Machine Learning Techniques Applied to Geoscience Information System and Remote Sensing

Machine Learning Techniques Applied to Geoscience Information System and Remote Sensing

Special Issue Editors

Saro Lee
Hyung-Sup Jung

MDPI • Basel • Beijing • Wuhan • Barcelona • Belgrade

Special Issue Editors
Saro Lee
Korea Institute of Geoscience and
Mineral Resources (KIGAM)
Korea

Hyung-Sup Jung
University of Seoul
Korea

Editorial Office
MDPI
St. Alban-Anlage 66
4052 Basel, Switzerland

This is a reprint of articles from the Special Issue published online in the open access journal *Applied Sciences* (ISSN 2076-3417) from 2018 to 2019 (available at: https://www.mdpi.com/journal/applsci/special_issues/Machine_Learning_Techniques_GIS_RS)

For citation purposes, cite each article independently as indicated on the article page online and as indicated below:

LastName, A.A.; LastName, B.B.; LastName, C.C. Article Title. *Journal Name* **Year**, *Article Number*, Page Range.

ISBN 978-3-03921-215-6 (Pbk)
ISBN 978-3-03921-216-3 (PDF)

Contents

About the Special Issue Editors

Saro Lee, Ph.D., received his B.Sc. degree in Geology in 1991 and Ph.D. in landslide susceptibility mapping using GIS in 2000. He is currently a Principal Researcher at Geological Research Division, KIGAM. He is also a Full Professor at University of Science and Technology (UST). He started his professional career in 1995 as a Researcher in the KIGAM and has carried out many international cooperative research projects in the field of mineral potential and geological hazard. He has also managed and delivered lectures in the KOICA International Training Program (Mineral Exploration and GIS/RS). His research interests include geospatial predictive mapping with GIS and RS, including landslide susceptibility, ground subsidence hazard, groundwater potential, mineral potential, and habitat mapping. He has published around 100 peer-reviewed SCI(E) papers and has a citation rate of about 9000 and h-index of 50 in Google. He is also Associate Editor and an Editorial Board member of many international journals, including *Landslide, Arabian Journal of Geosciences, and Geosciences Journal*.

Hyung-Sup Jung received his M.Sc. degree in Geophysics in 1998 and Ph.D. degree in Remote Sensing in 2007 from Yonsei University, Seoul, Korea. He is currently a Professor with the Department of Geoinformatics at the University of Seoul. He is also serving as Editor-in-Chief of the *Korean Journal of Remote Sensing (KJRS)* and as an Associate Editor of IEEE *Geoscience and Remote Sensing Letters*. His primary research interests cover the development of algorithms related to 1) synthetic aperture radar (SAR), SAR interferometry (InSAR), multiple-aperture InSAR (MAI) and multi-temporal InSAR (MTInSAR); 2) automated geometric correction of multi-sensor images; and 3) multi-sensor image processing and fusion, thermal remote sensing, and multi-temporal optical and thermal sensing. He has been developing algorithms for remote sensing applications related to 1) 3D deformation mapping by combining MAI and InSAR; 2) 2D surface velocity estimation by combining MAI and along-track interferometry (ATI); 3) MAI-based ionospheric correction of radar interferograms; 4) multi-sensor fusion by the integration of optic and SAR, SAR and thermal and optic and thermal images; 5) automated geometric correction for optic and SAR images; and 6) Earth's surface variation monitoring, such as urban subsidence monitoring, glacier monitoring, volcano monitoring, deforestation monitoring, forest mapping, forest fire mapping, and snow depth estimation.

Preface to "Machine Learning Techniques Applied to Geoscience Information System and Remote Sensing"

As computer and space technologies have been developed, geoscience information systems (GIS) and remote sensing (RS) technologies, which deal with the geospatial information, have been rapidly maturing. Moreover, over the last few decades, machine learning techniques including artificial neural network (ANN), deep learning, decision tree, and support vector machine (SVM) have been successfully applied to geospatial science and engineering research fields. The machine learning techniques have been widely applied to GIS and RS research fields and have recently produced valuable results in the areas of geoscience, environment, natural hazards, and natural resources.

This Special Issue of *Applied Sciences* on the machine learning techniques applied to geoscience information systems and remote sensing aims to attract novel contributions. We have invited original research papers addressing the state-of-the-art in the following areas:

1) Application of machine learning techniques combined with GIS;

2) Application of machine learning techniques to remote sensing;

3) Application of machine learning techniques to global positioning system (GPS);

4) Spatial analysis and geocomputation based on machine learning techniques;

5) Spatial prediction using machine learning techniques;

6) Data processing of geoinformation using machine learning techniques;

7) Comparative analysis among several machine learning techniques applied to GIS and RS;

8) Application of machine learning techniques on geosciences, environments, natural hazards, and natural resources as case studies.

Twenty-one papers have been selected which reflect the topics of interest for this Special Issue. This Special Issue would not have been possible without the contributions of professional authors and reviewers, and the excellent editorial team of *Applied Sciences*.

Saro Lee, Hyung-Sup Jung
Special Issue Editors

Editorial

Special Issue on Machine Learning Techniques Applied to Geoscience Information System and Remote Sensing

Hyung-Sup Jung [1,*] and Saro Lee [2,3,*]

1 Department of Geoinformatics, University of Seoul, 163 Seoulsiripdaero, Dongdaemun-gu, Seoul 02504, Korea
2 Geoscience Platform Research Division, Korea Institute of Geoscience and Mineral Resources (KIGAM), 124 Gwahak-ro, Yuseong-gu, Daejeon 34132, Korea
3 Geophysical Exploration, Korea University of Science and Technology, 217 Gajeong-ro, Yuseong-gu, Daejeon 34113, Korea
* Correspondence: hsjung@uos.ac.kr (H.-S.J.); leesaro@kigam.re.kr (S.L.)

Received: 10 June 2019; Accepted: 11 June 2019; Published: 14 June 2019

1. Introduction

As computer and space technologies have been developed, geoscience information systems (GIS) and remote sensing (RS) technologies, which deal with the geospatial information, have been maturing rapidly. Moreover, over the last few decades, machine learning techniques, including artificial neural network (ANN), deep learning, decision tree, and support vector machine (SVM), have been successfully applied to geospatial science and engineering research fields. The machine learning techniques have been widely applied to GIS and RS research fields and have recently produced valuable results in the areas of geoscience, environment, natural hazards and natural resources.

This special issue of applied sciences on machine learning techniques applied to geoscience information system and remote sensing aims to attract novel contributions. We have invited original research papers addressing the state-of-the-art in the following:

(1) Application of machine learning techniques combined with GIS;
(2) Application of machine learning techniques to remote sensing;
(3) Application of machine learning techniques to Global Positioning System (GPS);
(4) Spatial analysis and geocomputation based on machine learning techniques;
(5) Spatial prediction using machine learning techniques;
(6) Data processing of geoinformation using machine learning techniques;
(7) Comparison analysis among several machine learning techniques applied to GIS and RS;
(8) Application of machine learning techniques on geosciences, environments, natural hazards, and natural resources as case studies.

Twenty-one papers have been selected, which reflect the topics of interest for this special issue.

2. Machine Learning Techniques and Their Applications

Truong et al. [1] in their paper entitled "Enhancing Prediction Performance of Landslide Susceptibility Model Using Hybrid Machine Learning Approach of Bagging Ensemble and Logistic Model Tree" performed landslide modeling via proposing a new machine learning ensemble method that integrates logistic model trees (LMTree) algorithm and bagging ensemble (BE). The proposed method was named as BE-LMtree, and the proposed method enhanced the performance of the landslide model.

Seo et al. [2] in their paper entitled "Learning-Based Colorization of Grayscale Aerial Images Using Random Forest Regression" exploited the random forest (RF) regression for aerial imagery colorization, developed an efficient algorithm to establish color relationships based on unchanged regions, and performed visual and quantitative analyses.

Arabameri et al. [3] in their paper entitled "Spatial Modelling of Gully Erosion Using GIS and R Programing: A Comparison among Three Data Mining Algorithms" determined the relationship between gully occurrence and conditioning factors using weights-of-evidence (WoE) Bayes theory; assessed the capability of RF, multivariate adaptive regression spline (MARS), and boosted regression tree (BRT) machine learning models to predict gully erosion (GE) susceptibility; and validated the models using the area under the curve (AUC) and seed cell area index (SCAI) methods.

Deng and Pu [4] in their paper entitled "Single-Class Data Descriptors for Mapping *Panax notoginseng* through P-Learning" mapped *Panax notoginseng* fields through a stack of single-class data descriptors (SCDDs) as the future technical milestone for planting pattern analysis, evaluated the abilities of SCDDs in identifying small *Panax notoginseng* fields in the complex agricultural landscapes, and provided the potential possibilities for monitoring the planting pattern changes of *Panax notoginseng* fields, further giving us new insights into the planting pattern transitions of the perennial ginseng in macrocosm.

Wiratama et al. [5] in their paper entitled "Dual-Dense Convolution Network for Change Detection of High-Resolution Panchromatic Imagery" proposed a dual-dense convolutional network to recognize pixel-wise change that is based on dissimilarity analysis of neighborhood pixels on panchromatic (PAN) images with high spatial resolution. The proposed method exploits two fully convolutional neural networks employed to measure dissimilarity of neighborhood pixels, and hence it showed a better performance in qualitative and quantitative evaluation.

Zhang et al. [6] in their paper on "Convolutional Neural Network-Based Remote Sensing Images Segmentation Method for Extracting Winter Wheat Spatial Distribution" proposed a new method to map winter wheat field areas using GF-2 high-resolution PAN images. A deep learning model named as a Hybrid Structure Convolutional Neural Network (HSCNN) was successfully applied to map the winter wheat field areas.

Liu et al. [7] in their paper titled "A New Weighting Approach with Application to Ionospheric Delay Constraint for GPS/GALILEO Real-Time Precise Point Positioning" adopted a weighting approach in the precise point positioning with integer and zero-difference ambiguity resolution demonstrator (PPP-WIZARD). The weighting method integrates a weight factor searching method with a moving-window average filter. The proposed method can significantly reduce convergence time as well as improve the reliability of positioning solutions in real-time precise point positioning.

Chen et al. [8] in their paper titled "Landslide Susceptibility Modeling Using Integrated Ensemble Weights of Evidence with Logistic Regression and Random Forest Models" employed the integrated ensemble WoE with logistic regression (LR) and RF models to map landslide susceptibility and quantitatively compared and analyzed by receiver operating characteristic (ROC) curves and AUC.

Azeez et al. [9] in their paper entitled "Modeling of CO Emissions from Traffic Vehicles Using Artificial Neural Networks" proposed a hybrid model to generate microscale prediction maps with toll gate locations. The proposed model combines the metaheuristic optimization technique and ANN model to predict traffic emissions. The achieved performance of the method was about 80.6%. The authors said that the developed model can be a promising tool for vehicular CO simulations in highly congested areas.

Kwak and Park [10] in their paper entitled "Impact of Texture Information on Crop Classification with Machine Learning and UAV Images" focused on the evaluation of the effectiveness of texture information for crop classification with unmanned aerial vehicle (UAV) images. The classification performance was compared between a single-date UAV image and a time-series UAV image set. The used classification algorithms were RF and SVM.

Park and Kim [11] in their paper entitled "Landslide Susceptibility Mapping Based on Random Forest and Boosted Regression Tree Models, and a Comparison of Their Performance" analyzed and compared the performance between the RF and boosted regression tree (BRT) models for landslide susceptibility analysis. The performance of the RF model was about 0.865 and that of the BRT model was about 0.851. The performance of the two ensemble models were very similar.

Li et al. [12] in their paper on "A Single Point-Based Multilevel Features Fusion and Pyramid Neighborhood Optimization Method for ALS Point Cloud Classification" proposed (i) two local features including the normal angle distribution (NAD) histogram and latitude sampling histogram (LSH), (ii) a multilevel single-point features fusion method based on a multi-neighborhood space and multi-resolution, and (iii) a fast classification optimization method based on a multi-scale pyramid. They validated the proposed method using large-scale airborne laser scanning (ALS) point clouds.

Choung and Kim [13] in their paper titled "Study of the Relationship between Urban Expansion and PM_{10} Concentration Using Multi-Temporal Spatial Datasets and the Machine Learning Technique: Case Study for Daegu, South Korea" assessed a possible relation between urban expansion and PM_{10} concentration in Daegu, Korea, from ten-year monitoring data acquired from 2007 to 2017 using the SVM method. The experiment result showed no relation between the urban expansion and the PM_{10} concentrations.

Wang et al. [14] in their paper entitled "Deep Fusion Feature Based Object Detection Method for High Resolution Optical Remote Sensing Images" proposed a novel transfer deep learning method to detect objects in high-resolution remote-sensed images. In addition, they improved the candidate window selection process and designed a deep feature extraction method with context scene feature fusion and detection. They validated the proposed method using high-resolution remote-sensed images.

Oh et al. [15] in their paper entitled "Land Subsidence Susceptibility Mapping Using Bayesian, Functional, and Meta-Ensemble Machine Learning Models" investigated the achieved performance of several models that have never been applied to land subsidence prediction. They produced land subsidence susceptibility (LSS) maps in abandoned subsurface coal mining areas using machine learning techniques such as the logit boost meta-ensemble model, Bayes net model, naïve Bayes (NB) model, logistic model, and multilayer perceptron model. The reliability and accuracy of the models were performed by the area under the receiver operating characteristic (ROC) curves.

Wiratama and Sim [16] in their paper entitled "Fusion Network for Change Detection of High-Resolution Panchromatic Imagery" proposed a fusion network by combining front- and back-end networks to perform the low- and high-level differential detection in one structure and a combining loss function between contrastive loss and binary cross entropy loss to accomplish fusion of the proposed networks in training stage. In addition, the two-stage decision as a post-processing is presented to validate and ensure the changes prediction at the inference stage to better obtain the final change map.

Mao et al. [17] in their paper entitled "Comparison of Machine Learning Regression Algorithms for Cotton Leaf Area Index Retrieval Using Sentinel-2 Spectral Bands" compared the algorithm performance of five advanced machine learning regression algorithms, including ANN, support vector regression (SVR), Gaussian process regression (GPR), RF, and gradient boosting regression tree (GBRT), to retrieve cotton leaf area index (LAI) in a relatively comprehensive manner. Although the five models showed different performance, all of the models showed a potential for cotton LAI retrieval.

Liu et al. [18] in their paper entitled "Spatial Data Reconstruction via ADMM and Spatial Spline Regression" proposed a novel constrained spatial smoothing (CSS) algorithm to reconstruct a spatial field of densities. They evaluated the proposed method from the problem of reconstructing the spatial distribution of cellphone traffic volumes based on aggregate volumes recorded at sparsely scattered base stations.

Utomo et al. [19] in their paper entitled "Landslide Prediction with Model Switching" provided a total solution in the form of an early warning system. The system is called the Model Switch-based Landslide Prediction System (MoSLaPS). To address the data imbalance problem, they also adapted the popular adaptive synthetic sampling (ADASYN) method to landslide prediction. Moreover, to address the low true-positive rate (TPR) problem, they proposed a novel event-class model switch predictor design that significantly improves TPR.

Li [20] in the paper entitled "A Critical Review of Spatial Predictive Modeling Process in Environmental Sciences with Reproducible Examples in R" assisted spatial modelers and scientists by critically reviewing the spatial predictive modeling process, developing guidelines for selecting the most appropriate spatial predictive methods, and identifying and developing the most accurate predictive model to generate spatial predictions.

Xu et al. [21] in their paper entitled "Mapping Areal Precipitation with Fusion Data by ANN Machine Learning in Sparse Gauged Region" showed an efficient method to map areal precipitation with the data fused from the remote-sensing precipitation acquired from Tropical Precipitation Measurement Satellite (TRMM) product and ground gauge precipitation using the ANN method.

Funding: This work was supported by the National Research Foundation of Korea funded by the Korea government under Grant NRF-2018M1A3A3A02066008 for H.J. and it was also supported by the Basic Research Project of the Korea Institute of Geoscience and Mineral Resources (KIGAM) funded by the Ministry of Science and ICT for S.L.

Acknowledgments: This special issue would not be possible without the contributions of professional authors and reviewers, and the excellent editorial team of Applied Sciences.

References

1. Truong, X.; Mitamura, M.; Kono, Y.; Raghavan, V.; Yonezawa, G.; Truong, X.; Do, T.; Tien Bui, D.; Lee, S. Enhancing Prediction Performance of Landslide Susceptibility Model Using Hybrid Machine Learning Approach of Bagging Ensemble and Logistic Model Tree. *Appl. Sci.* **2018**, *8*, 1046. [CrossRef]

2. Seo, D.; Kim, Y.; Eo, Y.; Park, W. Learning-Based Colorization of Grayscale Aerial Images Using Random Forest Regression. *Appl. Sci.* **2018**, *8*, 1269. [CrossRef]

3. Arabameri, A.; Pradhan, B.; Pourghasemi, H.; Rezaei, K.; Kerle, N. Spatial Modelling of Gully Erosion Using GIS and R Programing: A Comparison among Three Data Mining Algorithms. *Appl. Sci.* **2018**, *8*, 1369. [CrossRef]

4. Deng, F.; Pu, S. Single-Class Data Descriptors for Mapping Panax notoginseng through P-Learning. *Appl. Sci.* **2018**, *8*, 1448. [CrossRef]

5. Wiratama, W.; Lee, J.; Park, S.; Sim, D. Dual-Dense Convolution Network for Change Detection of High-Resolution Panchromatic Imagery. *Appl. Sci.* **2018**, *8*, 1785. [CrossRef]

6. Zhang, C.; Gao, S.; Yang, X.; Li, F.; Yue, M.; Han, Y.; Zhao, H.; Zhang, Y.; Fan, K. Convolutional Neural Network-Based Remote Sensing Images Segmentation Method for Extracting Winter Wheat Spatial Distribution. *Appl. Sci.* **2018**, *8*, 1981. [CrossRef]

7. Liu, T.; Wang, J.; Yu, H.; Cao, X.; Ge, Y. A New Weighting Approach with Application to Ionospheric Delay Constraint for GPS/GALILEO Real-Time Precise Point Positioning. *Appl. Sci.* **2018**, *8*, 2537. [CrossRef]

8. Chen, W.; Sun, Z.; Han, J. Landslide Susceptibility Modeling Using Integrated Ensemble Weights of Evidence with Logistic Regression and Random Forest Models. *Appl. Sci.* **2019**, *9*, 171. [CrossRef]

9. Azeez, O.; Pradhan, B.; Shafri, H.; Shukla, N.; Lee, C.; Rizeei, H. Modeling of CO Emissions from Traffic Vehicles Using Artificial Neural Networks. *Appl. Sci.* **2019**, *9*, 313. [CrossRef]

10. Kwak, G.; Park, N. Impact of Texture Information on Crop Classification with Machine Learning and UAV Images. *Appl. Sci.* **2019**, *9*, 643. [CrossRef]

11. Park, S.; Kim, J. Landslide Susceptibility Mapping Based on Random Forest and Boosted Regression Tree Models, and a Comparison of Their Performance. *Appl. Sci.* **2019**, *9*, 942. [CrossRef]

12. Li, Y.; Tong, G.; Du, X.; Yang, X.; Zhang, J.; Yang, L. A Single Point-Based Multilevel Features Fusion and Pyramid Neighborhood Optimization Method for ALS Point Cloud Classification. *Appl. Sci.* **2019**, *9*, 951. [CrossRef]

13. Choung, Y.; Kim, J. Study of the Relationship between Urban Expansion and PM10 Concentration Using Multi-Temporal Spatial Datasets and the Machine Learning Technique: Case Study for Daegu, South Korea. *Appl. Sci.* **2019**, *9*, 1098. [CrossRef]

14. Wang, E.; Li, Y.; Nie, Z.; Yu, J.; Liang, Z.; Zhang, X.; Yiu, S. Deep Fusion Feature Based Object Detection Method for High Resolution Optical Remote Sensing Images. *Appl. Sci.* **2019**, *9*, 1130. [CrossRef]
15. Oh, H.; Syifa, M.; Lee, C.; Lee, S. Land Subsidence Susceptibility Mapping Using Bayesian, Functional, and Meta-Ensemble Machine Learning Models. *Appl. Sci.* **2019**, *9*, 1248. [CrossRef]
16. Wiratama, W.; Sim, D. Fusion Network for Change Detection of High-Resolution Panchromatic Imagery. *Appl. Sci.* **2019**, *9*, 1441. [CrossRef]
17. Mao, H.; Meng, J.; Ji, F.; Zhang, Q.; Fang, H. Comparison of Machine Learning Regression Algorithms for Cotton Leaf Area Index Retrieval Using Sentinel-2 Spectral Bands. *Appl. Sci.* **2019**, *9*, 1459. [CrossRef]
18. Liu, B.; Mavrin, B.; Kong, L.; Niu, D. Spatial Data Reconstruction via ADMM and Spatial Spline Regression. *Appl. Sci.* **2019**, *9*, 1733. [CrossRef]
19. Utomo, D.; Chen, S.; Hsiung, P. Landslide Prediction with Model Switching. *Appl. Sci.* **2019**, *9*, 1839. [CrossRef]
20. Li, J. A Critical Review of Spatial Predictive Modeling Process in Environmental Sciences with Reproducible Examples in R. *Appl. Sci.* **2019**, *9*, 2048. [CrossRef]
21. Xu, G.; Wang, Z.; Xia, T. Mapping Areal Precipitation with Fusion Data by ANN Machine Learning in Sparse Gauged Region. *Appl. Sci.* **2019**, *9*, 2294. [CrossRef]

Article

Enhancing Prediction Performance of Landslide Susceptibility Model Using Hybrid Machine Learning Approach of Bagging Ensemble and Logistic Model Tree

Xuan Luan Truong [1,*], Muneki Mitamura [2], Yasuyuki Kono [3], Venkatesh Raghavan [2], Go Yonezawa [2], Xuan Quang Truong [4], Thi Hang Do [1,2], Dieu Tien Bui [5,*] and Saro Lee [6,7,*]

[1] Faculty of Information Technology, Hanoi University of Mining and Geology, No.14 Vien Street, Bac Tu Liem, Hanoi 10000, Vietnam; dohanghumg@gmail.com

[2] Graduate School for Creative Cities, Osaka City University, Osaka 558-8585, Japan; mitamura@sci.osaka-cu.ac.jp (M.M.); raghavan@media.osaka-cu.ac.jp (V.R.); yonezawa@media.osaka-cu.ac.jp (G.Y.)

[3] Center for Southeast Asian Studies, Kyoto University, Kyoto 606-8502, Japan; kono@cseas.kyoto-u.ac.jp

[4] Faculty of Information Technology, Hanoi University of Natural Resources and Environment, No. 14 Phu Dien, Bac Tu Liem, Hanoi 10000, Vietnam; txquang@hunre.edu.vn

[5] Geographic Information System Group, Department of Business and IT, University College of Southeast Norway, Gulbringvegen 36, N-3800 Bø i Telemark, Norway

[6] Geological Research Division, Korea Institute of Geoscience and Mineral Resources (KIGAM), 124, Gwahak-ro, Yuseong-gu, Daejeon 34132, Korea

[7] Department of Geophysical Exploration, Korea University of Science and Technology, 217 Gajeong-ro Yuseong-gu, Daejeon 305-350, Korea

* Correspondence: truongxuanluan@humg.edu.vn (X.-L.T.); Dieu.T.Bui@usn.no (D.T.B.); leesaro@kigam.re.kr (S.L.)

Received: 29 May 2018; Accepted: 23 June 2018; Published: 27 June 2018

Abstract: The objective of this research is introduce a new machine learning ensemble approach that is a hybridization of Bagging ensemble (BE) and Logistic Model Trees (LMTree), named as BE-LMtree, for improving the performance of the landslide susceptibility model. The LMTree is a relatively new machine learning algorithm that was rarely explored for landslide study, whereas BE is an ensemble framework that has proven highly efficient for landslide modeling. Upper Reaches Area of Red River Basin (URRB) in Northwest region of Viet Nam was employed as a case study. For this work, a GIS database for the URRB area has been established, which contains a total of 255 landslide polygons and eight predisposing factors i.e., slope, aspect, elevation, land cover, soil type, lithology, distance to fault, and distance to river. The database was then used to construct and validate the proposed BE-LMTree model. Quality of the final BE-LMTree model was checked using confusion matrix and a set of statistical measures. The result showed that the performance of the proposed BE-LMTree model is high with the classification accuracy is 93.81% on the training dataset and the prediction capability is 83.4% on the on the validation dataset. When compared to the support vector machine model and the LMTree model, the proposed BE-LMTree model performs better; therefore, we concluded that the BE-LMTree could prove to be a new efficient tool that should be used for landslide modeling. This research could provide useful results for landslide modeling in landslide prone areas.

Keywords: landslide; bagging ensemble; Logistic Model Trees; GIS; Vietnam

---

1. Introduction

The problem of rainfall-induced landslides, which are triggered by high intense and long lasting precipitation, seems to be more serious in recent years in many regions around the world due to the effects of climate changes i.e., extreme rainfall events [1–8]. The rainfall-triggered landslide is especially exacerbated in countries that are located in storm centers of the world, such as Vietnam [9], Philippines [10], and China [11]. For example, the tropical typhoon of Rasmussen caused various floods and landslides with the total damages were estimated at $7 billion [12]. It anticipates that the number of landslides in the future will continue to rise due to effects of extreme rainfall events and changes of hydrological cycles [13]. Thus, landslide has become one of the hottest subject of the research community, however, accurately prediction of landslide still is a challenging real-world problem [14]. Therefore, more researches on landslide are still urgently required for deriving better detailed knowledge of slope failure and its mechanisms for designing remedial measures.

The development of a hazard map that provides detailed dimensional information of spatial distributions, temporal predictions, and destructive power of landslide is considered as an efficient tool for designing mitigation measures and management policies. However, the hazard map at the regional scale requires very detailed temporal landslide inventories that are hardly available, especially in developing countries [15]. For this context, a landslide susceptibility map (LS-map) could be alternatively employed since it helps to identify areas with high landslide probability. According to Ciampalini, et al. [16], LS-map is a valuable decision-support tool that assists local authorities in land use infrastructural planning and management

To produce susceptibility map, a variety of studying approaches has been introduced because the accuracy of the susceptibility map at regional analysis scale is controlled not only by the quality of the input maps, but also the algorithms and techniques that are employed [17]. These approaches vary from expert weighting methods to deterministic and statistical models. Evaluation of these approaches has been well presented i.e., in Chacon et al. [18] and Van Westen, et al. [19]. In recent years, new approaches that are based on advanced statistical and machine learning methods have been proposed i.e., fuzzy k-Nearest Neighbor [17]; fuzzy rule based models [20–23]; neural networks [24–30]; support vector machines [31–38]; Random Forests; metaheuristic optimized least squares support vector machines [39,40]; Cuckoo optimized relevance vector machines [41]; Chi-squared automatic interaction detection (CHAID) [42]; tree-based algorithms [43–47]; and, gene expression programming [48]. The main advantage of these methods is that they are capable of involving several to a large number of variables for reliable results, and overall, these methods are able to provide better performance models when compared to those of conventional methods [43,49,50].

In the last years, the integration of advanced machine learning algorithms and homogeneous ensemble frameworks has been explored for landslide susceptibility modeling with promising results. For example, Tien Bui, et al. [51] show that the landslide model based on a combination of functional trees with Bagging performs better than the neural network models. Pham et al. [23] concluded that the hybridization of Fuzzy Unordered Rules Induction Algorithm and Rotation forest ensemble has increased the prediction performance of the landslide model when compared to the benchmark of support vector machines model. Pham et al. [26] reported that the landslide model derived from a combinations of MultiBoost and Dagging with neural networks has significantly improved the prediction power of the landslide model using only the neural network. Thus, it could be concluded that homogeneous ensembles of machine learning are promising and should be further investigated aiming to improve the prediction capability of landslide susceptibility model.

Based on the mentioned motivation, this research aim is to expand the body knowledge of landslide modeling through introducing a new machine learning ensemble approach that combines the Logistic Model Trees (LMTree) algorithm [52] and Bagging Ensemble (BE) [53], named as BE-LMtree, for enhancing the performance of the landslide model. LMTree is a relative new and promising machine learning algorithm that was rarely explored for the landslide study, whereas Bagging ensemble is an framework that has proven efficient in landslide modeling [51,54]. Consequently, a combination of

BE and LMTree has resulted in a new powerful prediction method, and to the best of our knowledge, this is the first time that the BE-LMTree is studied for landslide susceptibility.

2. Theoretical Background of the Methods

2.1. Logistic Model Tree

Logistic Model Trees (LMTree), which is a relatively new machine learning algorithm, is developed based on the integration of tree induction algorithm and additive logistic regression [52]. The difference of LMTree when compared to the other decision tree algorithms is that the tree growing process is carried out using the LogitBoost algorithm [52,55] and the tree pruning is performed using Classification And Regression Tree (CART) [56].

Given a training dataset $T = (x_i, y_i)_{i=1}^{ds}$ with $x_i \in R^D$ is the input vector, ds is the number of data samples, D is the dimension of the training dataset, and $y_i \in (1, 0)$ is the label class. In this research context, the input vector consists of eight variables (slope, aspect, elevation, land cover, soil type, lithology, distance to fault, and distance to river), whereas the label class contains two classes, landslide (LS) and non-landslide (Non-LS). The landslide class is coded as "1" and the non-landslide is coded as "0". The objective of LMTree is to construct a tree-like structure model that is capable of classifying the training dataset into the two above classes in term of probability. The predicted numeric value to the landslide class of sample is used as susceptibility index.

Structurally, the LMTree model consists of a root node, a set of inner nodes, and a set of leaves. The aim of the training phase that includes the tree growing and the tree pruning processes is to determine the best tree structure with numbers of inner nodes and leaves. Accordingly, first, a logistic regression model Equation (1) is built at the root note using the binary LogitBoost algorithm [57] and the training dataset. In the next step, the training dataset at the root is split using the C4.5 splitting rule [58] in order to sort appropriate sub-datasets for the inner nodes, and then, logistic regression models Equation (1) for these inner nodes is built using their associated sorted datasets and the binary LogitBoost. The tree continues growing in the same procedure until it meets the stopping criterion of less than 15 samples at nodes. Finally, to prevent the LMTree model from over-fitting, the tree pruning is performed using the CART algorithm that is based on a combination of the model error and the model complexity [52].

In the LMTree building process, the binary LogitBoost algorithm [57] is used to generate logistic regression models Equation (1) for all of the inner nodes and leaves, as follows.

$$f_{\text{LS,Non}-\text{LS}}(x) = \sum_{i=1}^{D} \beta_i x_i + \beta_0 \tag{1}$$

where D is the total number of landslide input factors and β_i is the logistic coefficient.

The membership probability [52] of the landslide class at the leaves of the LMTree model is posterior probabilities derived using Equation (2) and is used as landslide susceptibility index.

$$p((\text{LS}, \text{Non} - \text{LS})\,|x) = \frac{\exp f_{\text{LS,Non}-\text{LS}}(x)}{\exp f_{\text{LS}}(x) + \exp f_{\text{Non}-\text{LS}}(x)} \tag{2}$$

The complexity of the LMTree model could be estimated using the following equation [52]:

$$\text{MC} = \text{O}(dept * ds * \log n + ds * D^2 * dept + nt^2) \tag{3}$$

where MC is the model complexity; *dept* is the depth of the initial unpruned tree; *nt* is the number of nodes in the LMTree; *ds* is the number of training samples; and, D is the number of landslide predisposing factors.

2.2. Bagging Ensemble

Ensemble learning is a machine learning paradigm where multiple classifiers are trained and combined to enhance the prediction capability of a model. Different from popular machine learning approaches where one model is built from the training data, ensemble frameworks try to generate a set of sub-datasets from the training data, and then, each sub-dataset is used to construct a classifier, which is also called a based learner. At last, all of the based learners are combined to form the final prediction model using combination techniques i.e., averaging or majority voting [59].

Different ensemble techniques have been successfully proposed i.e., Bagging, AdaBoost, Multiboost, Stacking, and Rotation forest [60]; however, in landslide modeling, Bagging ensemble has proven robust and better than other ensembles [26,51,54], therefore, it is selected for this study.

Bagging also called Bootstrap aggregating in the full name is one of the earliest procedure for generating sub-datasets and combining based learners proposed by Breiman [53]. Using the training dataset, this technique generates bootstrap samples in which some of the samples are replicated and some samples are omitted. These bootstrap samples, which are called bootstrapped sub-datasets, are used to construct based learners using the same classification algorithm i.e., the LMTree in this work. These based learners are then combined using the majority voting strategy.

3. The Study Area and Spatial Datasets

3.1. Description of the Upper Reaches Area of Red River Basin

The study area is the Upper reaches area of the Red River Basin (URRB) ($103°33'36''$–$104°30'50''$ E, $22°05'40''$–$22°47'52''$ N) that belongs to the Lao Cai, a north-western mountainous province in Vietnam (Figure 1). The URRB covers an area of 3273.5 km^2 with complex topography, steep slopes, and narrow valleys. The topography is highly fragmented with high mountains ranges, wide valleys, and deep streams, which result in high relief amplitudes [40]. The altitude varies from 48.1 m to 2812.6 m above sea level, with the mean and the standard deviation of 528.6 m and 484.9 m, respectively. Topographically, 61.8% of the URRB is occupied by slope angles that are higher than 15°, whereas areas with slopes less than 5° cover approximately 7.3% the total area of the URRB. The remaining 30.9% are areas located in the slope group 5–15°.

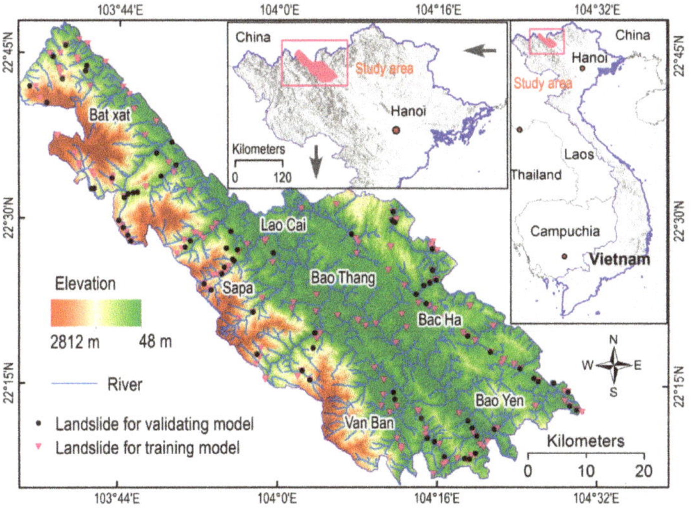

Figure 1. Location of the Upper Reaches Area of Red River Basin (Vietnam).

Hydrologically, due to the fragmentation of the terrain, the river system in the study area is dense and evenly distributed (Figure 1). These rivers are characterized by being narrow and steep, which are favorable conditions for the occurrence of flash flood and landslides. The Red River, which is the second largest river in Vietnam, is the major channel system of the URRB. This river originates from Yunnan province (China) and flows south-eastward to the study area [61].

The climate of URRB is divided into two seasons: the rainy season begins from April to October and the dry season lasts from November to March next year. The average temperature ranges range from 23 °C to 29 °C [62] and the average annual rainfall is from 1400 mm to 1900 mm [63].

The URRB is located in an active tectonic region with the relatively fast movement of the Red River fault zone that results in continuously landslide occurrences over the years [40]. It should be noted that the Red River fault zone is one of the four main tectonic features in north Vietnam that begins from Tibetan plateau (China) and extends to the Red River area of Vietnam [64,65]. Twenty seven geological formations outcrop in the basin with varied area and space distribution (Figure 2). Quaternary deposits, which consist of mainly granule, grit, breccia, pebble, boulder, and sand, cover 7.04% of the total area of the basin. Whereas, 86.68% of the basin is covered by nine geological formations, Suoi Chieng (23.62%), Ha Giang (10.96%), Nui Con Voi (10.54%), Sinh Quyen (10.43%), Ngoi Chi (8.44%), Cam Duong (8.29%), Ye Yen Sun (6.23%), Po Sen (5.96%), and Muong Hum (2.21%). The main lithologies are biotite schist, garnet-biotite gneiss, coaly shale, marble cherty shale, quartz-plagioclase-biotite schist, and two-mica schist. Detailed distribution of the lithological formations in the basin is shown in Figure 2.

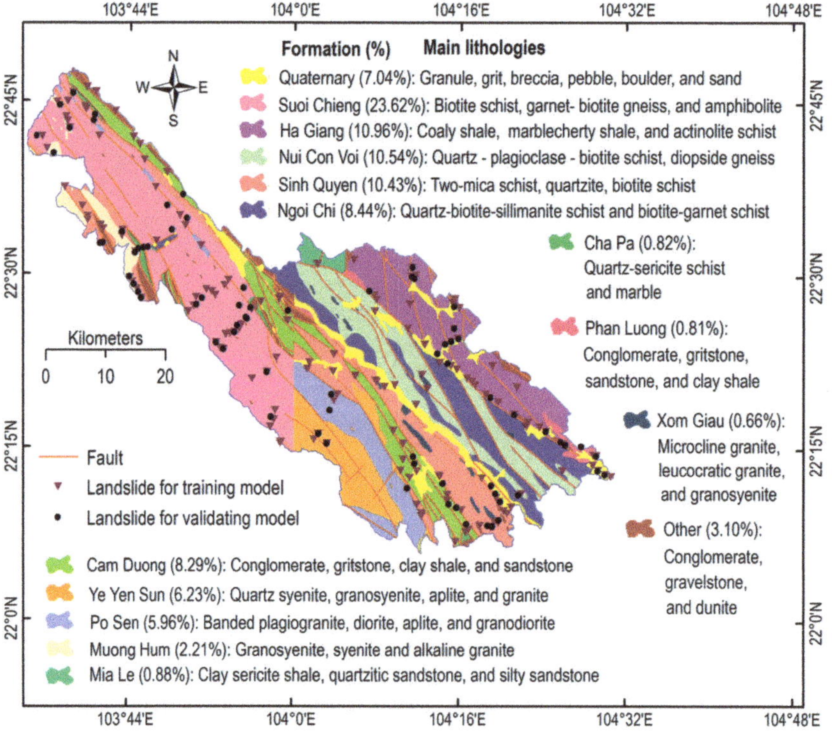

Figure 2. Geological map of the study area.

3.2. Geospatial Data

Landslide inventory map for the URRB was constructed from two main sources: (i) historic landslides from the project VAST05.02/14-15 in 2015, which was prepared by Tien Bui et al. [40]; and, (ii) landslide polygons from the State-Funded Landslide Project (SFLP) 2016 [9], a national landslide program that is carrying out in Vietnam. The SFLP project has systematically investigated and collected historic landslides for all northwest mountainous provinces in Vietnam, including the study area. Accordingly, these landslides were mainly interpreted and mapped using aerial photos and field investigations. Detailed descriptions of methods and techniques for obtaining these historic landslides in the SFLP project are present in [9].

As result, a total of 255 historic soil-mixed-boulder slides that occurred during the last two decades were registered for the landslide inventory map (Figure 1). It is noted that many rock falls were excluded out of this research because their falling mechanism are very different when compared to that of the soil-mixed-boulder slides. Analysis of the landslide inventory map showed that these slides occurred due to rainfall during tropical rainstorms [40]. Our statistical analysis of these slides showed that the largest and the smallest landslides are 116627.9 m^2 and 6.2 m^2, respectively, with the mean is 3742.5 m^2 and the standard deviation is 11467.3 m^2. Approximately 9.1% of the landslide inventories are large landslides (>10,000 m^2), whereas 9.1% of the landslide inventories are medium landslides (1000–10,000 m^2), and the remaining are landslides less than 1000 m^2. Two examples of landslide photos in the study area are shown in Figure 3.

Figure 3. Two photos of landslides in the study area: (**a**) Landslide at the Mong Sen area and (**b**) Landslide at Km 7 Lao Cai. The two photos were taken by Xuan-Luan Truong in August 2014.

Because the rainfall-trigged landslides in this study area occurred due to interactions of various geo-environmental factors, including topography, land cover, lithology, soil type, and river network [9,40,66,67], these factors were selected for this analysis. Digital elevation model (DEM) with resolution of 25 × 25 m for the URRB area was constructed using digital topographic maps 1:50,000 scale provided by the Ministry of Natural Resource and Environment of Vietnam. Using this DEM, three morphometric factors, slope, elevation, and aspect, were generated. To build the slope map (Figure 4a), seven categories were used. For the elevation map (Figure 4b), eight categories were considered. These categories were determined using Jenks natural break available in ArcGIS. For the aspect map, nice facing slopes were used (Figure 4c).

Land cover map (Figure 4d) at scale of 1:50,000 with nine classes for the URRB area was derived from the project No.02/2012/ HD-HTSP funded by Ministry of Education and Training of Vietnam. The nine classes were obtained through the classification of Landsat 8 OLI imagery in 2013 using ENVI software. Soil type map (Figure 4e) at 1:100,000 scale with 13 soil types for the URRB area was provided by Department of Agriculture and Rural Development of the Lao Cai province.

Lithological map for the URRB area was constructed based on National Geological and Mineral Resources Maps at scale of 1: 200,000, as provided by the Ministry of Natural Resource and Environment of Vietnam. Our analysis showed that more than 15 formations outcrop in the URRB area (see Figure 2). For this research, the lithological map with seven categories was constructed (Figure 4f) and these categories were separated based on clay composition, weathering characteristics, and material strength [24,68,69]. Detailed characteristics of the seven categories could be found in Tien Bui, et al. [70]. Fault is an popular factor for landslide susceptibility that was used various works i.e., in [71–73], and especially, it is an important factor for landslide modeling in areas that are affected by tectonic activities [74]. In this research, distance to fault map (Figure 4g) with seven classes [40] for the URRB area was constructed by buffering the fault lines extracted from the National Geological and Mineral Resources Maps above.

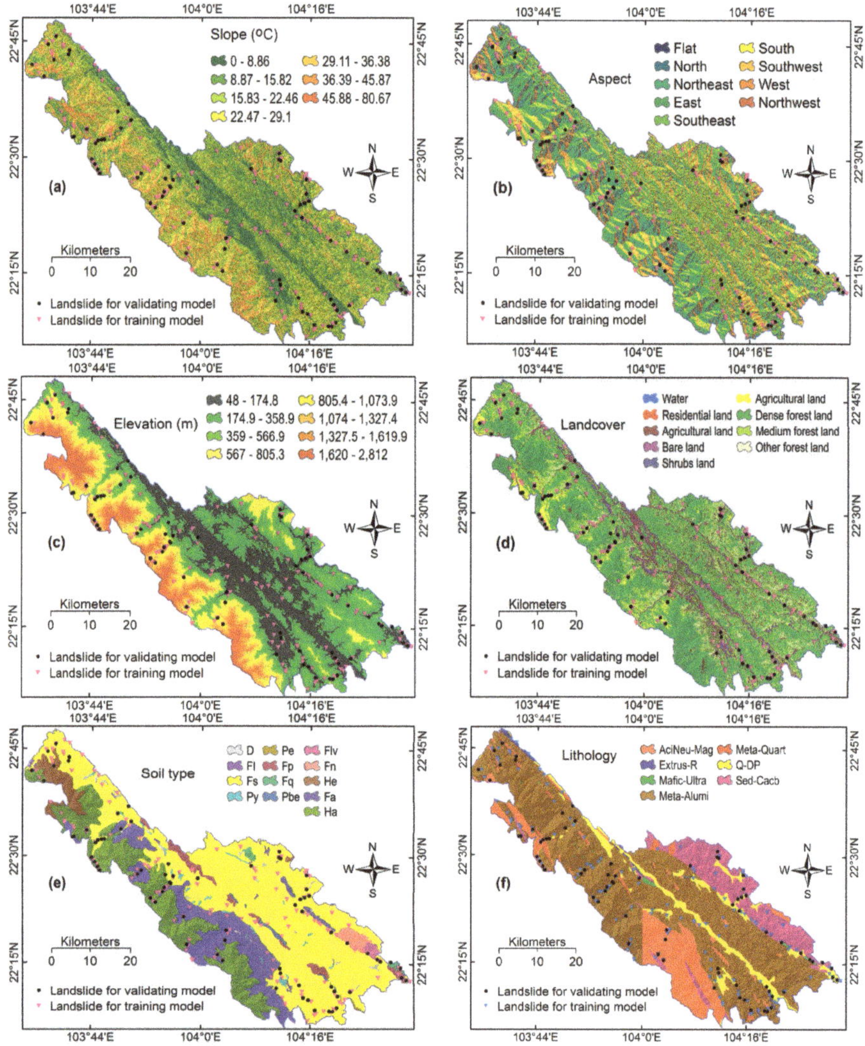

Figure 4. *Cont.*

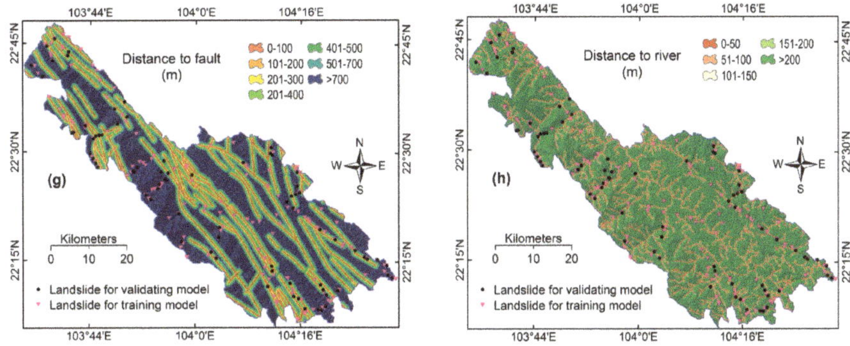

Figure 4. Landslide predisposing factors used in this study: (**a**) Slope; (**b**) Aspect; (**c**) Elevation; (**d**) Land cover; (**e**) Soil type ; (**f**) Lithology; (**g**) Distance to fault; and, (**h**) Distance to river.

Soil type (e) legend: **D**: Sloping soil; **Fl**: Cultivated rice yellowish red soil; **Fs**: Yellowish red soil on claystone and metamorphic rocks; **Py**: Alluvial soil deposited by river; **Pe**: neutral-less acidic and light texture alluvial soil; **Fp**: Brown-yellowish soil on old alluvium; **Fq**: Light yellowish soil on sandstone; **Pbe**: Neutral and less acidic alluvial soil; **Flv**: Red soil on limestone; **Fn**: Brown-yellowish soil on limestone; **He**: Humus yellow red soil on claystone and metamorphic rocks; **Fa**: Yellowish red soil on acid magmatic rock; and **Ha**: Humus yellow red soil on acid igneous rock. **Lithology (f) legend**: **AciNeu-Mag**: Acid-neutral magmatic rocks; **Extrus-R**: Extrusive rocks; **Mafic-ultra**: Mafic-ultramafic rocks; **Meta-Alumi**: Metamorphic rock with aluminosilicate components; **Meta-Quart**: Metamorphic rock with rich quarts components; **Q-DP**: Quaternary deposits; and, **Sed-Cacb**: Sedimentary carbonate rocks.

4. Proposed a Hybrid Machine Learning Approach of Bagging Ensemble (BE) and Logistic Model Tree (LMTree)

In this section, the proposed hybrid machine learning approach for Landslide Susceptibility Modeling at Upper Reaches Area of Red River Basin (Viet Nam) is described and presented in the first time. Methodological concept of the proposed BE-LMT model used in this study is shown in Figure 5.

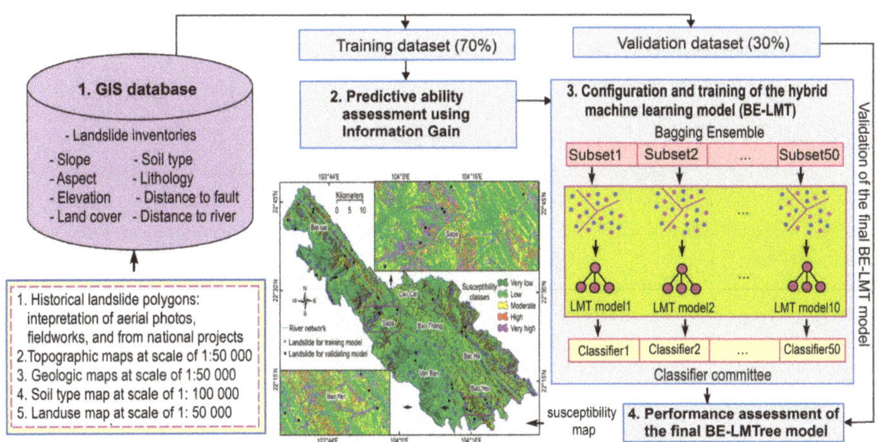

Figure 5. Methodological concept of the proposed Bagging ensemble (BE)-Logistic Model Trees (LMTree) model used in this study.

The proposed approach is a hybridization of LMTree and BE and is named as BE-LMTree. It should be noted that the data processing and coding were conducted using IDRISI Selva 17.0 (Clark University, Worcester, MA, USA, 2017) and ArcGIS 10.4 (ESRI Inc., Redlands, CA, USA 2017). The BE code is from Kuncheva [59] whereas the Logistic Model Tree algorithm is available at Weka's API [75]. The proposed BE-LMTree model was programmed by the authors in the Matlab environment.

4.1. Establishment of GIS Database, the Training Dataset and the Validation Dataset

In the first step, a GIS database for this project was designed and established using ArcCatalog software. Accordingly, the File Geodatabase format was used due to the ability to host and process very large geographic datasets with their different data types in a only one file system [76]. Accordingly, the GIS database consists of 255 landslide polygons and eight predisposing factors (slope, aspect, elevation, land cover, soil type, lithology, distance to fault, and distance to river). These landslide polygons and factors were converted to raster format with a resolution of 25 m. In this research, the categories of the eight predisposing factors were coded and normalized, as suggested in [24,77], to avoid the imbalance of categorical magnitudes [78].

In landslide modeling, cross validation [79] that has proven efficient for evaluating the model performance should be used. Accordingly, in this research, 179 landslide polygons (70%, 1006 pixels) were randomly extracted [80] and used for training the landslide models, whereas the other 76 landslides (30%, 441 pixels) were used for assessing the prediction capability of the models. Because the proposed approach in this study employs "on-off" classification, the equal amount of non-landslide pixels were also randomly sampled in the not-yet landslide areas of the basin, area with slope angles less than 5°, as suggested in [32]. Detailed discussions on sampling strategies can be found at [81]. In the next step, values of the eight predisposing factors for all of the aforementioned pixels were extracted to build the training dataset and the validation dataset. Finally, the coding process that was proposed in [17] was performed, in which the landslide pixels were assigned "1" and the non-landslide pixels were assigned "0".

Because the aforementioned partition of the landslide dataset into the training and validation datasets was randomly generated only once; therefore, a further cross validation was additionally used to ensure that the modeling result is the objective. Accordingly, 10-fold cross validation was employed in the training phase with the training dataset to build landslide models. Thus, the training dataset was randomly partitioned into 10 equally sized subsets; nine subsets were used for building the landslide model, whereas the remaining subset was used for testing the landslide model. This procedure was repeated 10 times where each subset was being used once as the testing dataset. Once the model was successfully trained using the training dataset with the 10-fold cross validation procedure, the model was again validated using the validation dataset.

4.2. Merit Evaluation of Factor

Identification of relevant features is an essential task when employing machine learning techniques for landslide susceptibility [82]. This is because landslide is a typical real-world problem that is influenced by various factors, but the contribution of these factors to the prediction model is different. If non-contribution factors are included in the model, then they may cause noises that reduce the prediction power of the final model; therefore, these factors should be excluded.

To detect non-relevant factors in this study, Pearson technique was employed to quantify the predictive power of all landslide predisposing factors. Accordingly, the meritof these features were estimated using Pearson correlation values [83] of the predisposing factors and the output using the following equation:

$$Merit_i = \frac{covr(IF_i, y)}{\sqrt{varr(IF_i) * varr(y)}} \tag{4}$$

where $Merit_i$ is the correlation value of landslide predisposing factor *IFi* and the label class y; $covr(.)$ is the covariance; and, $varr(.)$ is the variance.

4.3. Configuration and Training of the BE-LMTree Model

Configuration of the BE-LMTree model consists of two steps: (i) Determining the minimum number of samples (NS) that are used for growing the LMTree; and, (ii) Determining the number of bootstrap subsets (BS) used for BE. Because at least five samples are required to build a logistic regression model at a tree node [52], we varied NS from 5 to 100 with a step size of 1, and then, estimating the classification rate of the corresponding LMTree model on both the training dataset and the validation dataset. As a result, minimum of 10 samples is the best for the data at hand; therefore, NS of 10 was selected. For the case of determining the number of the bootstrap subsets, since no thumb rule is available, an empirical test was carried out by varying BS from 2 to 100, and then, compute their classification rates of the LMTree model both on the training dataset and the validation dataset. The test result revealed that the BE-LMTree with 50 tree-based classifiers provided the highest classification accuracy for the data at hand; therefore, BS of 50 is selected. Once the BE-LMTree model had been configured, the training process was carried out to derive the final BE-LMTree model.

4.4. Performance Assessment of the Final BE-LMTree Model

Because the landslide modeling in this research is considered to be a binary form of pattern recognition, therefore the performance of the final BE-LMTree model could be assessed using confusion matrix (Figure 6) [40], both on the training dataset and the validation dataset. Based on the matrix, several model measures are further derived i.e., sensitivity (SEN), specificity (SPE), positive predictive power (PP2), and negative predictive power (NP2), Kappa statistics, and classification accuracy (CLA) for the assessment, as suggested in [50]. It should be noted that a perfect landslide model will have 100% for SEN, SPE, PP2, NP2, and CLA.

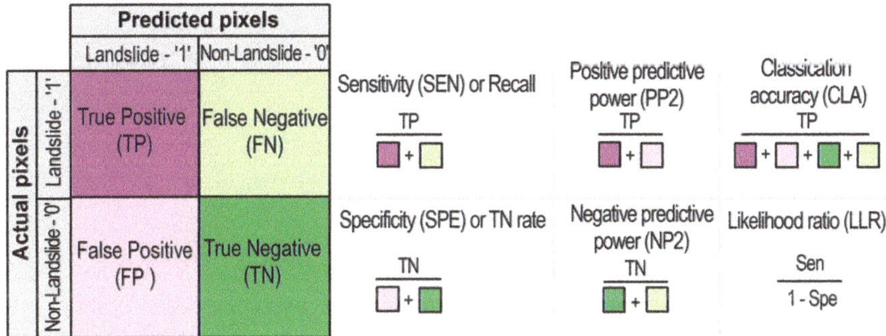

Figure 6. Confusion matrix and model measures used in this research.

For the case of CLA, although CLA provides the overall performance of the landslide model, however, a landslide model with a high CLA value may not classify the landslide pixels well. Therefore, the likelihood ratio (LLR) is additionally used [84]. LLR is a metric that assesses the trade-off of both SEN and SPE of landslide models. The higher the LLR value, the better the landslide model.

Global performance of the BE-LMTree model is summarized and assessed using the Receiver Operating Characteristic (ROC) Curve and Area Under the curve (AUC) [40,41,85]. In general, the closer the curve to the upper left corner, the better performance of the landslide model. Once the ROC curve is constructed, AUC for the model is computed and used to quantify the quality of the model. Accordingly, the performance of the model is excellent (AUC belong to 0.9–1), good (AUC belong to 0.8–0.9), fair (AUC belong to 0.7–0.8), and poor (AUC is less than 0.7) [86].

4.5. Computing Landslide Susceptibility Index

When the final BE-LMTree model is satisfied in the performance assessment check, the model is used to compute susceptibility index for all the pixels of the study area. These susceptibility indices are then converted to the ASCII raster format in ArcGIS using a Python application that was developed by the authors. Finally, the landslide susceptibility map is classified by five susceptibility classes: very high, high, moderate, low, and very low [87].

5. Results and Discussion

5.1. Predictive Ability Assessment

Result of the predictive ability evaluation of the eight predisposing factors is shown in Table 1. It is noted that the 10-fold cross validation was used to ensure the stable assessment result, as suggested in [88]. It could be seen that slope the highest predictive with the average merit (AM) is 0.225, followed by distance to river (AM of 0.171), lithology (AM of 0.148), aspect (AM of 0.129), and elevation (AM of 0.102). In contrast, soil type (AM of 0.038), distance to fault (AM of 0.055), and land cover (AM of 0.077) have low predictive ability values (Table 1).

The findings are reasonable because slope is widely recognized as the most important factor for landslide in various projects [89,90]. From the above results, it could be seen that all predisposing factors revealed predictive values to landslide model; therefore, we concluded that they are all relevant factors and are included in this analysis.

Table 1. Predictive ability of eight landslide predisposing factors using Pearson technique and 10-fold cross validation techniques.

No.	Predisposing Factors	Average Merit	Standard Deviation
1	Slope	0.225	0.008
2	Distance to river	0.171	0.008
3	Lithology	0.148	0.008
4	Aspect	0.129	0.008
5	Elevation	0.102	0.006
6	Land cover	0.077	0.008
7	Distance to fault	0.055	0.005
8	Soil type	0.038	0.005

5.2. Model Training and Evaluation

Using the eight predisposing factors, the BE-LMTree model was trained using the training dataset with the 10-fold cross validation technique. The training result is shown in Figure 7. It could be seen that the CLA of the BE-LMTree model is 93.81%, indicating a high degree of fit of the model with the dataset. Kappa statistics of 0.876 indicates the high agreement of the model and the training dataset. SEN of the BE-LMTree model is 93.02%, indicating that the proportion of the landslide pixels is correctly classified to the landslide class is 93.02%. Whereas, SPE is 94.63%, indicating that the proportion of the non-landslide pixels is correctly classified to the non-landslide class is 94.63%. PP2 is 94.72%, indicating that the probability that the BE-LMTree model correctly classifies pixels to the landslide class is 94.72%. NP2 is 92.89% indicating that the probability the BE-LMTree model correctly classifies pixels to the non-landslide class is 92.89%. Overall, these above measures have demonstrated that the BE-LMTree model performed very well with the training dataset.

To assess the contribution of landslide factors to the BE-LMT model, each factor was removed, and then, the classification accuracy (CLA) was estimated. The reduction of CLA of the BE-LMT model when one or more factors were removed indicates the contribution of these factors to the model. The result is shown in Table 2. It could be seen that when Distance to Fault and Soil type were removed from the LMT model, the CLA was reduced 2.12%. Therefore, although the average merit of Distance to fault (0.055) and Soil type (0.038) are small (see Table 1), the two factors contributed to 2.12% increasing classification accuracy of the BE-LMT model. An even larger accuracy decrease (4.3%, see Table 2) occurred when the four most significant variables (Slope, Distance to river, Lithology, and Aspect) are used into the BE-LMT model. Overall, it is reasonable of the to keep all factors in this research.

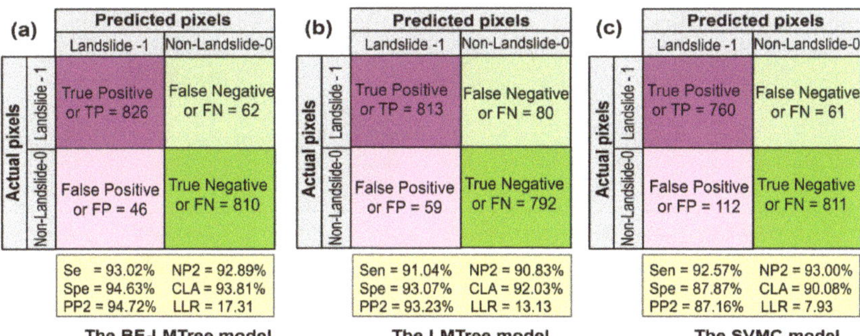

Figure 7. Confusion matrices and performance measures of the three landslide models using the training dataset: (a) the BE-LMTree model; (b) the LMTree model; and (c) the SVMC model.

Table 2. Contribution of the landslide predisposing factors to the BE-LMT model.

No.	Removing Factor	Classification Accuracy-CLA (%)
1	Slope	91.74
2	Aspect	92.31
3	Elevation	92.49
4	Land cover	93.60
5	Soil type	93.59
6	Lithology	91.97
7	Distance to fault	92.83
8	Distance to river	93.35
9	Distance to Fault and Soil type	91.69
10	Elevation, Land cover, Distance to fault and Soil type	89.51

The prediction performance of the BE-LMTree model is assessed using the validation dataset and the result is shown in Figure 8. It could be observed that the CLA is 87.89%, indicating a high prediction result. Kappa statistics of 0.759 indicates that the prediction performance of the model is 75.9% better than random. SEN of the BE-LMTree model is 92.25%, indicating that the proportion of the landslide pixels, which is accurately predicted, is 92.25%. SPE of the BE-LMTree model is 84.35%, indicating that the proportion of the non-landslide pixels is accurately predicted is 84.35%. PP2 of the model is 82.73%, indicating that the probability that the BE-LMTree model accurately predicts pixels to the landslide class is 82.73%. NP2 is 93.05%, indicating that the probability that the BE-LMTree model accurately predicts pixels to the non-landslide class is 93.05%.

Figure 9 shows 72 mispredicted landslide pixels (false positive) and 29 mispredicted non-landslide pixels (false negative) for the study area. We see that the 76.4% and 20.8% of the mispredicted landslide pixels were located in areas with slope angles <8.86° or slope angles from 36.39° to 5.87°, respectively. The mispredicted landslide pixels were also mainly located in elevation 174.78–358.94 m (76.4%), the lithology of sedimentary carbonate rocks (73.6%), the yellowish red soil on claystone and metamorphic rocks (87.5%), distance to fault >700 m (76.4%), and distance to river >200 m (79.2%). Distribution of the mispredicted landslide pixels in the classes in the other factors was more even. Regarding the mispredicted non-landslide pixels, they were mainly located in the distance to river >200 m (79.3%), the dense forest land (69.0%), and the yellowish red soil on claystone and metamorphic rocks (62.1%). For the other factors, the distribution of the mispredicted non-landslide pixels in their classes was quite even.

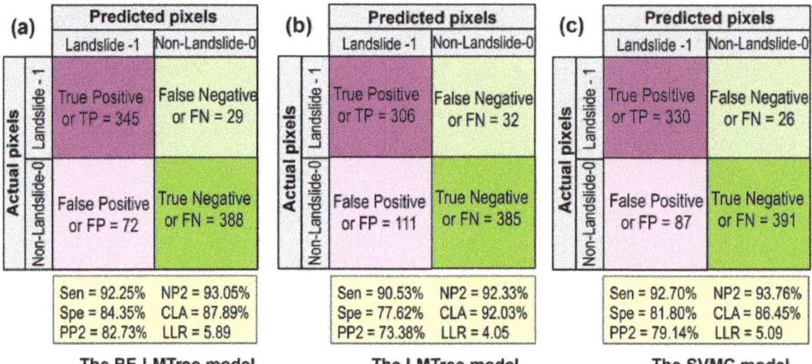

Figure 8. Confusion matrices and prediction measures of the three landslide models using the validation dataset: (**a**) the BE-LMTree model; (**b**) the LMTree model; and (**c**) the SVMC model.

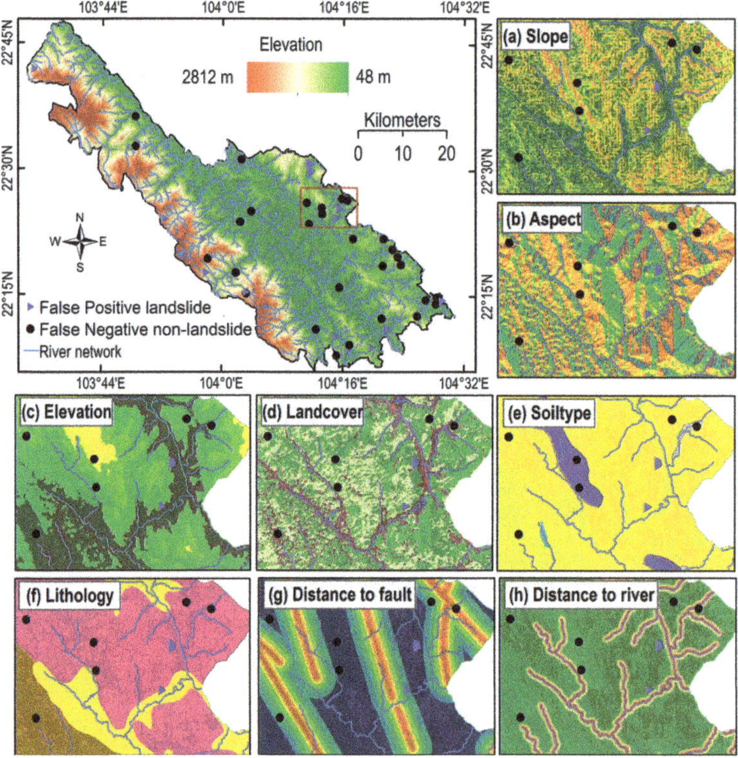

Figure 9. Mispredicted landslide pixels (false positive) and mispredicted non-landslide pixels in the validation dataset versus the eight landslide predisposing factors (legend for the eight factors was the same as in Figure 4). (**a**) Slope; (**b**) Aspect; (**c**) Elevation; (**d**) Landcover; (**e**) Soil type; (**f**) Lithology; (**g**) Distance to fault; and (**h**) Distance to river.

The global prediction capability of the BE-LMTree model is summarized and presented using the ROC curve and AUC (Figure 10). It can be seen that AUC is 0.834, indicating that the prediction capability of the proposed model is 83.4%, which is a high prediction capability.

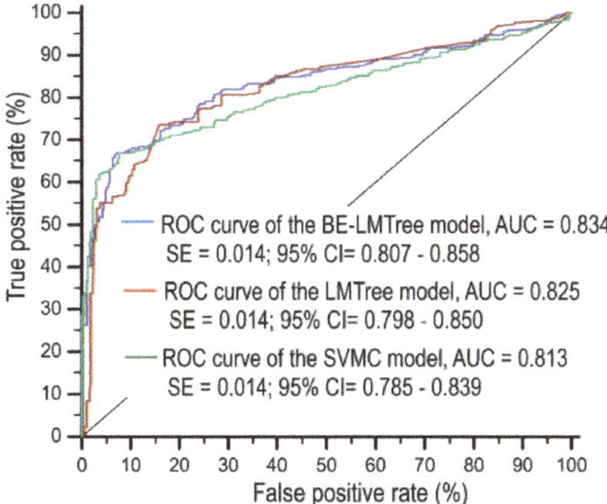

Figure 10. Receiver Operating Characteristic (ROC) curve and Area Under the curve (AUC) of the BE-LMTree model, the LMTree model, and the SVMC model using the validation dataset. SE: Standard Error; CI: Confidence Interval.

5.3. Comparison of the BE-LMTree Model with Benchmark

Because this is the first time that the BE-LMTree model is investigated for landslide modeling, the validity of the proposed model therefore was evaluated and compared with the benchmark. We select support vector machine (SVMC) as a benchmark because SVMC has proven efficient and outperforms other conventional methods [38,91]. For constructing the SVMC model, the radial basic function (RBF) kernel [41,92,93] was selected and the grid-search method [94–96] was used to derive the best the regularization (C = 9) and kernel width (γ = 0.245). In addition, the performance of the LMTree model was also included to present the merit of the proposed BE-LMTree model that is an integration of the Bagging ensemble and the LMTree.

The result is shown in Figures 7, 8, and 10. Using the training dataset, the CLA of the SVMC model (90.08%) and the LMTree model (92.03%) is slightly lower than CLA (93.81%) of the BE-LMTree model. Regarding LLR, the SVMC model (7.93) and the LMTree model (13.13) have lower values when compared to that of the BE-LMTree model (17.31). The other detailed metrics of the two models are shown in Figure 7. Overall, the BE-LMTree model performs better than the SVMC model and the LMTree model in the training dataset.

Using the validation dataset, the prediction performance of the SVMC model and the LMTree model is evaluated (Figure 8). It could be seen that the proposed BE-LMTree model (CLA = 87.98, LLR = 5.89) has a higher prediction performance when compared to those of the SVMC model (CLA = 86.45%, LLR = 5.09) and the LMTree model (CLA = 82.85%, LLR = 4.05). The global prediction capabilities of the three landslide models are assessed using the ROC curve and AUC (Figure 10). It could be been that the proposed BE-LMTree model (AUC = 0.834) is slightly higher than those of the SVMC model (AUC = 0.825) and the LMTree model (AUC = 0.813). Other detailed prediction performances of the three models are presented in Figure 8. Based on the aforementioned analysis, it could be concluded that the proposed BE-LMTree model is capable of producing the best landslide susceptibility result for this study area.

5.4. The Landslide Susceptibility Map

The final BE-LMTree model derived from the training step above was then used to compute landslide susceptibility indices for the Upper Reaches Area of Red River Basin (URRB), Vietnam. Accordingly, all of the predisposing factors in the raster maps were converted into ASCII format, and then fed to the BE-LMTree model to generate susceptibility indices. Distribution of these susceptibility indices is shown in Figure 11.

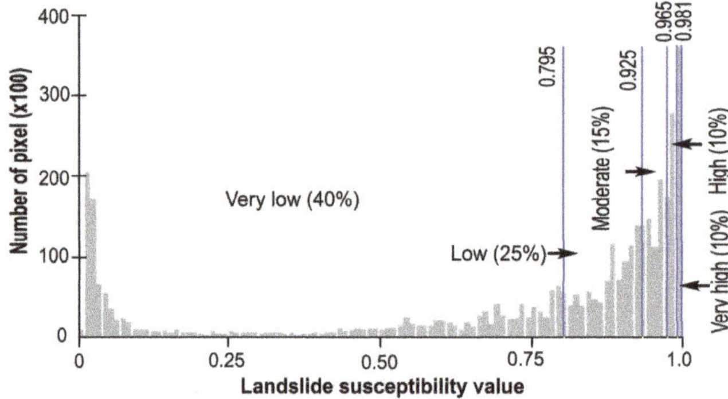

Figure 11. Distribution of these susceptibility indices versus of the five susceptibility classes.

These landslide susceptibility indices were then transformed to the raster format to manage in ArcGIS software using a python application that was programmed by the authors. Finally, the landslide susceptibility map (Figure 12) for the URRB was cartographically presented by five classes: very high (10%), high (10%), moderate (15%), low (25%), and very low (40%). To determine the thresholds for these classes, the extensively used graphic curve method has been considered to be the most suitable; a detailed explanation of it is available in [87,97,98]. The thresholds for these classes were determined based on an analysis of the susceptibility index map and the landslide inventory map, and then, the percentage of the landslide pixel versus the percentage of the susceptibility indices was calculated. At last, the four thresholds for the five classes were obtained.

Characteristics of the five landslide susceptibility classes that were derived from the BE-LMTree model the study area are shown in Table 3. Accordingly, the overall landslide frequency (OLF) proposed in [99] for the five classes was derived, and theoretically, the overall frequency should gradually grow from the very low class to the very high class [87]. It can be seen that the very high occupied only 10% of the study area, but it has the highest OLF value (4.40), followed by the high class (OLF = 1.59), the moderate class (OLF = 0.86), the low class (OLF = 0.43), and the very low class (OLF = 0.41). These confirm that the BE-LMTree model performed well with the URRB area.

Table 3. Characteristics of the landslide susceptibility classes derived from the BE-LMTree model the study area.

No.	Index Interval	Landslide Susceptibility (%)	Expression	Overall Landslide Frequency (OLF)	Areas (km²)
1	1.000–0.981	90–100	Very high	4.40	327.4
2	0.965–0.980	80–90	High	1.59	327.4
3	0.925–0.964	65–80	Moderate	0.86	491.0
4	0.795–0.924	40–65	Low	0.43	818.4
5	0.000–0. 794	0–50	Very low	0.41	1309.4

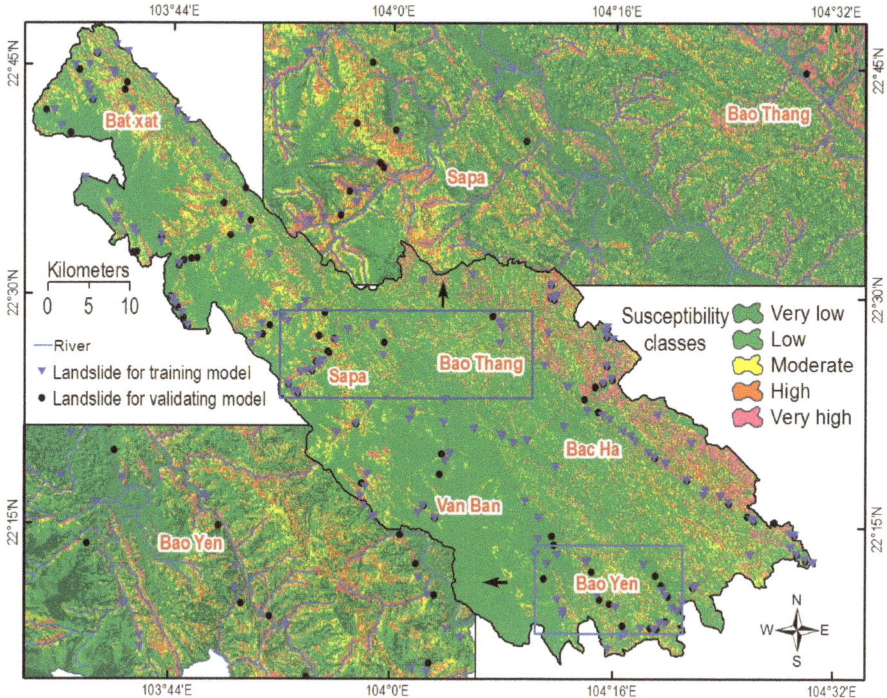

Figure 12. Landslide susceptibility map for the study area using the proposed BE-LMTree model.

Visual interpretation of the map (Figure 12) shows that the high probability of landslide is for areas i.e., Sapa, Bat Xat, and Bao Yen, therefore these areas should receive more attention in the development of remedial measures for the landslide prevention. Inversely, the low probability of landslide is for the Van Ban area. In fact, this area belongs to the Hoang Lien National Park, which is covered by the protected and dense tropical forest [100], therefore, having a low probability of landslide.

6. Concluding Remarks

This paper proposes a new modeling approach that is a hybrid intelligence of BE-LMTree for landslide susceptibility mapping with a case study at URRB. According to current literature, the BE-LMTree model has not been used for landslide modeling. For this purpose, the GIS database for the URRB area has been established, which contains a total of 255 historic soil-mixed-boulder slides and eight geo-environmental factors. These factors checked their merits to landslide using the Pearson correlation. The GIS database was then used to construct and verify the BE-LMTree model. Quality of the final BE-LMTree model was checked using confusion matrices and several model measures.

The results in this study point out that the new approach of the BE-LMTree could help to model landslide susceptibility with desirable prediction capability. When compared to the support vector machines (SVMC), a recognized benchmark in landslide modeling, the proposed BE-LMTree model presents a better performance. Therefore, the BE-LMTree is a new promising tool that could be used to enhance the quality of landslide susceptibility mapping.

For the case of the LMTree, this technique has been recently investigated for landslide susceptibility mapping with promising results i.e., in [50], the performance of the LMTree model in this research is lower than that of the SVMC model and the BE-LMTree model (Figures 4 and 5). Therefore, it could be concluded that the integration of the BE and the LMTree has significantly improved the quality of the LMTree model. This is due to the stability and robustness of the BE procedures itself with the ability to reduce variances [101]. This finding agrees with [51], who concluded that the performance of the landslide model is enhanced with the use of ensemble frameworks.

The main disadvantage of the proposed approach is that the quality of the BE-LMTree model is heavily controlled by the minimum number of samples (NS) that is used for growing the LMTree and the number of bootstrap subsets (BS) used in the BE. In this research, NS and BS were determined using an empirical test. Although the NS and the BS found results in the high performance BE-LMTree model, however these do not warrant them being the optimal parameters. Therefore, the performance of the BE-LMTree model may be further enhanced if optimization algorithms are considered to integrate in the model. In addition, the BE-LMTree may create a complex forest trees i.e., 50 trees in this research. Therefore, the interpretation of the BE-LMTree model may be complicated. Despite the aforementioned limitations, the BE-LMTree can be considered as a new and valid tool for landslide susceptibility modeling.

Author Contributions: X.L.T., M.M., Y.K., V.R., G.Y., X.Q.T., and T.H.D collected data and processed input data. X.L.T., X.Q.T., D.T.B., and S.L. carried out the modeling process and wrote the paper.

Funding: This work is supported by: (i) the Scientific Research Project 02/2012/HD—HTQTSP funded by Ministry of Education and Training, Vietnam; and (ii) by the Scientific Research Project DTNCCB-DHUD.2012-G/01 funded by NAFOSTED, Ministry of Science and Technology, Vietnam. This work is supported by the Basic Research Project of the Korea Institute of Geoscience and Mineral Resources (KIGAM) funded by the Minister of Science and ICT of Korea.

Acknowledgments: We would like to thank three anonymous reviewers for their valuable and constructive comments on the earlier version of the manuscript.

Conflicts of Interest: The authors declare no conflict of interest.

References

1. Huggel, C.; Clague, J.J.; Korup, O. Is climate change responsible for changing landslide activity in high mountains? *Earth Surf. Process. Landf.* **2012**, *37*, 77–91. [CrossRef]

2. Uchida, T.; Sakurai, W.; Okamoto, A. Historical Patterns of Heavy Rainfall Event and Deep-Seated Rapid Landslide Occurrence in Japan: Insight for Effects of Climate Change on Landslide Occurrence. In *Advancing Culture of Living with Landslides, Proceedings of the World Landslide Forum WLF 2017, Ljubljana, Slovenia, 29 May–2 June 2017*; Springer: Cham, Switzerland, 2017; pp. 251–257.

3. Ciervo, F.; Rianna, G.; Mercogliano, P.; Papa, M. Effects of climate change on shallow landslides in a small coastal catchment in southern Italy. *Landslides* **2017**, *14*, 1043–1055. [CrossRef]

4. Sewell, R.; Parry, S.; Millis, S.; Wang, N.; Rieser, U.; DeWitt, R. Dating of debris flow fan complexes from Lantau Island, Hong Kong, China: The potential relationship between landslide activity and climate change. *Geomorphology* **2015**, *248*, 205–227. [CrossRef]

5. Gallina, V.; Torresan, S.; Critto, A.; Sperotto, A.; Glade, T.; Marcomini, A. A review of multi-risk methodologies for natural hazards: Consequences and challenges for a climate change impact assessment. *J. Environ. Manag.* **2016**, *168*, 123–132. [CrossRef] [PubMed]

6. Montz, B.E.; Tobin, G.A.; Hagelman, R.R., III. *Natural Hazards: Explanation and Integration*; Guilford Publications: New York, NY, USA, 2017.

7. Maes, J.; Kervyn, M.; de Hontheim, A.; Dewitte, O.; Jacobs, L.; Mertens, K.; Vanmaercke, M.; Vranken, L.; Poesen, J. Landslide risk reduction measures: A review of practices and challenges for the tropics. *Prog. Phys. Geogr.* **2017**, *41*, 191–221. [CrossRef]

8. Gian, Q.A.; Tran, D.-T.; Nguyen, D.C.; Nhu, V.H.; Tien Bui, D. Design and implementation of site-specific rainfall-induced landslide early warning and monitoring system: a case study at Nam Dan landslide (Vietnam). *Geomat. Nat. Hazards Risk* **2017**, *8*, 1978–1996. [CrossRef]

9. Hung, L.Q.; Van, N.T.H.; Son, P.V.; Ninh, N.H.; Tam, N.; Huyen, N.T. Landslide Inventory Mapping in the Fourteen Northern Provinces of Vietnam: Achievements and Difficulties. In *Advancing Culture of Living with Landslides: Volume 1 ISDR-ICL Sendai Partnerships 2015–2025*; Sassa, K., Mikoš, M., Yin, Y., Eds.; Springer International Publishing: Cham, Switzerland, 2017; pp. 501–510.

10. Acosta, L.A.; Eugenio, E.A.; Macandog, P.B.M.; Magcale-Macandog, D.B.; Lin, E.K.-H.; Abucay, E.R.; Cura, A.L.; Primavera, M.G. Loss and damage from typhoon-induced floods and landslides in the Philippines: Community perceptions on climate impacts and adaptation options. *Int. J. Glob. Warm.* **2016**, *9*, 33–65. [CrossRef]

11. Shan, W.; Hu, Z.; Guo, Y.; Zhang, C.; Wang, C.; Jiang, H.; Liu, Y.; Xiao, J. The impact of climate change on landslides in southeastern of high-latitude permafrost regions of China. *Front. Earth Sci.* **2015**, *3*, 7. [CrossRef]

12. LeComte, D. International weather highlights 2014: Winter storms, typhoons, hurricanes, and flooding. *Weatherwise* **2015**, *68*, 20–26. [CrossRef]

13. Jiménez-Perálvarez, J.; El Hamdouni, R.; Palenzuela, J.; Irigaray, C.; Chacón, J. Landslide-hazard mapping through multi-technique activity assessment: An example from the Betic Cordillera (southern Spain). *Landslides* **2017**, *4*, 1975–1991. [CrossRef]

14. Pham, B.; Tien Bui, D.; Pourghasemi, H.; Indra, P.; Dholakia, M.B. Landslide susceptibility assesssment in the Uttarakhand area (India) using GIS: A comparison study of prediction capability of naïve bayes, multilayer perceptron neural networks, and functional trees methods. *Theor. Appl. Climatol.* **2015**, *128*, 255–273. [CrossRef]

15. Corominas, J.; van Westen, C.; Frattini, P.; Cascini, L.; Malet, J.P.; Fotopoulou, S.; Catani, F.; Van Den Eeckhaut, M.; Mavrouli, O.; Agliardi, F.; et al. . Recommendations for the quantitative analysis of landslide risk. *Bull. Eng. Geol. Environ.* **2014**, *73*, 209–263. [CrossRef]

16. Ciampalini, A.; Raspini, F.; Lagomarsino, D.; Catani, F.; Casagli, N. Landslide susceptibility map refinement using PSInSAR data. *Remote Sens. Environ.* **2016**, *184*, 302–315. [CrossRef]

17. Tien Bui, D.; Nguyen, Q.-P.; Hoang, N.-D.; Klempe, H. A Novel Fuzzy K-Nearest Neighbor Inference model with Differential Evolution for Spatial Prediction of Rainfall-Induced Shallow Landslides in a Tropical Hilly Area using GIS. *Landslides* **2017**, *14*, 1–17. [CrossRef]

18. Chacon, J.; Irigaray, C.; Fernandez, T.; El Hamdouni, R. Engineering geology maps: Landslides and geographical information systems. *Bull. Eng. Geol. Environ.* **2006**, *65*, 341–411. [CrossRef]

19. Van Westen, C.J.; Van Asch, T.W.J.; Soeters, R. Landslide hazard and risk zonation—Why is it still so difficult? *Bull. Eng. Geol. Environ.* **2006**, *65*, 167–184. [CrossRef]

20. Akgun, A.; Sezer, E.A.; Nefeslioglu, H.A.; Gokceoglu, C.; Pradhan, B. An easy-to-use MATLAB program (MamLand) for the assessment of landslide susceptibility using a Mamdani fuzzy algorithm. *Comput. Geosci.* **2012**, *38*, 23–34. [CrossRef]

21. Meng, Q.; Miao, F.; Zhen, J.; Wang, X.; Wang, A.; Peng, Y.; Fan, Q. GIS-based landslide susceptibility mapping with logistic regression, analytical hierarchy process, and combined fuzzy and support vector machine methods: A case study from Wolong Giant Panda Natural Reserve, China. *Bull. Eng. Geol. Environ.* **2016**, *75*, 923–944. [CrossRef]

22. Gheshlaghi, H.A.; Feizizadeh, B. An integrated approach of analytical network process and fuzzy based spatial decision making systems applied to landslide risk mapping. *J. Afr. Earth Sci.* **2017**, *133*, 15–24. [CrossRef]

23. Pham, B.T.; Tien Bui, D.; Prakash, I.; Dholakia, M.B. Rotation forest fuzzy rule-based classifier ensemble for spatial prediction of landslides using GIS. *Natl. Hazards* **2016**, *83*, 97–127. [CrossRef]

24. Tien Bui, D.; Pradhan, B.; Lofman, O.; Revhaug, I.; Dick, O.B. Landslide susceptibility assessment in the Hoa Binh province of Vietnam: A comparison of the Levenberg-Marquardt and Bayesian regularized neural networks. *Geomorphology* **2012**, *171–172*, 12–29. [CrossRef]

25. Yilmaz, I. Landslide susceptibility mapping using frequency ratio, logistic regression, artificial neural networks and their comparison: A case study from Kat landslides (Tokat-Turkey). *Comput. Geosci.* **2009**, *35*, 1125–1138. [CrossRef]

26. Pham, B.T.; Tien Bui, D.; Prakash, I.; Dholakia, M.B. Hybrid integration of Multilayer Perceptron Neural Networks and machine learning ensembles for landslide susceptibility assessment at Himalayan area (India) using GIS. *Catena* **2017**, *149 Pt 1*, 52–63. [CrossRef]

27. Gorsevski, P.V.; Brown, M.K.; Panter, K.; Onasch, C.M.; Simic, A.; Snyder, J. Landslide detection and susceptibility mapping using LiDAR and an artificial neural network approach: A case study in the Cuyahoga Valley National Park, Ohio. *Landslides* **2016**, *13*, 467–484. [CrossRef]

28. Oh, H.-J.; Lee, S. Shallow Landslide Susceptibility Modeling Using the Data Mining Models Artificial Neural Network and Boosted Tree. *Appl. Sci.* **2017**, *7*, 1000. [CrossRef]

29. Conforti, M.; Pascale, S.; Robustelli, G.; Sdao, F. Evaluation of prediction capability of the artificial neural networks for mapping landslide susceptibility in the Turbolo River catchment (northern Calabria, Italy). *Catena* **2014**, *113*, 236–250. [CrossRef]

30. Pascale, S.; Parisi, S.; Mancini, A.; Schiattarella, M.; Conforti, M.; Sole, A.; Murgante, B.; Sdao, F. Landslide susceptibility mapping using artificial neural network in the Urban area of Senise and San Costantino Albanese (Basilicata, Southern Italy). In *International Conference on Computational Science and Its Applications*; Springer: Berlin, Germany, 2013; pp. 473–488.

31. Yao, X.; Tham, L.G.; Dai, F.C. Landslide susceptibility mapping based on Support Vector Machine: A case study on natural slopes of Hong Kong, China. *Geomorphology* **2008**, *101*, 572–582. [CrossRef]

32. Kavzoglu, T.; Sahin, E.; Colkesen, I. Landslide susceptibility mapping using GIS-based multi-criteria decision analysis, support vector machines, and logistic regression. *Landslides* **2014**, *11*, 425–439. [CrossRef]

33. Kumar, D.; Thakur, M.; Dubey, C.S.; Shukla, D.P. Landslide susceptibility mapping & prediction using Support Vector Machine for Mandakini River Basin, Garhwal Himalaya, India. *Geomorphology* **2017**, *295*, 115–125.

34. Colkesen, I.; Sahin, E.K.; Kavzoglu, T. Susceptibility mapping of shallow landslides using kernel-based Gaussian process, support vector machines and logistic regression. *J. Afr. Earth Sci.* **2016**, *118*, 53–64. [CrossRef]

35. Pham, B.T.; Bui, D.T.; Prakash, I.; Nguyen, L.H.; Dholakia, M. A comparative study of sequential minimal optimization-based support vector machines, vote feature intervals, and logistic regression in landslide susceptibility assessment using GIS. *Environ. Earth Sci.* **2017**, *76*, 371. [CrossRef]

36. Hong, H.; Pradhan, B.; Bui, D.T.; Xu, C.; Youssef, A.M.; Chen, W. Comparison of four kernel functions used in support vector machines for landslide susceptibility mapping: A case study at Suichuan area (China). *Geomat. Natl. Hazards Risk* **2016**, *8*, 544–569. [CrossRef]

37. Pham, B.T.; Jaafari, A.; Prakash, I.; Bui, D.T. A novel hybrid intelligent model of support vector machines and the MultiBoost ensemble for landslide susceptibility modeling. *Bull. Eng. Geol. Environ.* **2018**. [CrossRef]

38. Pham, B.T.; Tien Bui, D.; Prakash, I. Bagging based Support Vector Machines for spatial prediction of landslides. *Environ. Earth Sci.* **2018**, *77*, 146. [CrossRef]

39. Tien Bui, D.; Pham, T.B.; Nguyen, Q.-P.; Hoang, N.-D. Spatial Prediction of Rainfall-Induced Shallow Landslides Using Hybrid Integration Approach of Least Squares Support Vector Machines and Differential Evolution Optimization: A Case Study in Central Vietnam. *Int. J. Dig. Earth* **2016**, *9*, 1077–1097. [CrossRef]

40. Tien Bui, D.; Anh Tuan, T.; Hoang, N.-D.; Quoc Thanh, N.; Nguyen, B.D.; Van Liem, N.; Pradhan, B. Spatial Prediction of Rainfall-induced Landslides for the Lao Cai area (Vietnam) Using a Novel hybrid Intelligent Approach of Least Squares Support Vector Machines Inference Model and Artificial Bee Colony Optimization. *Landslides* **2017**, *14*, 447–458. [CrossRef]

41. Hoang, N.-D.; Tien Bui, D. A Novel Relevance Vector Machine Classifier with Cuckoo Search Optimization for Spatial Prediction of Landslides. *J. Comput. Civ. Eng.* **2016**, *30*, 1–10. [CrossRef]

42. Althuwaynee, O.F.; Pradhan, B.; Lee, S. A novel integrated model for assessing landslide susceptibility mapping using CHAID and AHP pair-wise comparison. *Int. J. Remote Sens.* **2016**, *37*, 1190–1209. [CrossRef]

43. Youssef, A.M.; Pourghasemi, H.R.; Pourtaghi, Z.S.; Al-Katheeri, M.M. Landslide susceptibility mapping using random forest, boosted regression tree, classification and regression tree, and general linear models and comparison of their performance at Wadi Tayyah Basin, Asir Region, Saudi Arabia. *Landslides* **2015**, *13*, 839–856. [CrossRef]

44. Lagomarsino, D.; Tofani, V.; Segoni, S.; Catani, F.; Casagli, N. A Tool for Classification and Regression Using Random Forest Methodology: Applications to Landslide Susceptibility Mapping and Soil Thickness Modeling. *Environ. Model. Assess.* **2017**, *22*, 201–214. [CrossRef]

45. Tsangaratos, P.; Ilia, I. Landslide susceptibility mapping using a modified decision tree classifier in the Xanthi Perfection, Greece. *Landslides* **2015**, *13*, 305–320. [CrossRef]

46. Kim, J.-C.; Lee, S.; Jung, H.-S.; Lee, S. Landslide susceptibility mapping using random forest and boosted tree models in Pyeong-Chang, Korea. *Geocarto Int.* **2017**, *33*, 1000–1015. [CrossRef]

47. Hong, H.; Liu, J.; Bui, D.T.; Pradhan, B.; Acharya, T.D.; Pham, B.T.; Zhu, A.X.; Chen, W.; Ahmad, B.B. Landslide susceptibility mapping using J48 Decision Tree with AdaBoost, Bagging and Rotation Forest ensembles in the Guangchang area (China). *CATENA* **2018**, *163*, 399–413. [CrossRef]

48. Hoang, N.-D.; Tien Bui, D. Spatial prediction of rainfall-induced shallow landslides using gene expression programming integrated with GIS: A case study in Vietnam. *Natl. Hazards* **2018**, *92*, 1871–1887. [CrossRef]

49. Pradhan, B. A comparative study on the predictive ability of the decision tree, support vector machine and neuro-fuzzy models in landslide susceptibility mapping using GIS. *Comput. Geosci.* **2013**, *51*, 350–365. [CrossRef]

50. Tien Bui, D.; Tuan, T.A.; Klempe, H.; Pradhan, B.; Revhaug, I. Spatial prediction models for shallow landslide hazards: A comparative assessment of the efficacy of support vector machines, artificial neural networks, kernel logistic regression, and logistic model tree. *Landslides* **2016**, *13*, 361–378. [CrossRef]

51. Tien Bui, D.; Ho, T.-C.; Pradhan, B.; Pham, B.-T.; Nhu, V.-H.; Revhaug, I. GIS-Based Modeling of Rainfall-Induced Landslides Using Data Mining Based Functional Trees Classifier with AdaBoost, Bagging, and MultiBoost Ensemble Frameworks. *Environ. Earth Sci.* **2016**, *75*, 1101–1123. [CrossRef]

52. Landwehr, N.; Hall, M.; Frank, E. Logistic Model Trees. *Mach. Learn.* **2005**, *59*, 161–205. [CrossRef]

53. Breiman, L. Bagging Predictors. *Mach. Learn.* **1996**, *24*, 123–140. [CrossRef]

54. Tien Bui, D.; Ho, T.C.; Revhaug, I.; Pradhan, B.; Nguyen, D. Landslide Susceptibility Mapping Along the National Road 32 of Vietnam Using GIS-Based J48 Decision Tree Classifier and Its Ensembles. In *Cartography from Pole to Pole*; Buchroithner, M., Prechtel, N., Burghardt, D., Eds.; Springer: Berlin/Heidelberg, Germany, 2013; pp. 303–317.

55. Pham, T.D.; Bui, D.T.; Yoshino, K.; Le, N.N. Optimized rule-based logistic model tree algorithm for mapping mangrove species using ALOS PALSAR imagery and GIS in the tropical region. *Environ. Earth Sci.* **2018**, *77*, 159. [CrossRef]

56. Breiman, L.; Friedman, J.H.; Olshen, R.A.; Stone, C.J. *Classification and Regression Trees*; Chapman and Hall/CRC: New York, NY, USA, 1984.

57. Doetsch, P.; Buck, C.; Golik, P.; Hoppe, N.; Kramp, M.; Laudenberg, J.; Oberdörfer, C.; Steingrube, P.; Forster, J.; Mauser, A. Logistic Model Trees with AUC Split Criterion for the KDD Cup 2009 Small Challenge. In Proceedings of the 2009 International Conference on KDD-Cup 2009, Paris, France, 28 June–1 July 2009; pp. 77–88.

58. Quinlan, J.R. *C4.5: Programs for Machine Learning*; Morgan Kaufmann: San Mateo, CA, USA, 1993.

59. Kuncheva, L.I. *Combining Pattern Classifiers: Methods and Algorithms*, 2nd ed.; John Wiley & Sons: Hoboken, NJ, USA, 2014.

60. Kotsiantis, S. Combining bagging, boosting, rotation forest and random subspace methods. *Artif. Intell. Rev.* **2011**, *35*, 223–240. [CrossRef]

61. Lu, X.X.; Oeurng, C.; Le, T.P.Q.; Thuy, D.T. Sediment budget as affected by construction of a sequence of dams in the lower Red River, Viet Nam. *Geomorphology* **2015**, *248*, 125–133. [CrossRef]

62. Do, T.; Nguyen, C.; Phung, T. *Assessment of Natural Disasters in Vietnam's Northern Mountains*; Munich University Library: Munich, Germany, 2013; p. 57.

63. Tran, T. Climate change adaptation from small and medium scale hydropower plants: A case study for Lao Cai province. *VNU J. Sci. Earth Environ. Sci.* **2011**, *27*, 32–38.

64. Jolivet, L.; Beyssac, O.; Goffe, B.; Avigad, D.; Lepvrier, C.; Maluski, H.; Thang, T.T. Oligo-Miocene midcrustal subhorizontal shear zone in Indochina. *Tectonics* **2001**, *20*, 46–57. [CrossRef]

65. Duan, B.V. The relation between fault movement potential and seismic activity of major faults in Northwestern Vietnam. *Vietnam J. Earth Sci.* **2017**, *39*, 240–255. [CrossRef]

66. Hue, T.T.; Duong, T.V.; Toan, D.V.; Nghinh, L.T.; Minh, V.C.; Pho, N.V.; Xuan, P.T.; Hoan, L.T.; Huyen, N.X.; Pha, P.D.; et al. *Investigation and Assessment of the Types of Geological Hazard in the Territory of Vietnam and Recommendation of Remedial Measures. Phase II: A Study of the Northern Mountainous Province of Vietnam*; Institute of Geological Sciences, Vietnam Academy of Science and Technology: Hanoi, Vietnam, 2004; p. 361.

67. Yem, N.T.; Thanh, N.Q.; Anh, P.L.; Chi, C.T.; Du, C.D.; Dung, N.P.; Dung, P.D.; Hai, N.P.; Hien, T.T.; Hoang, N.V.; et al. *Assessment of Landslides and Debris Flows at Some Prone Mountainouns Areas Vietnam and Recommendation of Remedial Measures. Phase I: A Study of the East Side of the Hoang Lien Son Mountainous Area of Vietnam*; Institute of Geological Sciences, Vietnam Academy of Science and Technology: Hanoi, Vietnam, 2006; p. 361.

68. Van, T.T.; Tuy, P.K.; Giap, N.X.; Ke, T.D.; Thai, T.N.; Giang, N.T.; Tho, H.M.; Tuat, L.T.; San, D.N.; Hung, L.Q.; et al. *Assessment and Prediction of Geological Hazards in the 8 Coastal Provinces of Central Vietnam from Quang Binh to Phu Yen—Current Status, Causes, Prediction and Recommendation of Remedial Measures*; Vietnam Institude of Geosciences and Mineral Resourses: Hanoi, Vietnam, 2002; p. 215.

69. Van, T.T.; Anh, D.T.; Hieu, H.H.; Giap, N.X.; Ke, T.D.; Nam, T.D.; Ngoc, D.; Ngoc, D.T.Y.; Thai, T.N.; Thang, D.V.; et al. *Investigation and Assessment of the Current Status and Potential of Landslides in Some Sections of the Ho Chi Minh Road, National Road 1A and Proposed Remedial Measures to Prevent Landslides from Threat of Safety of People, Property, and Infrastructure*; Vietnam Institute of Geosciences and Mineral Resources: Hanoi, Vietnam, 2006; p. 249.

70. Tien Bui, D.; Pradhan, B.; Lofman, O.; Revhaug, I.; Dick, O.B. Landslide susceptibility mapping at Hoa Binh province (Vietnam) using an adaptive neuro-fuzzy inference system and GIS. *Comput. Geosci.* **2012**, *45*, 199–211. [CrossRef]

71. Cevik, E.; Topal, T. GIS-based landslide susceptibility mapping for a problematic segment of the natural gas pipeline, Hendek (Turkey). *Environ. Geol.* **2003**, *44*, 949–962. [CrossRef]

72. Conforti, M.; Pascale, S.; Pepe, M.; Sdao, F.; Sole, A. Denudation processes and landforms map of the Camastra River catchment (Basilicata–South Italy). *J. Maps* **2013**, *9*, 444–455. [CrossRef]

73. Yilmaz, I. A case study from Koyulhisar (Sivas-Turkey) for landslide susceptibility mapping by artificial neural networks. *Bull. Eng. Geol. Environ.* **2009**, *68*, 297–306. [CrossRef]

74. Pachauri, A.; Pant, M. Landslide hazard mapping based on geological attributes. *Eng. Geol.* **1992**, *32*, 81–100. [CrossRef]

75. Witten, I.H.; Frank, E.; Hall, M.A.; Pal, C.J. *Data Mining: Practical Machine Learning Tools and Techniques*; Morgan Kaufmann: San Mateo, CA, USA, 2016.

76. Zeiller, M. *Modeling Our World: The ESRI Guide to Geodatabase Concepts*; ESRI Press: Redlands, CA, USA, 2010.

77. Tien Bui, D.; Hoang, N.-D. A Bayesian framework based on a Gaussian mixture model and radial-basis-function Fisher discriminant analysis (BayGmmKda V1. 1) for spatial prediction of floods. *Geosci. Model Dev.* **2017**, *10*, 3391. [CrossRef]

78. Dang, V.-H.; Dieu, T.B.; Tran, X.-L.; Hoang, N.-D. Enhancing the accuracy of rainfall-induced landslide prediction along mountain roads with a GIS-based random forest classifier. *Bull. Eng. Geol. Environ.* **2018**. [CrossRef]

79. Goetz, J.N.; Brenning, A.; Petschko, H.; Leopold, P. Evaluating machine learning and statistical prediction techniques for landslide susceptibility modeling. *Comput. Geosci.* **2015**, *81*, 1–11. [CrossRef]

80. Micheletti, N.; Foresti, L.; Robert, S.; Leuenberger, M.; Pedrazzini, A.; Jaboyedoff, M.; Kanevski, M. Machine learning feature selection methods for landslide susceptibility mapping. *Math. Geosci.* **2014**, *46*, 33–57. [CrossRef]

81. Erener, A.; Sivas, A.A.; Selcuk-Kestel, A.S.; Düzgün, H.S. Analysis of training sample selection strategies for regression-based quantitative landslide susceptibility mapping methods. *Comput. Geosci.* **2017**, *104*, 62–74. [CrossRef]

82. Nguyen, Q.-K.; Tien Bui, D.; Hoang, N.-D.; Trinh, P.T.; Nguyen, V.-H.; Yilmaz, I. A Novel Hybrid Approach Based on Instance Based Learning Classifier and Rotation Forest Ensemble for Spatial Prediction of Rainfall-Induced Shallow Landslides using GIS. *Sustainability* **2017**, *9*, 813. [CrossRef]

83. Guyon, I.; Elisseeff, A. An introduction to variable and feature selection. *J. Mach. Learn. Res.* **2003**, *3*, 1157–1182.

84. Lagomarsino, D.; Segoni, S.; Rosi, A.; Rossi, G.; Battistini, A.; Catani, F.; Casagli, N. Quantitative comparison between two different methodologies to define rainfall thresholds for landslide forecasting. *Natl. Hazards Earth Syst. Sci.* **2015**, *15*, 2413–2423. [CrossRef]

85. Lucà, F.; Conforti, M.; Robustelli, G. Comparison of GIS-based gullying susceptibility mapping using bivariate and multivariate statistics: Northern Calabria, South Italy. *Geomorphology* **2011**, *134*, 297–308. [CrossRef]

86. Cantor, S.B.; Kattan, M.W. Determining the area under the ROC curve for a binary diagnostic test. *Med. Decis. Mak.* **2000**, *20*, 468–470. [CrossRef] [PubMed]

87. Pradhan, B.; Lee, S. Landslide susceptibility assessment and factor effect analysis: Backpropagation artificial neural networks and their comparison with frequency ratio and bivariate logistic regression modelling. *Environ. Model. Softw.* **2010**, *25*, 747–759. [CrossRef]

88. Fushiki, T. Estimation of prediction error by using K-fold cross-validation. *Stat. Comput.* **2011**, *21*, 137–146. [CrossRef]

89. Van Den Eeckhaut, M.; Vanwalleghem, T.; Poesen, J.; Govers, G.; Verstraeten, G.; Vandekerckhove, L. Prediction of landslide susceptibility using rare events logistic regression: A case-study in the Flemish Ardennes (Belgium). *Geomorphology* **2006**, *76*, 392–410. [CrossRef]

90. Costanzo, D.; Rotigliano, E.; Irigaray, C.; Jiménez-Perálvarez, J.D.; Chacón, J. Factors selection in landslide susceptibility modelling on large scale following the gis matrix method: Application to the river Beiro basin (Spain). *Natl. Hazards Earth Syst. Sci.* **2012**, *12*, 327–340. [CrossRef]

91. Tien Bui, D.; Pradhan, B.; Lofman, O.; Revhaug, I.; Dick, O. Regional prediction of landslide hazard using probability analysis of intense rainfall in the Hoa Binh province, Vietnam. *Natl. Hazards* **2013**, *66*, 707–730. [CrossRef]

92. Hoang, N.-D.; Tien Bui, D. Predicting earthquake-induced soil liquefaction based on a hybridization of kernel Fisher discriminant analysis and a least squares support vector machine: A multi-dataset study. *Bull. Eng. Geol. Environ.* **2018**, *77*, 191–204. [CrossRef]

93. Hoang, N.-D.; Tien Bui, D.; Liao, K.-W. Groutability estimation of grouting processes with cement grouts using Differential Flower Pollination Optimized Support Vector Machine. *Appl. Soft Comput.* **2016**, *45*, 173–186. [CrossRef]

94. Ngoc-Thach, N.; Ngo, D.B.-T.; Xuan-Canh, P.; Hong-Thi, N.; Thi, B.H.; NhatDuc, H.; Dieu, T.B. Spatial pattern assessment of tropical forest fire danger at Thuan Chau area (Vietnam) using GIS-based advanced machine learning algorithms: A comparative study. *Ecol. Inform.* **2018**, *46*, 74–85. [CrossRef]

95. Vafaei, S.; Soosani, J.; Adeli, K.; Fadaei, H.; Naghavi, H.; Pham, T.D.; Tien Bui, D. Improving Accuracy Estimation of Forest Aboveground Biomass Based on Incorporation of ALOS-2 PALSAR-2 and Sentinel-2A Imagery and Machine Learning: A Case Study of the Hyrcanian Forest Area (Iran). *Remote Sens.* **2018**, *10*, 172. [CrossRef]

96. Tien Bui, D.; Pradhan, B.; Lofman, O.; Revhaug, I.; Dick, O.B. Application of support vector machines in landslide susceptibility assessment for the Hoa Binh province (Vietnam) with kernel functions analysis. In *iEMSs 2012—Managing Resources of a Limited Planet, Proceedings of the 6th Biennial Meeting of the International Environmental Modelling and Software Society, Leipzig, Germany, 1 July 2012*; Brigham Young University: Provo, UT, USA, 2012; pp. 382–389.

97. Chung, C.-J.; Fabbri, A.G. Predicting landslides for risk analysis—Spatial models tested by a cross-validation technique. *Geomorphology* **2008**, *94*, 438–452. [CrossRef]

98. Tien Bui, D.; Pradhan, B.; Lofman, O.; Revhaug, I.; Dick, O.B. Spatial prediction of landslide hazards in Hoa Binh province (Vietnam): A comparative assessment of the efficacy of evidential belief functions and fuzzy logic models. *Catena* **2012**, *96*, 28–40. [CrossRef]

99. Sarkar, S.; Kanungo, D.P. An integrated approach for landslide susceptibility mapping using remote sensing and GIS. *Photogramm. Eng. Remote Sens.* **2004**, *70*, 617–625. [CrossRef]

100. Kieu, Q.L.; Nguyen, T.T. Study on the distribution characteristics of the vegetation in high levations in Hoang Lien National park of Vietnam. *J. Vietnam. Environ.* **2015**, *6*, 84–88.

101. Mert, A.; Kılıç, N.; Akan, A. Evaluation of bagging ensemble method with time-domain feature extraction for diagnosing of arrhythmia beats. *Neural Comput. Appl.* **2014**, *24*, 317–326. [CrossRef]

Article

Learning-Based Colorization of Grayscale Aerial Images Using Random Forest Regression

Dae Kyo Seo [1], Yong Hyun Kim [2], Yang Dam Eo [3,*] and Wan Yong Park [4]

[1]   Department of Advanced Technology Fusion, Konkuk University, Seoul 05029, Korea; tjeory@konkuk.ac.kr
[2]   Department of Civil and Environmental Engineering, Seoul National University, Seoul 08826, Korea; yhkeen@gmail.com
[3]   Department of Technology Fusion Engineering, Konkuk University, Seoul 05029, Korea
[4]   Agency for Defense Development, Daejeon 34060, Korea; wypark@add.re.kr
*   Correspondence: eoandrew@konkuk.ac.kr; Tel.: +82-2-450-3078

Received: 18 April 2018; Accepted: 27 July 2018; Published: 31 July 2018

Abstract: Image colorization assigns colors to a grayscale image, which is an important yet difficult image-processing task encountered in various applications. In particular, grayscale aerial image colorization is a poorly posed problem that is affected by the sun elevation angle, seasons, sensor parameters, etc. Furthermore, since different colors may have the same intensity, it is difficult to solve this problem using traditional methods. This study proposes a novel method for the colorization of grayscale aerial images using random forest (RF) regression. The algorithm uses one grayscale image for input and one-color image for reference, both of which have similar seasonal features at the same location. The reference color image is then converted from the Red-Green-Blue (RGB) color space to the CIE L*a*b (Lab) color space in which the luminance is used to extract training pixels; this is done by performing change detection with the input grayscale image, and color information is used to establish color relationships. The proposed method directly establishes color relationships between features of the input grayscale image and color information of the reference color image based on the corresponding training pixels. The experimental results show that the proposed method outperforms several state-of-the-art algorithms in terms of both visual inspection and quantitative evaluation.

Keywords: colorization; random forest regression; grayscale aerial image; change detection

---

1. Introduction

Image colorization can be described as the process of assigning colors to the pixels of a grayscale image in order to increase the image's visual appeal [1]. This application is often utilized in the image processing community to colorize old grayscale images or movies [2]. Particularly in aerial and satellite imagery, the problem with colorization occurs in a multitude of scenarios that seek to replace topographic maps with vivid, photorealistic renderings of terrain models [3]. There are three main reasons for colorizing aerial and satellite images: (1) grayscale satellite images are available at higher spatial resolutions than their color counterparts are; (2) there are many old grayscale aerial and satellite images that should be represented by color images, typically for monitoring purposes; and (3) grayscale aerial–satellite images can be obtained for approximately one-tenth the cost of color images of the same resolution [3].

In the case of satellite images, modern systems acquire a panchromatic (grayscale) image, with high spatial and low spectral resolutions, and a multispectral (color) image that has complementary properties [4]. In other words, grayscale and color images with different resolutions over the same time period are provided. In order to perform colorization through the color information of the multispectral image while maintaining the high resolution of the panchromatic image, the two components are fused, which is called pansharpening [5]. The fused images provide increased interpretation capabilities and more reliable results [6]. However, this colorization method is confined to satellite images that provide panchromatic and multispectral images of the same time periods, making this type of colorization of aerial images impossible.

On the other hand, grayscale aerial image colorization is a poorly posed problem with more than one solution [7]. As mentioned above, the satellite images fuse the grayscale and color images from the same time period to perform colorization, whereas aerial images do not usually have the two types of imagery available. Furthermore, in contrast to the natural images, this colorization solution depends on the sun elevation angle, season, sensor parameters, etc. It is also problematic that the same intensity may represent different colors, so there is no exact solution [8]. In general, existing colorization methods can be divided into three main categories, all of which have limitations: user-scribbled methods, example-based methods, and those that employ a large number of training images [9]. User-scribbled techniques [10–13] require a user to manually add colored marks to a grayscale image [13]. The colors from these marks are then smoothly propagated across the entire image, based on an optimization framework. A key weakness is that such methods require users to provide a considerable number of scribbles on the grayscale image, which is time-consuming and requires expertise [14]. Moreover, it is almost impossible to add such markings to large volumes (gigabytes) of aerial imagery. In the case of the example-based method [1,8,9,14–16], it typically transfers the color information from a similar reference image to the input grayscale image rather than obtaining chromatic values from the user, thereby reducing the burden on users. However, as feature matching is critical to the quality of the results, satisfactory results cannot be obtained if feature matching is not performed correctly [15]. Moreover, the procedure is very sensitive to image brightness and contrast, whereas real aerial images always include large areas of shadow and low contrast, due to relief, vignetting, and so on. An alternative approach is to employ a large number of training images [17–19], which is a recent example of deep learning. These methods use multiple color images to automatically transfer the color information to the grayscale image, and deep neural networks are used to solve the colorization problem. A large database of color images comprising all kinds of objects is used for training the neural networks. The trained model can then be used to efficiently colorize grayscale images. However, this approach is computationally expensive, and the training is significantly slow [15].

In order to overcome these limitations, this study presents a new, fast learning-based technique for the colorization of grayscale aerial images that colorizes them without any user intervention. The algorithm uses one grayscale image as the input and one-color image for reference, both of which have similar seasonal features at the same location. Then, the reference color image is converted from the Red-Green-Blue (RGB) color space to a CIE L*a*b (Lab) color space in which luminance and two-dimensional (2D) color information are stored. Change detection between the input grayscale image and the luminance of the reference color image is performed, and the unchanged region is selected as training pixels, which allows for the extraction of meaningful training data. For colorization, the relationships are established through learning between features of the input grayscale image and the 2D color information of the reference color image based on training pixels. In other words, for the corresponding unchanged region, the color relationships between the two images are directly established, and the colors for the changed region are predicted. At this time, the study's main technical framework is random forest (RF) regression. Random forest is a data-mining method that has some advantages over most statistical modeling methods [20], including: the ability to model highly nonlinear dimensional relationships; resistance to overfitting; relative robustness with respect to the presence of noise in the data; and the capacity to determine the relevance of the variables used.

The main contributions of this paper can be summarized as follows: (1) to the best of our knowledge, this is the first work that exploits RF regression for aerial imagery colorization, although it has been used for natural image colorization [21–23]; (2) this paper develops a novel algorithm that establishes color relationships based on unchanged regions, which predict the color values of the changed regions; (3) this paper establishes color relationships by directly mapping the features of the input grayscale image with the color information of the reference color image; and (4) this paper performs visual and quantitative analyses that show that our method outperforms the current state-of-the-art methods. The rest of this paper is organized as follows. Section 2 describes the materials used in detail, the background of RF regression, and the proposed algorithm. Section 3 presents the colorization results and a detailed comparison with other state-of-the-art colorization algorithms. Section 4 presents the conclusions of the study.

2. Materials and Methods

2.1. Study Site and Data

The study sites are located in Gwangjin-gu, in the central–western part of South Korea. The input grayscale images were acquired on 10 June 2013, and the reference color images were acquired on 2 June 2016; both are aerial images at 1 m resolution. Coordinate definition and geometric correction were performed on the images using Environment for Visualizing Images (ENVI) geospatial analytical software (version 4.7, HARRIS Geospatial Solutions, Broomfield, CO, USA). The coordinate system of each image was projected as World Geodetic System (WGS) 84 Universal Transverse Mercator Coordinate System (UTM) 52N, and 30 Ground Control Points (GCPs) were selected for image registration. The 30 GCPs returned a total root mean square error of 0.4970, satisfying values within 0.5 m. Then, based on these GCPs, image registration was performed using the "Warp from GCPs: Image to Image" tool in ENVI. Finally, to achieve reasonable computational time, only a portion of the image was extracted prior to conducting the experiments, which was selected to be 1200 × 1200 pixels. A total of four sites were extracted and are shown in Figures 1–4.

(a)   (b)

Figure 1. Experimental area of site 1: (**a**) input grayscale image acquired on 10 June 2013, (**b**) reference color image acquired on 2 June 2016.

(a)   (b)

Figure 2. Experimental area of site 2: (**a**) input grayscale image acquired on 10 June 2013, (**b**) reference color image acquired on 2 June 2016.

(**a**)　　　　　　　　　　　(**b**)

Figure 3. Experimental area of site 3: (**a**) input grayscale image acquired on 10 June 2013, (**b**) reference color image acquired on 2 June 2016.

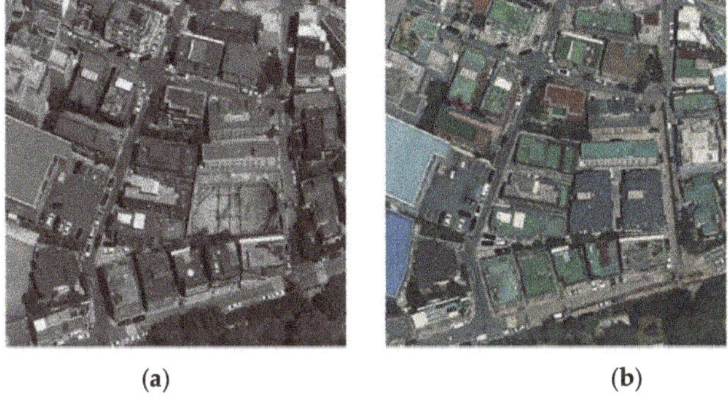

(**a**)　　　　　　　　　　　(**b**)

Figure 4. Experimental area of site 4: (**a**) input grayscale image acquired on 10 June 2013, (**b**) reference color image acquired on 2 June 2016.

2.2. Random Forest

Random forest is a highly versatile ensemble of decision trees that performs well for linear and non-linear prediction by finding a balance between bias and variance [20]. This ensemble learning method constructs and subsequently averages a large number of decision trees for classification or regression purposes [24,25]. At this time, to avoid correlation among the trees, RF increases the diversity of the trees by forcing them to grow from different training data created through a procedure called bootstrap aggregating (bagging) [26]. Bagging refers to aggregating base learners trained through bootstrapping, which creates training data subsets by randomly resampling a given original dataset [27]. That is, as a process of de-correlating trees to train different datasets, it increases stability and makes it more robust when facing slight variations in the input data [28]. Furthermore, approximately one-third of the samples are excluded for the tree training in the bagging process, which is known as the "out-of-bags" (OOB) samples. These OOB samples can be used to evaluate performance, which allows the RF to compute an unbiased estimation of the generalization error without using an external data subset [29]. From the predictions of the OOB samples for every tree in the forest, the mean square error (MSE) is calculated, and the overall MSE is obtained by aggregation, as shown in Equation (1):

$$MSE_{OOB} = \frac{1}{n} \sum_{i=1}^{n} (y_i - \overline{\hat{y}_{i_{OOB}}})^2 \tag{1}$$

where n is the number of trees, y_i denotes the prediction for the ith observation, and $\overline{\hat{y}_{i_{OOB}}}$ denotes the average of the OOB predictions for the ith observation.

Furthermore, the use of RF requires the specification of two standard parameters: the number of variables to be selected and tested for the best split at each node (mtry), and the number of trees to be grown (ntree). At each node per tree, the number of mtry variables from the total variables in the model is selected at random, the variable that best splits the input space and the corresponding split are computed, and the input space is split at this point [30]. In a regression problem, the standard value for mtry is one-third of the total number of variables due to computational benefits [31,32]. In the case of ntree, the majority of the studies set the ntree value to 500 since the errors are stabilized before this number of regression trees is achieved [33]. However, recent studies have found that the number of trees does not significantly affect performance improvement, which allows the selection of ntree to consider the performance and training time together [34–36].

Random forest can also be used to assess the importance of each variable during modeling. To determine the importance of input variables, a variable is randomly permuted, and regression trees are grown on the modified dataset. The measure of each variable's importance is then calculated as the difference in the MSE between the original OOB dataset and the modified dataset [37,38]. A key advantage of RF variable importance is that it not only deals with the impact of each variable individually but also looks at multivariate interactions with other variables [39].

2.3. Method

The proposed colorization framework can be decomposed into four steps: (1) pre-processing, (2) feature extraction, (3) colorization, and (4) post-processing, all of which are shown in Figure 5. The first step is to convert the color space and to select the pixels to be used in training for colorization. The second step is to extract feature descriptors of the input grayscale image to be used in learning for color prediction. In the third step, color relationships are established through the proposed method, and colorization is performed on the input grayscale image. The fourth step improves the colorization result by adjusting the histogram. Each of these steps is described below.

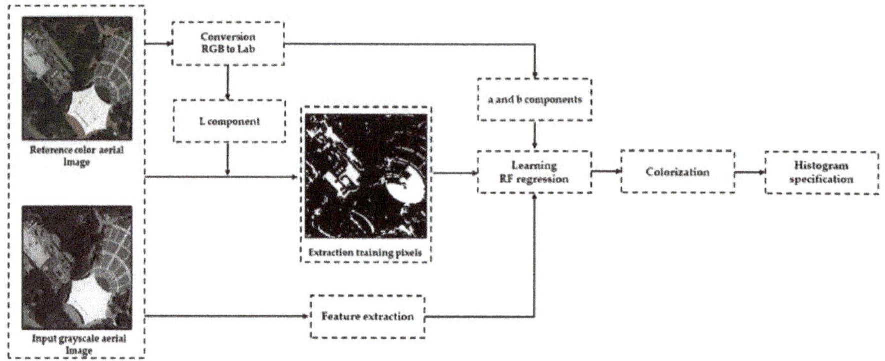

Figure 5. The flowchart of the proposed method. RGB: Red-Green-Blue color space, Lab: CIE L*a*b color space, L component: grayscale axis, a and b components: two-color axes, RF: random forest.

2.3.1. Preprocessing

As mentioned above, the proposed preprocessing step is divided into color-space conversion and extraction of training pixels. First, in this study, a Lab color space is selected since its underlying metric has been designed to express color coherency. Furthermore, Lab has three coordinates: L is the luminance, or lightness, which consequently represents the grayscale axis, whereas a and b represent the two-color axes [16]. In other words, the L component can be known in advance through the input grayscale image, and only the remaining 2D color information, a and b, can be predicted [23]. Thus, the color space of the reference color image is converted from RGB to Lab, and, in the colorization step, only the color relationships for a and b are established through regression. Then, in order to extract a meaningful set of training data, change detection between the input grayscale image and the L component of the reference color image is performed. The change detection method used here comprises two steps for accurate extraction. The first step is a pixel-based method that uses principal component analysis (PCA) [40], and the threshold for distinguishing between changed and unchanged pixels is selected using Otsu's method. However, this process will result in fragmentation and incomplete expression of the change [41]. Therefore, the second step applies the object-based method, which consists of four sub-steps: (1) the morphological closing operation, (2) gap filling, (3) the morphological opening operation, and (4) elimination of small patches.

The morphological closing operation is the combination of dilation followed by erosion, which is used to remove holes in the image [42]. Thus, the closing operation is applied to the image to fill the spaces. Then, gaps within the changed regions that are not filled by the closing operation are additionally filled, which makes the changed information more complete [41]. The morphological opening is then applied, in which erosion is conducted on the image, and it is followed by a dilation operation. The aim of the opening is to remove unnecessary portions. For the structure elements used in the morphological operation, the closing and the opening are set to 3×3 and 5×5, respectively, as selected in Xiao et al. [41]. Small, insignificant patches persist following the opening processing, which can be removed by applying an area threshold. The area threshold is set based on the minimum object size and is acquired through a zero-parameter version of simple linear iterative clustering (SLICO). The SLICO is a spatially localized version of the k-means [43]. To initialize it, the k-cluster centers, which are located on a regular grid and spaced S pixels apart, are sampled [44]. Then, an iterative procedure assigns each pixel to a cluster center using the distance measure D, as defined in Equation (2), which combines the distance of color proximity (Equation (3)) and the distance of spatial proximity (Equation (4)) [45]:

$$D = \sqrt{\left(\frac{d_c}{m}\right)^2 + \left(\frac{d_s}{S}\right)^2} \tag{2}$$

$$d_c = \sqrt{\sum_{s_i \in S} \left(I(x_1, y_1, s_i) - I(x_2, y_2, s_i)\right)^2} \tag{3}$$

$$d_s = \sqrt{(x_1 - x_2)^2 + (y_1 - y_2)^2} \tag{4}$$

where d_c and d_s represent the color and spatial distance between pixels $I(x_1, y_1, s_i)$ and $I(x_2, y_2, s_i)$ in the spectral band S_i, respectively; and m controls the compactness of the superpixels, which is adaptively chosen for each superpixel. In this study, the number of iterations is set to ten, which is sufficient for error convergence [46], and the initial size of the superpixels is set to 10×10, which represents optimal accuracy and computational time [45]. Finally, the unchanged regions are used for training, which consists of establishing color relationships in the unchanged regions from which those of the changed regions can be predicted.

2.3.2. Feature Extraction

The gray level of one pixel is not informative for color prediction, so additional information such as texture and local context is necessary [15]. In order to extract the maximum information for color prediction, features that describe the textural information of local neighborhoods of pixels are considered. Previous automatic colorization methods used Speed-Up Robust Features, Gabor features, or a histogram of oriented gradients as base tools for textural analysis [14,15]. These descriptors are known to be discriminative but also computationally and memory intensive due to their high number of features. Moreover, recent approaches have separated texture from structure using relative total variation, but their descriptors are not sufficiently accurate to discriminate textures among themselves. Consequently, statistical features are utilized here, as this approach is simple, easy to implement, and has strong adaptability and robustness, among which the gray-level co-occurrence matrix (GLCM) is used [47,48]. The GLCM is used extensively in texture description, and the co-occurrence matrices provide better results than do other forms of texture discrimination [49,50]. For remote sensing images, four types of statistics—angular second moment, contrast, correlation, and entropy—are better suited to texture feature extraction, so they have been selected for statistics in this study [51]. Also, in order to calculate the GLCM values, the window size should be set. The present study has ultimately selected a 5 × 5 window size, which better reflects coarse and fine textures [48]. Furthermore, the intensity value and the mean and standard deviation of the intensity within a 5 × 5 neighborhood are included as supplementary components.

2.3.3. Colorization

Like other learning-based approaches, this step consists of two components: (1) a training component and (2) a prediction component, which are described below.

In the training component, image colorization is formulated as a regression problem and is solved using RF. The training data employ seven features of the input grayscale image corresponding to the unchanged regions extracted from the preprocessing step. At this time, these features are trained with the a and b components of the reference color image at the same pixel location. In other words, rather than establishing the color relationships between the L component and the a and b components of the reference color image like other colorization methods do [15,18,19,23], this study establishes color relationships directly between the input grayscale image and the a and b components of the reference color image. Then, the RF regression parameters—mtry and ntree values—are defined. The mtry value is generally set to one-third of the number of features in the regression problem due to computational advantages, but, in this study, the total number of features is utilized, as in the original classification and regression trees procedure, since the number of features is not large enough to affect the computational time [32]. Furthermore, the ntree value is set to 32, which takes into account the performance and training time of the RF regression [35].

The prediction portion of this step uses the RF regression obtained from the training component to predict the colors for the a and b components of the input grayscale image. Then, the input grayscale image is used as the L component, and it is combined with the predicted a and b components. Finally, the Lab image is converted to the RGB color space.

2.3.4. Post-Processing

Finally, the colorized image acquired above is adjusted to reflect global properties based on histogram specification. Histogram specification is a useful technique for modifying histograms via image enhancement without losing the original histogram characteristics [52,53]. Essentially, the input histogram is transformed into a histogram of the specified image to highlight specific ranges. The histogram specification procedure is defined below.

First, the cumulative density functions (CDFs) of the input and specified images are acquired, as shown in Equations (5) and (6):

$$s_k = C_r(r_k) = \sum_{i=0}^{k} P(r_i) = \sum_{i=0}^{k} \frac{n_i}{n} \tag{5}$$

$$v_k = C_z(z_k) = \sum_{i=0}^{k} P(z_i) \tag{6}$$

where s_k and v_k are the respective histograms of the input and specified images, $C_r(r_k)$ and $C_z(z_k)$ are the CDFs of the respective input and specified images, $P(r_i)$ and $P(z_i)$ are the probability density functions of the respective input and specified images, $k = 0, 1, 2, \ldots, L-1$, L is the total number of gray levels, and n_i is the total number of gray levels r_i [53]. Then, the value of z_k, which satisfies Equation (7), is identified:

$$\{ (C_z(z_k) - s_k) = (v_k - s_k) \} \rightarrow 0 \tag{7}$$

In other words, the smallest integer between v_k and s_k should be determined. Finally, the mapping table of z_k will be the output of Equation (8):

$$z_k = C_z^{-1}(s_k) = C_z^{-1}(C_r(r_k)) \tag{8}$$

In this study, the colorization image that is converted to the RGB color space is selected as the input image, and the reference color image is selected as the specified image; histogram specification is carried out for the red, green, and blue bands.

3. Results and Discussion

3.1. Implementation and Performance

This section presents the colorization results of the proposed algorithm and compares these with the results of other state-of-the-art colorization algorithms. To ensure a fair comparison, colorization algorithms that use only a single-color image as the reference (exemplar-based method) and that are based on RF regression are compared, so that the methods of Welsh et al. [1], Bugeau et al. [14], Gupta et al. [15], Gupta et al. [21], and Deshpande et al. [22] are selected. When performing the colorization using these other methods, we used the same parameter settings suggested by their respective authors. In addition, two of the latest deep learning-based methods [18,19] are included for visual comparison, using the codes provided by the authors. The results of the various algorithms are compared by visual inspection (see Figures 6–9) and quantitative evaluation with ground-truth images, which are the actual color images of the input grayscale images.

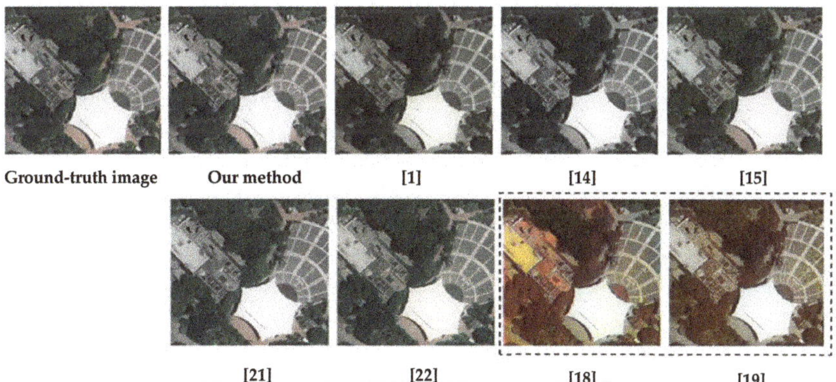

Figure 6. Comparison with existing state-of-the-art colorization methods at site 1.

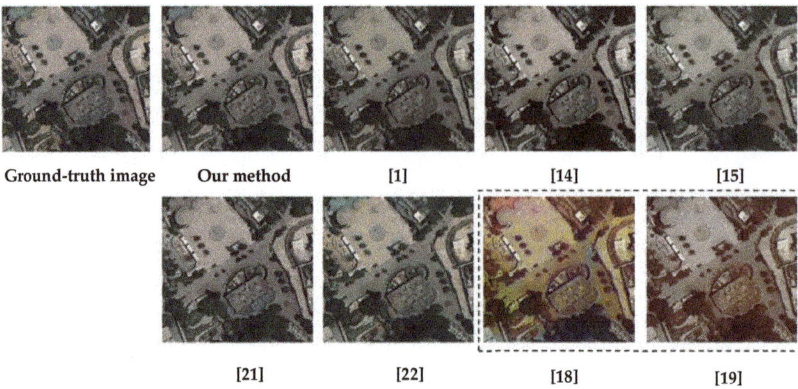

Figure 7. Comparison with existing state-of-the-art colorization methods at site 2.

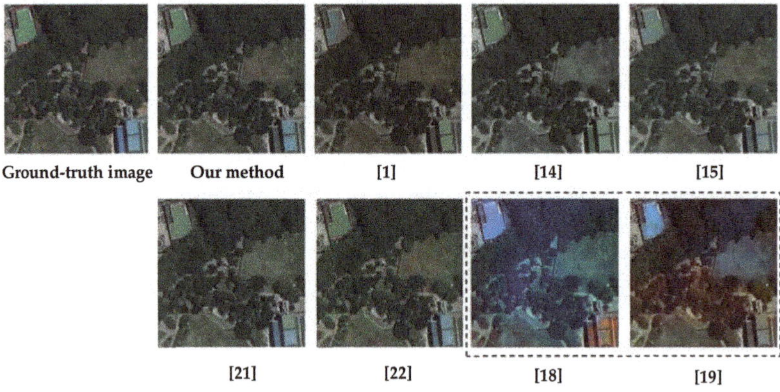

Figure 8. Comparison with existing state-of-the-art colorization methods at site 3.

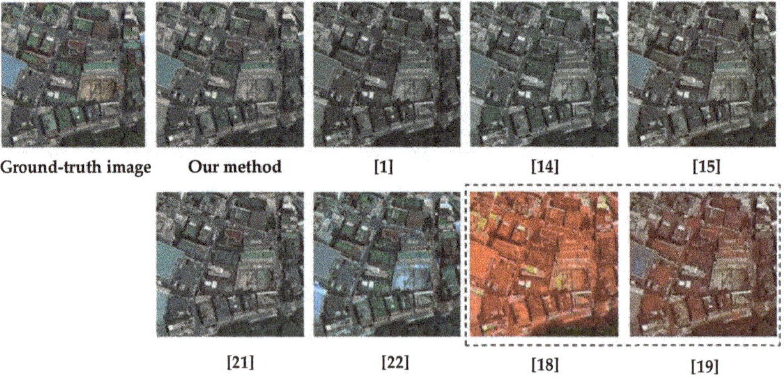

Figure 9. Comparison with existing state-of-the-art colorization methods at site 4.

From the overall visual inspection, the proposed algorithms appear to show better results than do the five existing methods for most cases. The Welsh et al. method [1] is based on transferring color from one initial color image considered as an example. Color prediction is performed by minimizing the distance in the simple statistics of a luminance image. However, as the results show, only a few colors are dealt with, and the results include many artifacts due to the lack of any spatial coherency criteria. In the case of the Bugeau et al. method [14], image colorization is solved by computing the colors using different features and associated metrics. Then, the best colors are automatically selected via a variational framework. Overall, the results fail to select the best colors and have a desaturating effect, confirming the limitations of the variational framework. The Gupta et al. method [15] performs image colorization by automatically exploiting multiple image features. This method transfers color information by performing feature-matching between a reference color image and an input grayscale image—a process that is critical to the quality of the results. This achieves better colorization than the other exemplar-based methods but still only deals with a few colors, resulting in incorrect matches in challenging scenarios.

The other Gupta et al. method [21] is a learning-based method in which learning is performed in superpixel units to enforce higher color consistency. Superpixels are extracted, features are computed for each superpixel, and learning is performed based on an RF regression. The results of this method contain more color information than do the results of the other exemplar-based methods, but the approach still does not retrieve certain colors such as blue or red (Figure 7), and it sometimes predicts completely different colors (as in Figures 8 and 9). The Deshpande et al. method [22] is also based on RF regression, which is performed within the LEArning to seaRCH (LEARCH) framework. Furthermore, histogram correction is performed on the colorization image to improve the visual appeal of the results. In the state-of-the-art-methods used for comparison, this method predicts sufficient color information. However, halo effects exist in the object boundaries, especially in Figure 7. Moreover, as shown in Figure 9, the more complex the structure, the more halo effects are added, leading to many artifacts.

In addition to exemplar-based and RF regression-based methods, deep learning-based methods [18,19] are used for comparison. These colorization algorithms use millions of images for training neural networks, which are based on ImageNet and convolutional neutral networks (CNN). Both results contain color information that is completely different from the ground-truth or reference color images, as shown in Figures 6–9. For example, although the colors of the tree are somewhat predicted, the colors of the buildings or the roads are not predicted at all, which suggests that artifacts are more obvious when the structure is complex. In other words, it is impossible to colorize aerial images through a model that is trained with natural images.

As can be seen in Figures 6–9, our approach more accurately predicts colors than do the other methods, producing results with fewer artifacts. In Figure 6, the color determined by our method is much clearer, especially in the red portion of the ground that is correctly recovered without artifacts. Figure 7 is a site with many human-made objects, and our method demonstrates remarkably high performance in color prediction, while other methods completely fail to correctly predict colors or contain halo effects. Figure 8 also has many human-made objects that contain multiple colors in the area of vegetation, which makes it more difficult to predict the color values at this site. Except for our method and that of Deshpande et al. [22], the colors of human-made objects are not correctly predicted, which indicates that our method is robust for color prediction. Figure 9, the urban area, contains the most complex structure and the greatest variety of colors for prediction. Although the results of the proposed method also contain slightly turbid colors, our method produces significantly better results than do the other methods. In other words, our method retrieves more color values with fewer artifacts from the reference image; these details can be confirmed in Figures 10–13.

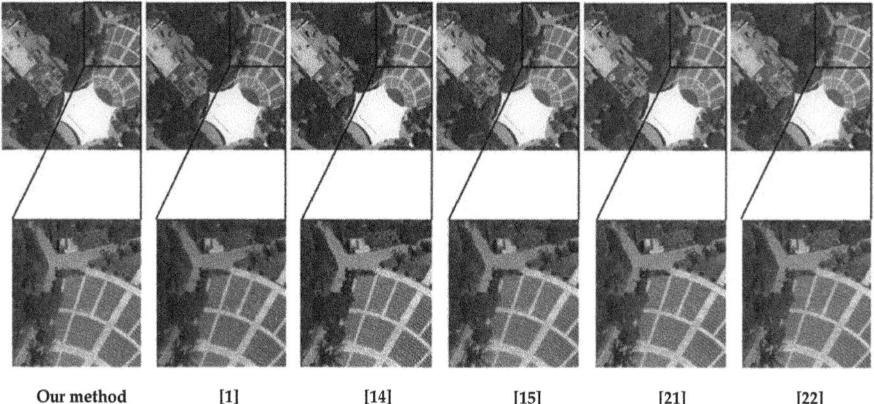

Figure 10. Enlargement of site 1: our method retrieves color values well in this region, compared with other methods.

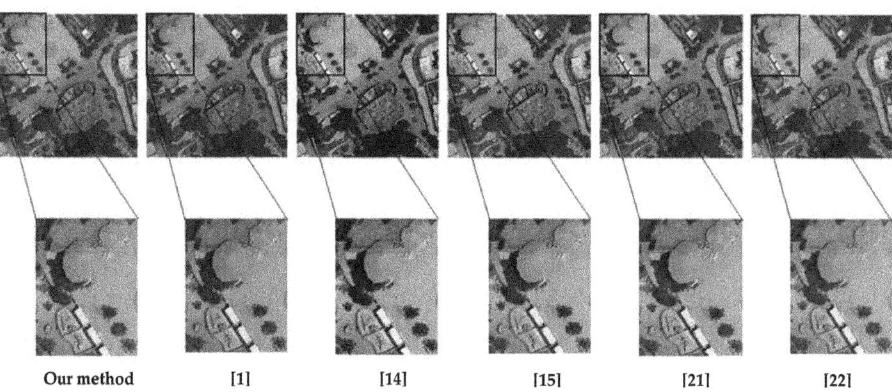

Figure 11. Enlargement of site 2: our method retrieves color values well in this region, compared with other methods.

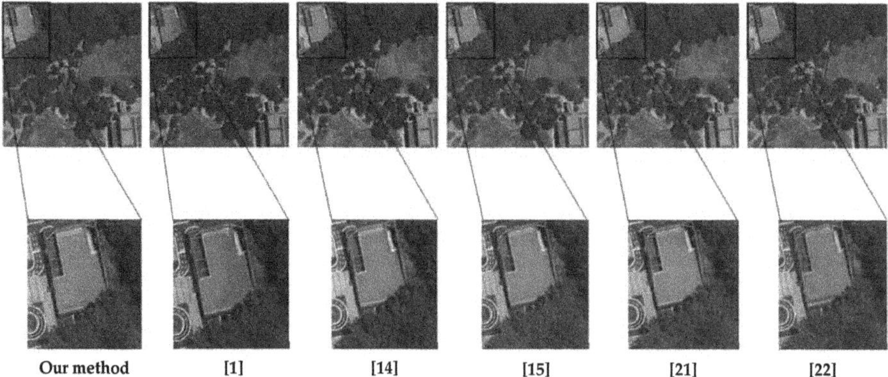

Figure 12. Enlargement of site 3: our method retrieves color values well in this region, compared with other methods.

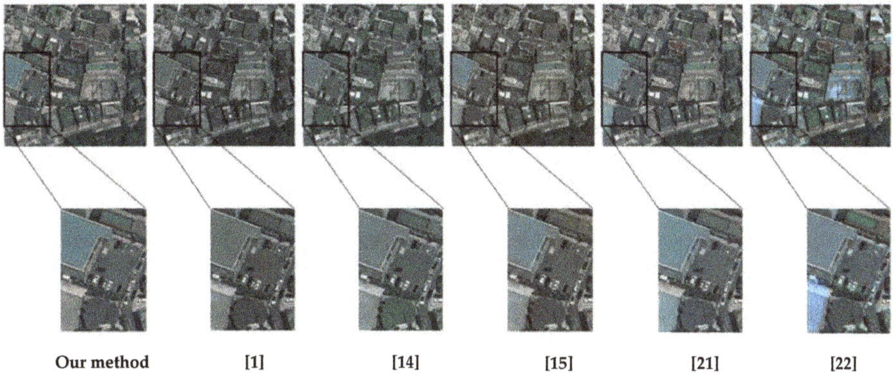

Our method   **[1]**   **[14]**   **[15]**   **[21]**   **[22]**

Figure 13. Enlargement of site 4: our method retrieves color values well in this region, compared with other methods.

Although visual inspection is a simple and direct way of appreciating the quality of the colorization results, it is highly subjective and cannot accurately evaluate the results of the various colorization methods. Therefore, for quantitative evaluation of the results, we employ the standard peak signal-to-noise ratio (PSNR) and normalized color difference (NCD). The PSNR, which is expressed in terms of the decibel (dB), is an estimate of the quality of the reconstructed (colorization) image compared with the ground-truth color image [54]. Given an $m \times n$ ground-truth color image u_0 and a colorization result u, PSNR is defined as:

$$PSNR = 10 \cdot \log\left(\frac{3mn \, (MAX)^2}{\sum_{RGB} \sum_{i=0}^{m-1} \sum_{j=0}^{n-1} [u(i,j) - u_0(i,j)]^2}\right) \tag{9}$$

where MAX_I is the maximum possible pixel value of the image (i.e., 255 with standard 8-bit samples), and $\sum_{RGB}()$ denotes summation over the red, green, and blue bands. The higher the PSNR value, the better the reconstruction process. Normalized color difference is used to measure the color quality degradation in color images [55]. For the NCD calculation, Lab space is used and is defined as:

$$NCD = \frac{\sum_{i=0}^{m-1} \sum_{i=0}^{n-1} [\, (\Delta L)^2 + (\Delta a)^2 + (\Delta b)^2]^{1/2}}{\sum_{i=0}^{m-1} \sum_{i=0}^{n-1} [\, (L_{u0})^2 + (a_{u0})^2 + (b_{u0})^2]^{1/2}}, \tag{10}$$

where ΔL, Δa and Δb are the differences between the components of the ground-truth color image and the colorization result, and L_{u0}, a_{u0}, and b_{u0} are each component values of the ground-truth color image [56]. The lower the NCD value, the better the color quality.

The PSNRs and NCDs of the various algorithms are shown in Tables 1 and 2. In the case of PSNR, the proposed method (Max: 35.0906, Min: 29.6542, Average (Avg): 32.8773) significantly outperforms those of Welsh et al. [1] (Max: 27.1564, Min: 23.5535, Avg: 25.9514), Bugeau et al. [14] (Max: 29.1172, Min: 26.2594, Avg: 27.5596), Gupta et al. [15] (Max: 30.6962, Min: 27.1006, Avg: 29.3068), Gupta et al. [21] (Max: 32.6280, Min: 28.0476, Avg: 30.6694), and Deshpande et al. [22] (Max: 31.5879, Min: 24.2211, Avg: 29.6766). The performance difference from the state-of-the-art methods ranges from 2.4626 to 7.9342 for the maximum PSNR, 1.6066–6.1007 for the minimum PSNR, and 2.1779–6.9259 for the average PSNR, which indicates high performance for all results, regardless of site. That is, it is possible to colorize images stably regardless of the object included in the image or the complexity of the included structure.

Table 1. Quantitative evaluation of algorithm performance using standard peak signal-to-noise ratio (PSNR). dB: decibel.

	PSNR (dB)					
Method	**Our Method**	**[1]**	**[14]**	**[15]**	**[21]**	**[22]**
Site 1	35.0906	26.9364	29.1172	30.6962	32.628	31.5879
Site 2	34.8936	27.1564	26.2594	29.9741	31.6331	31.8681
Site 3	31.8709	26.1595	28.0316	29.4565	30.369	31.0295
Site 4	29.6542	23.5535	26.8302	27.1006	28.0476	24.2211

Table 2. Quantitative evaluation of algorithm performance using normalized color difference (NCD).

	NCD					
Method	**Our Method**	**[1]**	**[14]**	**[15]**	**[21]**	**[22]**
Site 1	0.0707	0.1472	0.127	0.0963	0.0858	0.1042
Site 2	0.0716	0.1408	0.1573	0.1334	0.0929	0.1069
Site 3	0.1098	0.2094	0.1722	0.1472	0.1386	0.1373
Site 4	0.1244	0.2304	0.1677	0.1679	0.1422	0.1801

The NCD of the proposed method (Max: 0.0707, Min: 0.1244, Avg: 0.0941) also show better performance than do Welsh et al. [1] (Max: 0.1408, Min: 0.2304, Avg: 0.1819), Bugeau et al. [14] (Max: 0.1270, Min: 0.1722, Avg: 0.1561), Gupta et al. [15] (Max: 0.0963, Min: 0.1679, Avg: 0.1362), Gupta et al. [21] (Max: 0.0858, Min: 0.1422, Avg: 0.1148), and Deshpande et al. [22] (Max: 0.1042, Min: 0.1801, Avg: 0.1231), in which the range of the improved performance difference is 0.0151–0.0701 for the maximum NCD, 0.0178–0.1244 for the minimum NCD, and 0.0207–0.0878 in the average NCD. This means that the degradation in color quality is lowest when performing colorization through the proposed method. In other words, both visual and quantitative evaluations confirm the superiority of the method proposed herein.

3.2. Limitations

The results show that the proposed algorithm can realize better results than can the existing methods; however, there remain several limitations. Firstly, if there are errors in orthorectification or image registration, incorrect extraction can be performed during selection of the training pixels in the preprocessing step. Although RF regression is robust to training data, some color relationships can be established incorrectly. Secondly, our method retrieves more color values than do the other methods, but, if the structure is complex, it contains somewhat turbid colors. The histogram specification for the reference color image is performed by post-processing, but there are still limitations. Thirdly, in this study, aerial images three years apart are used. However, further verification is needed to determine the extent of the period in which the colorization can properly be performed. Finally, our method is established by directly correlating color relationships between the input grayscale image and the reference color image, making it dependent on the availability of reference color aerial imagery of the same input area with matching seasonal characteristics. Consequently, when suitable color aerial images are unavailable, colorization may fail.

4. Conclusions

This paper presents a colorization algorithm for aerial imagery. The proposed method uses a reference color image with similar seasonal features at the same location as an input grayscale image. The color space of the reference color image is converted to Lab, and unchanged regions are selected by applying change detection to the input grayscale image and the L component of the reference color image, which serves as meaningful training data. Moreover, color relationships are established in direct correspondence between the feature descriptors of the input grayscale image and the a and b components of the reference color image based on the RF regression. Finally, histogram specification is applied to the colorization image to improve the results and is compared with state-of-the-art methods. Experimental results for multiple sites show that our method achieves visually appealing colorizations with significantly improved quantitative performance. In other words, the proposed algorithm performs well and outperforms existing colorization approaches for aerial images.

Future work will include other complex descriptors or features in order to retrieve more color values for complex structures. In particular, we intend to find the combination of features that best describes the characteristics of the aerial images for colorization. We will also extend our application of the technique by applying satellite images obtained from various sensors other than aerial images. Furthermore, to overcome the limitations that may prevent colorization from being performed when reference color images are unavailable, a method of using reference color images that are obtained with different sensors or contain different seasons or resolutions will be sought out. Finally, we plan to colorize past grayscale aerial images using a time-series approach, possibly by incorporating monitoring frameworks.

Author Contributions: All authors contributed to the writing of the manuscript. D.K.S. carried out the experiments and analyzed the results; Y.H.K. designed the experiments and presented the direction of this study; Y.D.E. supervised this study; W.Y.P. provided scientific counsel.

Funding: This research was supported by the Basic Science Research Program through the NRF (National Research Foundation of Korea) funded by the Ministry of Education [No. 2016R1D1A1B03933562].

References

1. Welsh, T.; Ashikhmi, M.; Mueller, K. Transferring Color to Grayscale Images. *ACM Trans. Graph.* **2002**, *21*, 277–280. [CrossRef]
2. Bugeau, A.; Ta, V. Patch-Based Image Colorization. In Proceedings of the 21st International Conference on Pattern Recognition, Tsukuba, Japan, 11–15 November 2012; pp. 3058–3061.
3. Lipowezky, U. Grayscale Aerial and Space Image Colorization Using Texture Classification. *Pattern Recogn. Lett.* **2005**, *27*, 275–286. [CrossRef]
4. Yang, Y.; Wan, W.; Huang, S.; Lin, P.; Que, Y. A Novel Pan-Sharpening Framework Based on Matting Model and Multiscale Transform. *Remote Sens.* **2017**, *9*, 391. [CrossRef]
5. Li, S.; Kang, X.; Fang, L.; Hu, J.; Yin, H. Pixel-Level Image Fusion: A Survey of the State of the Art. *Inf Fusion.* **2017**, *33*, 100–112. [CrossRef]
6. Ghassemian, H. A Review of Remote Sensing Image Fusion Methods. *Inf Fusion.* **2016**, *32*, 75–89. [CrossRef]
7. Horiuchi, T. Estimation of Color for Gray-Level Image by Probabilistic Relaxation. In Proceedings of the Object Recognition Supported by User Interaction for Service Robots, Quebec, Canada, 11–15 August 2002; pp. 867–870.
8. Arbelot, B.; Vergne, R.; Hurtut, T.; Thollot, J. Automatic Texture Guided Color Transfer and Colorization. In Proceedings of the Joint Symposium on Computational Aesthetics and Sketch Based Interfaces and Modeling and Non-Photorealistic Animation and Rendering, Lisbon, Portugal, 7–9 May 2016; pp. 21–32.
9. Li, B.; Lai, Y.K.; Rosin, P.L. Example-Based Image Colorization via Automatic Feature Selection and Fusion. *Neurocomputing* **2017**, *266*, 687–698. [CrossRef]
10. Levin, A.; Lischinski, D.; Weiss, Y. Colorization Using Optimization. *ACM Trans. Graph.* **2004**, *23*, 689–694. [CrossRef]
11. Huang, Y.C.; Tung, Y.S.; Chen, J.C.; Wang, S.W.; Wu, J.L. An Adaptive Edge Detection Based Colorization Algorithm and Its Applications. In Proceedings of the 13th ACM International Conference on Multimedia, Hilton, Singapore, 6–11 November 2005; pp. 351–354.
12. Irony, R.; Cohen-Or, D.; Lischinski, D. Colorization by Example. In Proceedings of the Sixteen Eurographics Conference on Rendering Techniques, Konstanz, Germany, 29 June–1 July 2005; pp. 201–210.
13. Yatziv, L.; Sapiro, G. Fast Image and Video Colorization Using Chrominance Blending. *IEEE Trans. Image Process.* **2006**, *15*, 1120–1129. [CrossRef] [PubMed]

14. Bugeau, A.; Ta, V.T.; Papadakis, N. Variational Exemplar-Based Image Colorization. *IEEE Trans. Image Process.* **2014**, *23*, 298–307. [CrossRef] [PubMed]

15. Gupta, R.L.; Chia, A.Y.S.; Rajan, D.; Ng, E.S.; Zhiyoung, H. Image Colorization Using Similar Images. In Proceedings of the 20th ACM International Conference on Multimedia, Nara, Japan, 29 October–2 November 2012; pp. 369–378.

16. Charpiat, G.; Hofmann, M.; Scholkopf, B. Automatic Image Colorization via Multimodal Predictions. In Proceedings of the 10th European Conference on Computer Vision, Marseille, France, 12–18 October 2008; pp. 126–139.

17. Cheng, Z.; Yang, Q.; Sheng, B. Deep Colorization. In Proceedings of the IEEE International Conference on Computer Vision, Santiago, Chile, 7–13 December 2015; pp. 415–423.

18. Zhang, R.; Isola, P.; Efros, A.A. Colorful Image Colorization. In Proceedings of the Computer Vision—ECCV 2016—14th European Conference, Amsterdam, The Netherlands, 11–14 October 2016; pp. 649–666.

19. Larrson, G.; Maire, M.; Shakhnarovich, G. Learning Representations for Automatic Colorization. In Proceedings of the Computer Vision—ECCV 2016—14th European Conference, Amsterdam, The Netherlands, 11–14 October 2016; pp. 577–593.

20. Brieman, L. Random Forests. *Mach. Learn.* **2001**, *45*, 5–32. [CrossRef]

21. Gupta, R.K.; Chia, A.Y.; Rajan, D.; Zhiyong, H. A Learning-based Approach for Automatic Image and Video Colorization. In Proceedings of the Computer Graphics International, Bournemouth, UK, 12–15 June 2012; pp. 1–10.

22. Deshpande, A.; Rock, J.; Forsyth, D. Learning Large-Scale Automatic Image Colorization. In Proceedings of the 2015 IEEE International Conference on Computer Vision, Santiago, Chile, 7–13 December 2015; pp. 567–575.

23. Mohn, H.; Caebelein, M.; Hansch, R.; Hellwich, O. Towards Image Colorization with Random Forests. In Proceedings of the 13th International Joint Conference on Computer Vision, Funchal, Madeira, Portugal, 27–29 January 2018; pp. 270–278.

24. Culter, D.R.; Edwards, T.C.; Beard, K.H.; Culter, A.; Hess, K.T.; Gibson, J.C. Random Forests for Classification in Ecological Society of America. *Ecology* **2007**, *88*, 2783–2792. [CrossRef]

25. Yang, Y.; Cao, C.; Pan, X.; Li, X.; Zhu, X. Downscaling Land Surface Temperature in an Arid Area by Using Multiple Remote Sensing Indices with Random Forest Regression. *Remote Sens.* **2017**, *9*, 789. [CrossRef]

26. Shataee, S.; Kalbi, S.; Fallah, A.; Pelz, D. Forest Attribute Imputation Using Machine Learning Methods and ASTER Data: Comparison of K-NN, SVR, Random Forest Regression Algorithms. *Int. J. Remote Sens.* **2012**, *33*, 6254–6280. [CrossRef]

27. Peters, J.; Baets, B.D.; Verhoest, N.E.C.; Samson, R.; Degroeve, S.; Becker, P.D.; Huybrechts, W. Random Forests as a Tool for Ecohydrological Distribution Modeling. *Ecol. Model.* **2007**, *207*, 304–318. [CrossRef]

28. Rodriguez-Galiano, V.; Sanchez-Castillo, M.; Olmo-Chica, M.; Chica-Rivas, M. Machine Learning Predictive Models for Mineral Prospectivity: An Evaluation of Neural Networks, Random Forest, Regression Trees and Support Vector Machines. *Ore Geol. Rev.* **2015**, *71*, 804–818. [CrossRef]

29. Prasad, A.M.; Iverson, L.R.; Liaw, A. Newer Classification and Regression Tree Techniques: Bagging and Random Forests for Ecological Prediction. *Ecosystems* **2006**, *9*, 181–199. [CrossRef]

30. Hutengs, C.; Vohland, M. Downscaling Land Surface Temperature at Regional Scales with Random Forest Regression. *Remote Sens. Environ.* **2016**, *178*, 127–141. [CrossRef]

31. Changas, C.S.; Junior, W.C.; Behring, S.B.; Filho, B.G. Spatial Prediction of Soil Surface Texture in a Semiarid Region Using Random Forest and Multiple Linear Regressions. *Catena* **2016**, *139*, 232–240. [CrossRef]

32. Scornet, E. Tuning Parameters in Random Forests. In Proceedings of the ESAIM: Proceedings and Surveys, Grenoble, France, 29–31 August 2017; pp. 144–162.

33. Lawrence, R.L.; Wood, S.D.; Sheley, R.L. Mapping Invasive Plants Using Hyperspectral Imagery and Breiman Cutler Classification (RandomForest). *Remote Sens. Environ.* **2016**, *100*, 356–362. [CrossRef]

34. Belgiu, M.; Dragut, L. Random Forest in Remote Sensing: A Review of Applications and Future Directions. *ISPRS J. Photogramm. Remote Sens.* **2015**, *114*, 24–31. [CrossRef]

35. Seo, D.K.; Kim, Y.H.; Eo, Y.D.; Park, W.Y.; Park, H.C. Generation of Radiometric, Phenological Normalized Image Based on Random Forest Regression for Change Detection. *Remote Sens.* **2017**, *9*, 1163. [CrossRef]

36. Sug, H. Applying Randomness Effectively Based on Random Forest for Classification Task of Datasets of Insufficient Information. *J. Appl. Math.* **2012**, *2012*, 1–13. [CrossRef]

37. Palmer, D.S.; O'Boyle, N.M.; Glen, R.C.; Mitchell, J.B.O. Random Forest Models to Predict Aqueous Solubility. *J. Chem. Inf. Model.* **2007**, *47*, 150–158. [CrossRef] [PubMed]

38. Dye, M.; Mutanga, O.; Ismail, R. Combining Spectral and Textural Remote Sensing Variables Using Random Forests: Predicting the Age of Pinus Forests in KwaZulu-Natal, South Africa. *J. Spat Sci.* **2012**, *57*, 193–211. [CrossRef]

39. Quintana, D.; Saez, Y.; Isasi, P. Random Forest Prediction of IPO Underpricing. *Appl. Sci.* **2017**, *7*, 636. [CrossRef]

40. Hussain, M.; Chen, D.; Cheng, A.; Wei, H.; Stanley, D. Change Detection from Remotely Sensed Images: From Pixel-Based to Object-Based Approaches. *ISPRS J. Photogramm. Remote Sens.* **2013**, *80*, 91–106. [CrossRef]

41. Xiao, P.; Zhang, X.; Wang, D.; Yuan, M.; Feng, X.; Kelly, M. Change Detection of Built-up Land: A Framework of Combining Pixel-Based Detection and Object-Based Recognition. *ISPRS J. Photogramm. Remote Sens.* **2016**, *119*, 402–414. [CrossRef]

42. Wang, J.; Qin, Q.; Gao, Z.; Zhao, J.; Ye, X. A New Approach to Urban Road Extraction Using High-Resolution Aerial Image. *Int. J. Geo-Inf.* **2016**, *5*, 114. [CrossRef]

43. Crommelinck, S.; Bennett, R.; Gerke, M.; Koeva, M.N.; Yang, M.Y.; Vosselman, G. SLIC Superpixels for Object Delineation from UAV Data. In Proceedings of the International Conference on Unmanned Aerial Vehicles in Geomatics, Bonn, Germany, 4–7 September 2017; pp. 9–16.

44. Mei, T.; An, L.; Li, Q. Supervised Segmentation of Remote Sensing Image Using Reference Descriptor. *IEEE Geosci. Remote Sens. Lett.* **2015**, *12*, 938–942. [CrossRef]

45. Csillik, O. Fast Segmentation and Classification of Very High Resolution Remote Sensing Data Using SLIC Superpixels. *Remote Sens.* **2017**, *9*, 243. [CrossRef]

46. Achanta, R.; Shaji, A.; Smith, K.; Luchi, A.; Fua, P.; Susstrunk, S. SLIC Superpixels Compared to State-of-the-Art Superpixels Methods. *IEEE Trans. Pattern Anal. Mach. Intell.* **2012**, *34*, 2274–2281. [CrossRef] [PubMed]

47. Zainal, Z.; Ramli, R.; Mustafa, M.M. Grey-Level Cooccurence Matrix Performance Evaluation for Heading Angle Estimation of Movable Vision System in Static Environment. *J. Sens.* **2013**, *2013*, 1–15. [CrossRef]

48. Zhang, X.; Cui, J.; Wang, W.; Lin, C. A Study for Texture Feature Extraction of High-Resolution Satellite Images Based on a Direction Measure and Gray Level Co-Occurrence Matrix Fusion Algorithm. *Sensors* **2017**, *17*, 1474. [CrossRef] [PubMed]

49. Huang, X.; Liu, X.; Zhang, L. A Multichannel Gray Level Co-Occurrence Matrix for Multi/Hyperspectral Image Texture Representation. *Remote Sens.* **2014**, *6*, 8424–8445. [CrossRef]

50. Jia, B.; Wang, W.; Yoon, S.C.; Zhuang, H.; Li, Y.F. Using a Combination of Spectral and Texture Data to Measure Water-Holding Capacity in Fresh Chicken Breas Fillets. *Appl. Sci.* **2018**, *8*, 343. [CrossRef]

51. Zheng, S.; Zheng, J.; Shi, M.; Li, X. Classification of Cultivated Chinese Medicinal Plants Based on Fractal Theory and Gray Level Co-Occurrence Matrix Textures. *J. Remote Sens.* **2014**, *18*, 868–886. [CrossRef]

52. Sun, C.C.; Ruan, S.J.; Shie, M.C.; Pai, T.W. Dynamic Contrast Enhancement Based on Histogram Specification. *IEEE Trans. Consum. Electron.* **2005**, *51*, 1300–1305. [CrossRef]

53. Xie, L.; Wang, G.; Zhang, X.; Xiao, B.; Zhou, B.; Zhang, F. Remote Sensing Image Enhancement Based on Wavelet Analysis and Histogram Specification. In Proceedings of the 2014 IEEE 3rd International Conference on Cloud Computing and Intelligence Systems, Shenzhen, China, 27–29 November 2014; pp. 55–59.

54. Al-Najjar, Y.; Chen, S.D. Comparison of Image Quality Assessment: PSNR, HVS, SSIM, UIQI. *Int. J. Sci. Eng. Res.* **2012**, *3*, 1–5.

55. Senthilkumaran, N.; Saromary, J. Detailed Performance Evaluation of Bilateral Filters for De-noising Chromosome Image. *Int. J. Inf. Technol.* **2017**, *3*, 64–70.

56. Szczepanski, M.; Smolka, B.; Plataniots, K.N.; Ventesanopoulos, A.N. On the Distance Function Approach to Color Image Enhancement. *Discret.Appl. Math.* **2004**, *139*, 283–305. [CrossRef]

 *applied sciences*

Article

Spatial Modelling of Gully Erosion Using GIS and R Programing: A Comparison among Three Data Mining Algorithms

Alireza Arabameri [1], Biswajeet Pradhan [2,*], Hamid Reza Pourghasemi [3], Khalil Rezaei [4] and Norman Kerle [5]

[1] Department of Geomorphology, Tarbiat Modares University, Tehran 36581-17994, Iran; alireza.ameri91@yahoo.com
[2] Centre for Advanced Modelling and Geospatial Information Systems (CAMGIS), Faculty of Engineering and IT, University of Technology Sydney, Ultimo, NSW 2007, Australia
[3] Department of Natural Resources and Environmental Engineering, College of Agriculture, Shiraz University, Shiraz 71441-65186, Iran; hm_porghasemi@yahoo.com
[4] Faculty of Earth Sciences, Kharazmi University, Tehran 14911-15719, Iran; kh.rezaei@gmail.com
[5] Department of Earth Systems Analysis (ESA), Faculty of Geo-Information Science and Earth Observation (ITC), University of Twente, 7522 Enschede, The Netherlands; kerle@itc.nl
* Correspondence: biswajeet24@gmail.com or Biswajeet.Pradhan@uts.edu.au

Received: 13 July 2018; Accepted: 10 August 2018; Published: 14 August 2018

Abstract: Gully erosion triggers land degradation and restricts the use of land. This study assesses the spatial relationship between gully erosion (GE) and geo-environmental variables (GEVs) using Weights-of-Evidence (WoE) Bayes theory, and then applies three data mining methods—Random Forest (RF), boosted regression tree (BRT), and multivariate adaptive regression spline (MARS)—for gully erosion susceptibility mapping (GESM) in the Shahroud watershed, Iran. Gully locations were identified by extensive field surveys, and a total of 172 GE locations were mapped. Twelve gully-related GEVs: Elevation, slope degree, slope aspect, plan curvature, convergence index, topographic wetness index (TWI), lithology, land use/land cover (LU/LC), distance from rivers, distance from roads, drainage density, and NDVI were selected to model GE. The results of variables importance by RF and BRT models indicated that distance from road, elevation, and lithology had the highest effect on GE occurrence. The area under the curve (AUC) and seed cell area index (SCAI) methods were used to validate the three GE maps. The results showed that AUC for the three models varies from 0.911 to 0.927, whereas the RF model had a prediction accuracy of 0.927 as per SCAI values, when compared to the other models. The findings will be of help for planning and developing the studied region.

Keywords: gully erosion; environmental variables; data mining techniques; SCAI; GIS

---

1. Introduction

Today, reducing natural resources, especially soil and water, is one of the major problems and major threats to human life and is one of the most important environmental problems worldwide that has intensified in recent years, with increasing population and the alternation of human activities [1]. According to the data from United Nations research, the world's population is growing at a rate of 1.8% per year and it is expected to rise from 8 billion in 2025 to 9.4 billion in 2050 [2]. This increase in world population would demand the need for food, water, forage, and others, which consequently would add huge pressure on land exploitation, non-standard exploitation, and eventually lead to an increase in erosion rates [1,3]. Soil erosion is one of the factors that endangers water and soil [1]. Soil erosion by water, such as GE, is considered as a major cause of land degradation around the world [4,5]. It leads to a range of problems, such as desertification, flooding and sediment deposition in reservoirs [6,7], the destructive effects on the ecosystem reducing soil fertility, and imposes huge economic costs [8]. GE is typically defined as a deep channel that has been eroded by concentrated water flow, removing surface soils and materials [9,10]. The amount of moisture and its changes as a result of the dry and wet seasons is a main parameter in creating cracks and grooves in fine-grained clay formations containing clay and silt, and ultimately developing rilled erosion and gullies [10]. The alternation of warm and dry seasons makes it possible to create cracks, in the formation of fine grains, in warm seasons with the drying of the land and the wilting of the vegetation, and these cracks at the time of the first sudden rainfall concentrate the runoff and therefore cause rill and GE to emerge [11]. GE occurs when the erosion of the water flow or the erodibility of the sediments or the formation of the area is higher than the geomorphological threshold of the area [11]. Mapping gully erosion systems is essential for implementing soil conservation measures [6]. GEVs that influence gully occurrence are rainfall, topography-derived factors such as elevation, slope degree, slope aspect, and plan curvature, lithology [12], soil properties [13], and LU/LC [14]. The distribution of precipitation affects the hydraulic flow and moisture content of the soil, and the erosion strength of the flow and soil resistance to erosion is different before and after erosion [11]. Generally, the amount and volume of flow are controlled by the topographic features of the area including slope, aspect, and drainage area of the area. Depth and morphology of the cross section of the gullies are controlled by soil erodibility features of the geological layers of the area. The characteristics of the region's soil affect the subsurface flow and the phenomenon of piping erosion, and the pipes cause a gully when their ceiling collapses [10].

Susceptibility maps of GE are essential for conservation of natural resources, and for evaluating the relationship among gully occurrence and relevant GEVs [12]. Several models have been applied to assess soil erosion and GE rate in a quantitative and qualitative way, such as the Universal Soil Loss Equation (USLE) [1,15], Erosion Potential Method, Modified Pacific Southwest Interagency Committee Model (MPSIAC) [16], Water Erosion Prediction Project (WEEP) [17], European Soil Erosion Model (EUROSEM) [18], Ephemeral Gully Erosion Model (EGEM) [19], and Chemicals, Runoff, and Erosion from Agricultural Management Systems (CREAMS) [20].

Within the soil conservation research field, the distribution of soil erosion is one of the primary sources of information. This is also relevant for GE; however, in the above mentioned methods, spatial distribution of gullies has not been addressed. Remote sensing-based methods to identify GE have been developed [21], including with RF machine learning, though they serve more to validate susceptibility models and to explain the actual erosion presence and distribution. In recent years, scientific research for susceptibility analysis of GE, and work on the statistical relationships between GEVs and the spatial distribution of gullies, have been addressed using various statistical and machine learning methods including bivariate statistics (BS) [1], weights-of-evidence (WoE) [13], index of entropy (IofE) [8], logistic regression (LR) [22–26], information value (IV) [24,25], random forest (RF) [27], bivariate statistical models [28,29], maximum entropy (ME) [30,31], frequency ratio (FR) [28], analytical hierarchy processes (AHP) [29], artificial neural network (ANN) [12,31], support vector machine (SVM) [31], and boosted regression trees (BRT) [12]. For this purpose, various GEVs such as topography (e.g., elevation, slope, aspect, plan curvature, profile curvature, slope length), lithology, land use, soil properties (e.g., soil texture, soil type, erosivity, soil water content), land use, climate (rainfall intensity, rainfall period, and spatial distribution of rainfall), infrastructures (road, bridge) and hydrology (e.g., TWI, SPI, drainage density) were used.

A comprehensive literature review shows that there are still dimensions that require further research, and that a large number of potentially useful methods have not yet been fully implemented to provide GE susceptibility maps. The main objectives of this study are: (i) To determine the relationship between gully occurrence and conditioning factors using Weights-of-Evidence Bayes theory, (ii) assessing the capability of RF, MARS, and BRT data mining/machine learning models to predict GE susceptibility; and (iii) validation of models using the AUC curve and SCAI methods. Study of the research background showed that using MARS, BRT, and RF data mining models in GE zonation is very new. It will help managers in future planning to prevent human intervention in sensitive areas.

2. Materials and Methods

2.1. Study Area

The Shahroud watershed, with an area of about 848 km^2 and elevation range from 1084 to 2131 m a.s.l., is located in the northeastern part of Semnan Province, Iran (Figure 1). The study area receives an average rainfall of less than 250 mm has an arid and semi-arid climate [32]. Various types of lithological formations cover this watershed, and the landforms are mainly low level pediment fans and valley terrace deposits. The dominant land use is rangelands, but irrigation farming and bare lands are also present.

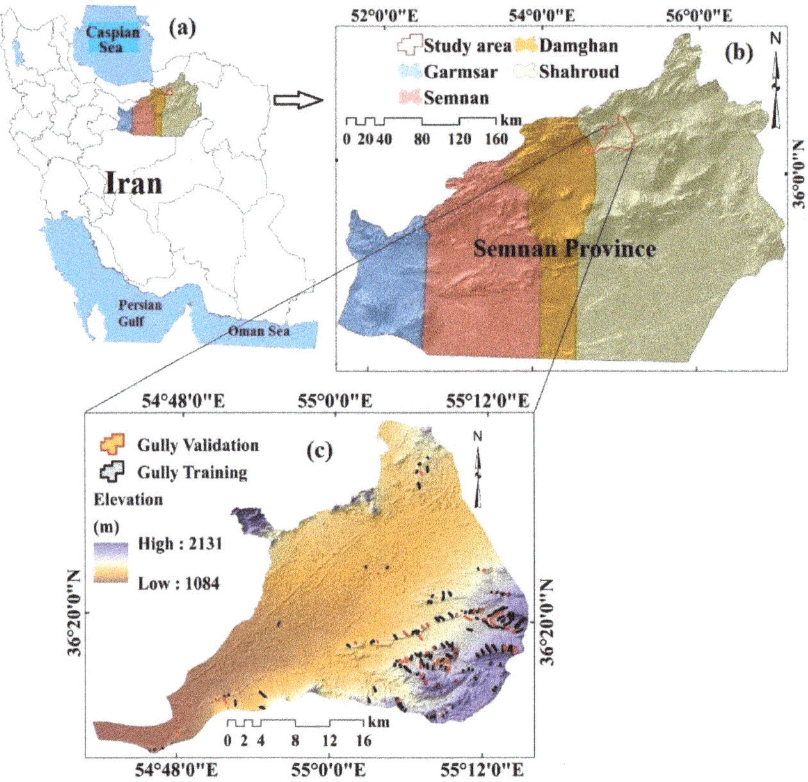

Figure 1. (**a**) Location of the Semnan provinces in Iran, (**b**) location of study area, and (**c**) gully erosion locations with the digital elevation model map of the Shahroud watershed.

2.2. Data and Method

Figure 2 shows the methodological approach applied to map GE susceptibility in the Shahroud watershed using BRT, MARS, and RF models. For preparing an accurate and reliable gully inventory map, extensive field surveys with a DGPS device were performed in the study area to determine the location of the Gullies [27,28]. Then, among 172 detected gully locations, randomly (70/30 ratio), 121 gully locations (70%) and 51 gully locations (30%) in the polygon format were used for training the testing models [28]. The locations of training and testing gullies are shown in Figure 1. Interventionary studies involving animals or humans, and other studies require ethical approval must list the authority that provided approval and the corresponding ethical approval code.

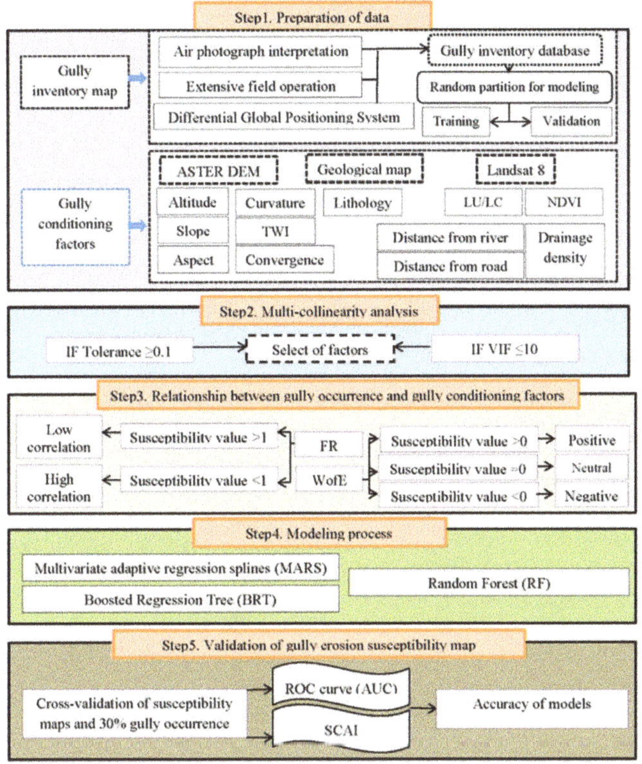

Figure 2. Flowchart of research methodology.

The tools used in present study are ArcGIS10.5, ENVI 4.8, SAGA-GIS 2.1.1, and a DGPS. The basic maps used were geological maps [33], at a scale of 1:100,000, topographic maps, at a scale of 1:50,000, satellite images acquired by Landsat8, and ASTER GDEM with spatial resolution of 30 m [34]. In this study, based on literature review [24,26,31] and local conditions of the study area, twelve factors were selected. Elevation map was divided into six classes: <1200 m, 1200–<1350 m, 1350–<1450 m, 1450–1600 m, and >1600 m (Figure 3a). Slope degree affects surface runoff [35], soil erosion, and pattern of drainage density. Slope degree map was classified into six classes [24,26]: <5°, 5–<10°, 10–<15°, 15–<20°, 20–<25°, 25–30°, and >30° (Figure 3b).

The aspect map was classified into nine classes (Figure 3c). Positive and negative values of plan curvatures define convexity and concavity of slope curvature, whereas zero is flat surface. The plan curvature map was divided into 3 categories: Concave, Flat, and Convex. The TWI indicator is important for identifying prone areas to GE [36]. TWI is calculated by Equation (1):

$$\mathrm{TWI} = \ln\left(\frac{S}{\tan \propto}\right) \tag{1}$$

TWI map of study area is divided into four classes [24,26,37] including <5, 5–<7.5, 7.5–11, and >11 (Figure 3e). The convergence index (CI) gives a measure of how flow in a cell diverges (convergence index in negative and positive values) [38]. The CI map was prepared in SAGA-GIS 2.1.1 and divided into 3 classes: <0, 0–10, and >10 (Figure 3f). In this research, for the computation of the effect of drainage network and infrastructures on GE, the distance from rivers and roads was considered [14] and divided into four classes: <170 m, 170–<370 m, 370–650 m, and >650 m for rivers (Figure 3g) and <500 m, 500–<1500 m, 1500–3000 m, and >3000 m for roads (Figure 3h). The line density tool in ArcGIS 10.5 was used for calculating drainage density and then its map was divided into four categories: <1.4, 1.4–<2.4, 2.4–3.7, and >3.7 km/km^2 (Figure 3i). A geological map at a 1:100,000 scale was used to prepare the lithological unit layer. The lithological units were classified into ten categories based on their sensitivity to gully occurrence using expert knowledge method (Figure 3j and Table 1). The advantage of this method is it is easy to use, however this method has certain disadvantages, such as the possibility of a mistake by the expert.

Table 1. Lithology of the study area.

Code	Lithology	Geological Age
Murmg	Gypsiferous marl	Miocene
Qft2	Low level piedment fan and vally terrace deposits	Quaternary
Ku	Upper cretaceous, undifferentiated rocks	Cretaceous
Jd	Well—bedded to thin—bedded, greenish—grey argillaceous limestone with intercalations of calcareous shale (DALICHAI FM)	Jurassic
PeEz	Reef-type limestone and gypsiferous marl (ZIARAT FM)	Paleocene-Eocene
PlQc	Fluvial conglomerate, Piedmont conglomerate and sandstone.	Pliocene-Quaternary
Jl	Light grey, thin—bedded to massive limestone (LAR FM)	Jurassic-Cretaceous
E2c	Conglomerate and sandstone	Eocene
PlQc	Fluvial conglomerate, Piedmont conglomerate and sandstone.	Pliocene-Quaternary
E1c	Pale-red, polygenic conglomerate and sandstone	Paleocene-Eocene

The LU/LC map was obtained using Landsat 8 images [39–41]. The main LU/LC types identified in the study area were range, irrigation farming, and bare lands (Figure 3k). The NDVI map was also produced using Landsat 8 images and classified into 3 categories: <0.11, 0.11–0.25, and >0.25 (Figure 3l).

For multi-collinearity checking, the tolerance (TOL) and variance inflation factor (VIF) were used. If during modeling there is collinearity among the variables, the accuracy of the model's prediction decreases. Values of TOL and VIF were ≤0.1 and ≥10, respectively, indicating that multi-collinearity among parameters [28].

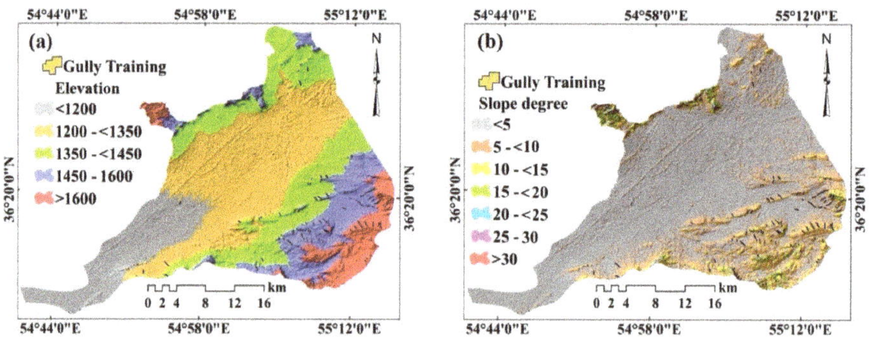

Figure 3. *Cont.*

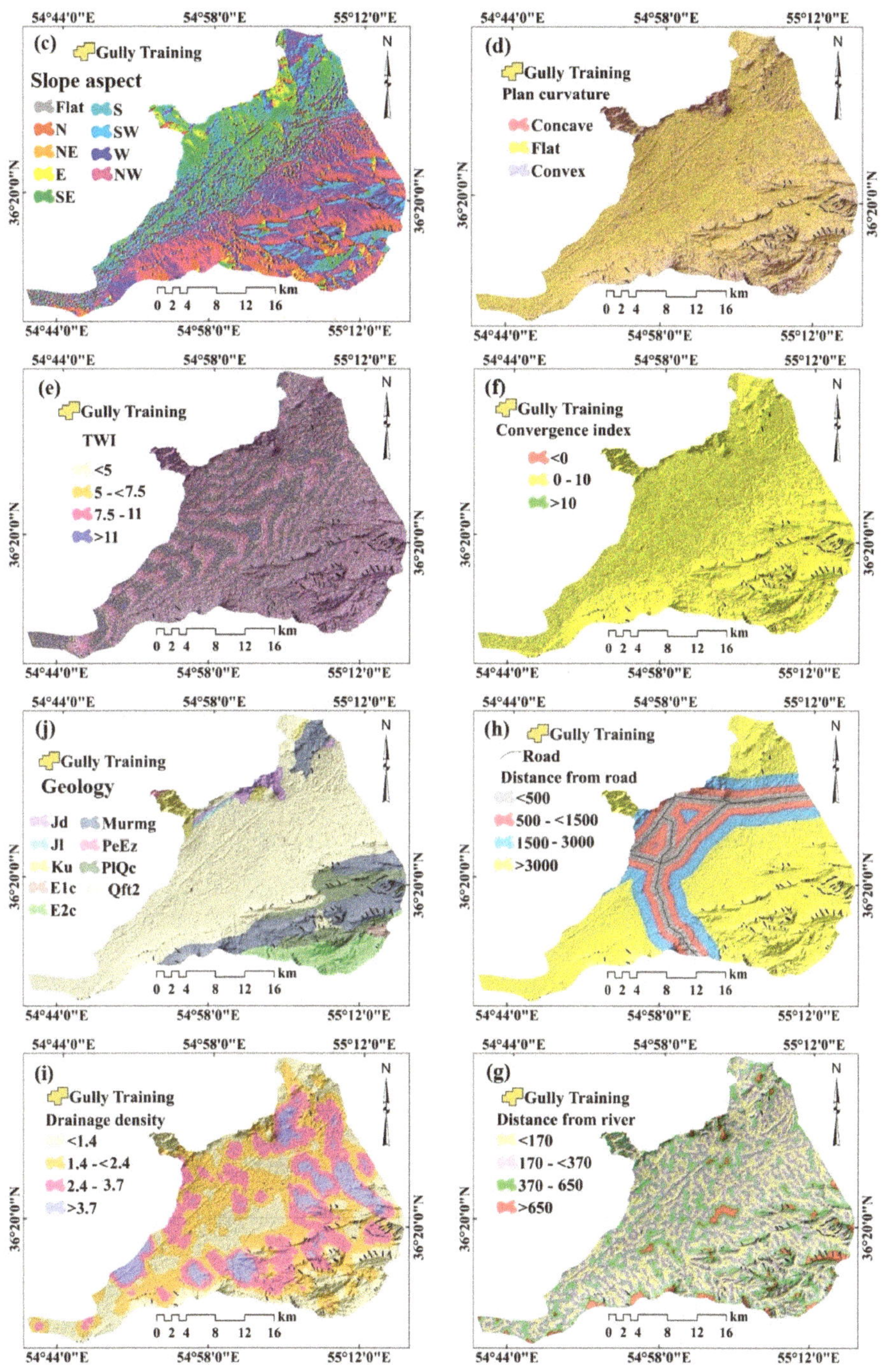

Figure 3. *Cont.*

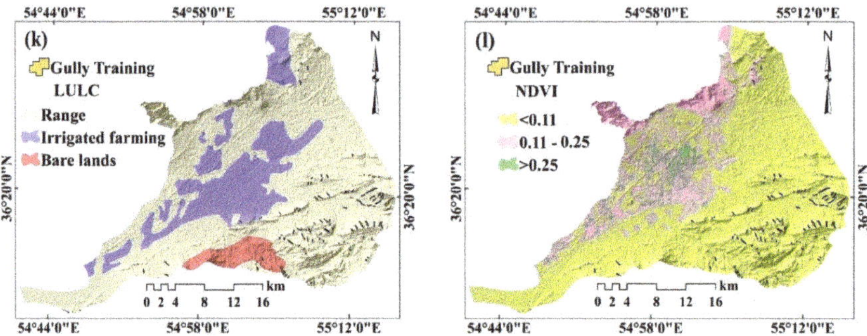

Figure 3. Gully erosion conditioning factors: (**a**) Elevation, (**b**) slope, (**c**) aspect, (**d**) plan curvature, (**e**) TWI, (**f**) convergence index, (**j**) geology, (**h**) distance from road, (**i**) drainage density, (**g**) distance from road, (**k**) land use/land cover (LU/LC), and (**l**) NDVI.

2.3. Gully Erosion Modelling

2.3.1. WoE Model

WoE is according to the Bayesian probability framework, to predict the significance of effective factor classes through a statistical approach [42–51]. In this method, the spatial relationship between GE areas and GEVs are identified. The WoE model is based on the calculation of positive (W^+) and negative (W^-) weights. This model computes the weight of each GEVs according to the existence or absence of the gully inventory [52] as follows:

$$W_i^+ = \ln \left(\frac{P\{B|L\}}{P\{B|\overline{L}\}} \right) \tag{2}$$

$$W_i^- = \ln \left(\frac{P\{\overline{B}|L\}}{P\{\overline{B}|\overline{L}\}} \right) \tag{3}$$

$$C = W^+ + W^- \tag{4}$$

$$S(C) = \sqrt{S^2(W^+) + S^2(W^-)} \tag{5}$$

$$S^2(W^+) = \frac{1}{N\{B \cap L\}} + \frac{1}{\{B \cap L\}} \tag{6}$$

$$S^2(W^-) = \frac{1}{\{\overline{B} \cap L\}} + \frac{1}{\{\overline{B} \cap \overline{L}\}} \tag{7}$$

$$W = \left(\frac{C}{S(C)} \right) \tag{8}$$

where ln is the natural log function and P is the probability, B and $\overline{B}$ indicate the presence and absence of the gully geo-environmental factor, respectively, L is the presence of gully, and $\overline{L}$ is the absence of a gully. W^+ and W^- are positive and negative weights, with W^+ indicating that a geo-environmental factor is present in the gully inventory. $S^2(W^+)$ is the variance of the W^+ and $S^2(W^-)$ is the variance of W^-. C indicates the overall association between GEVs and gully occurrence. S (C) is the standard deviation of the contrast and W is final weight of each class factor.

2.3.2. RF Model

RF is a controlled learning method that uses multiple trees in the classification [21]. The RF algorithm, by replacing and continuously changing the factors that affect the target, leads to the creation of a large number of decision trees, then all trees are combined to make decisions [21]. The RF consists of 3 user-defined parameters, which include: (1) The number of variables used in the construction of each tree, which expresses the power of each independent tree; (2) number of trees in RF; and (3) minimum number of nodes [43]. RF prediction power increases with the increasing strength of independent trees and reducing the correlation between them [44]. This algorithm uses 66% of the data to grow a tree called Bootstrap, and then a predictor variable is introduced randomly during the growing process to split a node in the tree construction. The remaining 33% of the data is also used to evaluate the fitted tree [45]. This process is repeated several times and the average of all predicted values is used as the final prediction of the algorithm. In this model, two factors, including the mean decrease accuracy and mean decrease Gini, are used to prioritize of each of the effective factors. The use of the mean decrease accuracy in comparison to mean decrease Gini index is more effective in determining the priority of effective factors, especially in the context of the relationship between environmental factors [46]. The RF analyses were carried out in R 3.3.1, using the "Randomforest" package [21].

2.3.3. BRT Model

BRT is one of several techniques that can help improve the performance of a single model by combining multiple models [47]. BRT uses two algorithms for modeling: Boosting and regression [48]. Boosting is a way to increase the accuracy of the model, and based on this, the construction, combination, and averaging of a large number of models are better and more accurate than an individual model on its own [49]. BRT overcomes the greatest weakness of the single decision tree, which is relatively weak in data processing. In BRT, only the first tree of all the training data is constructed, the next trees are grown on the remaining data from the tree before it; trees are not built on all data and only use a number of data [50]. The main idea in this method is to combine a set of weak predictor models (high predictive error) to arrive at strong prediction (low predictive error) [51]. Thus, in this study, BRT was used for GE spatial modeling using GMB (Generalized Boosted Models) and dismo (Species Distribution Modeling) packages in R 3.3.1.3.

2.3.4. MARS Model

The MARS model is a form of regression algorithm that was introduced by Friedman in 1991 to predict continuous numerical outputs [52]. This technique generates flexible regression models for predicting the target variable by means of dividing the problem space into intervals of input variables and processing a basic function in each interval.

The base function represents information in relation to one or more independent variables. A base function is defined in a given interval, in which the primary and end points are called knote. The knote is the key concept in this method and represents the point at which the behavior of the function changes at that point. The base function expresses the relationship between the input variables and the target variable and is in the form of *Max* $(0, X - c)$ or *Max* $(0, c - X)$, in which c is threshold value and X is the impute variable. The general form of the MARS model is as follows:

$$f(x) = \beta_0 + \sum_{j=1}^{P} \sum_{b=1}^{B} \left[\beta_{jb}(+) Max\left(0, x_i - H_{bj}\right) + \beta_{jb}(-) \, Max\left(0, H_{bj} - x_J\right) \right] \tag{9}$$

where x = input, $f(x)$ = output, P = predictor variables, and B = basis function. *Max* $(0, x - H)$ and *Max* $(0, H - x)$ are basis function and do not have to be present if their coefficients are 0. β_0 is constant, β_{jb} is the coefficient of the *j*th base function (BF), and the H values are called knots. The MARS model includes three main steps: (1) A forward stepwise algorithm to select certain spline basis functions, (2) a backward stepwise algorithm to delete base functions (BFs) until the best set is found, and (3) a smoothing method which gives the final MARS approximation a certain degree of continuity [52]. First, the MARS model estimates the value of the target function with a constant value, and then generates the best processing in the forward direction by searching among the variables.

The search process continues as long as all possible (BFs) are added to the model. At this stage, a very complex model with a large number of knotes is obtained. In the next step, through the process of pruning backward, BFs that are less important are identified and deleted by using the generalized cross-validation (*GCV*) criterion [27]. *GCV* is a criterion for data fitting and eliminates a large number of BFs and reduces the probability of overfitting. This indicator is obtained by using Equation (10):

$$GCV = \frac{1}{2} \sum_{i=1}^{N} [y_i - f(x_i)]^2 / \left[1 - \frac{C(B)}{N}\right]^2 \tag{10}$$

where N is the number of data and $C(B)$ is a complexity penalty that increases with the number of BF in the model and which is defined as:

$$C(B) = (B = 1) + dB \tag{11}$$

where d is a penalty for each BF included into the model. This process continues until a complete review of all the basic functions, and at the end of the optimal model is obtained by applying base functions [52]. MARS model is an adaptive approach, since the selection of BFs and node locations is based on the data and type of purpose. After determining the optimal MARS model, the analysis of variance (ANOVA) method can be used to estimate the participation rate of each of the input variables and BFs. A detailed description of the MARS model can be found in [45]. MARS was run with R 3.3.1 and the "Earth" package [53].

2.4. Validation of GESMs Using Three Data Mining Models

A single criterion is not enough to select the best model among a large number of models, and judging about choosing a superior model by one criteria. It is not a suitable approach and it raises the chance of mistake in choosing the suitable model [27,37,54]. In this study, to compare the performance between data mining models and select the appropriate model, AUC and SCAI were used [28,36,55]. For calculating AUC, different thresholds were considered from 0 to 1, and for each threshold, the number of cells detected by the model as gully erosion was compared with observed gully erosion cells and positive and negative ratio indicators was calculated. After calculating these two indicators, we arranged them in ascending order, then they were plotted to calculate AUC. The AUC values range from 0.5 to 1. If a model cannot estimate the occurrence of an event better than a probable or random viewpoint, its AUC is 0.5 and therefore it will have the least accuracy, while if the AUC is equal to one, the model will have the highest accuracy [56,57]. The quantitative–qualitative relationship between AUC value and prediction accuracy can be classified as follows: 0.5–0.6, poor; 0.6–0.7, average; 0.7–0.8, good; 0.8–0.9, very good; and 0.9–1, excellent. SCAI is the ratio of the percentage area of each of the zoning classes to the percentage of gullies occurring on each class. Based on the SCAI indicator, the values of SCAI in very high sensitivity class are lower than very low sensitivity class.

3. Results

3.1. Multi-Collinearity Analysis

Multicollinearity is a condition of very high inter-correlations or inter-associations among the independent variables. Therefore, it is a type of disturbance in the data, and if present in the data, the statistical conclusions of the data may not be reliable [27]. A TOL value less than 0.1 or a VIF value larger than 10 indicates a high multicollinearity [56]. The outcomes of the coherent analysis among the 12 GEVs are shown in Table 2. The outcomes showed that the TOL and VIF of all GEVs were $\geq$0.1 and $\leq$5, respectively. As a result, no multi-collinearity is seen among the GEVs.

Table 2. Multi-collinearity of effective factors using tolerance (TOL) and variance inflation factor (VIF).

Conditioning Factors	Collinearity Statistics	
	Tolerance	VIF
Constant Coefficient	-	-
Slope degree	0.998	1.002
Distance from road	0.672	1.489
Distance from river	0.323	3.094
Plan curvature	0.674	1.483
Lithology	0.945	1.058
LU	0.864	1.158
Drainage density	0.826	1.211
Elevation	0.920	1.087
Convergence index	0.666	1.503
Aspect	0.299	3.343
TWI	0.942	1.062
NDVI	0.941	1.063

3.2. Spatial Relationship Using WoE Model

The outcomes of WoE model are shown in Table 3. In elevation, the results of WoE indicate that there is a direct correlation between classes of elevation and GE, and with an increase in elevation, GE also increases. Therefore, the class of >1600 m with WoE 47.95 had the greatest impact on gully occurrence. The result of slope degree indicate that classes 5–<10 with WoE 34.96 had a strong relation with GEIM. For slope aspect, NE–facing slopes with a value of 19.46 show high probability of gully occurrence. In the case of plan curvature, among the three classes of concave, flat, and convex, the concave class had the highest value (78.04), and thus a positive correlation with GE. This result is in line with [11,50]. In TWI, the class of >11 has the strongest relationship with GE with the highest value (78.04). In the case of the convergence index, the class of 0–10 with values of 13.18 has a positive relation with gully occurrence. With respect to distance from river, class of >650 with value of 25.86 and regarding distance from road the class of >3000 m with values of 16.25 had the greatest effect on gully occurrence. For the drainage density factor, the class of <1.4 km/km^2 showed the highest value (14.23) and thus high correlation with gully occurrence. According to the lithology factor, Gypsiferous marl with greatest value (51.23) is more prone to GE than other lithology units. Concerning LU/LC, most gullies are located in the range land use type and this class with the highest value (21.02) has the strongest relationship with gully occurrence. In NDVI, results indicated that all gullies are located in the class of <0.11, showing that very low vegetation density renders slopes susceptible to GE.

Table 3. Relationship between conditioning factors and gully erosion using weights-of-evidence (WoE) model.

Factor	Class	Number of Pixels in Domain	Pixels of Gullies	Weights-of-Evidence (WoE)				
				C	S2 (w$^+$)	S2 (w$^-$)	S	W
	<1200	144,200	21	−3.16	0.05	0.00	0.22	−14.41
	1200–<1350	348,463	89	−2.87	0.01	0.00	0.11	−26.60
1	1350–<1450	230,735	502	−0.37	0.00	0.00	0.05	−7.52
	1450–1600	133,305	1057	0.33	0.00	0.00	0.00	0.00
	>1600	85,376	1074	1.88	0.00	0.00	0.04	47.95
	<5	705,163	896	−1.83	0.00	0.00	0.04	−44.90
	5–<10	171,923	1259	1.34	0.00	0.00	0.04	34.96
	10–<15	38,854	397	1.38	0.00	0.00	0.05	25.36
2	15–<20	13,936	121	1.13	0.01	0.00	0.09	12.13
	20–<25	6223	50	1.03	0.02	0.00	0.14	7.22
	25–30	3396	15	0.42	0.07	0.00	0.26	1.62
	>30	2584	5	−0.41	0.20	0.00	0.45	−0.92

Table 3. *Cont.*

Factor	Class	Number of Pixels in Domain	Pixels of Gullies	Weights-of-Evidence (WoE)				
				C	S2 (w^+)	S2 (w^-)	S	W
	Flat	16,770	2	−3.22	0.50	0.00	0.71	−4.55
	N	72,345	208	−0.01	0.00	0.00	0.07	−0.19
	NE	79,383	209	−0.11	0.00	0.00	0.07	−1.52
	E	72,794	43	−1.66	0.02	0.00	0.15	−10.81
3	SE	91,567	54	−1.68	0.02	0.00	0.14	−12.24
	S	114,731	246	−0.34	0.00	0.00	0.07	−5.13
	SW	119,263	396	0.15	0.00	0.00	0.05	2.81
	W	142,533	459	0.12	0.00	0.00	0.05	2.35
	NW	232,693	1126	0.76	0.00	0.00	0.04	19.46
	Concave	54,613	1493	2.99	0.00	0.00	0.04	78.04
4	Flat	574,180	749	−1.43	0.00	0.00	0.04	−33.33
	Convex	313,286	501	−0.80	0.00	0.00	0.05	−16.27
	<7	24,272	63	0.00	0.00	0.00	0.02	−0.15
5	5–<7.5	42,453	91	−0.32	0.01	0.00	0.11	−3.00
	7.5–11	89,328	225	−0.16	0.00	0.00	0.07	−2.29
	>11	786,026	2364	0.21	0.00	0.00	0.06	3.87
	<0	75,370	26	−2.21	0.04	0.00	0.20	−11.21
6	0–10	776,920	2534	0.95	0.00	0.00	0.07	13.18
	>10	89,789	183	−0.39	0.01	0.00	0.08	−5.08
	<170	382,383	522	−1.07	0.00	0.00	0.05	−21.99
7	170–<370	329,586	835	−0.21	0.00	0.00	0.04	−4.99
	370–650	179,671	914	0.75	0.00	0.00	0.04	18.62
	>650	50,444	472	1.31	0.00	0.00	0.05	25.86
	<500	90,285	0	−0.10	0.00	0.00	0.02	−5.29
8	500–<1500	102,453	0	−0.12	0.00	0.00	0.02	−6.05
	1500–3000	113,685	12	−3.44	0.08	0.00	0.29	−11.91
	>3000	635,661	2731	4.70	0.00	0.08	0.29	16.25
	<1.4	277,251	1150	0.55	0.00	0.00	0.04	14.23
9	1.4–<2.4	353,215	1000	−0.04	0.00	0.00	0.04	−1.12
	2.4–3.7	231,503	573	−0.21	0.00	0.00	0.05	−4.49
	>3.7	80,115	20	−2.54	0.05	0.00	0.22	−11.32
	Murmg	144,412	1544	1.97	0.00	0.00	0.04	51.23
	Qft2	617,176	417	−2.36	0.00	0.00	0.05	−44.47
	Ku	23,972	0	−0.03	0.00	0.00	0.02	−1.35
	Jd	18,232	0	−0.02	0.00	0.00	0.02	−1.03
10	PeEz	1449	0	0.00	0.00	0.00	0.02	−0.08
	PlQc	71,058	600	1.24	0.00	0.00	0.05	26.85
	Jl	3,274	0	0.00	0.00	0.00	0.02	−0.18
	E2c	58,380	174	0.03	0.01	0.00	0.08	0.33
	E1c	4,820	8	−0.56	0.13	0.00	0.35	−1.60
	Range	708,879	2669	2.48	0.00	0.01	0.12	21.02
11	Farming	193,682	33	−3.06	0.03	0.00	0.18	−17.47
	Bare land	39,523	41	−1.06	0.02	0.00	0.16	−6.75
	<0.11	863,198	2743	0.09	0.00	0.00	0.02	4.59
12	0.11–0.25	56,745	0	−0.06	0.00	0.00	0.02	−3.26
	>0.25	22,140	0	−0.02	0.00	0.00	0.02	−1.25

1. Elevation, 2. Slope degree, 3. Slope aspect, 4. Plan curvature, 5. topographic wetness index (TWI), 6. Convergence index, 7. Distance from river, 8. Distance from road, 9. Drainage density, 10. Lithology, 11. land use (LU), 12. NDVI.

3.3. Applying RF Model

The outcomes of the confusion matrix for RF model are shown in Table 4. The result shows that the model predicted 2487 non-gully pixels as non-gullies and 256 non-gullies as gully. On the other hand, the RF model predicted 2677 gullies as gullies and 66 gullies as non-gullies. Moreover, the out-of-bag error (OOB) for RF was 5.82%. This means that the model has a precision of 94.18%, which expresses the excellent accuracy of the model in predicting gully erosion.

Table 4. Confusion matrix from the random forest (RF) model (0 = no gully, 1 = gully).

	0	1	**Class Error**
0	2487	256	0.0933
1	66	2677	0.0240

Prioritization results of RF are shown in Table 5 and Figure 4. The results show that the distance from roads (381.67, 22%), elevation (335.06, 19%), and lithology (234.21, 14%) had the highest values, followed by slope degree, drainage density, distance from river, NDVI, convergence index, slope aspect, TWI, plan curvature, and LU/LC.

Table 5. Relative influence of effective conditioning factors in the RF model.

Conditioning Factors	**Weight**
Distance from road	381.67
Elevation	335.06
Lithology	234.21
Slope degree	153.85
Drinage density	126.72
Distance from river	106.84
NDVI	105.26
Convergence index	73.97
Slope aspect	72.41
TWI	71.3
Plan curvature	42.43
LU	25.38

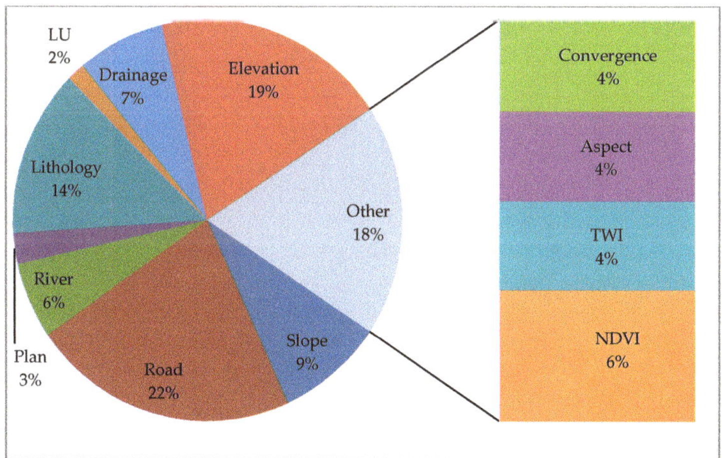

Figure 4. Relative influence of effective conditioning factors in the random forest (RF) model.

Finally, the GESM by the RF model was prepared in ArcGIS 10.5 and divided into five classes from very low to very high (Figure 5a), using a natural break classification [8]. According to the results, of the entire study area (847.87 km^2), 525.97 km^2 (62.03%) are located in the very low susceptibility class, 148.28 km^2 (17.49%) in the low susceptibility, 79.42 km^2 (9.37%) in the moderate class, 56.34 km^2 (6.64%) in the high class, and 37.88 km^2 (4.47%) are located in the very high susceptibility class. Of the total area of GE (0.729 km^2) in the study area, 0.86% (0.01 km^2) are located in the very low susceptibility class,

5.67% (0.04 km^2) in the low susceptibility, 14.80% (0.11 km^2) in the moderate susceptibility, 21.95% (0.16 km^2) in the high susceptibility, and 56.72% (0.41 km^2) in the very high susceptibility classes.

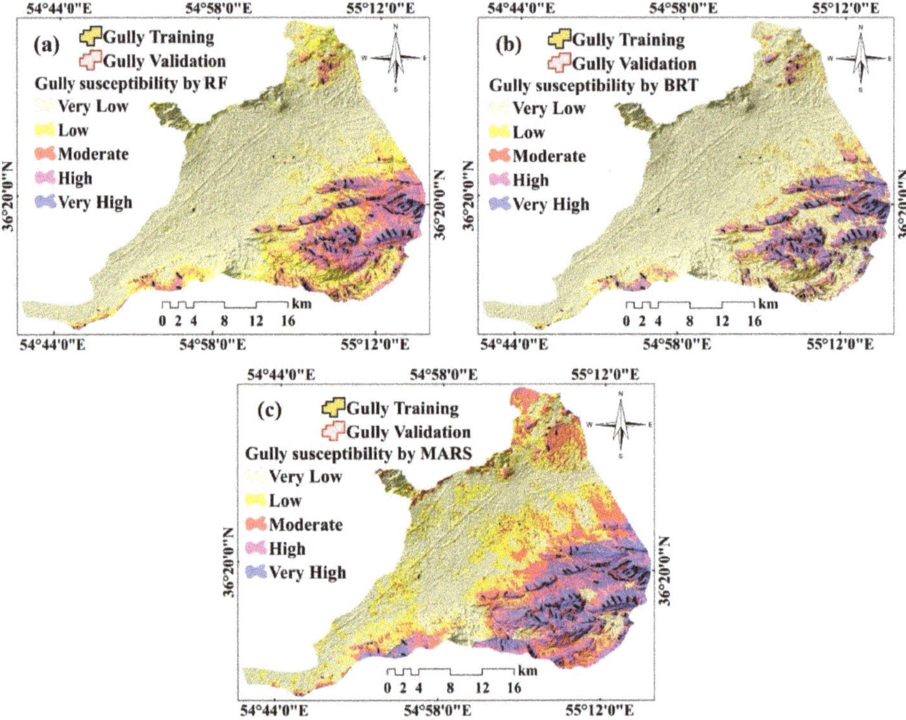

Figure 5. Gully erosion susceptibility maps using: (**a**) RF model, (**b**) BRT model, and (**c**) multivariate adaptive regression spline (MARS) model.

3.4. Applying BRT Model

The BRT model was used to reveal the spatial correlation between the existing GE and the GEVs in the study area. The results of the model are shown in Figure 6. They indicate that the factors distance from roads (31.1%), elevation (27.2%), and lithology (11%) had the highest importance on GE, mirroring the outcomes of the RF model, followed by slope degree (7%), drainage density (6.7%), distance from river (5.1%), slope aspect (3.8%), convergence index (2.4%), NDVI (2.2%), plan curvature (1.6%), TWI (1.6%), and LU/LC (0.3%). The gully susceptibility map by the BRT model was also prepared in ArcGIS 10.5 and divided into five classes of very low to very high (Figure 6c). The results of the GE susceptibility class by the BRT model covered 847.87 km^2 of the study area an area distribution in the very low, low, moderate, high, and very high susceptibility classes are 605.37 km^2, 88.38 km^2, 52.01 km^2, 34.13 km^2, and 67.98 km^2, and percentage distribution in the susceptibility classes of are 71.40, 10.42, 6.13, 4.03, and 8.02, respectively. Of the actual GE area of 0.729 km^2, 0.04 (5.55%), 0.03 (4.56%), 0.06 (8.26%), 0.08 (11.34%), and 0.51 km^2 (70.28%) are located in the very low to very high susceptibility classes, respectively.

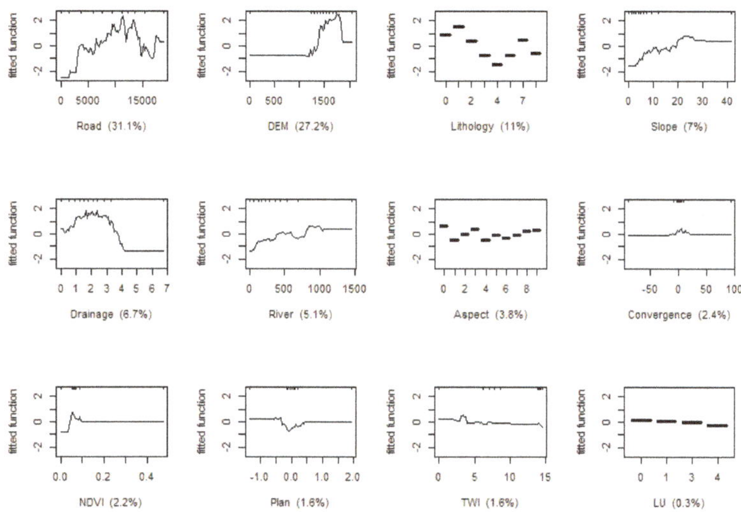

Figure 6. Relative influence of effective conditioning factors in boosted regression tree (BRT) model.

3.5. Applying MARS Model

The optimal MARS model included 28 terms, and the GCV was 0.157. MARS model provides the optimal model only by selecting the necessary parameters. In this research, nine GEVs including lithology, distance from road, distance from river, drainage density, elevation, aspect, convergence index, slope, and NDVI were used to construct the optimal model from the 12 GEVs. The GESM by the MARS model was implemented in ArcGIS 10.5 using Equation (12). According to Equation (12), distance from roads, elevation, and lithology were the most important variables. Values of GESM by MARS model varies from −9.8 to 7.3. At first, GESM classified using quantile, equal interval, natural break, and geometrical interval classification techniques, then, by comparatively analyses of the distribution of training and validation gullies in high and very high classes, the natural break classification technique was most accurate. As a result, GESM by MARS were classified into very low (−9.86–−6.24), low (−6.24–−2.31), moderate (−2.3–0.04), high (0.04–0.38), and very high (0.38–7.32) gully erosion susceptibility zones by natural break classification technique (Figure 5c). The results indicate that 0.02 km^2 (2.10%) of GE in the study area are located in the very low susceptibility class, with 339.01 km^2 (39.98% of total study area) and 0.58 km^2 (79.16%) located in the very high susceptibility class with 105.50 km^2 (0.58%) (Table 6). In general, the results indicate that for all three models with increasing susceptibility (from very low to very high), the area of the respective classes decreased, while in contrast the areas of GE increased. These results is in line with Youssef et al. (2015).

Table 6. Area under the curve (AUC) values of RF, MARS, and BRT data mining models.

Models	AUC	Standard Error	Asymptotic Significant	Asymptotic 95% Confidence Interval	
				Lower Bound	Upper Bound
RF	0.927	0.007	0.000	0.914	0.941
MARS	0.911	0.008	0.000	0.896	0.926
BRT	0.919	0.007	0.000	0.905	0.933

3.6. Validation of Models

The results of the validation of the models using the AUC curve and SCAI indicator are shown in Figure 7, and in Tables 6 and 7. The results show that the values of the AUC for the three models vary from 0.911 to 0.927, indicating very good prediction accuracy for all models, with RF resulting in the highest value. In addition, the SCAI values for the three models, RF (61.08–0.00), MARS (10.45–0.03), BRT (12.59–0.01), show that the RF model has higher SCAI values compared to the other models in the very low, low, and very high susceptibility classes (Figure 7). In spite of the high efficiency and accuracy of the RF model for GE sensitivity mapping, so far this model has not been used by the research community.

$$
\begin{aligned}
GESP_{MARS} = \;& 0.74 + (0.659 \times Lithology1) + (0.656 \times Lithology7) - 0.0001 \\
& \times max(0, 13445 - Distance\ from\ road) + 0.0001 \\
& \times max(0, Distance\ from\ road - 13445) - 0.0002 \\
& \times max(0, 2907.97 - Distance\ from\ River) - 0.087 \\
& \times max(0, 2.377 - Drainage\ density) - 0.106 \\
& \times max(0, Drainage\ density - 2.377) + 0.001 \times max(0, 1793 \\
& - Elevation) - 0.002 \times max(0, Elevation - 1793) - 0.605 \\
& \times Lithology7 \times Aspect4 - 0.0001 \times max(0, 7355.32 \\
& - Distance\ from\ road) \times Lithology1 - 0.0001 \\
& \times max(0, 11249.2 - Distance\ from\ road) \times Lithology7 \\
& - 0.00002 \times max(0, Distance\ from\ road - 11249.2) \\
& \times Lithology7 + 0.0001 \times max(0, 13445 \\
& - Distance\ from\ road) \times Lithology10 - 0.005 \\
& \times max(0, 84.853 - Distance\ from\ River) \times Lithology1 \\
& - 0.0003 \times max(0, Distance\ from\ River - 84.853) \\
& \times Lithology1 + 0.001 \times Lithology2 \times max(0, Elevation \\
& - 1249) - 0.001 \times Lithology2 \times max(0, 1249 - Elevation) \\
& - 0.019 \times Lithology7 \times max(0, 0.772 - Convergence) - 23.54 \\
& \times Lithology7 \times max(0, NDVI - 0.055) - 22.23 \times Lithology7 \\
& \times max(0, 0.055 - NDVI) - 0.00001 \times max(0, 7.65 - Slope) \\
& \times max(0, Distance\ from\ road - 13445) + 0.00001 \\
& \times max(0, Slope - 7.65) \times max(0, Distance\ from\ road - 13445) \\
& - 0.0001 \times max(0, 8.186 - Slope) \times max(0, 1793 - Elevation) \\
& + 0.00004 \times max(0, Slope - 8.19) \times max(0, 1793 - Elevation) \\
& - 0.0000001 \times max(0, Distance\ from\ road - 3877.78) \\
& \times max(0, 907.97 - Distance\ from\ River) + 0.00000004 \\
& \times max(0, 8861.03 - Distance\ from\ road) \times max(0, 907.97 \\
& - Distance\ from\ River) + 0.0000001 \times max(0, Road \\
& - 8861.03) \times max(0, 907.97 - Distance\ from\ River) - 0.001 \\
& \times max(0, Distance\ from\ road - 13445) \\
& \times max(0, Drainage\ density - 1.821) - 0.000001 \\
& \times max(0, Distance\ from\ road - 11435.4) \times max(0, 1793 \\
& - Elevation) - 0.000001 \times max(0, Distance\ from\ road \\
& - 13238.3) \times max(0, 1793 - Elevation)
\end{aligned}
\tag{12}
$$

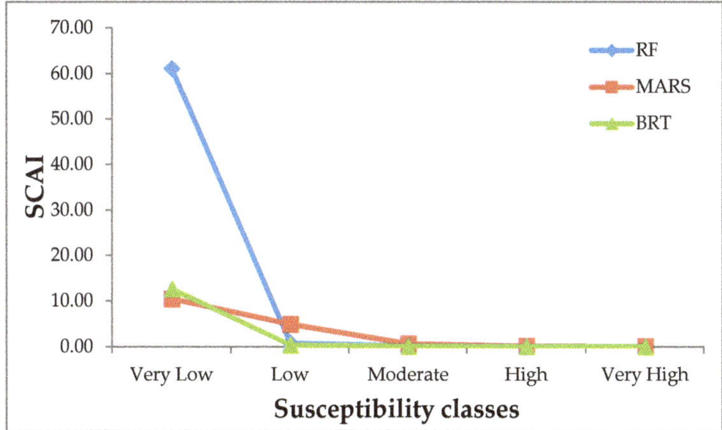

Figure 7. Seed cell area index (SCAI) values for different susceptibility classes in RF, MARS, and BRT data mining models.

Table 7. Seed cell area index (SCAI) values in RF, multivariate adaptive regression spline (MARS), and boosted regression tree (BRT) data mining models.

Model	Susceptibility Classes	Total Area of Classes		Gully in Classes		No Gully Area (km)	Seed Cell (%)	SCAI
		Area (km)	%	Area (km)	%			
RF	Very Low	525.97	62.03	0.01	0.86	525.96	0.01	61.08
	Low	148.28	17.49	0.04	5.67	148.24	0.24	0.74
	Moderate	79.42	9.37	0.11	14.80	79.31	1.15	0.08
	High	56.34	6.64	0.16	21.95	56.18	2.41	0.03
	Very High	37.88	4.47	0.41	56.72	37.46	9.27	0.00
MARS	Very Low	339.01	39.98	0.02	2.10	339.00	0.04	10.45
	Low	194.83	22.98	0.01	1.48	194.82	0.05	4.89
	Moderate	131.17	15.47	0.04	5.67	131.13	0.27	0.58
	High	77.35	9.12	0.08	11.59	77.26	0.93	0.10
	Very High	105.50	12.44	0.58	79.16	104.92	4.64	0.03
BRT	Very Low	605.37	71.40	0.04	5.55	605.33	0.06	12.59
	Low	88.38	10.42	0.03	4.56	88.34	0.32	0.33
	Moderate	52.01	6.13	0.06	8.26	51.95	0.98	0.06
	High	34.13	4.03	0.08	11.34	34.05	2.06	0.02
	Very High	67.98	8.02	0.51	70.28	67.46	6.40	0.01

4. Discussion

Determining effective parameters in GE and providing a GESM are the first steps in risk management. In regards to this, prediction of areas susceptible to erosion is associated with uncertainty, various models can be used to predict it accurately. Over the past decades, numerous statistical and empirical models have been developed to predict environmental hazards, such as GE, by various researchers around the world [12,14,28,30,31,45]. Due to some of the limitations of the aforementioned models such as time consuming, complexity, costly, and need a lot of data, in recent years data mining methods have been presented. Data mining is a process of discovery of relationships, patterns, and trends that consider the vast amount of information stored in databases with template recognition technology [51,58,59]. The most important applications of data mining are categorization, estimation, forecasting, group dependency, clustering, and descriptions. The results of data mining models show that in RF, BRT, and MARS mode, distance from roads had the highest impact in the occurrence of gully erosion in the study area. This result is in line with [10,49]. If the engineering measures are not considered in site selection and construction of roads as anthropogenic structures in nature, they can act as a causative factor in environmental hazards such as landslide and gully erosion. The construction of roads in bare lands with erosion-sensitive formations has led to the expansion of gully erosion in the study area, so that the construction of a road without proper culverts causes disrupted of natural drainage and runoff concentrations, thus eroding the bare lands and resulting in the formation of a gully. The results of the validation of data mining models showed that the RF model more accurately predicted areas that are sensitive to gully erosion. These results are consistent with the results of [36,43,46,59], which introduced the RF model as a strong and high-performance model. One of the most widely used data mining methods is the RF model. The advantages of the RF method over other models is that this model can apply several input factors without eliminating any factors, and return a very small set of categories that support high prediction accuracy [6]. The classification accuracy of this model is affected by many factors such as the number, scale, type, and precision of input data. Thus, in the processing, the use of all suitable factors causes the accuracy of the model to increase. Compared with other models, RF has higher sufficiency to apply a very high number of datasets [6]. The RF model has the potential as a tool of spatial model for assessing environmental issues and environmental hazards. The RF model combines several tree algorithms to generate a repeated prediction of each phenomenon. This method can learn complicated patterns and consider the nonlinear relationship between explanatory variables and dependent variables. It can also incorporate and combine different types of data in the analysis, due to the lack of distribution of assumptions about the data used. This model can use and apply thousands of input variables without deleting one of them. This method is less sensitive to artificial neural networks, in case of noise data, and can better estimate the parameters [60]. The greatest advantages of RF model are high predictive accuracy, the ability to learn nonlinear relationships, the ability to determine the important variables in prediction, its nonparametric nature, and in dealing with distorted data, it works better than other algorithms for categorization. The main disadvantages of this algorithm include high memory occupation, hard and time-consuming in implementation for large datasets, high cost of pruning, high number of end nodes in case of overlap, and the accumulation of layers of errors in the case of the tree growing. [15,61] stated that the CART, BRT, and RF models showed better accuracy compared to bivariate and multivariate methods. Pourghasemi et al. concluded that the RF and maximum entropy (ME), models have high performance and precision in modeling [31]. Mojaddadi et al. showed that BRT, CART, and RF methods are suitable for modelling [55]. Chen et al. indicates that the MARS and RF models are good estimators for mapping [36]. Lai et al. indicated that the RF model has significant potential for weight determination on landslide modelling [62]. Kuhnert et al stated that RF with AUC = 97.0 is suitable for landslide susceptibility [27]. Lee et al stated that the prediction accuracy of RF model is high (90.8) and that this model had a high capability for landslide prediction [43]. They applied RF and boosted-tree models for spatial prediction of flood susceptibility in Seoul metropolitan city, Korea [43]. They stated that the RF model has better performance compared to boosted-tree. As a scientific achievement, the methodology framework used in this research has shown that the proper selection of effective variables in gully erosion, along with the use of modern data mining models and Geography Information System (GIS) technique, are able to successfully identify areas susceptible to gully erosion. The susceptibility map prepared using this methodology is a suitable tool for sustainable planning to protect the land against gully erosion processes. Therefore, this methodology can be used to assess gully erosion in other similar areas, especially in arid and semi-arid regions.

5. Conclusions

GE is one of the main processes causing soil degradation and there is a need to improve methods to predict susceptible areas and responsible environmental factors, to allow early intervention to prevent, limit, or reverse gully formation. The utility of three data mining models, RF, BRT, and MARS, to predict GE in the Shahroud watershed, Iran, was assessed. For this purpose, twelve causative factors and 121 gully locations (70%) are used for applying the models. In addition, 51 gully locations (30%) are used for validation of models. The correlation between GE and conditioning factor classes was researched with a WoE Bayes theory. Distance to roads, elevation, and lithology were the key factors. Validation of the models showed that all three models have high accuracy for GE mapping. Data mining/machine learning methods have a unique ability and accuracy for GESM. The results also showed that the southwestern part of the study region has a high susceptibility to GE.

Therefore, it is recommended that the following suggestions should be made to prevent and reduce soil erosion and its subsequent risks in the Sharoud watershed: (1) Control of gullies by restoration of vegetation adaptable with the natural conditions of the area; (2) gully controlling by building dams that could prevent soil erosion by slowing down the flow of water and aggravation of sedimentation; (3) awareness of farmers by environmental officials of the region, in terms of the type and principles of proper cultivation and prevention of overgrazing and destruction of vegetation; (4) correction of land use based on natural ability and restrictions related to geomorphologic and physiographic soil characteristics of the area.

Author Contributions: Conceptualization, A.A., B.P., and H.R.P.; Data curation, A.A.; Formal analysis, A.A., and H.R.P.; Investigation, A.A., B.P., and H.R.P.; Methodology, B.P., A.A., H.R.P., and K.R.; Resources, B.P. and A.A.; Software, H.R.P., and A.A.; Supervision, B.P., N.K. and H.R.P.; Validation, H.R.P., and A.A.; Writing–original draft, A.A.; Writing–review and editing, B.P., A.A., H.R.P., N.K. and K.R.

Funding: This research was supported by the UTS under grant number 321740.2232335 and 321740.2232357.

Conflicts of Interest: The authors declare no conflict of interest.

References

1. Magliulo, P. Assessing the susceptibility to water-induced soil erosion using a geomorphological, bivariate statistics-based approach. *Environ. Earth Sci.* **2012**, *67*, 1801–1820. [CrossRef]

2. UNEP. The Emissions Gap Report. United Nations Environment Programme (UNEP). Nairobi, 2017. Available online: www.unenvironment.org/resources/emissions-gap-report (accessed on 13 January 2018).

3. Haregeweyn, N.; Tsunekawa, A.; Poesen, J.; Tsubo, M.; Meshesha, D.T.; Fenta, A.A.; Nyssen, J.; Adgo, E. Comprehensive assessment of soil erosion risk for better land use planning in river basins: Case study of the Upper Blue Nile River. *Sci. Total Environ.* **2017**, *574*, 95–108. [CrossRef] [PubMed]

4. Nampak, H.; Pradhan, B.; Mojaddadi Rizeei, H.; Park, H.-J. Assessment of Land Cover and Land Use Change Impact on Soil Loss in a Tropical Catchment by Using Multi-Temporal SPOT-5 Satellite Images and RUSLE model. *Land Degrad. Dev.* **2018**. [CrossRef]

5. Rizeei, H.M.; Saharkhiz, M.A.; Pradhan, B.; Ahmad, N. Soil erosion prediction based on land cover dynamics at the Semenyih watershed in Malaysia using LTM and USLE models. *Geocarto Int.* **2016**, *31*, 1158–1177. [CrossRef]

6. Zhang, X.; Fan, J.; Liu, Q.; Xiong, D. The contribution of gully erosion to total sediment production in a small watershed in Southwest China. *Phys. Geogr.* **2018**, *39*, 246–263. [CrossRef]

7. Mojaddadi, H.; Habibnejad, M.; Solaimani, K.; Ahmadi, M.; Hadian-Amri, M. An Investigation of Efficiency of Outlet Runoff Assessment. *J. Appl. Sci.* **2009**, *9*, 105–112.

8. Zabihi, M.; Mirchooli, F.; Motevalli, A.; Darvishan, A.K.; Pourghasemi, H.R.; Zakeri, M.A.; Sadighi, F. Spatial modelling of gully erosion in Mazandaran Province, northern Iran. *Catena* **2018**, *161*, 1–13. [CrossRef]

9. Kirkby, M.; Bracken, L. Gully processes and gully dynamics. *Earth Surf. Process. Landf. J. Br. Geomorphol. Res. Group* **2009**, *34*, 1841–1851. [CrossRef]

10. Torri, D.; Poesen, J.; Borselli, L.; Bryan, R.; Rossi, M. Spatial variation of bed roughness in eroding rills and gullies. *Catena* **2012**, *90*, 76–86. [CrossRef]

11. Mccloskey, G.; Wasson, R.; Boggs, G.; Douglas, M. Timing and causes of gully erosion in the riparian zone of the semi-arid tropical Victoria River, Australia: Management implications. *Geomorphology* **2016**, *266*, 96–104. [CrossRef]

12. Rahmati, O.; Tahmasebipour, N.; Haghizadeh, A.; Pourghasemi, H.R.; Feizizadeh, B. Evaluating the influence of geo-environmental factors on gully erosion in a semi-arid region of Iran: An integrated framework. *Sci. Total Environ.* **2017**, *579*, 913–927. [CrossRef] [PubMed]

13. Dube, F.; Nhapi, I.; Murwira, A.; Gumindoga, W.; Goldin, J.; Mashauri, D. Potential of weight of evidence modelling for gully erosion hazard assessment in Mbire District–Zimbabwe. *Phys. Chem. Earth Part A/B/C* **2014**, *67*, 145–152. [CrossRef]

14. Zakerinejad, R.; Maerker, M. An integrated assessment of soil erosion dynamics with special emphasis on gully erosion in the Mazayjan basin, southwestern Iran. *Nat. Hazards* **2015**, *79*, 25–50. [CrossRef]

15. Pham, T.G.; Degener, J.; Kappas, M. Integrated universal soil loss equation (USLE) and Geographical Information System (GIS) for soil erosion estimation in A Sap basin: Central Vietnam. *Int. Soil Water Conserv. Res.* **2018**, *6*, 99–110. [CrossRef]

16. Pournader, M.; Ahmadi, H.; Feiznia, S.; Karimi, H.; Peirovan, H.R. Spatial prediction of soil erosion susceptibility: An evaluation of the maximum entropy model. *Earth Sci. Inform.* **2018**, *11*, 389–401. [CrossRef]

17. Althuwaynee, O.F.; Pradhan, B.; Park, H.-J.; Lee, J.H. A novel ensemble bivariate statistical evidential belief function with knowledge-based analytical hierarchy process and multivariate statistical logistic regression for landslide susceptibility mapping. *Catena* **2014**, *114*, 21–36. [CrossRef]

18. Morgan, R.; Quinton, J.; Smith, R.; Govers, G.; Poesen, J.; Auerswald, K.; Chisci, G.; Torri, D.; Styczen, M. The European Soil Erosion Model (EUROSEM): A dynamic approach for predicting sediment transport from fields and small catchments. *Earth Surf. Process. Landf. J. Br. Geomorphol. Res. Group* **1998**, *23*, 527–544. [CrossRef]

19. Barber, M.; Mahler, R. Ephemeral gully erosion from agricultural regions in the Pacific Northwest, USA. *Ann. Wars. Univ. Life Sci.-SGGW. Land Reclam.* **2010**, *42*, 23–29. [CrossRef]

20. Leonard, R.; Knisel, W.; Still, D. GLEAMS: Groundwater loading effects of agricultural management systems. *Trans. ASAE* **1987**, *30*, 1403–1418. [CrossRef]

21. Liaw, A.; Breiman, W.M. Cutler's Random Forests for Classification and Regression. Available online: https://www.rdocumentation.org/packages/randomForest (accessed on 1 April 2018).

22. Akgün, A.; Türk, N. Mapping erosion susceptibility by a multivariate statistical method: A case study from the Ayvalık region, NW Turkey. *Comput. Geosci.* **2011**, *37*, 1515–1524. [CrossRef]

23. Conoscenti, C.; Angileri, S.; Cappadonia, C.; Rotigliano, E.; Agnesi, V.; Märker, M. Gully erosion susceptibility assessment by means of GIS-based logistic regression: A case of Sicily (Italy). *Geomorphology* **2014**, *204*, 399–411. [CrossRef]

24. Conforti, M.; Aucelli, P.P.; Robustelli, G.; Scarciglia, F. Geomorphology and GIS analysis for mapping gully erosion susceptibility in the Turbolo stream catchment (Northern Calabria, Italy). *Nat. Hazards* **2011**, *56*, 881–898. [CrossRef]

25. Lucà, F.; Conforti, M.; Robustelli, G. Comparison of GIS-based gullying susceptibility mapping using bivariate and multivariate statistics: Northern Calabria, South Italy. *Geomorphology* **2011**, *134*, 297–308. [CrossRef]

26. Meyer, A.; Martınez-Casasnovas, J. Prediction of existing gully erosion in vineyard parcels of the NE Spain: A logistic modelling approach. *Soil Tillage Res.* **1999**, *50*, 319–331. [CrossRef]

27. Kuhnert, P.M.; Henderson, A.K.; Bartley, R.; Herr, A. Incorporating uncertainty in gully erosion calculations using the random forests modelling approach. *Environmetrics* **2010**, *21*, 493–509. [CrossRef]

28. Rahmati, O.; Haghizadeh, A.; Pourghasemi, H.R.; Noormohamadi, F. Gully erosion susceptibility mapping: The role of GIS-based bivariate statistical models and their comparison. *Nat. Hazards* **2016**, *82*, 1231–1258. [CrossRef]

29. Svoray, T.; Michailov, E.; Cohen, A.; Rokah, L.; Sturm, A. Predicting gully initiation: Comparing data mining techniques, analytical hierarchy processes and the topographic threshold. *Earth Surf. Proc. Land.* **2012**, *37*, 607–619. [CrossRef]

30. Zakerinejad, R.; Märker, M. Prediction of Gully erosion susceptibilities using detailed terrain analysis and maximum entropy modeling: A case study in the Mazayejan Plain, Southwest Iran. *Geogr. Fis. Din. Quat.* **2014**, *37*, 67–76.

31. Pourghasemi, H.R.; Yousefi, S.; Kornejady, A.; Cerdà, A. Performance assessment of individual and ensemble data-mining techniques for gully erosion modeling. *Sci. Total Environ.* **2017**, *609*, 764–775. [CrossRef] [PubMed]

32. I.R. of Iran Meteorological Organization. 2012. Available online: http://www.mazandaranmet.ir/ (accessed on 12 October 2017).

33. Geological Survey Department of Iran (GSDI). 2012. Available online: http://www.mazandaranmet.ir/ (accessed on 12 October 2017).

34. Althuwaynee, O.F.; Pradhan, B.; Lee, S. Application of an evidential belief function model in landslide susceptibility mapping. *Comput. Geosci.* **2012**, *44*, 120–135. [CrossRef]

35. Rizeei, H.M.; Pradhan, B.; Saharkhiz, M.A. Surface runoff prediction regarding LULC and climate dynamics using coupled LTM, optimized ARIMA, and GIS-based SCS-CN models in tropical region. *Arab. J. Geosci.* **2018**, *11*, 53. [CrossRef]

36. Chen, W.; Xie, X.; Wang, J.; Pradhan, B.; Hong, H.; Bui, D.T.; Duan, Z.; Ma, J. A comparative study of logistic model tree, random forest, and classification and regression tree models for spatial prediction of landslide susceptibility. *Catena* **2017**, *151*, 147–160. [CrossRef]

37. Tehrany, M.S.; Pradhan, B.; Mansor, S.; Ahmad, N. Flood susceptibility assessment using GIS-based support vector machine model with different kernel types. *Catena* **2015**, *125*, 91–101. [CrossRef]

38. Claps, P.; Fiorentino, M.; Oliveto, G. Informational entropy of fractal river networks. *J. Hydrol.* **1996**, *187*, 145–156. [CrossRef]

39. Aal-Shamkhi, A.D.S.; Mojaddadi, H.; Pradhan, B.; Abdullahi, S. Extraction and modeling of urban sprawl development in Karbala City using VHR satellite imagery. In *Spatial Modeling and Assessment of Urban Form*; Springer: Berlin, Germany, 2017; pp. 281–296.

40. Abdullahi, S.; Pradhan, B.; Mojaddadi, H. City compactness: Assessing the influence of the growth of residential land use. *J. Urban Technol.* **2018**, *25*, 21–46. [CrossRef]

41. Rizeei, H.M.; Shafri, H.Z.; Mohamoud, M.A.; Pradhan, B.; Kalantar, B. Oil palm counting and age estimation from WorldView-3 imagery and LiDAR data using an integrated OBIA height model and regression analysis. *J. Sensors* **2018**, *2018*, 2536327. [CrossRef]

42. Xie, Z.; Chen, G.; Meng, X.; Zhang, Y.; Qiao, L.; Tan, L. A comparative study of landslide susceptibility mapping using weight of evidence, logistic regression and support vector machine and evaluated by SBAS-InSAR monitoring: Zhouqu to Wudu segment in Bailong River Basin, China. *Environ. Earth Sci.* **2017**, *76*, 313. [CrossRef]

43. Lee, S.; Kim, J.-C.; Jung, H.-S.; Lee, M.J.; Lee, S. Spatial prediction of flood susceptibility using random-forest and boosted-tree models in Seoul metropolitan city, Korea. *Geomat. Nat. Hazards Risk* **2017**, *8*, 1185–1203. [CrossRef]

44. Cutler, D.R.; Edwards Jr, T.C.; Beard, K.H.; Cutler, A.; Hess, K.T.; Gibson, J.; Lawler, J.J. Random forests for classification in ecology. *Ecology* **2007**, *88*, 2783–2792. [CrossRef] [PubMed]

45. Simpson, G.L.; Birks, H.J.B. Statistical learning in palaeolimnology. In *Tracking Environmental Change Using Lake Sediments*; Springer: Berlin, Germany, 2012; pp. 249–327.

46. Nicodemus, K.K. Letter to the editor: On the stability and ranking of predictors from random forest variable importance measures. *Brief. Bioinform.* **2011**, *12*, 369–373. [CrossRef] [PubMed]

47. Bui, D.T.; Bui, Q.-T.; Nguyen, Q.-P.; Pradhan, B.; Nampak, H.; Trinh, P.T. A hybrid artificial intelligence approach using GIS-based neural-fuzzy inference system and particle swarm optimization for forest fire susceptibility modeling at a tropical area. *Agr. For. Meteorol.* **2017**, *233*, 32–44.

48. Elith, J.; Leathwick, J.R.; Hastie, T. A working guide to boosted regression trees. *J. Anim. Ecol.* **2008**, *77*, 802–813. [CrossRef] [PubMed]

49. Regmi, A.D.; Devkota, K.C.; Yoshida, K.; Pradhan, B.; Pourghasemi, H.R.; Kumamoto, T.; Akgun, A. Application of frequency ratio, statistical index, and weights-of-evidence models and their comparison in landslide susceptibility mapping in Central Nepal Himalaya. *Arab. J. Geosci.* **2014**, *7*, 725–742. [CrossRef]

50. Aertsen, W.; Kint, V.; Van Orshoven, J.; Özkan, K.; Muys, B. Comparison and ranking of different modelling techniques for prediction of site index in Mediterranean mountain forests. *Ecol. Model.* **2010**, *221*, 1119–1130. [CrossRef]

51. Krishnaiah, V.; Narsimha, G.; Chandra, N.S. Heart disease prediction system using data mining techniques and intelligent fuzzy approach: A review. *Heart Dis.* **2016**, *136*, 43–51. [CrossRef]

52. Oh, H.-J.; Pradhan, B. Application of a neuro-fuzzy model to landslide-susceptibility mapping for shallow landslides in a tropical hilly area. *Comput. Geosci.* **2011**, *37*, 1264–1276. [CrossRef]

53. Torgo, L. *Data Mining with R: Learning with Case Studies*; Chapman and Hall/CRC: Boca Raton, FL, USA, 2016.

54. Umar, Z.; Pradhan, B.; Ahmad, A.; Jebur, M.N.; Tehrany, M.S. Earthquake induced landslide susceptibility mapping using an integrated ensemble frequency ratio and logistic regression models in West Sumatera Province, Indonesia. *Catena* **2014**, *118*, 124–135. [CrossRef]

55. Mojaddadi, H.; Pradhan, B.; Nampak, H.; Ahmad, N.; Ghazali, A.H.B. Ensemble machine-learning-based geospatial approach for flood risk assessment using multi-sensor remote-sensing data and GIS. *Geomat. Nat. Hazards Risk* **2017**, *8*, 1080–1102. [CrossRef]

56. Pourghasemi, H.R.; Beheshtirad, M.; Pradhan, B. A comparative assessment of prediction capabilities of modified analytical hierarchy process (M-AHP) and Mamdani fuzzy logic models using Netcad-GIS for forest fire susceptibility mapping. *Geomat. Nat. Hazards Risk* **2016**, *7*, 861–885. [CrossRef]

57. Hong, H.; Naghibi, S.A.; Dashtpagerdi, M.M.; Pourghasemi, H.R.; Chen, W. A comparative assessment between linear and quadratic discriminant analyses (LDA-QDA) with frequency ratio and weights-of-evidence models for forest fire susceptibility mapping in China. *Arab. J. Geosci.* **2017**, *10*, 167. [CrossRef]

58. Mezaal, M.R.; Pradhan, B.; Shafri, H.; Mojaddadi, H.; Yusoff, Z. Optimized Hierarchical Rule-Based Classification for Differentiating Shallow and Deep-Seated Landslide Using High-Resolution LiDAR Data. In *Global Civil Engineering Conference*; Springer: Berlin, Germany, 2017.

59. Rizeei, H.M.; Pradhan, B.; Saharkhiz, M.A. An integrated fluvial and flash pluvial model using 2D high-resolution sub-grid and particle swarm optimization-based random forest approaches in GIS. *Complex Intell. Syst.* **2018**, 1–20. [CrossRef]

60. Kantardzic, M. *Data mining: Concepts, Models, Methods, and Algorithms*; John Wiley & Sons: Hoboken, NJ, USA, 2011.

61. Pham, B.T.; Pradhan, B.; Bui, D.T.; Prakash, I.; Dholakia, M. A comparative study of different machine learning methods for landslide susceptibility assessment: A case study of Uttarakhand area (India). *Environ. Model. Softw.* **2016**, *84*, 240–250. [CrossRef]

62. Lai, C.; Chen, X.; Wang, Z.; Xu, C.-Y.; Yang, B. Rainfall-induced landslide susceptibility assessment using random forest weight at basin scale. *Hydrol. Res.* **2017**, nh2017044. [CrossRef]

 *applied sciences*

Article

Single-Class Data Descriptors for Mapping *Panax notoginseng* through P-Learning

Fei Deng and Shengliang Pu *

School of Geodesy and Geomatics, Wuhan University, Wuhan 430079, China; fdeng@sgg.whu.edu.cn
* Correspondence: shengliangpu@163.com; Tel.: +86-027-6875-8467

Received: 16 August 2018; Accepted: 21 August 2018; Published: 24 August 2018

Abstract: Machine learning-based remote-sensing techniques have been widely used for the production of specific land cover maps at a fine scale. P-learning is a collection of machine learning techniques for training the class descriptors on the positive samples only. *Panax notoginseng* is a rare medicinal plant, which also has been a highly regarded traditional Chinese medicine resource in China for hundreds of years. Until now, *Panax notoginseng* has scarcely been observed and monitored from space. Remote sensing of natural resources provides us new insights into the resource inventory of Chinese materia medica resources, particularly of *Panax notoginseng*. Generally, land-cover mapping involves focusing on a number of landscape classes. However, sometimes a subset or one of the classes will be the only part of interest. In term of this study, the *Panax notoginseng* field is the right unit class. Such a situation makes single-class data descriptors (SCDDs) especially significant for specific land-cover interpretation. In this paper, we delineated the application such that a stack of SCDDs were trained for remote-sensing mapping of *Panax notoginseng* fields through P-learning. We employed and compared SCDDs, i.e., the simple Gaussian target distribution, the robust Gaussian target distribution, the minimum covariance determinant Gaussian, the mixture of Gaussian, the auto-encoder neural network, the k-means clustering, the self-organizing map, the minimum spanning tree, the k-nearest neighbor, the incremental support vector data description, the Parzen density estimator, and the principal component analysis; as well as three ensemble classifiers, i.e., the mean, median, and voting combiners. Experiments demonstrate that most SCDDs could achieve promising classification performance. Furthermore, this work utilized a set of the elaborate samples manually collected at a pixel-level by experts, which was intended to be a benchmark dataset for the future work. The measuring performance of SCDDs gives us challenging insights to define the selection criteria and scoring proof for choosing a fine SCDD in mapping a specific landscape class. With the increment of remotely sensed satellite data of the study area, the spatial distribution of *Panax notoginseng* could be continuously derived in the local area on the basis of SCDDs.

Keywords: mapping; single-class data descriptors; materia medica resource; *Panax notoginseng*; one-class classifiers; geoherb

---

1. Introduction

Traditional Chinese medicine (TCM) [1] originated in ancient China and has evolved over thousands of years as the only health care and disease healing [2]. A long time before the birth of modern Western medicine, traditional medicinal recipes were handed down orally generation by generation in many parts of the world [3]. Given that TCM is a practical medicine built on experience, and has been mainly practiced and researched in China [4,5], the essence of TCM has always been the most advanced and experienced medicine in the world [6]. Moreover, scientists proved that TCM can coexist with Western medicine [3,7]. Geoherb is a type of Chinese herb with a geographical indication corresponding to a specific geographical location or origin, which has a certification that the product possesses certain qualities, and its production will be protected by intellectual property rights law [8,9]. Compared to the herbal resources produced in other areas, the quality and efficacy of geoherbs are much better [10]. As a highly-regarded TCM resource and a rare kind of geoherb in China [11,12], *Panax notoginseng* (see Figure 1) has been cultivated for more than 400 years in the south-west region of China [13], especially in Wenshan Prefecture, Yunnan province. The conventional methods of TCM resource surveys mainly focus on the qualitative description of species rather than the natural storage or dynamic changes of the planting fields, resulting in a problematic situation that TCM resources appear difficult to monitor over time, which is not conducive to the sustainable development of TCM [14]. In the past few decades, a resource census of TCM has been carried out three times (e.g., 1960–1962, 1969–1973, and 1983–1987). Until 2009, the government of China proposed *"To carry out a nationwide census of TCM resources, strengthen the monitoring of TCM resources and the construction of an information network"* [15]. Furthermore, the State Council of China highlighted *"Strengthening the landscape-scale dynamic monitoring and protection of TCM"* [16] again. From 2011, the government of China planned to conduct the fourth national census of TCM resources, and remote-sensing techniques were regarded as the core-key technologies for surveying and monitoring TCM resources in a large area. The inheritance, innovation, modernization, and internationalization of TCM would be the four basic tasks for a considerable period of time [17,18].

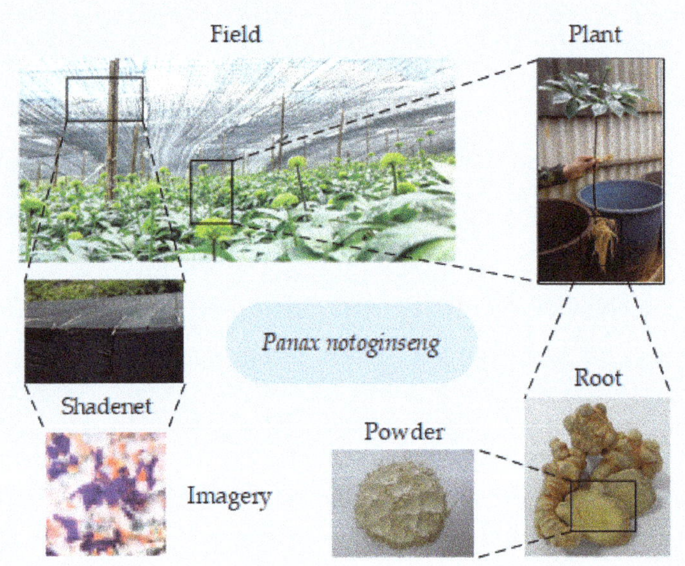

Figure 1. *Panax notoginseng* is a well-known geoherb, which has acquired a very favorable reputation for the treatment of blood disorders, including blood stasis, bleeding, and blood deficiency. Its root can be turned to powder as a medicinal material, and the shadenet cover can be observed from space by means of satellite imagery.

Remote-sensing techniques have been applied to monitor land cover at a range of spatial and temporal scales, in order to satisfy a range of scientific and practical requirements [19–21]. In particular, remote-sensing mapping is an efficient technique to acquire spatial and temporal cropland information repeatedly and consistently [22,23]. In many cases, remotely-sensed data are utilized to derive landscape information on a specific land-cover class of interest [24]. The ability to map and monitor land-cover types and their dynamics for diverse applications has been enhanced by the availability and constantly increased coverage, of satellite images [25,26]. Many problems are encountered in mapping land cover from remotely sensed data by a classification analysis or landscape class description [27] in order to quantify the relationships among all the pixels in an image, such as similarities or differences in spectra signature or spatial texture [22], and extract land cover classes from remote-sensing images [28–30]. The application of remote-sensing techniques plays a significant role in the quantitative resource survey of TCM, particularly in exploring monitoring abilities for the sustainable utilization and bio-diversity protection of Chinese materia medica resources in macrocosm. Under the prerequisite that remote-sensing techniques provide up-to-date landscape surveying at a fine scale [31], one of the motivations for this work comes from the fact that only a single class of interest is involved in a mapping task.

Recently, there has been a number of applications [32], i.e., geological products [33], vegetation indices [34], aerosol products [35], ocean data products [36], dust source identification [37], and crops identification [38] in remote-sensing based on machine learning techniques. In this study, we present an original innovation to apply most of the available single-class data descriptors (SCDDs) through P-learning to conduct mapping of *Panax notoginseng* fields. After that, the work of measuring performance will give us profound insights into defining the selection criteria and scoring proof for choosing a fine SCDD. As such, the introduction of SCDDs for remote-sensing mapping of *Panax notoginseng* fields will be helpful to promoting the development of a *Panax notoginseng* resource inventory and dynamic monitoring towards the quantitative direction. Attributing to the standardized cultivation technique of good agricultural practice (GAP) [39], or so-called controlled-environment agriculture using shade houses provide a distinct image texture to interpret *Panax notoginseng* fields visually [40]. Additionally, due to only *Panax notoginseng* fields being the target class, the task of mapping *Panax notoginseng* fields becomes a specific typology of land-cover classification, which could be regarded as a problem of single-class data description or a special type of one-class classification [41]. SCDDs are the appealing alternatives to the conventional supervised classifiers because they can be trained with only the target training samples [42]. These kinds of algorithms have emerged to only require training samples from the target class, which are referred to as P-learning [43]. Notice that the single class meant no more than one landscape class, and the P-learning based class description may depict the sight that no negative samples are used for training. Such a classification approach aims to identify only one landscape class of interest regardless of the other classes presented in the study area [44]. In the case of single-class data description, we always face an imbalanced binary classification [45,46] including (1) the positive class (i.e., *Panax notoginseng* fields); and (2) the negative class (i.e., the other classes). In this case, the positive class is assumed to be sampled well, while the other classes may be sampled very sparsely or totally absent. When no samples of the other classes are available, most classification errors (e.g., false negative) cannot be estimated [47]. In addition, the procedure that trains an accurate SCDD is challenging, particularly in the face of a large number of unlabeled samples; or say, only a small class or relatively few training samples are available [42]. Therefore, it might be very expensive to collect the negative samples which are so abundant that a good sampling seems elusive. Although this was an extreme case, we carefully designed the training and test sets, which were composed of the qualified positive and negative samples. Note that if we want to improve the overall performance of the numerous classifiers which may differ in complexity, a combination of these classifiers will always be a viable solution [48].

Regarding another motivation of this work with respect to TCM, the quality control of TCM remains a significant issue that affects medicinal herbs, formulations, and even TCM practice. Due more to the lax enforcement of standards [49], resulting in the diminishing popularity of TCM rather than a failure of remedies, particularly, the patchy regulation has led to inconsistent herb quality, unqualified practitioners, unsubstantiated claims for secret formulas, and both deliberate and inadvertent mislabeling and adulteration, sometimes with fatal consequences. Considering *Panax notoginseng* is a vulnerable crop which has a serious succession cropping obstacle [50], consequently the continuous planting of the same crop in the same field will lead to the decrease of yield and quality. In order to promote the quality of production, it is crucial to monitor the spatial planting patterns of *Panax notoginseng* fields, such as crop rotation [51,52]. The planting pattern implies standardized planting with the specific crop structure and spatio-temporal configuration in the same field for a specific region under the particular natural resource and socio-economic conditions [53,54] so as to realize the sustainable utilization of agricultural resources and crop yield. Until now, the concrete planting pattern changes of *Panax notoginseng* are still poorly known. To the best of our knowledge, until now no such work has been done which, on the one hand, enriches the approaches to monitor spatial planting pattern changes of the perennial ginseng from space; on the other hand, employs SCDDs for mapping *Panax notoginseng*.

Our studies on mapping *Panax notoginseng* aim to provide fruitful information for studies on the quality assurance of TCM production, precision farming, the construction of agro-ecosystems, sustainable development, and the protection of the biodiversity of *Panax notoginseng*. Furthermore, determining the planting area of *Panax notoginseng* is an important part of obtaining more accurate information about annual yield and natural storage, except for mapping the spatial distribution. The current study, which involves mapping the planting parcels of *Panax notoginseng* at a 30 m spatial resolution, has three aims: (1) mapping *Panax notoginseng* fields through a stack of SCDDs as the future technical milestone for planting pattern analysis; (2) evaluating the abilities of SCDDs in identifying small *Panax notoginseng* fields in the complex agricultural landscapes; and (3) providing the potential possibilities for monitoring the planting pattern changes of *Panax notoginseng* fields, further giving us new insights into the planting pattern transitions of the perennial ginseng in macrocosm. The case study area is located in Wenshan City of China, which is characterized by a distinctive crop rotation agricultural system. The highlights of this study include: (1) striving for the research of the landscape-scale remote-sensing interpretation of TCM resources for the first time; (2) employing a stack of SCDDs with a comparative perspective to conduct mapping of *Panax notoginseng* fields; (3) defining the selection criteria and scoring proof for choosing a most appropriate SCDD; and (4) evaluating the abilities of SCDDs in identifying the fragmented parcels of *Panax notoginseng* in the complex agricultural landscapes.

The rest of this paper is structured as follows. The description of materials and methods is introduced in Section 2, and the experiments and analysis are presented and discussed in Sections 3 and 4, respectively. Finally, the conclusions of this work are summarized in Section 5.

2. Materials and Methods

2.1. Study Area and Data

Our study area comprises two independent blocks which are situated in Wenshan City, Wenshan Prefecture, Yunnan Province, in the south-west region of China (see Figure 2). These places mainly lie between longitude 103.71° E–104.46° E and latitude 23.07° N–23.73° N. Wenshan Prefecture is on a plateau, where the temperatures are quite constant throughout the year, with more precipitation during the summer months. Due to its low latitude and tempered by its high elevation, Wenshan Prefecture has a mild, humid, and subtropical climate, particularly suitable for planting *Panax notoginseng*. This is the reason why Wenshan Prefecture is the specific geographical location and origin of *Panax notoginseng* (i.e., why it is called a geoherb).

The "sa" and "sb" are two typical square regions. There are 151×151 pixels in the image space, respectively, corresponding to an area 20.5209 km^2 or equivalent to 2052.09 ha for both. The two blocks have a typical representation of the dense *Panax notoginseng* fields, upon which mapping *Panax notoginseng* fields will be carried out using multiple SCDDs. Multi-spectral cloud-free images acquired by the Landsat-8 Operational Land Imager (OLI) at a 30 m spatial resolution were utilized in this study. Their acquisition date was 18 March 2015. Since only one scene (path/row: 128/044) was utilized for the analysis in this study, the atmosphere can be considered to be homogeneous, and therefore the atmospheric correction may be not necessary [41,55]. Note that the planting fields of *Panax notoginseng* are rather sparse in most cases, and the cloud-free satellite images are not easy to collect because of the special geographical environment (i.e., a mountainous area is often in heavy clouds, refer to Table 1).

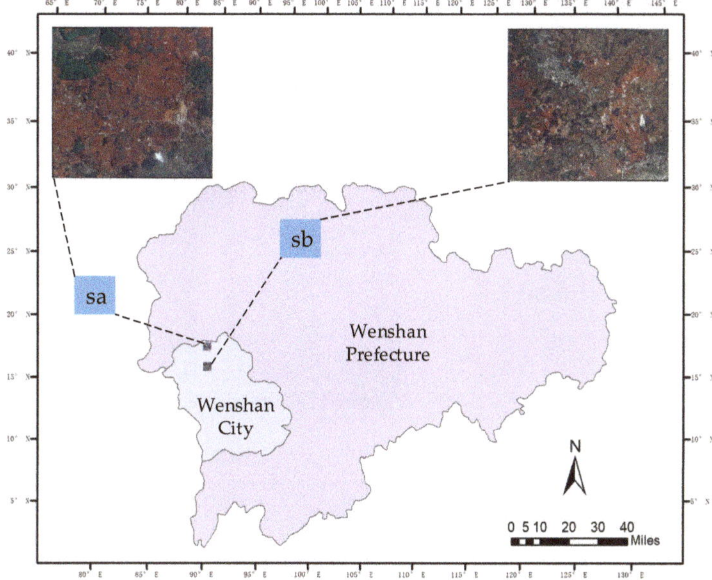

Figure 2. Study area in Wenshan City, Wenshan Prefecture, Yunnan Province, China.

Table 1. Cloud cover statistics of a 16-day revisited Landsat-8 OLI scene (path/row 128/044) until May 2017.

Cloud (%)	Number	Percentages
0–10	3	3.41
10–20	7	7.95
20–40	13	14.77
40–60	17	19.32
60–80	23	26.14
80–100	25	28.41

2.2. Shadenet Structures

Plastic sheets are used as materials to build the shadenet structure which can be regarded as a micro-scale planting environment and are relatively common [56], having unique characteristics [57], i.e., optical transparency, shade percentages, gas-tightness, and high-reflectivity. Agriculturalists have long known the importance of the planting environment for crop growth, always by manipulating the growing environment to provide a more conducive environment for crop growth. As for *Panax notoginseng*, sunlight is often modified by shading to provide the more optimal growing environments so as to enhance their production. Therefore, the production of *Panax notoginseng* takes place within an enclosed growing structure called a shade house. The shade house (see Figure 3) provides the protection and maintains the optimal growing conditions throughout the development of *Panax notoginseng*. The shade house of *Panax notoginseng* is a framed structure often covered with the black plastic nets which are made of the polyethylene thread with different shade percentages. It provides a partially-controlled atmosphere and environment by reducing the light intensity and effective heat during daytime to *Panax notoginseng* grown under the very large plastic sheets.

Figure 3. Shade house. Sub-figures are the snapshots and photos in the different phases of the construction of a shade house of *Panax notoginseng* including the materials for building, i.e., sticks (5 cm × 2 m), upright rods, and black plastic net over the poles; viewpoints for the observation, i.e., parallel to bed, perpendicular to bed, at close range, over a long distance, and from a satellite image.

Additionally, attributed to the standardized cultivation technique of GAP, the shade house provides a distinct spatial texture to interpret the spatial distribution of *Panax notoginseng* fields in reference to Google Earth historical images based on a calendar, i.e., there are 20 days when the satellite images were captured from 6 March 2013, to 12 December 2015, which is associated with the different sampling sites (see Figure 4), i.e., there are a total of 123 sites in Wenshan City. Polypropylene fabric shade is the dominant shade for field-cultivated *Panax notoginseng* in Wenshan City. Its black color maintains the proper shade and also forms the distinct texture of the shade net in satellite images, which can be visually interpreted and makes it easier to collect training samples using region of interest (ROI) tools.

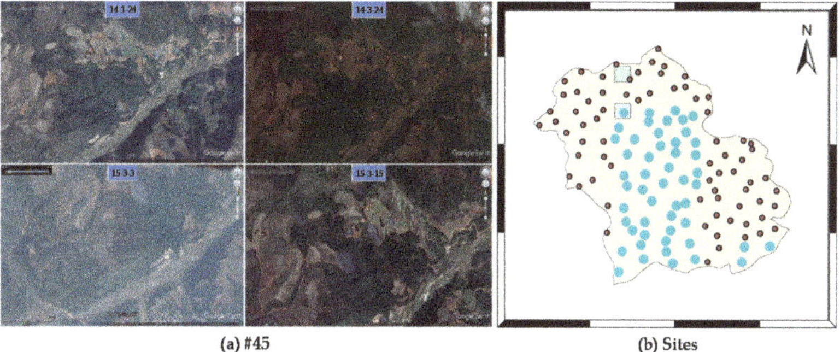

(a) #45   (b) Sites

Figure 4. Sampling sites. (**a**) The site #45; (**b**) the sites in Wenshan City.

2.3. Design Sets

Good classification depends not only on the factors associated with classifiers, but fine design sets also play a significant role in assessing the classification results objectively [58]. There is a non-negligible truth that, if only given positive samples, we cannot estimate most of the classification errors. Therefore, both positive and negative samples are supposed to be prepared for this study. Silva, et al. [41] utilized a manually-collected set of samples of the target class and a random sampling of samples of all classes, to keep the training effort low. In that case, they used the non-pure negative samples under the assumption of which few samples of the target class would be submerged.

In this study, we manually collected the positive (i.e., 1211) and negative (i.e., 8522) samples with a class-wise separability of 1.9918. Subsequently, we prepared three kinds of design sets, i.e., the training set (i.e., 727 positive samples, and 5114 negative samples at a 60% split); test set (484 positive samples, and 3408 negative samples, the remaining 40% split); and validation set (only 51 positive samples for the "sa", and 94 positive samples for the "sb"); as well as additional reference results (see Figure 5) obtained by the expert processing software. The training and test sets are random subsets of the raw collected samples by splitting operation. Note that once determined, they should be fixed so that all SCDDs could be fairly compared. For a good estimate, the test set should be labeled, randomly drawn from the class of interest, independent from the training set, and as large as possible. The validation set (i.e., only comprising true labels of the positive class) is used for validation and a nominal set. Note that the raw samples covering the whole of Wenshan City are specially designed for training and testing SCDDs. Meanwhile, the true labels for validation set are applied to calculate a representative accuracy (i.e., a correct rate) of the classification results as a final validation.

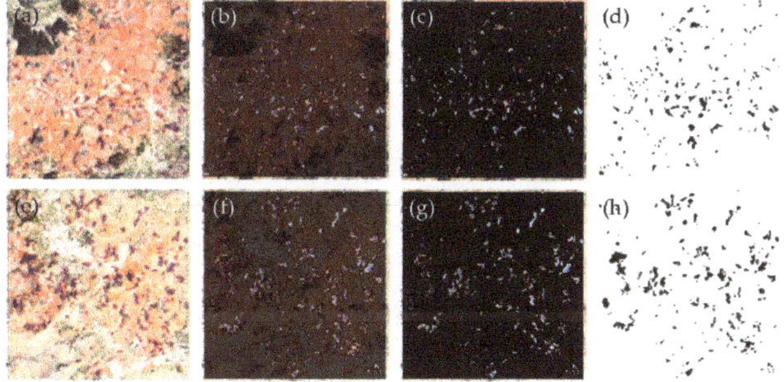

Figure 5. Reference results. (**a**) a regular subset, (**b**) the overlap, (**c**) an irregular subset of Landsat-8 OLI image, and (**d**) the reference result for the "sa"; (**e**) a regular subset, (**f**) the overlap, (**g**) an irregular subset of Landsat-8 OLI image, and (**h**) the reference result for the "sb".

2.4. Single-Class Data Descriptors (SCDDs)

The SCDD [47] is a trained mapping to predict classes. Additionally, they can be divided into several categories depending on the type of the training data and the discrimination function [59]. For example, the positive samples only (i.e., P-learning) or the positive and unlabeled samples (i.e., PU-learning) [31,59]. In general, for certain SCDDs, the corresponding model can be defined as

$$h(x) = \begin{cases} target, & f(x) \leq \theta \\ other, & f(x) > \theta \end{cases},$$

(1)

or vice versa in the opposite conditions.

The threshold θ is set according to the target error. Formally, each instance is mapped to one element of the set of the positive and negative class labels. In this study, due to only the target class (i.e., *Panax notoginseng* fields) is the class of interest, and the task of mapping *Panax notoginseng* fields can be regarded as a specific single-class data description problem. Hence, 10 SCDDs [47], i.e., the simple Gaussian target distribution data description (SGTD coded as c1) [60]; the robust Gaussian target distribution data description (RGTD coded as c2) [60]; the minimum covariance determinant Gaussian data description (MCDG coded as c3) [60]; the mixture of Gaussian data description (MoG coded as c4) [60]; the auto-encoder neural network data description (AENN coded as c6) [61]; the k-means clustering data description (k-means coded as c7) [62]; the self-organizing map data description (SOM coded as c10) [63]; the minimum spanning tree data description (MST coded as c11) [64]; the k-nearest neighbor data description (K-NN coded as c13) [65]; the incremental support vector data description (IncSVDD coded as c17) [66]; the Parzen density estimator data description (PDE coded as c5, which is a known underestimated descriptor) [67]; and the principal component analysis data description (PCA coded as c9, which is known as an overestimated descriptor) [68].

In addition to the SCDDs mentioned above, three ensemble classifiers, i.e., the mean combiner (meanc coded as cmea), the median combiner (medianc coded as cmed), and the voting combiner (votec coded as cvot), taking the mean, median, and vote strategies, respectively. The ensemble-based approach refers to the multiple-classifier system in which the outputs of all member classifiers are combined to derive an accurate classification. We combine the SCDDs which should be accurate but different in an ensemble strategy because each of them can focus on a specific feature or characteristic in the feature space [24]. Thus, a much more flexible and outstanding classifier can be obtained by combining all the strong points of the different SCDDs. There are three strategies of combining the different SCDDs, which are referred to as (1) sequential, (2) parallel, and (3) stacked [47]. Here, the above-enumerated SCDDs (i.e., c1–c17) are computed in the same feature space, and which are typically combined in a stacked way. Notice that, the combining procedure is computationally intensive in the face of many different base classifiers. Thus, the action that prunes the base classifiers according to their performance (i.e., underestimated or overestimated) is inevitable sometimes. In this study, ten SCDDs except for the underestimated and overestimated ones, which are regarded as the qualified member classifiers. The abovementioned SCDDs and combiners have been well implemented with a Matlab toolbox for data description developed and distributed by Dr. David Tax.

2.5. Performance Evaluation

In this study, the error computation and performance evaluation involve several accuracy metrics, i.e., the basic errors (see Table 2), F_1 measure, receiver operating characteristics (ROC) curve, area under the ROC curve (AUC) error, cost curve, confusion matrix, and correct rate, which are employed to evaluate the SCDDs in a more comprehensive way. In order to find a good SCDD, four basic errors can be calculated, and two of them have to be minimized, namely the false positives (FP) and false negatives (FN). Hence, we put forward and discuss several representative measures which can reflect the probability that the prediction is informed versus chance. The FN can be estimated on the positive set. In fact, the FN is much harder to estimate when no negative samples are available [47]. If only minimizing the FN will lead to the SCDD which may wrongly label a number of the negative samples as the positive class. In order to avoid such a degenerate solution, the negative samples have to be collected as well. This is the reason why we elaborately prepared the design sets and the reference results for training and testing. Moreover, two other measures, such as precision and recall (i.e., the true positive (TP) rate), are often used in performance evaluation. Finally, a derived performance indicator F_1 score can be computed. Note that a good SCDD should have both small rates of the FN and FP.

Table 2. Binary error matrix.

Types		Predicted Label	
		Target	**Other**
Actual Label	**Target**	true positive (TP)	false positive (FP)
	Other	false negative (FN)	true negative (TN)

Due to the error on the positive class can be estimated relatively well, assuming that, a threshold can be set beforehand on the target error for the SCDDs, and then a ROC curve can be obtained by varying this threshold and measuring the error on the negative samples. The ROC graph [69,70] shows how the FP varies for varying the FN, which is a technique for visualizing and selecting the best classifiers based on their performance. The smaller these fractions are, the more the SCDD is to be preferred. Although the ROC graph shows an intuitive metric of the performance of the SCDD, as one side, it is a bit difficult to compare the ROC curves; for another, we want to reduce the ROC curve to a single scalar value representing the expected performance. Thus, the AUC error [71,72] often is taken, and which is computed from the ROC curve, which integrates the fraction TP over varying thresholds (or by varying fraction FP equivalently). The larger the AUC value, the better the SCDD's performance. That is, the higher values may indicate a better separation between the positive and negative samples. As for the ROC graph [70], there are many thresholds that may be suboptimal. That is, there is another operating point for which at least one of the errors is lower. Nevertheless, this concern can be indicated by a cost curve [73], which will be another method for visualizing the performance of the SCDD. For a varying cost-ratio between both classes, the expected cost is computed. Additionally, once a trained mapping has been determined, we can obtain the sufficient site-specific metrics derived from a confusion matrix (i.e., a binary contingency table for the SCDD). The binary confusion matrix (see Table 2) is a specific table layout that allows the visualization of the performance of the SCDD [74]. Such an error matrix is constructed via classifying a predefined test set or comparing two sets of labels, and which combines the spatial position and quantitative information of the classification results to implement a performance evaluation. What is important is that the very small changes of labels are well reflected in a confusion matrix.

3. Results and Analysis

3.1. Resultant Maps

After the SCDDs have been performed, and then the final outputs associated with mapping *Panax notoginseng* fields can be derived. For one side, we conduct a statistical analysis from a perspective of quantitative evaluation. Besides, we want an intuitive comparison by showing the paired maps which are overlapped in false-color style between the classification and reference maps regarding the spatial distribution of *Panax notoginseng* fields.

As such, the classification map of each SCDD in comparison with the reference map that can be simultaneously graphed for the "sa" and "sb" (see Figures 6 and 7). By visual inspection, the obvious differences have been observed with respect to the underestimated PDE (see Figures 6 and 7) and the overestimated PCA (see Figures 6 and 7). The heavy magenta patches mean seriously underestimated, while a large number of green patches mean severely overestimated. The majority of the rest of the

SCDDs arise from the predominately magenta and slightly green patches, which mean underestimated. Meanwhile, AENN is different (see Figures 6 and 7). Notice that the error threshold on the positive class is set by a default float value of 0.1 for all SCDDs. In a number of clusters a default integer value of 2 will be acceptable because only two classes are available (i.e., the positive and negative class). Additionally, the number of components is set to 5 in terms of PCA.

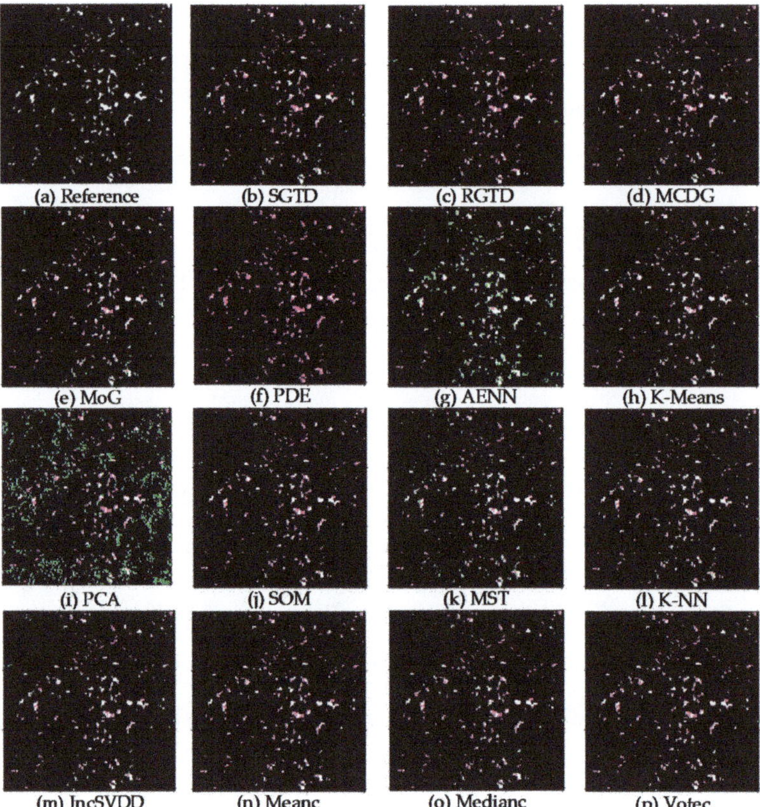

Figure 6. Classification results which are shown in false-color style in comparison with the reference result for the "sa". (**a**) Reference result; (**b**) simple Gaussian target distribution (SGTD); (**c**) robust Gaussian target distribution (RGTD); (**d**) minimum covariance determinant Gaussian (MCDG); (**e**) mixture of Gaussian (MoG); (**f**) Parzen density estimator (PDE); (**g**) auto-encoder neural network (AENN); (**h**) k-means; (**i**) principal component analysis (PCA); (**j**) self-organizing map (SOM); (**k**) minimum spanning tree (MST); (**l**) k-nearest neighbor (K-NN); (**m**) incremental support vector data description (IncSVDD); (**n**) mean combiner (meanc); (**o**) median combiner (medianc); and (**p**) voting combiner (votec). Here, the white pixels means the well estimated, the magenta means underestimated, and the green means overestimated.

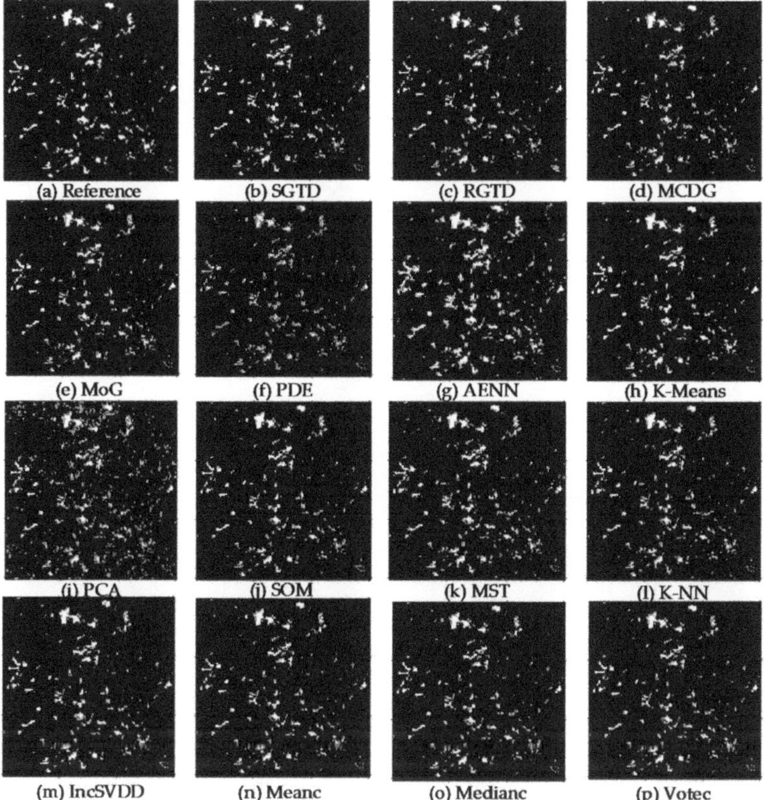

Figure 7. Classification results which are shown in false-color style in comparison with the reference result for the "sb". (**a**) Reference result; (**b**) SGTD; (**c**) RGTD; (**d**) MCDG; (**e**) MoG; (**f**) PDE; (**g**) AENN; (**h**) k-means; (**i**) PCA; (**j**) SOM; (**k**) MST; (**l**) K-NN; (**m**) IncSVDD; (**n**) meanc; (**o**) medianc; and (**p**) votec.

3.2. Measuring Performance

Performance evaluation is crucial in the assessment of a set of SCDDs. For Table 3, the false negative rate (FNR) gives the error on the positive class, while the false positive rate (FPR) shows the error on the negative class. Meanwhile, P is precision, and R denotes recall equivalent with TPR. The statistical metrics can be grouped into (1) a group that the smaller is the better, i.e., the FNR and FPR, and (2) another group that the greater is the better, i.e., P, R, F_1, and AUC. Although it is difficult to describe all SCDDs together, we attempt to make it possible by rating them on a rank table later. Table 3 illustrates that the FNR (e.g., the c5) and FPR (e.g., the c2 and c9) are prominently higher than two known inferior SCDDs (e.g., the c5 and c9), and a slightly inferior one (e.g., c2) is marked as well. Additionally, these unqualified SCDDs are again attention-catching in the second group. For all SCDDs, the precision (e.g., the c2 and c9), recall (e.g., the c5), and F_1 score (e.g., the c2, c5, and c9) support the analysis drawn from the first group, while the AUC error always appears mediocre. However, there are some differences owing to the measuring ability of statistical metrics and the intrinsic characteristics of the diverse SCDDs. The ROC curve (see Figure 8) gives a two-dimensional depiction of the performance of the SCDD. This is due to too many SCDDs with the approximate accuracies so that discriminating the individual ROC curve seems difficult. Thus, we plot them one by one, and the aforementioned inferior SCDDs still can be well reflected. Note that the c5 has the lowest operation point, while the c2 and c9 are found in terms of the ROC curve. In addition, the c6 deserves attention. Here, the ROC graphs are not going to be read significantly, as the purpose is achieved. The cost curve (see Figure 9) is a specific performance visualizer using the expected cost, another technique to measure the performance of SCDDs. Each operating point appears as a line in this plot, while the certain one of them is indicated by the dotted line. The combination of operating points that forms the lower hull is indicated by the thick curve and shows the best operating points over the range of costs. Here, the dotted line of c5, and the thick arcs of c2 and c9, again support the performance analysis drawn from the Figure 8. In particular, c6 has a representation that is the same as the ROC graph.

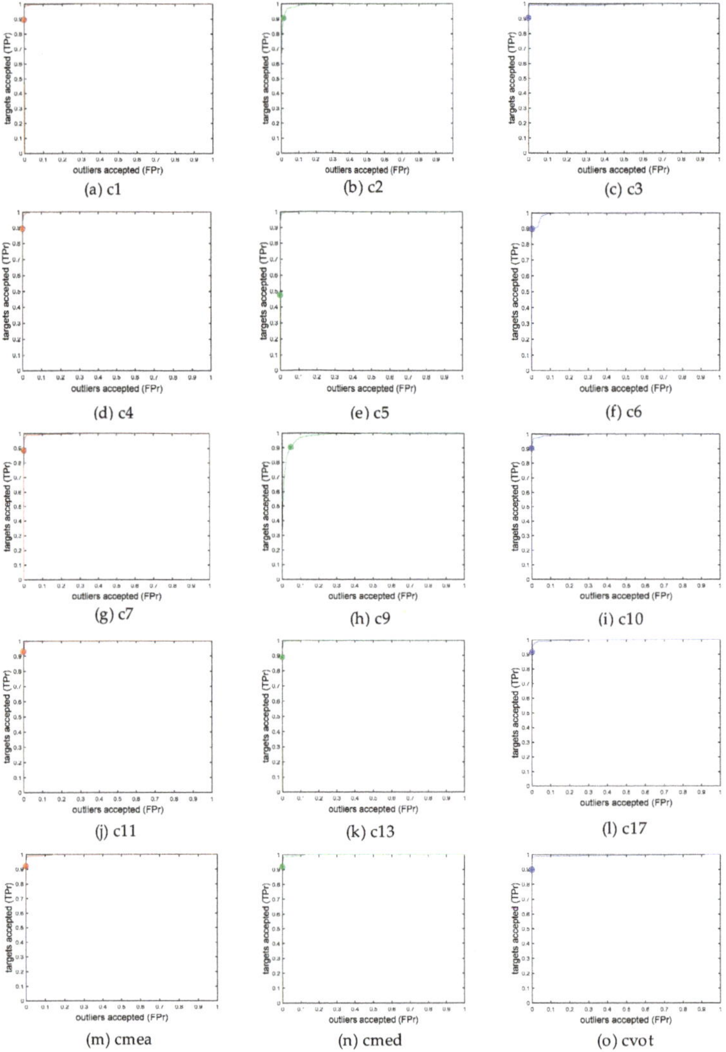

Figure 8. ROC curves. (**a**) c1: SGTD; (**b**) c2: RGTD; (**c**) c3: MCDG; (**d**) c4: MoG; (**e**) c5: PDE; (**f**) c6: AENN; (**g**) c7: k-means; (**h**) c9: PCA; (**i**) c10: SOM; (**j**) c11: MST; (**k**) c13: K-NN; (**l**) c17: IncSVDD; (**m**) cmea: meanc; (**n**) cmed: medianc; and (**o**) cvot: votec.

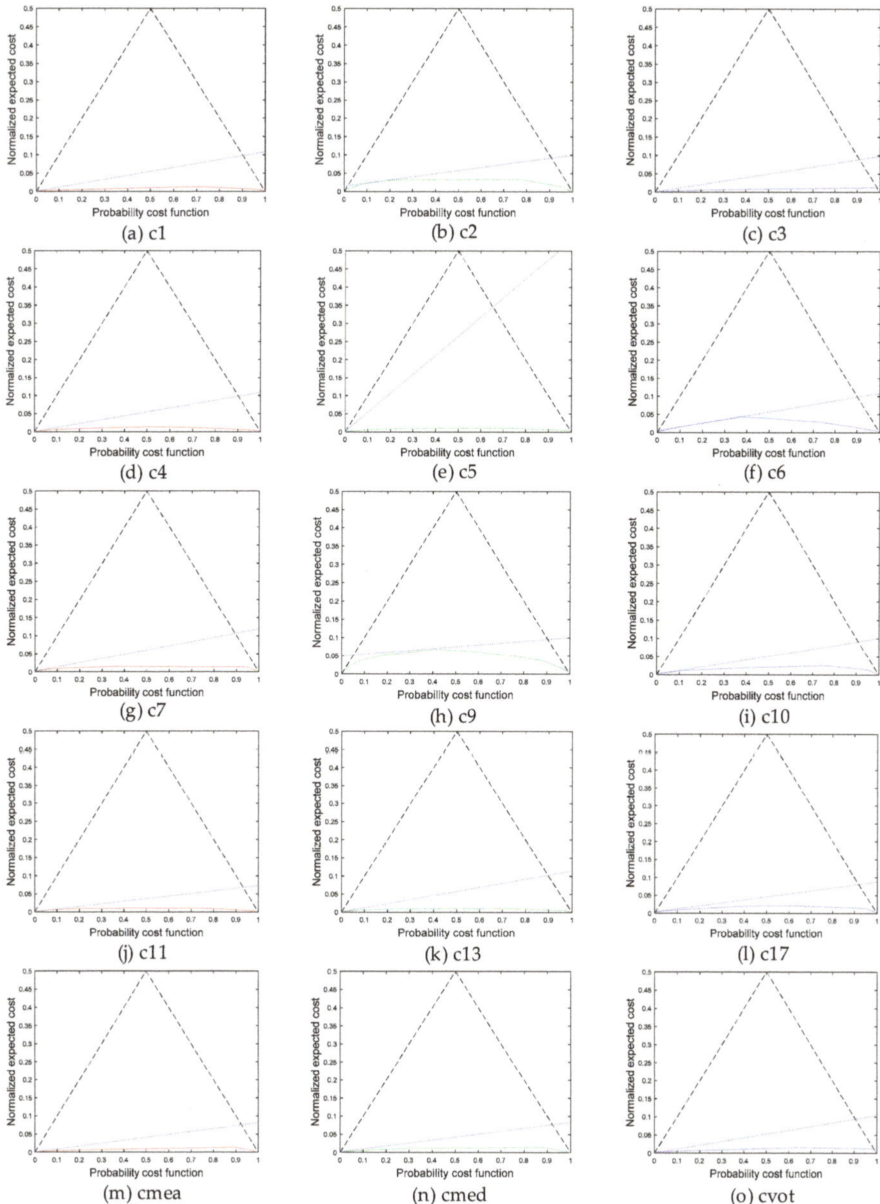

Figure 9. Cost curves. (**a**) c1: SGTD; (**b**) c2: RGTD; (**c**) c3: MCDG; (**d**) c4: MoG; (**e**) c5: PDE; (**f**) c6: AENN; (**g**) c7: k-means; (**h**) c9: PCA; (**i**) c10: SOM; (**j**) c11: MST; (**k**) c13: K-NN; (**l**) c17: IncSVDD; (**m**) cmea: meanc; (**n**) cmed: medianc; and (**o**) cvot: votec.

The confusion matrix is often applied to visualize the performance of the SCDD by a specific table layout. The OA means overall accuracy, K is the Kappa coefficient, PA denotes the producer's accuracy, and UA represents the user's accuracy. It is somewhat intricate that we structure two kinds of confusion matrices together (see Figure 10), which look similar but not identical. That is the reason why we design two types of confusion matrix (1) classifier-dependent (i.e., OAt, Kt, PAt, and UAt), which is generated using the test set; and (2) classifier-independent (i.e., OAa, Ka, PAa, UAa, OAb, Kb, PAb, and UAb), which is produced by utilizing the reference result for "sa" and "sb". For the two classifier-independent error matrices, which give an analogous presentation and a comforting result, though with a slight discrepancy in the amplitude, the inferior SCDDs, i.e., the c5 has a smaller K and PA while the c9 has a smaller K and UA. As for the classifier-dependent confusion matrix, all SCDDs have good performance except for the c5 has a lower K and PA while the c9 has a lower K and UA.

The summary can be drawn as (1) overall accuracy appears mediocre or fails in the face of the SCDD for the imbalanced data; and (2) two classifier-independent error matrices demonstrate that the SCDDs are, indeed, fixed by unchangeable splitting samples. Meanwhile, they seem worse and more unstable compared to the classifier-dependent matrices, which may be disturbed in the presence of more uncertainty. The correct rate (see Figure 10, and Table 4) is a custom performance measure, which is used for validation. Here, the true labels used are a subset of the positive samples drawn from a merged set for the "sa" and "sb". This measure is a simple attempt to repeat and verify the previous analysis.

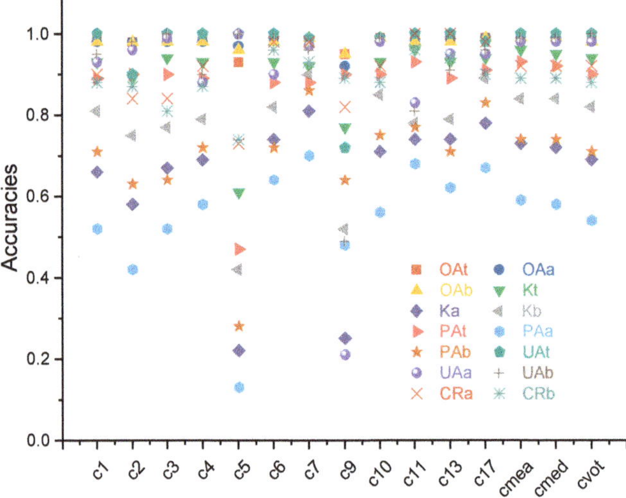

Figure 10. Accuracy metrics derived from the confusion matrix. Here, the postfix "t" means taking the test set as the true labels, while the "a" and "b" mean taking the reference results as the true labels, respectively. The CRa and CRb are the correct rates of the classification results in comparison with the true labels for the "sa" and "sb".

Table 3. Accuracy metrics, i.e. the FNR, FPR, P, R, F_1, and AUC.

	c1	c2	c3	c4	c5	c6	c7	c9	c10	c11	c13	c17	Cmea	Cmed	Cvot
FNR	0.11	0.10	0.10	0.11	0.53	0.12	0.12	0.10	0.10	0.07	0.11	0.09	0.07	0.08	0.10
FPR	0.00	0.01	0.00	0.00	0.00	0.00	0.00	0.05	0.00	0.00	0.00	0.00	0.00	0.00	0.00
P	1.00	0.90	1.00	1.00	1.00	1.00	0.99	0.72	0.99	1.00	1.00	0.98	1.00	1.00	1.00
R	0.89	0.90	0.90	0.89	0.47	0.88	0.88	0.90	0.90	0.93	0.89	0.91	0.93	0.92	0.90
F_1	0.94	0.90	0.95	0.94	0.64	0.94	0.93	0.80	0.94	0.96	0.94	0.94	0.96	0.96	0.95
AUC	1.00	0.99	0.99	1.00	1.00	1.00	1.00	0.98	1.00	1.00	1.00	1.00	1.00	1.00	0.99

Table 4. Rank table. Here, the FN and FP are the false negative rate and false positive rate, respectively. Signs "+" and "−" are the rating directions for judging the ascending or descending order. Here, the bold codes mean the inferior single-class data descriptors (SCDDs). Note that the accuracy metrics derived from confusion matrix for the "sb" have not been included herein.

C.	FN+	FP+	P−	R−	F1−	AUC−	OAt−	Kt−	PAt−	UAt−	OAa−	Ka−	PAa−	UAa−	CRa−	Rank−
c1	c5	c9	c9	c5	c5	c9	c5	c5	c5	c9	c9	c5	c5	c9	c5	15
c2	c6	c2	c2	c6	c9	cvot	c9	c9	c6	c2	c5	c9	c2	c11	c9	14
c3	c7	c17	c17	c7	c2	c2	c2	c2	c7	c17	c2	c2	c9	c4	c2	13
c4	c13	c10	c10	c13	c7	c3	c7	c7	c13	c10	c1	c1	c3	c6	c3	12
c5	c4	c7	c7	c4	c6	c17	c6	c6	c4	c7	c4	c3	c1	c1	c1	11
c6	c1	c1	c6	c1	c13	c10	c13	c13	c1	c6	c3	cvot	cvot	c13	c4	10
c7	c10	c3	c13	c10	c10	c7	c10	c10	c10	c13	cvot	c4	c10	c17	c10	9
c9	c2	c4	c1	c2	c4	cmea	c4	c4	c2	c1	c11	c10	cmed	c2	cmea	8
c10	c9	c6	c4	c9	c1	cmed	c1	c1	c9	c4	c10	cmed	c4	c7	cmed	7
c11	c3	c11	c3	c3	c17	c6	c17	c17	c3	c3	c6	cmea	cmea	cmea	cvot	6
c13	cvot	c13	cvot	cvot	c3	c1	c3	c3	cvot	cvot	cmed	c6	c13	cmed	c6	5
c17	c17	cmea	cmed	c17	cvot	c13	cvot	cvot	c17	cmed	cmea	c13	c6	c10	c7	4
cmea	cmed	cmed	c11	cmed	cmed	c5	cmed	cmed	cmed	c11	c13	c11	c17	cvot	c17	3
cmed	c11	cvot	cmea	c11	c11	c11	c11	c11	c11	cmea	c17	c17	c11	c3	c11	2
cvot	cmea	c5	c5	cmea	cmea	c4	cmea	cmea	cmea	c5	c7	c7	c7	c5	c13	1
C.	FN−	FP−	P+	R+	F1+	AUC+	OAt+	Kt+	PAt+	UAt+	OAa+	Ka+	PAa+	UAa+	CRa+	Rank+

4. Discussion

4.1. Selection Criteria

Although sufficient statistical analyses have been conducted, it is not known how to recognize which SCDD looks good. In fact, it is not easy to determine which is a fine, or even the best in the face of so many SCDDs with multiple performance measures. Therefore, we want to set up a handful of naive selection criteria to achieve such a goal by means of a rank board (see Table 4). For this work, more empirical selection criteria are adopted. Intrinsically, most of the statistical metrics are derived from the basic errors (i.e., the true positive, the true negative, the false positive, and the false negative). The derived measures (i.e., the precision, recall, F_1, AUC, OA, KC, PA, and UA) could be quantitatively analyzed with actions such as rating and scoring. Note that the AUC measure appears mediocre herein. The OA may not be a reliable metric for the real performance of the SCDD in this study, because it yields misleading results supposing the training data are imbalanced when the numbers of observations in different classes vary greatly. The ROC curve and cost graph are used for supporting numerical indicators, which are especially suitable for classification problems in which there are only two classes (i.e., the positive and negative classes). The limitation [70] of both ROC analysis and cost curve is the lack of any effective method to show the performance results obtained from several different data sets in a single plot. This difficulty follows the fact that only two dimensions are used to present the performance of a single data set.

It is important to realize the optimal selection criteria for the hybrid classifiers, such as comparing the performance of an ensemble classifier with a member classifier, which is also presented in this study. The fixed combination strategies or so-called rules (i.e., the mean rule, median rule, and voting rule), are more likely to obtain better classification results, just as the inferior classifiers will reduce the whole performance. In particular, the time taken will be an insufferable problem. It is crucial to address the question of under what criteria does one classifier outperform another. Additionally, a decision needs to be made to determine which classifier should be selected over others. That is if, given the current operating conditions, a set of selection criteria can be derived. It is often easy, by varying the parameter setting, such as a threshold or the variables of the mathematical model, or by varying the class distribution in the training set, to create a whole set of SCDDs. One commonly used selection criterion is to select the SCDD whose parameter settings and training conditions most closely agree with the current operating conditions, which is called the performance-independent criterion [75]. This is the reason why we try to fix all irrelevant conditions prior to developing the performance-dependent selection criteria. A plain criterion is to choose the qualified SCDDs regardless of their training conditions or parameter settings.

4.2. Scoring Model

For SCDDs with multiple performance measures, it would be expected for them to be scored. Then, there is always one possibility that all SCDDs can be quantitatively evaluated and scored. Consequently, a score-oriented method is presented here to clarify this concern. In this study, we put forward a kind of scoreboard on the basis of the rank board to give each SCDD an explicit score so that we can determine which SCDD is optimal.

According to Table 5, we use the rows (M.) to denote the measures and the columns (C.) to represent the different SCDDs. Signs are used to identify the metric belonging to which group: "−" denotes the error metric (i.e., the smaller the better), while "+" is the accuracy metric (i.e., the greater the better). The score variable Sc_j is calculated by the following equation:

$$s_{ij}(x) = \begin{cases} n - \dfrac{x_{ij} - \mathrm{Min}(x_i)}{(\mathrm{Max}(x_i) - \mathrm{Min}(x_i))/n} & \text{if sign} = \text{``-''} \\ \dfrac{x_{ij} - \mathrm{Min}(x_i)}{(\mathrm{Max}(x_i) - \mathrm{Min}(x_i))/n} & \text{if sign} = \text{``+''} \end{cases}, \tag{2}$$

or

$$s_{ij}(x) = \begin{cases} \dfrac{\text{Max}(x_i)-x_{ij}}{(\text{Max}(x_i)-\text{Min}(x_i))/n} & \text{if sign} = \text{``}-\text{''} \\[2ex] n - \dfrac{\text{Max}(x_i)-x_{ij}}{(\text{Max}(x_i)-\text{Min}(x_i))/n} & \text{if sign} = \text{``}+\text{''} \end{cases}, \tag{3}$$

and

$$Sc_j = \sum s_j, \tag{4}$$

where x_i represents the measures in the ith row, x_j represents SCDDs in the jth column, and x_{ij} denotes the measured value of the jth SCDD with ith metric. The s_i represents the scores in the ith row, the s_j represents the scores in the jth column, and the s_{ij} denotes the score value of the jth SCDD with the ith metric. Sc_j represents the total score of the jth SCDD. For simplicity, we assume that there are five SCDDs and five measures herein to facilitate the illustration of the scoreboard and the derivation of Equations (2)–(4). The n is a key scale to slice a certain metric for all SCDDs so that each SCDD can be assigned a normalized float value (ranging from 0 to n) associated with this metric. In the end, the gross score of every SCDD can be obtained and plotted by performing the summation by column. Figure 11 illustrates that two inferior SCDDs, i.e., the c5 is underestimated, and the c9 is overestimated, which are prominently identified. Meanwhile, two slightly inferior SCDDs, i.e., the c2 and c6, are apt to be observed again.

Table 5. Score table. Here, we assume five SCDDs with five metrics.

M./C.	j1	j2	j3	j4	j5	Sign–
i1	x11 | s11	x12 | s12	x13 | s13	x14 | s14	x15 | s15	–
i2	x21 | s21	x22 | s22	x23 | s23	x24 | s24	x25 | s25	–
i3	x31 | s31	x32 | s32	x33 | s33	x34 | s34	x35 | s35	+
i4	x41 | s41	x42 | s42	x43 | s43	x44 | s44	x45 | s45	+
i5	x51 | s51	x52 | s52	x53 | s53	x54 | s54	x55 | s55	+
Score	Sc1	Sc2	Sc3	Sc4	Sc5	Sign+

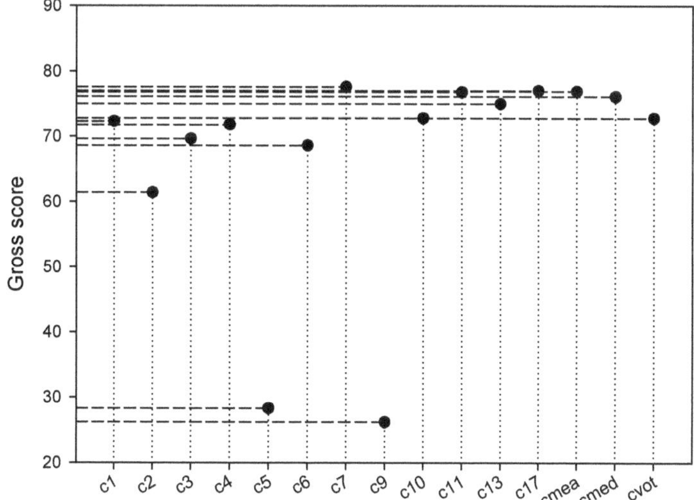

Figure 11. Scoring result, $n = 4$.

4.3. McNemar's Test

The four-cell confusion matrix is very intuitive to show the similarities and differences between the proportions (i.e., the true and false allocated parts) concerning two sets of specific labels. On the basis of the error table, we wish to achieve the statistical significance of the differences between the proportions using McNemar's test [76] so as to assess two allocated results, which are obtained by two SCDDs, or just given, under a position-specific comparison. Here, McNemar's test is specifically useful for comparing paired proportions derived from two sets of samples. In the formulas below, we use the notations:

$$pdisc = p_{F+} + p_{F-}$$
$$pdiff = p_{F+} - p_{F-} ,$$
(5)

where the p_{F+} is the proportion of testing samples that the first SCDD is true, and the second is false; meanwhile, the p_{F-} denotes the proportion of testing samples that the first SCDD is false while the second is true. Thus, McNemar's test focuses on the proportions of testing samples that one SCDD is true while another is false (and vice versa) [41].

$$SE_p = \sqrt{(pdisc - pdiff^2)/N},$$
(6)

where SE_p represents the standard error derived from the difference between the proportions, and N is the total number of the pairs of objects. McNemar's test will perform the evaluation of the $100(1 - \alpha)\%$ confidence interval for comparing the difference between two accuracy values based on the differences (D_a) between the proportions. Assuming a normal distribution z_α, the general expression of the confidence interval [76] can be expressed as:

$$D_a \pm z_\alpha SE_p.$$
(7)

For exploring in a straightforward manner, we split Equation (7) into a real image, then the image part may be more crucial to give the proper assessment with regard to a confusion matrix. In this way, the statistical assessment of the differences is carried out to determine if these are significantly different or not [76]. Since three kinds of error matrices are presented in this study, here we name them as CDt (i.e., the classifier-dependent using the test set), CIa (i.e., the classifier-independent using the reference map of the "sa"), and CIb (i.e., the classifier-independent using the reference map of the "sb").

The difference between the accuracies yielded by the paired SCDDs (or two sets of labels) is D_α, ranging from $D_a - |z_\alpha|SE_p$ to $D_a + |z_\alpha|SE_p$ at the $100 \times (1 - 0.05)\%$ confidence interval. In order to make all confidence intervals comparable, the one-sided absolute range ($|z_\alpha|SE_p$) around D_a is exhibited in Table 6. Two inferior SCDDs (i.e., the c5 and c9) have their appearances again. As for two slightly inferior SCDDs (i.e., the c2 and c6), only the c2 can be observed. Such results estimated by McNemar's test provide a useful back-up to previous analysis.

Table 6. Statistical significance ($\times 1000$).

	c1	c2	c3	c4	c5	c6	c7	c9	c10	c11	c13	c17	Cmea	Cmed	Cvot
CDt	3	4	3	3	7	3	3	6	3	3	3	3	2	3	3
CIa	1	2	1	1	2	1	1	3	1	1	1	1	1	1	1
CIb	1	2	1	1	2	1	1	3	1	2	1	1	1	1	1

4.4. Special Concerns and Limitations

Panax notoginseng is a rare kind of ginseng, and which also is an antique and endangered medicinal plant (i.e., a traditional Chinese geoherb). This paper aims to explore its potential and provides some insights into the application of SCDDs for the landscape-scale mapping of *Panax notoginseng*. We wish this work could be the referenced technical basis for exploring more novel points that make outstanding contributions and provide the fruitful information for studies on the quality assurance of the production of TCM, precision farming, the construction of agro-ecosystems, sustainable development, and the protection of biodiversity of *Panax notoginseng*. Special concerns and limitations of this study can be summarized as follows:

This work utilizes a manually-collected set of samples of the target class and grid-constraint uniformly-collected negative samples. The uncertainty exists that a few possible land-cover classes may be left out, even though the classification results look rather good.

Thirteen SCDDs are employed and compared, however, there may be many other algorithms and their variants. Anyhow, the available ones have been used in this study.

The comparison with the different SCDDs does not judge them to be good or not. Actually, we wish to extend the ability of SCDDs to achieve the expected experiences in a straightforward way to find the optimal approach to monitoring the plant pattern changes of *Panax notoginseng*.

The class imbalance is a non-negligible problem in terms of a real specific land-cover classification using SCDDs. We strive for trying to observe what influences it would cause, and find two mediocre-appearing measures, i.e., the OA and AUC.

The selection criteria and scoring model are presented to determine the optimal SCDD which is outstanding and deserves attention.

The division of the site-specific error matrices by discriminating if the SCDD is dependent on the training set or not provides a more comprehensive approach to assess the final results.

The combination of SCDDs which are taken as the base classifiers can reduce the error or improve the accuracy. However, lower computational efficiency would be an annoying problem. Additionally, the pruned ensembles can give better performance.

Some classification accuracies may not be the reliable indicators, particularly if the training data are imbalanced [41,77]. As single-class data description is a special type of one-class classification, there are difficulties that may exist when trying to fit a single-class learner using the positive samples only. If SCDDs are trained with the samples of the single target class, then only the sensitivity can be estimated. There is a possibility that using only the sensitivity to fine-tune an algorithm may result in a class descriptor with high sensitivity but low specificity and overestimating the true extension of the class of interest [41]. In terms of the scope of this study, the introduction of single-class data description regarding remote-sensing mapping of *Panax notoginseng* fields based on P-learning, which provides us the new insights to promote the development of the resource inventory and dynamic monitoring of *Panax notoginseng*.

5. Conclusions

Natural TCM resources have seldom been observed and monitored from space before. Due to the promoted GAP technique, the small or fragmented parcels covered by black plastic sheets create the opportunity and probability for us to recognize and analyze the eco-geographic characteristics of *Panax notoginseng* at a landscape-scale. This paper delineates an application whereby a stack of SCDDs is used for remote-sensing mapping of *Panax notoginseng* fields through P-learning. The measuring performance of SCDDs provides us the challenging insights to define the selection criteria and scoring proof for choosing an optimal SCDD for remote-sensing mapping a specific landscape class. Future work would involve (1) developing new algorithms to enrich the approaches of specific land cover mapping; (2) improving the design sets, updating the sampling strategy, and overcoming the imbalance issue; and (3) extending to the state-of-the-art SCDDs published, which have not been presented in this study.

Author Contributions: F.D. and S.P. conceived and designed the experiments; S.P. performed the experiments and analyzed the data; and all relevant co-authors participated in writing and editing the paper.

Funding: Research Fund of State Key Laboratory of Geohazard Prevention and Geoenvironment Protection (SKLGP2018Z006).

Acknowledgments: The authors thank the editors and the anonymous reviewers for their insightful comments and helpful suggestions, which highly improved the quality of the manuscript. We are grateful to D. Tax for the development and free use of the dd_tools.

Conflicts of Interest: The authors declare no conflict of interest. The funders had no role in the design of the study; in the collection, analyses, or interpretation of data; in the writing of the manuscript, and in the decision to publish the results.

References

1. Tsai, C. A brief introduction to traditional Chinese medicine. In *30 Years' Review of China's Science and Technology*; World Scientific: Singapore, 1981; pp. 125–138.
2. Li, X.; Yang, G.; Li, X.; Zhang, Y.; Yang, J.; Chang, J.; Sun, X.; Zhou, X.; Guo, Y.; Xu, Y.; et al. Traditional Chinese medicine in cancer care: A review of controlled clinical studies published in Chinese. *PLoS ONE* **2013**, *8*, e60338.
3. Stone, R. Lifting the veil on traditional Chinese medicine. *Science* **2008**, *319*, 709–710. [CrossRef] [PubMed]
4. Xiong, X. Integrating traditional Chinese medicine into Western cardiovascular medicine: An evidence-based approach. *Nat. Rev. Cardiol.* **2015**, *12*, 374. [CrossRef] [PubMed]
5. Harvey, A.L.; Edrada-Ebel, R.; Quinn, R.J. The re-emergence of natural products for drug discovery in the genomics era. *Nat. Rev. Drug Discov.* **2015**, *14*, 111–129. [CrossRef] [PubMed]
6. Dong, J. The relationship between traditional Chinese medicine and modern medicine. *Evid.-Based Complement. Altern.* **2013**, *2013*, 153148. [CrossRef] [PubMed]
7. Xue, T.; Roy, R. Studying traditional Chinese medicine. *Science* **2003**, *300*, 740–741. [CrossRef] [PubMed]
8. General Administration of Quality Supervision, Inspection and Quarantine of the People's Republic of China. Provisions for the Protection of Products of Geographical Indication. Available online: http://www.wipo.int/edocs/lexdocs/laws/en/cn/cn041en.pdf (accessed on 10 August 2018).
9. Addor, F.; Grazioli, A. Geographical indications beyond wines and spirits. *J. World Intellect. Prop.* **2002**, *5*, 865–897. [CrossRef]
10. Standing Committee of the National People's Congress. Law of the People's Republic of China on Traditional Chinese Medicine. Available online: http://www.gov.cn/xinwen/2016-12/26/content_5152773.htm (accessed on 10 August 2018).
11. Fan, Z.; Miao, C.; Qiao, X.; Zheng, Y.; Chen, H.; Chen, Y.; Xu, L.; Zhao, L.; Guan, H. Diversity, distribution, and antagonistic activities of rhizobacteria of *Panax notoginseng*. *J. Ginseng Res.* **2016**, *40*, 97–104. [CrossRef] [PubMed]
12. Park, H.J.; Kim, D.H.; Park, S.J.; Kim, J.M.; Ryu, J.H. Ginseng in traditional herbal prescriptions. *J. Ginseng Res.* **2012**, *36*, 225–241. [CrossRef] [PubMed]
13. Wei, J.X.; Du, Y.C. *Modern Science Research and Application of Panax Notoginseng*; Yunnan Science and Technology Press: Kunming, China, 1996.
14. Zhou, Y.Q.; Chen, S.L.; Zhang, B.G.; Zhang, J.S.; Zhang, J.; Chen, Z.J.; Cun, X.M. Studies on the resources survey methods of Panax notogingseng based on remote sensing. *China J. Chin. Mater. Med.* **2005**, *30*, 1902–1905.
15. The State Council of the People's Republic of China. Several Opinions of the State Council on Supporting and Promoting the Development of Traditional Chinese Medicine. Available online: http://www.gov.cn/zwgk/2009-05/07/content_1307145.htm (accessed on 10 August 2018).
16. The State Council Information Office of the Peoples Republic of China. Health Service Development Plan of Traditional Chinese Medicine (2015–2020). Available online: http://www.gov.cn/zhengce/content/2015-05/07/content_9704.htm (accessed on 10 August 2018).
17. The Ministry of Science and Technology of the People's Republic of China. Outline of Traditional Chinese Medicine Innovation and Development Plan (2006–2020). Available online: http://www.most.gov.cn/tztg/200703/t20070320_42240.htm (accessed on 10 August 2018).
18. Sun, X.; Lin, D.; Wu, W.; Lv, Z. Translational Chinese medicine: A way for development of traditional Chinese medicine. *Chin. Med.* **2011**, *2*, 186–190. [CrossRef]
19. Sanchez-Hernandez, C.; Boyd, D.S.; Foody, G.M. One-class classification for mapping a specific land-cover class: SVDD classification of fenland. *IEEE Trans. Geosci. Remote Sens.* **2007**, *45*, 1061–1073. [CrossRef]

20. Boyd, D.S.; Foody, G.M. *Changing Land Cover*; Global Environmental Issues; John Wiley & Sons: Hoboken, NJ, USA, 2004; pp. 65–94.

21. Cihlar, J. Land cover mapping of large areas from satellites: Status and research priorities. *Int. J. Remote Sens.* **2000**, *21*, 1093–1114. [CrossRef]

22. Wang, J.; Xiao, X.; Qin, Y.; Dong, J.; Zhang, G.; Kou, W.; Jin, C.; Zhou, Y.; Zhang, Y. Mapping paddy rice planting area in wheat-rice double-cropped areas through integration of Landsat-8 OLI, MODIS, and PALSAR images. *Sci. Rep.* **2015**, *5*, 10088. [CrossRef] [PubMed]

23. Thenkabail, P.S.; Knox, J.W.; Ozdogan, M.; Gumma, M.K.; Congalton, R.G.; Wu, Z.; Milesi, C.; Finkral, A.; Marshall, M.; Mariotto, I.; et al. Assessing future risks to agricultural productivity, water resources and food security: How can remote sensing help? *Photogramm. Eng. Rem. Sens.* **2012**, *78*, 773–782.

24. Foody, G.M.; Mathur, A.; Sanchez-Hernandez, C.; Boyd, D.S. Training set size requirements for the classification of a specific class. *Remote Sens. Environ.* **2006**, *104*, 1–14. [CrossRef]

25. Song, B.; Li, P.; Li, J.; Plaza, A. One-class classification of remote sensing images using kernel sparse representation. *IEEE J-STARS* **2016**, *9*, 1613–1623. [CrossRef]

26. Chen, C.H. An overview of recent progress on information processing for remote sensing. In *Information Processing for Remote Sensing*; World Scientific: Singapore, 1999; pp. 39–49.

27. Foody, G.M. Status of land cover classification accuracy assessment. *Remote Sens. Environ.* **2002**, *80*, 185–201. [CrossRef]

28. Wilkinson, G.G. Results and implications of a study of fifteen years of satellite image classification experiments. *IEEE Trans. Geosci. Remote Sens.* **2005**, *43*, 433–440. [CrossRef]

29. Chen, C.H. *Frontiers of Remote Sensing Information Processing*; World Scientific: Singapore, 2003.

30. Cristianini, N.; Shawe-Taylor, J. *An Introduction to Support Vector Machines and Other Kernel-Based Learning Methods*; Cambridge University Press: Cambridge, UK, 2000.

31. Wan, B.; Guo, Q.; Fang, F.; Su, Y.; Wang, R. Mapping US urban extents from MODIS data using one-class classification method. *Remote Sens.* **2015**, *7*, 10143–10163. [CrossRef]

32. Mathieu, P.P.; Aubrecht, C. *Earth Observation Open Science and Innovation*; Springer Open: Cham, Switzerland, 2018; pp. 165–218.

33. Tse, C.H.; Lam, E.Y. Geological applications of machine learning on hyperspectral remote sensing data. *Proc. SPIE Int. Soc. Opt. Eng.* **2015**, *9405*, 940512.

34. Brown, M.E.; Lary, D.J.; Vrieling, A.; Stathakis, D.; Mussa, H. Neural networks as a tool for constructing continuous NDVI time series from AVHRR and MODIS. *Int. J. Remote Sens.* **2008**, *29*, 7141–7158. [CrossRef]

35. Lary, D.J.; Remer, L.A.; MacNeill, D.; Roscoe, B.; Paradise, S. Machine Learning and Bias Correction of MODIS Aerosol Optical Depth. *IEEE Geosci. Remote Sens. Lett.* **2009**, *6*, 694–698. [CrossRef]

36. Aurin, D.A.; Mannino, A. A Database for Developing Global Ocean Color Algorithms for Colored Dissolved Organic Material, CDOM Spectral Slope, and Dissolved Organic Carbon. In Proceedings of the Ocean Optics XXI, Glasgow, Scotland, UK, 8–12 October 2012.

37. Lary, D.J.; Alavi, A.H.; Gandomi, A.H.; Walker, A.L. Machine learning in geosciences and remote sensing. *Geosci. Front.* **2016**, *7*, 3–10. [CrossRef]

38. Khobragade, A.N.; Raghuwanshi, M.M. *Contextual Soft Classification Approaches for Crops Identification Using Multi-sensory Remote Sensing Data: Machine Learning Perspective for Satellite Images*; Springer International Publishing: Cham, Switzerland, 2015; pp. 333–346.

39. FAO. Development of a Framework for Good Agricultural Practices. Available online: http://www.fao.org/docrep/meeting/006/y8704e.htm (accessed on 10 August 2018).

40. Davis, N. Controlled-environment agriculture-past, present and future. *Food Technol.* **1985**, *39*, 124–126.

41. Silva, J.; Bacao, F.; Caetano, M. Specific Land Cover Class Mapping by Semi-Supervised Weighted Support Vector Machines. *Remote Sens.* **2017**, *9*, 181. [CrossRef]

42. Mack, B.; Roscher, R.; Stenzel, S.; Feilhauer, H.; Schmidtlein, S.; Waske, B. Mapping raised bogs with an iterative one-class classification approach. *ISPRS J. Photogramm.* **2016**, *120*, 53–64. [CrossRef]

43. Liu, X.; Liu, H.; Gong, H.; Lin, Z.; Lv, S. Appling the one-class classification method of maxent to detect an invasive plant *Spartina alterniflora* with time-series analysis. *Remote Sens.* **2017**, *9*, 1120. [CrossRef]

44. Marconcini, M.; Fernández-Prieto, D.; Buchholz, T. Targeted land-cover classification. *IEEE Trans. Geosci. Remote Sens.* **2014**, *52*, 4173–4193. [CrossRef]

45. Sahare, M.; Gupta, H. A review of multi-class classification for imbalanced data. *Int. J. Adv. Comput. Res.* **2012**, *2*, 160–164.

46. Sun, Y.; Wong, A.K.C.; Kamel, M.S. Classification of imbalanced data: A review. *Int. J. Pattern Recogn. Artif. Intell.* **2009**, *23*, 687–719. [CrossRef]

47. Tax, D.M.J. One-Class Classification: Concept-Learning in the Absence of Counter-Examples. Ph.D. Thesis, Delft University of Technology, Delft, The Netherlands, 2001; p. 65.

48. Khan, S.S.; Madden, M.G. One-class classification: Taxonomy of study and review of techniques. *Knowl. Eng. Rev.* **2014**, *29*, 345–374. [CrossRef]

49. Normile, D. The new face of traditional Chinese medicine. *Science* **2003**, *299*, 188–190. [CrossRef] [PubMed]

50. Chen, L.; Dong, K.; Yang, Z.; Dong, Y.; Tang, L.; Zheng, Y. Allelopathy autotoxcity effect of successive cropping obstacle and its alleviate mechanism by intercropping. *Chin. Agric. Sci. Bull.* **2017**, *33*, 91–98.

51. Fu, G.; Zhang, Q.; Liang, C.; Cheng, Z. Stereoscopic planting pattern of kernel-used apricot and medicinal plants in the loess drought hilly region in West Henan Province. *Med. Plant* **2011**, *2*, 5–11.

52. Panigrahy, S.; Sharma, S.A. Mapping of crop rotation using multidate Indian remote rensing satellite digital data. *ISPRS J. Photogramm.* **1997**, *52*, 85–91. [CrossRef]

53. Pirkouhi, M.G.; Nobahar, A.; Dadashi, M.A. Effects of variety, planting pattern and density of plant phenology traits basil plants (*Ocimum basilicum* L.). *Int. J. Agric. Crop Sci.* **2012**, *4*, 1221–1227.

54. Yunusa, I.A.M. Effects of planting density and plant arrangement pattern on growth and yields of maize (*Zea mays* L.) and soya bean (*Glycine max* (L.) Merr.) grown in mixtures. *J. Agric. Sci.* **1989**, *112*, 1–8. [CrossRef]

55. Song, C.; Woodcock, C.E.; Seto, K.C.; Lenney, M.P.; Macomber, S.A. Classification and change detection using Landsat TM data: When and how to correct atmospheric effects? *Remote Sens. Environ.* **2001**, *75*, 230–244. [CrossRef]

56. Yang, D.; Chen, J.; Zhou, Y.; Chen, X.; Chen, X.; Cao, X. Mapping plastic greenhouse with medium spatial resolution satellite data: Development of a new spectral index. *ISPRS J. Photogramm.* **2017**, *128*, 47–60. [CrossRef]

57. Von Elsner, B.; Briassoulis, D.; Waaijenberg, D.; Mistriotis, A.; Von Zabeltitz, C.; Gratraud, J.; Russo, G.; Suay-Cortes, R. Review of structural and functional characteristics of greenhouses in European Union countries: Part I, design requirements. *J. Agric. Eng. Res.* **2000**, *75*, 1–16. [CrossRef]

58. Foody, G.M.; Boyd, D.S.; Sanchez-Hernandez, C. Mapping a specific class with an ensemble of classifiers. *Int. J. Remote Sens.* **2007**, *28*, 1733–1746. [CrossRef]

59. Mack, B.; Roscher, R.; Waske, B. Can I trust my one-class classification? *Remote Sens.* **2014**, *6*, 8779–8802. [CrossRef]

60. McLachlan, G.; Peel, D. *Finite Mixture Models*; John Wiley & Sons: Hoboken, NJ, USA, 2004.

61. Møller, M.F. A scaled conjugate gradient algorithm for fast supervised learning. *Neural Netw.* **1993**, *6*, 525–533. [CrossRef]

62. Arthur, D.; Vassilvitskii, S. In k-means++: The advantages of careful seeding. In Proceedings of the Eighteenth Annual ACM-SIAM Symposium on Discrete Algorithms, New Orleans, LA, USA, 7–9 January 2007; pp. 1027–1035.

63. Kohonen, T. The self-organizing map. *Neurocomputing* **1998**, *21*, 1–6. [CrossRef]

64. Gallager, R.G.; Humblet, P.A.; Spira, P.M. A distributed algorithm for minimum-weight spanning trees. *ACM Trans. Program. Lang. Syst. (TOPLAS)* **1983**, *5*, 66–77. [CrossRef]

65. Altman, N.S. An introduction to kernel and nearest-neighbor nonparametric regression. *Am. Stat.* **1992**, *46*, 175–185.

66. Tax, D.; Ypma, A.; Duin, R. Support vector data description applied to machine vibration analysis. In Proceedings of the 5th Annual Conference of the Advanced School for Computing and Imaging, Heijen, The Netherlands, 15–17 June 1999; pp. 15–23.

67. Parzen, E. On estimation of a probability density function and mode. *Ann. Math. Stat.* **1962**, *33*, 1065–1076. [CrossRef]

68. Jolliffe, I.T. Principal component analysis and factor analysis. In *Principal Component Analysis*; Springer: New York, NY, USA, 2002; pp. 150–166.

69. Powers, D.M. Evaluation: From precision, recall and F-measure to ROC, informedness, markedness and correlation. *J. Mach. Learn. Technol.* **2011**, *2*, 37–63.

70. Fawcett, T. An introduction to ROC analysis. *Pattern Recogn. Lett.* **2006**, *27*, 861–874. [CrossRef]

71. Bradley, A.P. The use of the area under the ROC curve in the evaluation of machine learning algorithms. *Pattern Recogn.* **1997**, *30*, 1145–1159. [CrossRef]

72. Hanley, J.A.; McNeil, B.J. The meaning and use of the area under a receiver operating characteristic (ROC) curve. *Radiology* **1982**, *143*, 29–36. [CrossRef] [PubMed]

73. Drummond, C.; Holte, R.C. Explicitly representing expected cost: An alternative to ROC representation. In Proceedings of the 6th ACM SIGKDD International Conference on Knowledge Discovery and Data Mining, Boston, MA, USA, 20–23 August 2000; pp. 198–207.

74. Stehman, S.V. Selecting and interpreting measures of thematic classification accuracy. *Remote Sens. Environ.* **1997**, *62*, 77–89. [CrossRef]

75. Ting, K.M. Matching model versus single model: A study of the requirement to match class distribution using decision trees. In *European Conference on Machine Learning*; Springer: Berlin/Heidelberg, Germany, 2004; pp. 429–440.

76. Foody, G.M. Classification accuracy comparison: Hypothesis tests and the use of confidence intervals in evaluations of difference, equivalence and non-inferiority. *Remote Sens. Environ.* **2009**, *113*, 1658–1663. [CrossRef]

77. Hwang, J.P. A new weighted approach to imbalanced data classification problem via support vector machine with quadratic cost function. *Expert Syst. Appl.* **2011**, *38*, 8580–8585. [CrossRef]

Article

Dual-Dense Convolution Network for Change Detection of High-Resolution Panchromatic Imagery

Wahyu Wiratama [1], **Jongseok Lee** [1], **Sang-Eun Park** [2] **and Donggyu Sim** [1,*]

[1]   Department of Computer Engineering, Kwangwoon University, Seoul 139701, Korea;
     wiratama@kw.ac.kr (W.W.); suk2080@kw.ac.kr (J.L.)

[2]   Department of Geoinformation Engineering, Sejong University, Seoul 143747, Korea; separk@sejong.ac.kr

*   Correspondence: dgsim@kw.ac.kr; Tel.: +82-2941-6470

Received: 10 September 2018; Accepted: 27 September 2018; Published: 1 October 2018

Abstract: This paper presents a robust change detection algorithm for high-resolution panchromatic imagery using a proposed dual-dense convolutional network (DCN). In this work, a joint structure of two deep convolutional networks with dense connectivity in convolution layers is designed in order to accomplish change detection for satellite images acquired at different times. The proposed network model detects pixel-wise temporal change based on local characteristics by incorporating information from neighboring pixels. Dense connection in convolution layers is designed to reuse preceding feature maps by connecting them to all subsequent layers. Dual networks are incorporated by measuring the dissimilarity of two temporal images. In the proposed algorithm for change detection, a contrastive loss function is used in a learning stage by running over multiple pairs of samples. According to our evaluation, we found that the proposed framework achieves better detection performance than conventional algorithms, in area under the curve (AUC) of 0.97, percentage correct classification (PCC) of 99%, and Kappa of 69, on average.

Keywords: change detection; convolutional network; deep learning; panchromatic; remote sensing

---

1. Introduction

Change detection is a challenging task in remote sensing for identifying changed areas between two images acquired at different times from the same geographical area. It has been widely used in both civil and military fields such as agricultural monitoring, urban planning, environment monitoring, and reconnaissance. In general, change detection is performed in three steps. First, a preprocessing step is commonly used to conduct registration of two images and to correct geometric and radiometric distortions. In the second step, a feature map is extracted, for example, a difference image is computed in order to generate change features with the assumption that two images are not perfectly registered for all of the pixels. Lastly, a classification or clustering algorithm is driven in order to distinguish changed pixels and unchanged pixels based on statistical characteristics.

For change detection, many manually designed features such as a difference image (DI) [1–7], local change vector [8], and texture vector [9–11] have been proposed. In further classification analysis, an unsupervised change detection was proposed based on fuzzy c-mean (FCM) clustering [12,13]. The optimization algorithm based on Markov random field (MRF) and genetic algorithm was employed so as to optimize the FCM. On the other hand, a supervised learning algorithm was presented based on an active learning and MRF in order to detect change areas [14]. In addition, a support vector machine (SVM) has widely been used to perform binary classification based on texture information and change vector analysis [9,15–17]. Since the classification process mainly depends on extracted features, the selection of handcrafted features for effective image representation is known to be crucial. In general, handcrafted features in change detection are sensitive due to geometric and radiometric distortions, as well as imperfect registration. All of those mentioned classification algorithms would be reasonably good for training data sets. However, those algorithms are not able to incorporate accurate and reliable statistical characteristics for a huge amount of data sets, and thus would not yield good detection performance for new data sets.

Recently, a deep convolution neural network (DCNN) was developed to produce a hierarchy feature-maps via learned filters, and it can automatically learn a complex feature space from a huge amount of image data. The DCNN can achieve superior performance compared to conventional classification algorithms with handcrafted features. Recently, several change detection methods using deep learning algorithms have been proposed [18–20]. A difference image is fed into the deep neural networks as input data [18]. In addition, the neighboring features on each pixel on the difference image are taken as inputs. The restricted Boltzmann machine (RBM) is used for pre-training and is then unrolled in order to create a deep neural network. On the other hand, the change detection is performed by combining a sparse autoencoder, convolutional neural network (CNN), and unsupervised clustering algorithm [19]. In addition, a log-ratio map was used and transformed by a sparse autoencoder into a suitable feature space. A change detection map is directly extracted from the two images using a pre-trained CNN [20]. A unique higher dimensional feature map is produced by the CNN through different convolutional layers. The change map is computed using pixel-wise Euclidean distance of hyper dimensional features. Another change detection algorithm has also been proposed that adopts a log-ratio difference [21]. It is used as a feature input for detecting changes between multi-temporal synthetic aperture radar (SAR) images. In addition, a deep neural network was developed by stacking RBMs to learn and recognize changed pixels and unchanged pixels. In addition, a combined algorithm with the deep belief networks (DBNs) and change analysis are presented to highlight changes [22]. The presented algorithm merges and vectorizes local pixel features into DBN inputs. Then, the DBN model is established in order to capture key information for discrimination and to suppress irrelevant variations. An unsupervised clustering algorithm is then used to classify changed and unchanged pixels. Another approach utilizes joint features for change detection [23]. This work proposed an efficient change rule with a reliable expression of difference information. It learns the reliable change rule by recording the change information for a long-term sequence of remote sensing data with long short-term memory (LSTM) model. As mentioned above, all of the deep learning-based change detection algorithms yield relatively good performance. However, most of them still rely on the difference image as a feature input of their networks, resulting in them being sensitive to noisy conditions caused by geometric, radiometric distortions, and different viewing angles. In order to solve these problems, an alternative approach for change detection was developed by measuring similarity. A Siamese convolutional network was proposed to detect changed areas for optical aerial images [24]. The Siamese convolutional network with shared weights learns to extract features directly from image pairs. This work uses shared weights that are dependent from those of two branch networks. The shared weights can reduce parameters to be optimized, resulting in faster convergence. However, this model is also less flexible, which leads to overfitting due to shared weights with some other neurons.

In order to overcome the problems described above, this paper proposes a dual-dense convolutional network for recognizing pixel-wise change based on dissimilarity analysis of neighborhood pixels on high resolution panchromatic (PAN) images. In this proposed algorithm, two fully convolutional neural networks are employed to measure dissimilarity of neighborhood pixels. Furthermore, dense connection in convolution layers is applied to reuse preceding feature maps by connecting them to all subsequent layers. It is proposed to enhance a feature-map representation. While the conventional change detection algorithm [24] and conventional Siamese network use shared weights, the proposed algorithm removes shared weights in order to obtain independent optimal weights for two points of input data. So, each network can independently learn for optimal weights, called the "dual-dense convolutional network (dual-DCN)". During its training, the dual-DCN is driven to learn more robust different representations to better distinguish different types of changes.

The proposed algorithm gives better performance compared to conventional methods in qualitative and quantitative evaluation. It yields AUC of 0.97, PCC of 99%, and Kappa of 69 on average.

The rest of this paper is organized into five sections: In Section 2, the conventional convolutional neural network and problem statements will be described. Section 3 presents the proposed algorithm in detail. Section 4 will present and analyze experiment results. Finally, we conclude it in the last section.

2. Convolutional Neural Network and Problem Statement

The convolutional neural networks (CNNs) are a category of neural networks which are very effective in image recognition, classification, and so on [25]. The CNN is one of the deep learning approaches that is composed of multiple convolutional and nonlinearity layers with optional pooling, followed by fully-connected layers, as shown in Figure 1.

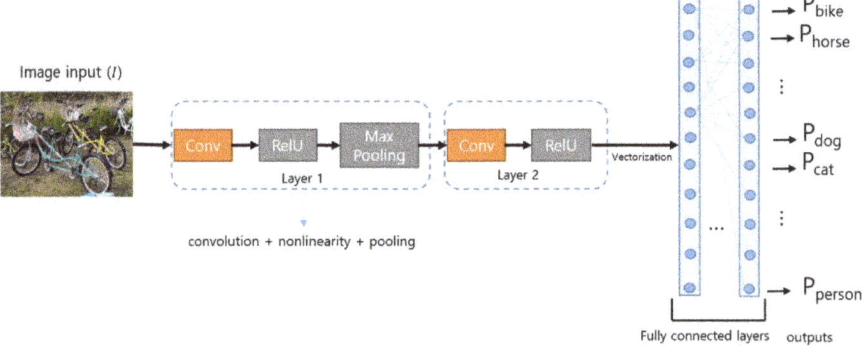

Figure 1. Traditional convolutional neural network (CNN) architecture.

Let I be an image ($m \times n \times c$) to be input, where m, n, and c are the height, width, and channel numbers of the image, respectively. In the convolutional layers, I is convolved by 2D k kernels and mapped by a nonlinearity function, called rectified linear units (ReLU), to build k feature-maps (F). The feature-map output of the lth layer is connected to the input of the ($l + 1$)th convolution and pooling layer. The final feature-maps are connected to a fully-connected layer. The last layer of fully-connected layer produces the class probability output (P_{class}). A cross-entropy function is then used to compute an objective loss. All of the weighting parameters of the network can be trained using the backpropagation algorithm.

Changes on remote-sensed images can be detected by analyzing two registered images over the same geographical area. For change detection, CNN could be employed to learn changed image characteristics and detect changed areas on remote-sensed imagery. However, the difference image (DI) or the feature fusion (FF) is widely used as an input feature of CNN, as shown in Figure 2. The DI is extracted by image subtraction or log ratio. Then, the FF is constructed by concatenating the two images. Note that these approaches are sensitive to noise as direct pixel-wise comparison features; thus, the traditional CNNs with DI or FF features could be weak to distorted data. In practice, distorted images and data are common in the remote sensing field. This distortion can be caused by not only radiometric but also geometric and viewing angle factors. In general, a geometric distortion is generated when satellites or aircrafts acquire images. In addition, image registration is required to align two images, even over the same geographical area, in a pre-processing stage. However, it is almost impossible to perfectly achieve distortion correction through automated methods. In addition, a viewing angle difference in acquisition is another challenging issue in registration and change detection. This problem cannot be resolved without precise 3-D building models, complicated algorithms, and manual intervention. For robust change detection, a robust and stable classification model is required that resolves all of the problems described above.

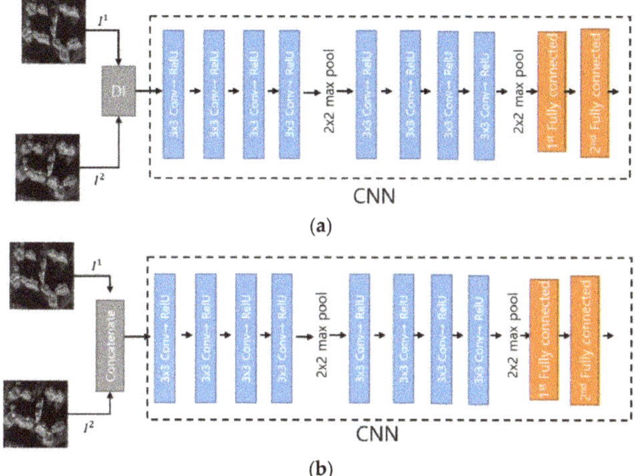

(a)

(b)

Figure 2. Conventional approaches. (**a**) Difference Image (DI) + CNN and (**b**) Feature Fusion (FF) + CNN.

3. Change Detection with the Proposed Dual-Dense Convolutional Neural Network

In general, generic change detection algorithms consist of two phases: pre-processing and change detection. Figure 3a depicts a general procedure of the conventional change detection system. The pre-processing stage performs radiometric correction, geometric rectification, and image registration. The registered images are then fed into a change detection algorithm in order to identify changed areas with feature vectors, for example, a difference image. In the general change detection systems, the radiometric correction and image registration stages are important and indispensable for better performance. The radiometric correction is performed in order to alleviate distortion for radiometric consistency. Then, the geometric correction is performed by aligning the global earth coordinates with the corresponding image points. Even though two images are compensated using multiple steps, they are still not perfectly registered, as they are independently processed with many error factors. Thus, an additional registration between two images is frequently required in order to reduce mis-alignment.

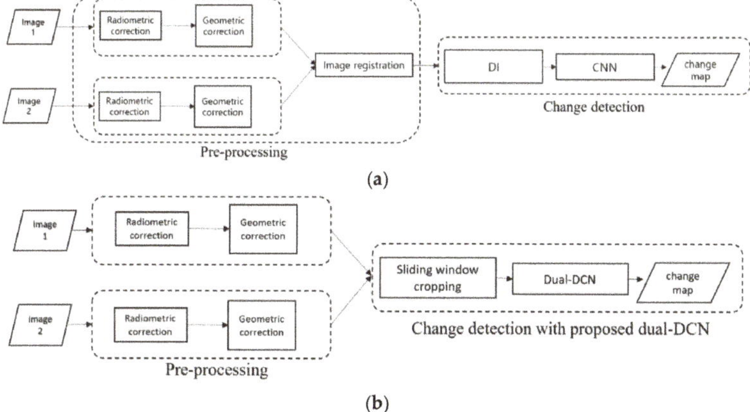

(a)

(b)

Figure 3. Change detection system schema. (**a**) Conventional change detection system and (**b**) the change detection system with the proposed dual-DCN.

In urban and mountainous areas in particular, most automatic image registration methods remain ineffective. It would degrade performance of change detection by direct pixel-wise comparison using the difference image. In order to resolve imperfect registration impacting on the change detection, a dual-DCN, as shown in Figure 3b is proposed by employing a dissimilarity distance in order to overcome the mis-alignment problem for better performance of change detection even without a perfect image registration. The proposed algorithm employs two deep convolution networks to keep all the information of the original data. The generic characteristics of the CNN handle some local distortion and alignment, thus, the proposed algorithm absorbs the misalignment problem. In addition, the dense connectivity in the convolution layer is introduced by reusing all preceding feature-maps to enhance the feature-map representation.

3.1. Pre-Processing for Change Detection

As mentioned previously, an atmospheric correction is required to remove scattering and absorption effects from the atmosphere to characterize the surface reflectance effects for a time-series image analysis. This work uses KOMPSAT-3 images with product level 1G. In these images, the radiometric correction has been done by converting the image pixel values (Digital Numbers/DNs) to surface reflectance values. It involves the conversion of DNs to a radiance value, and then to top-of-atmosphere (TOA) radiance. On the other hand, gain and offset values are provided by KOMPSAT-3 to derive the TOA reflectance values [26]. After the atmosphere errors are corrected, the geometric correction is performed in order to ensure that the pixels in the image are in their proper geometric positions on the Earth's surface. For our test images, geo-rectification and orthorectification are each conducted. For the geo-rectification, ground control points (GCPs) are identified in an unrectified image and correspond to their real coordinates to estimate the parameters (polynomial coefficients) of polynomial functions by the least square fitting. In addition, orthorectification can partly correct the image for image distortions caused by variations in the terrain topography in tandem with non-optimal satellite sensor viewing angle. Optical distortions are corrected, and terrain effects are corrected using coarse digital elevation model (DEM), namely shuttle radar topography mission DEM (SRTM DEM) for KOMPSAT-3 imagery [26].

In general change detection systems, an image registration is applied in order to ensure that two images become spatially aligned. Even though the correction of geometric distortion is performed, the spatial alignment of two images could contain a relatively large error of up to ±6 pixels. In order to overcome this distortion, automatic image registration is widely used. However, it requires high computational load, and is furthermore not easy to obtain perfect registration. They impact the performance of change detection algorithms, resulting in the possibility that a great deal of false change areas could occur. The proposed dual-DCN is proposed so as to handle distortion problems and simplify image registration. The dissimilarity distance of local characteristics is measured in order to identify a change with the dense dual-DCN model.

3.2. Dual-Dense Convolutional Neural Network for Change Detection

In order to achieve accurate change detection without a perfect registration, this paper proposes a dual-dense convolutional network (dual-DCN) with two deep convolutional networks, as shown in Figure 4. This proposed network identifies the change areas by measuring the dissimilarity distance of two inputs at the last stage for use of all the information of the two input images. Two branch networks, N^1 and N^2, handle two input images acquired at different time instances, respectively. The proposed network is based on CNN, thus, it can robustly conduct a pixel-wise change detection by inspecting the neighboring pixels.

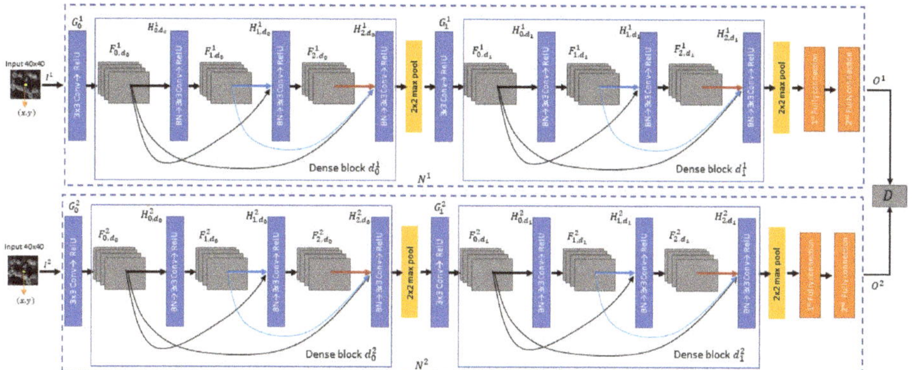

Figure 4. The proposed dual-DCN architecture for change detection.

A pair of images is cropped into two patches (40×40) by sliding in the raster scan order and two cropped patches (I^1 and I^2) are fed into the proposed network. The center pixel of the cropped patch is identified as changed or unchanged with the presence of a single dissimilar value between the cropped two patches. The Siamese network proposed by Reference [24] extracts features from an image pair. The pair of the convolutional networks is used to capture similarity characteristics by sharing the weights of the two network paths.

However, the shared weights of Siamese network reduce the parameters optimized during training for fast convergence. However, it is known to be frequently overfitted. Thus, the proposed network does not employ the shared weights to provide more flexible optimization than a restricted Siamese network. The parameters of each network branch of the proposed algorithm can be independently optimized in order to avoid early overfitted convergence. In addition, the proposed network employs dense connection [27] in the convolutional layers by reusing all the preceding feature-maps, in order to enhance representation capability of the feature-maps, as shown in Figure 4. The preceding feature maps are directly connected to all of the subsequent layers. The traditional CNN connects the output of the $(l-1)$th layer as input to the lth [28]. In the proposed dual-DCN model, the lth layer receives all of the preceding feature-maps. The feature map of the lth layer at the rth dense block and the ith network can be computed by

$$F_{l,d_r}^i = H_{l-1,d_r}^i \left(\left[F_{0,d_r}^i, F_{1,d_r}^i, \ldots, F_{l-1,d_r}^i \right] \right), r = 0, 1; i = 1, 2 \tag{1}$$

where $\left[F_{0,d_r}^i, F_{1,d_r}^i, \ldots, F_{l-1,d_r}^i \right]$ indicates concatenation of the feature-maps of all of the previous layers, layer $0, \ldots,$ layer $(l-1)$. Each dense block is a group of convolution layers with the dense connectivity to avoid variant sizes of the feature maps. $H(\cdot)$ plays a role in batch normalization (BN) [29], 3×3 convolution, and ReLU. The BN is used to normalize parameters change of the preceding layers. The ReLU is used by thresholding at zero following 3×3 convolution. The convergence of the stochastic gradient descent algorithm can be accelerated. G including 3×3 convolution followed by ReLU is employed before a dense block in order to generate the feature-map F_0. In the proposed architecture, each dense block contains three $H(\cdot)$, including 64 feature maps of each layer. After a dense block is performed, a down-sampling operation is applied to produce variety scales with 2×2 maximum pooling. Furthermore, the feature maps at the last convolutional layer are vectorized and fed into the fully-connected layer consisting of 64 neurons and 0.5 drop-out. The probability output, O^i, at the last stage is computed by the sigmoid function. Furthermore, Euclidean distance (D) is employed in order to measure the dissimilarity between I^1 and I^2 computed by

$$D = {\left\| O^1 - O^2 \right\|}_2 \tag{2}$$

When the value of D approaches 1, the center pixel of the 40×40 patch is set to a changed one, otherwise, it is set to unchanged.

3.3. Training of the Proposed Dual-DCN for Change Detection

Given a training set consisting of image pairs, the proposed network can be end-to-end trained by the backpropagation algorithm. For each image pair, let Y be a binary label of the ground truth in which $Y = 0$ if both inputs are similar, and $Y = 1$ if both inputs are dissimilar. The proposed dual-DCN is trained based on dissimilarity by computing the contrastive loss $L(D, Y)$ as an objective function [30]. This loss function employs a partial loss function for similar and dissimilar of a pair image. It produces a low value of D for unchanged pixels pair and high value for a pair of change pixels.

This proposed network is optimized using the stochastic gradient descent (SGD) optimizer. Each mini-batch arises from a single image pair that contains many changes and many absences of changes. The proposed algorithm randomly initializes all new layers by drawing weights from Glorot uniform [31]. The learning rate, decay rate, and momentum are set to 0.01, 1×10^{-6} and 0.9, respectively. The epoch number is set to 30.

4. Experimental Evaluation and Discussion

This paper uses a KOMPSAT-3 image dataset that was captured over South Korea. The KOMPSAT-3 image data set is provided by the Korea Aerospace Research Institute. Note that panchromatic band images which provide 0.7 m GSD are used for change detection. Figure 5 shows the example of an overlapped panchromatic images (1214×886) of the training dataset. These images were cropped by 40×40 sliding patch. The labels for the dataset were manually constructed for all of the center pixels of cropped patch pair.

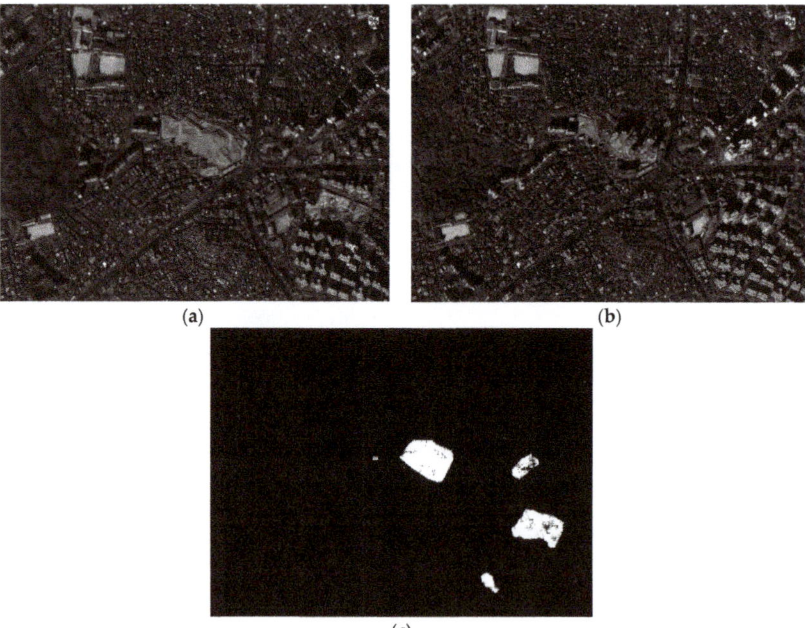

Figure 5. Seoul training data set: (**a**) image acquired in March 2014, (**b**) image acquired in December 2015, and (**c**) ground truth.

Figure 6 shows the two panchromatic images of $(29,368 \times 27,388)$ and $(29,188 \times 28,140)$ used in our experiments, which were acquired by KOMPSAT-3 on March 2014 and October 2015, respectively. These two images were acquired not only at different time instances, but also with different viewing angles. They have geometric misalignment of approximately ± 6 pixels for overlapped area.

Figure 6. PAN images of Seoul area, overlapped area denoted by blue lines. (Left image: March 2014, right image: October 2015).

Figure 7 shows four selected urban areas from Figure 6, and they contain changed areas due to building construction. In Area 1, there are two types of construction changes, under construction changes and completed construction changes. Moreover, in certain areas, there are tall buildings, which could lead to false changes in change detection due to differing viewing angles. Rather than construction changes and tall buildings, we can find a forest area in Area 2. There are many tall buildings in Area 3, and accurate detection is not easy due to a large different viewing angle.

Area 4 is used to assess the influence of change detection due to differing seasons for a forest area. This case is challenging because the change due to the season should be disregarded for practical applications. Note that the labels for four areas were manually obtained, as shown in Figure 7.

In order to evaluate the change detection performance of the proposed algorithm and conventional algorithms, several metrics are used in this study, including receiver operating characteristic (ROC) curve, area under the curve (AUC), percentage correct classification (PCC), and Kappa coefficient [32]. For existing algorithms, DI + CNN, FF + CNN, and Siamese network were implemented. This CNN architecture includes 8 depth convolutional, 2 pooling, and 2 fully connected layers. For fair comparison, the same parameters of training parameters, the number feature maps, and training dataset were used in our evaluation.

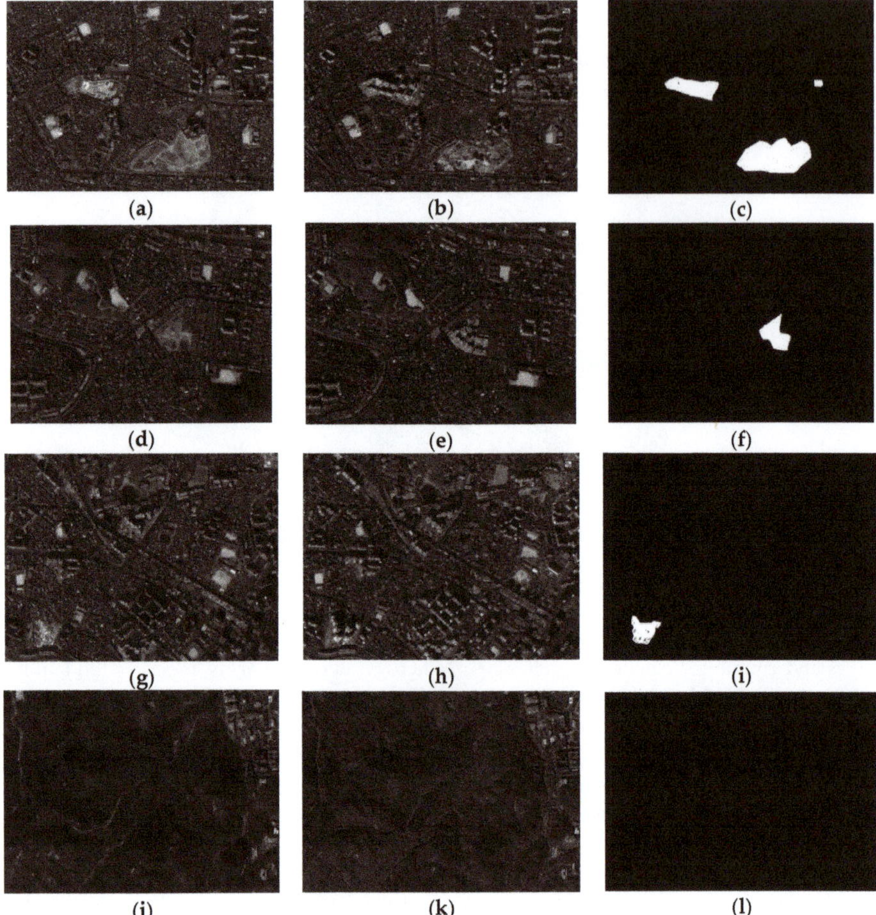

Figure 7. Four areas of Figure 6. (**a**) Input image for Area 1 (March 2014). (**b**) Input image for Area 1 (October 2015). (**c**) Ground truth for Area 1. (**d**) Input image for Area 2 (March 2014). (**e**) Input image for Area 2 (October 2015). (**f**) Ground truth for Area 2. (**g**) Input image for Area 3 (March 2014). (**h**) Image input for Area 3 (October 2015). (**i**) Ground truth for Area 3. (**j**) Input image for Area 4 (March 2014) (**k**) Input image for Area 4 (October 2015). (**l**) Ground truth for Area 4.

Figure 8 shows detection results for four areas with the exiting algorithms and the proposed algorithm. As shown in Figure 8, the proposed algorithm and FF + CNN generate better detection accuracy for Area 1. On the other hand, DI + CNN and Siamese network produce many false positives for the area. For urban surfaces, it is relatively difficult to handle the misalignment and the different viewing angle impacts because there exists tall buildings and complex constructions, resulting in the fact that false detections are likely to be performed. For Area 2, FF + CNN and DI + CNN yield more false positives, particularly in forest and urban areas. Moreover, Siamese net achieves a better detection result than other conventional algorithms. However, many false positives are still detected in certain areas. Overall, the proposed dual-DCN gives proper change detection performance, even in different viewing angle conditions. For Area 3, the proposed algorithm is still able to properly detect the changes.

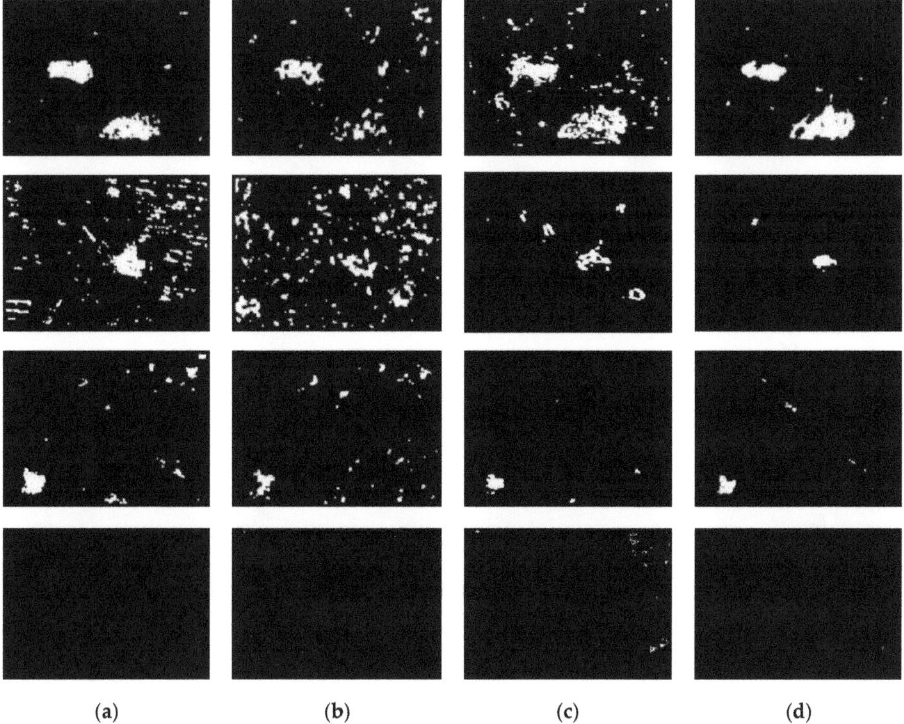

(a)   (b)   (c)   (d)

Figure 8. Detection results for four areas with the existing algorithms and proposed algorithm. (a) FF+ CNN, (b) DI + CNN, (c) Siamese network, and (d) the proposed dual-DCN.

The other algorithms result in more false positives. Note that input images for Area 4 were acquired in difference seasons for a forest area. For the test data, Siamese net produces some false positives. As shown in Figure 8, the proposed algorithm yields a better detection result with the proposed dual-DCN. The proposed algorithm can alleviate the impacts of distortions caused by imperfect geometric correction and different viewing angles. As mentioned previously, the proposed dual-DCN was designed to learn the dissimilarity of two local images in order to avoid false changes. That is why the false positive rate is relatively lower by the proposed algorithm. In contrast, DI + CNN and FF + CNN yield higher false positive rates, particularly for Areas 2 and 3. Moreover, the Siamese network produces higher false positives in Area 1, due to less optimized parameters. Figure 9 shows that the proposed algorithm can give better ROC than the conventional algorithms. According to ROC curves, the proposed dual-DCN shows better quantitative detection performance in AUC of 0.97, on average, as tabulated in Table 1. FF + CNN is slightly better in AUC than the proposed dual-DCN for Area 2, because it has better true positive for this case. However, the proposed algorithm has a lower false positive rate than FF + CNN. Table 1 summarizes the PCC and Kappa values of different methods for the three areas. As shown in Table 1, the proposed algorithm achieves higher PCC and Kappa values. We can say that the proposed dual dense convolutional network architecture has the ability to identify both changed and unchanged areas by disregarding irrelevant variations and false changes, even in cases of complicated urban surfaces, geometric distortion, and different viewing angles.

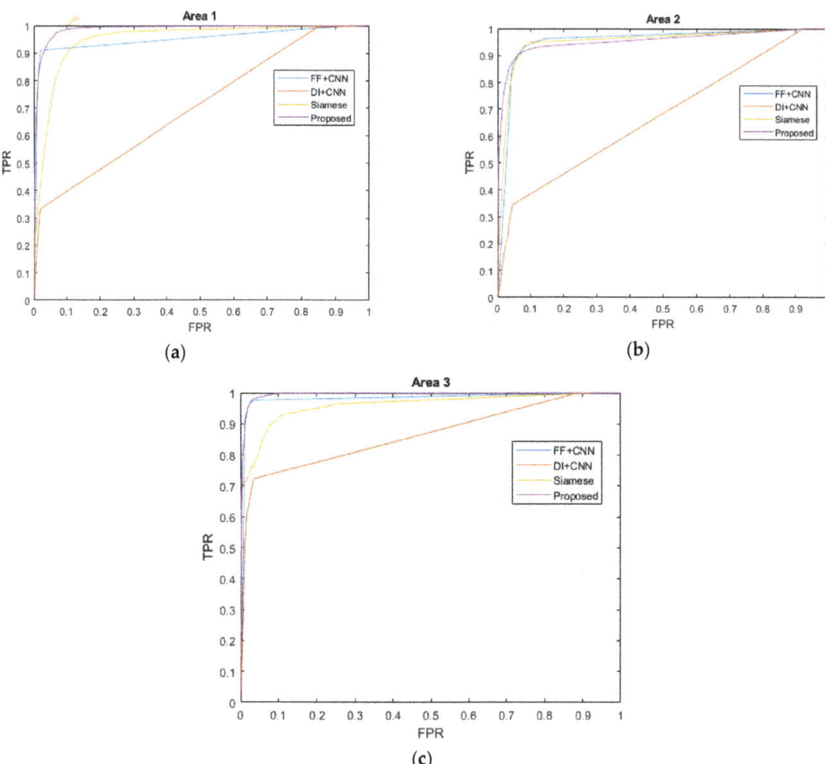

Figure 9. ROC for three areas. (**a**) ROC for Area 1, (**b**) ROC for area 2, and (**c**) ROC for Area 3.

Table 1. Quantitative assessments of the existing and proposed algorithms.

Metrics	Algorithms	Area 1	Area 2	Area 3	Avg
AUC	FF + CNN	0.95	0.95	0.98	0.96
	DI + CNN	0.70	0.68	0.88	0.75
	Siamese net	0.96	0.92	0.91	0.93
	The proposed	0.99	0.93	0.99	0.97
PCC (%)	FF + CNN	97	92	98	96
	DI + CNN	94	97	97	96
	Siamese net	96	98	99	98
	The proposed	98	99	99	99
Kappa	FF + CNN	78	19	47	48
	DI + CNN	30	32	28	30
	Siamese net	52	35	68	52
	The proposed	78	60	69	69

Regarding time complexity, the proposed DCN requires more computational complexity than the single architecture using FF + CNN and DI + CNN by a factor of approximately two with sequential machines. However, the proposed dual-DCN can work in parallel, thus, throughput can be enhanced with a parallel machine such as GPU. In addition, the proposed also takes about 20% more running time than the Siamese network because it includes additional preceding of feature maps.

5. Conclusions

In this paper, we presented a robust change detection algorithm for high-resolution panchromatic imagery. The proposed algorithm learns and analyzes the dissimilarity of two input images with the densely convolutional network by incorporating local information. We found that the proposed algorithm achieves higher detection accuracy, even with noisy conditions such as geometric distortion and different viewing angles in qualitative and quantitative analysis. Further work can be conducted to extend the framework for other modalities such as multi-spectrum images and SAR data.

Author Contributions: All authors contributed to the writing of the manuscript. W.W. and J.L. conceived and designed the experiments; W.W. performed the experiments and analyzed the data; S.-E.P. and D.S. supervised this study.

Funding: This research was supported by the MSIT (Ministry of Science and ICT), Korea, under the ITRC (Information Technology Research Center) support program (IITP-2018-2016-0-00288) supervised by the IITP (Institute for Information & communications Technology Promotion) and Basic Science Research Program through the National Research Foundation of Korea (NRF) funded by the Ministry of Science, ICT & Future Planning (NRF-2018R1A2B2008238).

Conflicts of Interest: The authors declare no conflict of interest.

References

1. Coppin, P.R.; Bauer, M.E. Digital change detection in forest ecosystems with remote sensing imagery. *Remote Sens. Rev.* **1996**, *13*, 207–234. [CrossRef]
2. Bazi, Y.; Bruzzone, L.; Melgani, F. Automatic identification of the number and values of decision thresholds in the log-ratio image for change detection in SAR images. *IEEE Geosci. Remote Sens. Lett.* **2006**, *3*, 349–353. [CrossRef]
3. Singh, K.K.; Mehrotra, A.; Nigam, M.J.; Pal, K. Unsupervised change detection from remote sensing using hybrid genetic FCM. In Proceedings of the IEEE 2013 Students Conference on Engineering and Systems (SCES), Allahabad, India, 12–14 April 2013; pp. 1–5.
4. Bi, C.; Wang, H.; Bao, R. SAR image change detection using regularized dictionary learning and fuzzy clustering. In Proceedings of the 2014 IEEE 3rd International Conference on Cloud Computing and Intelligence Systems (CCIS), Shenzhen, China, 27–29 November 2014; pp. 327–330.
5. Gong, M.; Zhou, Z.; Ma, J. Change detection in synthetic aperture radar images based on image fusion and fuzzy clustering. *IEEE Trans. Image Process.* **2012**, *21*, 2141–2151. [CrossRef] [PubMed]
6. Gong, M.; Su, L.; Jia, M.; Chen, W. Fuzzy clustering with a modified MRF energy function for change detection in synthetic aperture radar images. *IEEE Trans. Fuzzy Syst.* **2014**, *22*, 98–109. [CrossRef]
7. Gong, M.; Zhao, J.; Liu, J.; Miao, Q.; Jiao, L. Change detection in synthetic aperture radar images based on deep neural networks. *IEEE Trans. Neural Netw. Learn. Syst.* **2016**, *27*, 125–138. [CrossRef] [PubMed]
8. Johnson, R.D.; Kasischke, E.S. Change vector analysis: A technique for the multispectral monitoring of land cover and condition. *Int. J. Remote Sens.* **1998**, *19*, 411–426. [CrossRef]
9. Gao, F.; Zhang, L.; Wang, J.; Mei, J. Change Detection in Remote Sensing Images of Damage Areas with Complex Terrain Using Texture Information and SVM. In Proceedings of the International Conference on Circuits and Systems (CAS 2015), Paris, France, 9–10 August 2015.
10. Guo, Z.; Du, S. Mining parameter information for building extraction and change detection with very high-resolution imagery and GIS data. *GISci. Remote Sens.* **2017**, *54*, 38–63. [CrossRef]
11. Huang, S.; Ramirez, C.; Kennedy, K.; Mallory, J.; Wang, J.; Chu, C. Updating land cover automatically based on change detection using satellite images: Case study of national forests in Southern California. *GISci. Remote Sens.* **2017**, *54*, 495–514. [CrossRef]
12. Hao, M.; Zhang, H.; Shi, W.; Deng, K. Unsupervised change detection using fuzzy c-means and MRF from remotely sensed images. *Remote Sens. Lett.* **2013**, *4*, 1185–1194. [CrossRef]
13. Hao, M.; Hua, Z.; Li, Z.; Chen, B. Unsupervised change detection using a novel fuzzy c-means clustering simultaneously incorporating local and global information. *Multimed. Tools Appl.* **2017**, *76*, 20081–20098. [CrossRef]

14. Yu, H.; Yang, W.; Hua, G.; Ru, H.; Huang, P. Change detection using high resolution remote sensing images based on active learning and Markov random fields. *Remote Sens.* **2017**, *9*, 1233. [CrossRef]

15. Habib, T.; Inglada, J.; Mercier, G.; Chanussot, J. Support vector reduction in SVM algorithm for abrupt change detection in remote sensing. *IEEE Geosci. Remote Sens. Lett.* **2009**, *6*, 606–610. [CrossRef]

16. Volpi, M.; Tuia, D.; Bovolo, F.; Kanevski, M.; Bruzzone, L. Supervised change detection in VHR images using contextual information and support vector machines. *Int. J. Appl. Earth Obs. Geoinf.* **2013**, *20*, 77–85. [CrossRef]

17. Bovolo, F.; Bruzzone, L.; Marconcini, M. A novel approach to unsupervised change detection based on a semisupervised SVM and a similarity measure. *IEEE Trans. Geosci. Remote Sens.* **2008**, *46*, 2070–2082. [CrossRef]

18. Zhao, J.; Gong, M.; Liu, J.; Jiao, L. Deep learning to classify difference image for image change detection. In Proceedings of the IEEE 2014 International Joint Conference on Neural Networks (IJCNN), Beijing, China, 61–1 July 2014; pp. 411–417.

19. Gong, M.; Yang, H.; Zhang, P. Feature learning and change feature classification based on deep learning for ternary change detection in SAR images. *ISPRS J. Photogramm. Remote Sens.* **2017**, *129*, 212–225. [CrossRef]

20. El Amin, A.M.; Liu, Q.; Wang, Y. Convolutional neural network features-based change detection in satellite images. In Proceedings of the First International Workshop on Pattern Recognition, Tokyo, Japan, 11–13 May 2016.

21. Liu, J.; Gong, M.; Zhao, J.; Li, H.; Jiao, L. Difference representation learning using stacked restricted Boltzmann machines for change detection in SAR images. *Soft Comput.* **2016**, *20*, 4645–4657. [CrossRef]

22. Zhang, H.; Gong, M.; Zhang, P.; Su, L.; Shi, J. Feature-level change detection using deep representation and feature change analysis for multispectral imagery. *IEEE Geosci. Remote Sens. Lett.* **2016**, *13*, 1666–1670. [CrossRef]

23. Lyu, H.; Lu, H.; Mou, L. Learning a transferable change rule from a recurrent neural network for land cover change detection. *Remote Sens.* **2016**, *8*, 506. [CrossRef]

24. Zhan, Y.; Fu, K.; Yan, M.; Sun, X.; Wang, H.; Qiu, X. Change Detection Based on Deep Siamese Convolutional Network for Optical Aerial Images. *IEEE Geosci. Remote Sens. Lett.* **2017**, *14*, 1845–1849. [CrossRef]

25. Yoo, H.-J. Deep convolution neural networks in computer vision. *IEIE Trans. Smart Process. Comput.* **2015**, *4*, 35–43. [CrossRef]

26. KOMPSAT-3 Product Specifications Version 2.0. Available online: http://www.si-imaging.com/resources/?pageid=2&uid=232&mod=document (accessed on 25 June 2018).

27. Huang, G.; Liu, Z.; Van Der Maaten, L.; Weinberger, K.Q. Densely Connected Convolutional Networks. In Proceedings of the 2017 IEEE Conference on Computer Vision and Pattern Recognition (CVPR), Honolulu, HI, USA, 21–26 July 2017; pp. 4700–4708.

28. Krizhevsky, A.; Sutskever, I.; Hinton, G.E. Imagenet classification with deep convolutional neural networks. *Adv. Neural Inf. Process. Syst.* **2012**, *25*, 1097–1105. [CrossRef]

29. Ioffe, S.; Szegedy, C. Batch normalization: Accelerating deep network training by reducing internal covariate shift. In Proceedings of the International Conference on Machine Learning, Lille, France, 6–11 July 2015.

30. Hadsell, R.; Chopra, S.; Le Cun, Y. Dimensionality reduction by learning an invariant mapping. In Proceedings of the 2006 IEEE Computer Society Conference on Computer Vision and Pattern Recognition, New York, NY, USA, 17–22 June 2006.

31. Glorot, X.; Bengio, Y. Understanding the difficulty of training deep feedforward neural networks. In Proceedings of the Thirteenth International Conference on Artificial Intelligence and Statistics, Sardinia, Italy, 13–15 May 2010; pp. 249–256.

32. Fitz, R.W.; Lees, B.G. Assesing the classification accuracy of multisource remote sensing data. *Remote Sens. Environ.* **1994**, *47*, 362–368.

 *applied sciences*

Article

Convolutional Neural Network-Based Remote Sensing Images Segmentation Method for Extracting Winter Wheat Spatial Distribution

Chengming Zhang [1,2,*], Shuai Gao [3,*], Xiaoxia Yang [1,2], Feng Li [4], Maorui Yue [5], Yingjuan Han [6], Hui Zhao [6], Ya'nan Zhang [1] and Keqi Fan [1]

[1] College of Information Science and Engineering, Shandong Agricultural University, 61 Daizong Road, Taian 271000, China; yangxx@sdau.edu.cn (X.Y.); zyn980113@hotmail.com (Y.Z.); fkq980810@hotmail.com (K.F.)
[2] Shandong Technology and Engineering Center for Digital Agriculture, 61 Daizong Road, Taian 271000, China
[3] Chinese Academy of Sciences, Institute of Remote Sensing and Digital Earth, 9 Dengzhuangnan Road, Beijing 100094, China
[4] Shandong Climate Center, Mountain Road, Jinan 250001, China; lifengsd@outlook.com
[5] Taian Agriculture Bureau, Naihe Road, Taian 271000, China; yuemaoruita@outlook.com
[6] Key Laboratory for Meteorological Disaster Monitoring and Early Warning and Risk Management of Characteristic Agriculture in Arid Regions, CMA, 71 Xinchangxi Road, Yinchuan 750002, China; yjhan_nx@outlook.com (Y.H.); zhaohui_cau@outlook.com (H.Z.)
* Correspondence: chming@sdau.edu.cn (C.Z.); gaoshuai@radi.ac.cn (S.G.); Tel.: +86-139-5382-3659 (C.Z.); +86-010-64806258 (S.G.)

Received: 26 September 2018; Accepted: 16 October 2018; Published: 19 October 2018

Featured Application: In Gaofen-2 images, it is difficult to accurately extract winter wheat spatial distribution using traditional methods. Because our approach can better solve this problem, it has played an important role in agricultural surveys and improved the efficiency of agricultural surveys. Our approach has been utilized by the Department of Agriculture and the Meteorological Bureau of Shandong Province, China.

Abstract: When extracting winter wheat spatial distribution by using convolutional neural network (CNN) from Gaofen-2 (GF-2) remote sensing images, accurate identification of edge pixel is the key to improving the result accuracy. In this paper, an approach for extracting accurate winter wheat spatial distribution based on CNN is proposed. A hybrid structure convolutional neural network (HSCNN) was first constructed, which consists of two independent sub-networks of different depths. The deeper sub-network was used to extract the pixels present in the interior of the winter wheat field, whereas the shallower sub-network extracts the pixels at the edge of the field. The model was trained by classification-based learning and used in image segmentation for obtaining the distribution of winter wheat. Experiments were performed on 39 GF-2 images of Shandong province captured during 2017–2018, with SegNet and DeepLab as comparison models. As shown by the results, the average accuracy of SegNet, DeepLab, and HSCNN was 0.765, 0.853, and 0.912, respectively. HSCNN was equally as accurate as DeepLab and superior to SegNet for identifying interior pixels, and its identification of the edge pixels was significantly better than the two comparison models, which showed the superiority of HSCNN in the identification of winter wheat spatial distribution.

Keywords: remote sensing image segmentation; convolutional neural networks; Gaofen-2; hybrid structure convolutional neural networks; winter wheat spatial distribution; classification-based learning

1. Introduction

Winter wheat is the most important food crop in China, comprising 21.38% of the gross cropped area of the domestic food crops in 2017 according to the data released by the National Bureau of Statistics, with its output accounting for 21.00% of the total food crop production [1]. For national food security, the Chinese government has assigned a minimum area of arable land in each region that needs to be safeguarded (the "red line") [2]. Timely and accurate acquisition of the size and spatial distribution of winter wheat fields assists the relevant government departments in guiding the farming activities, estimating the yield, and adjusting the agricultural structure for ensuring food security [3].

Remote sensing is capable of imaging and large-area monitoring, making it a good data source for rapid and accurate extraction of winter wheat planting information. Researchers have successfully extracted winter wheat spatial distribution information from MODIS (moderate-resolution imaging spectroradiometer) and ETM/TM (enhanced thematic mapper plus/thematic mapper), achieving accuracies of 85.5% and 89.1%, respectively [4,5]. This exhibits the advantage of remote sensing in this application. However, owing to limitations in the spatial resolution of the data source, the spatial resolution of the extraction results is also rather coarse and unable to satisfy the requirement of the application [6–10]. With the development of high-resolution remote sensing satellites, a crop planting area can be monitored more accurately using the corresponding images as the data source [11,12]. The winter wheat cultivation information is extracted from the remote-sensing images captured by Gaofen-1 of the Gaofen series of Chinese satellites, yielding satisfactory results, with maximum accuracy reaching about 89% [13–18]. Most researchers still use traditional methods, such as decision trees and textures features. These methods can only take advantage of low-level features, which make it easy to make mistakes in identifying pixels at the edge of winter wheat planting area.

Image segmentation has been successfully used in the processing of camera images and applied by researchers to high-resolution remote sensing images, achieving significantly more accurate classification by a pixel-by-pixel segmentation [19–21]. Feature extraction is the key step in remote sensing image segmentation. In high-resolution remote sensing images, as the spectral difference between the same type of objects is increased, and between different types of objects is diminished, the former has more probability of exhibiting different spectral properties, whereas the latter tends to be spectrally similar, which makes feature extraction increasingly difficult [22,23]. Traditional methods including k-nearest neighbors and maximum entropy can only identify low-level image features such as color, shape, and texture. They are not capable of visually providing a semantic description. This hinders the extraction of higher-level features and limits the use of these methods in the segmentation of high-resolution remote sensing images [24,25].

With the development of machine learning, algorithms such as neural networks (NNs) [26] and support vector machine (SVM) [27,28] are being used in the segmentation of high-resolution images [29–31]. In some studies, when compared with traditional statistical methods and object-oriented methods, machine learning algorithms yielded better image segmentation results [32, 33]. Both SVM and NNs are shallow-learning algorithms [34–36], which do not express complex functions well owing to the limitations in their network structure. Therefore, these models cannot adapt to the continuously increasing complexity caused by the increasing sample size and diversity [37,38].

Progress in deep learning has facilitated solving these problems by using deep neural networks (DNNs) [39–42]. As an important branch of deep learning, a convolutional NN (CNN) is widely used with visual data because of its excellent feature learning ability [43–45]. A CNN is a deep learning network, composed of several layers, capable of nonlinear mapping. Its strength in learning is exemplified by the good image segmentation results achieved [46–52]. Further, the capacity of many large CNNs can be scaled according to the size of the training data and complexity and processing ability of the model, and their performance in image segmentation has improved significantly [53–60].

A fully convolutional network (FCN) is a deep learning network for image segmentation, which was proposed in 2015. Taking advantage of convolution computation in its feature organization and extraction abilities, an FCN realizes pixel-by-pixel segmentation of camera images by constructing a multi-layer convolutional structure and setting appropriate deconvolutional layers [61–63]. Accordingly, a series of convolution-based segmentation models has been developed including SegNet [64], UNet [65], DeepLab [66], multi-scale FCN [67], and ReSeg [68]. Of these models, SegNet and UNet are clearly structured, and it is easy to understand the convolution structure of the model. The processing speed is fast. DeepLab uses a method called "Atrous Convolution", which has a strong advantage in processing detailed images. multi-scale FCN is designed to address the huge scale gap between different classes of targets, i.e., sea/land and ships. ReSeg exploits the local generic features extracted by Convolutional Neural Networks and the capacity of Recurrent Neural Networks (RNN) to retrieve distant dependencies. Each model has its own strengths and is adept at dealing with certain image types.

In the work of extracting the spatial distribution of crops with high GF-1 as the data source, in addition to methods such as decision trees, textures features, and maximum entropy, research has also been carried out using deep learning. However, most of these studies directly use the existing deep learning model as a tool, and seldom consider the influence of characteristics difference of edge pixels and inner pixels in the crop planting area are large.

On board the Gaofen-2 satellite is a panchromatic camera with a spatial resolution of 1 m, and a multi-spectral camera with a spatial resolution of 4 m, which provides ideal data for extracting winter wheat plantation information. Before the application of a CNN to GF-2 remote-sensing images for this purpose, trial extraction is performed with classical network architectures (such as SegNet) where misidentified pixels are categorized, of which approximately 90% are found at the edge of the crop field. Further analysis indicates the structure of the convolutional layer as the source of this problem. The outcome produced by operating the convolution kernel in the pixel block is treated as the eigenvalue of the central pixel of the pixel block. As such, for the pixels at the edge, 50% of the pixels involved in each convolution are from negative samples, whereas, for the pixels at the corner, this number is 75% or higher. This results in a significant difference between the eigenvalues of the pixels at these locations and those at the center of the image, and an increase in the probability of the recognition results being placed in a wrong category. To avoid these problems, a new method is herein proposed for the extraction of the winter wheat field information from the GF-2 remote sensing images. The main procedures are as follows.

1. First, a CNN consisting of two independent sub-networks of different depths is established. The deep and shallow sub-networks are trained to be sensitive only to the pixels at the interior and edge of a winter wheat planting field, respectively, and only these pixels are extracted. This model is named as a Hybrid Structure Convolutional Neural Network (HSCNN).

2. A classification algorithm is adopted in the model training. For initial training of the sub-network used for the edge pixel extraction, edge pixels are considered as positive samples, with the pixels at other locations being treated as negative samples. The inner pixels are then designated as positive samples, with the pixels at other locations as negative samples, for training the sub-network used for the inner pixel extraction. After the successful completion of the training, the neural network is able to extract the winter wheat field from the GF-2 images accurately.

3. Finally, a GF-2 image is segmented by the trained model. Because SegNet and DeepLab are classic semantic segmentation models of images, and, the working principles of these two models are very similar to our work, we choose these two models as the comparison model, and a comparison is performed with them to evaluate the accuracy of the segmentation results.

2. Data Sources and Methods

2.1. Data Sources

2.1.1. Study Region

The whole study region is Shandong province, China. Shandong is located along the eastern coast of China (in the lower stream of the Yellow river), within 34°22′ N–38°24′ N and 114°47.5′ E–122°42′ E. It measures 721.03 km from east to west, and 437.28 km from north to south. The land area of the province is 155,800 km^2, of which 14.59% is mountainous, 5.56% is water (such as lakes), 15.98% is forest, and 53.82% is cultivated land. The annual total planting area of crops in the province is approximately 162 million mu. The main food crops of this region are wheat and maize. In 2016, the wheat planting area was 57.45405 million mu, and in 2017 it was 57.6435 million mu [69].

In this paper, we used the ground data and remote sensing data of Feicheng county, Ningyang county and Zhangqiu county, Shandong province. The three counties are similar in topography, all relatively flat, which can eliminate the influence of topographic fluctuations on the experimental results.

2.1.2. Ground-Based Data

For manufacturing sample to train our model, we conducted a field survey in Feicheng county, Ningyang county and Zhangqiu county in 2017 and 2018, and obtained the land use data of 369 sample points, among which 257 were winter wheat sample points and 112 were bare land. The survey results include the time, location and type of land use.

2.1.3. Remote Sensing Data

We selected 39 GF-2 remote sensing image, size of each image is 7300 × 6900. Of these images, 15 were captured on 17 February 2017, 11 were captured on 21 March 2018 and 13 were captured on 12 April 2018. We select images from different periods to increase the anti-interference abilities of the HSCNN. These remote sensing data cover Feicheng county, Ningyang county and Zhangqiu county, and are matched with ground investigation time. At the same time, the selected remote sensing data have fewer clouds and better clarity.

The Environment for Visualizing Images (ENVI) software was used for preprocessing the tasks, including fusion of panchromatic spectrum and multispectral band to obtain 1-m spatial resolution multispectral data, and the contrast stretch to generate a color-enhanced color composite image.

2.2. Network Architecture of Our Method

The HSCNN model is divided into five functional groups of components, input (a), inner-CNN (b), edge-CNN (f), vote function (j), and output (k), as shown in Figure 1. Both the edge-CNN and inner-CNN have convolution layers, an encoder layer, and a classifier layer. In the training stage, the inputs are original images and artificial classification labels. In the classification stage, the inputs are original GF-2 images, output is a single-band file, and content of each pixel in the output is the category number of the corresponding original image pixel. The HSCNN indicates the winter wheat area using category number 100, and category number 200 distinguishes other land use. The reason for adopting the two numbers is to fit with the coding value table we are working on to obtain detailed land use information.

2.2.1. Inner-Layers and Edge-Layers

The operational characteristics of the pixel block-based convolution for image segmentation are described in Section 1, in addition to the effect of the pixel block location on the convolution results. Based on this analysis, two convolution sub-structures of different depths are setup for the feature extraction of the winter wheat field. The deep convolution sub-network is used to extract the features of the pixels in the interior of the winter wheat plantation, shown as inner-layers (c) in Figure 1.

The shallower sub-network is used to extract the features of the pixels at the edge of the winter wheat plantation, shown as edge-layers (g) in Figure 1. The benefits of this design are discussed in Section 4 based on the experimental results.

In our approach, an inner pixel refers to the pixel that only contains winter wheat pixels in the pixel block when convolution operation is carried out with the pixel as the center pixel. An edge pixel refers to the pixel that contains winter wheat pixels and other pixels when computing the feature of the pixel.

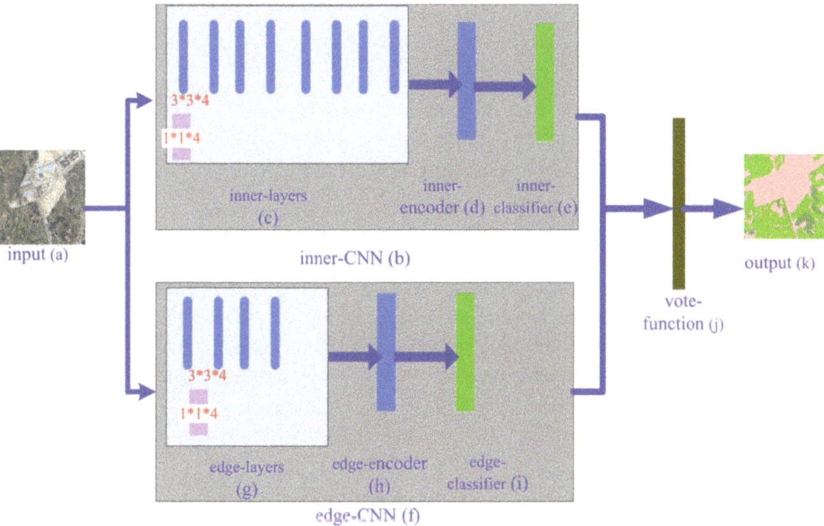

Figure 1. Network architecture of the Hybrid Structure Convolutional Neural Network (HSCNN): (**a**) input; (**b**) inner-CNN; (**c**) inner-layers; (**d**) inner-encoder; (**e**) inner-classifier; (**f**) edge-CNN; (**g**) edge-layers; (**h**) edge-encoder; (**i**) edge-classifier; (**j**) vote function; (**k**) output.

All kernels of the HSCNN take the form, $w \times h \times c$, where w is the width, h is the height, and c is the number of channels of a kernel. Two types of kernels are used in the first convolutional layers of inner-layers (c) and edge-layers (g). For one type w and h are set to 1, and for the other type the values are set to 3. In both cases, c is set to 4 because the data in the four multi-spectral bands of GF-2 are used. Kernels of the form $1 \times 1 \times 4$ are used to extract the features of the pixels. The generated feature map is used instantaneously as the input of the encoder, and does not participate in the subsequent convolution. Convolution kernels of the form $3 \times 3 \times 4$ are used to extract the spatial relation between the pixels and generate the spatial semantics by multi-level convolution.

After the operation of first convolution layer on the original image, we obtain a feature map which has only one channel. Because the input of convolution layer is the feature map calculated by the previous convolution layer, so the w and h values of the kernels used in all other convolutional layers are set to 3, and c is 1 from the second layer. To extract more features from the edge pixels of the crop field, the number of kernels used in each convolutional layer of edge-layers (g) is twice that used in the corresponding layer of inner-layers (c).

In the HSCNN, each convolution layer has only one activation layer attached, and there is no pool layer. Accordingly, the convolution result of each pixel block can be used directly as the feature of its central pixel, without the need to determine the position of the pixel that the feature corresponds to through deconvolution. As such, the HSCNN does not utilize a deconvolutional layer. This reduces the extent of computation and positioning error of the deconvolution, thereby improving the accuracy of the segmentation.

2.2.2. Inner-Encoder and Edge-Encoder

The inner-encoder and edge-encoder are used to encode the eigenvector extracted by the convolution layers on the pixel, ensuring that the classifier can establish the relationship between the eigenvector and pixel type. In the HSCNN model, the inner- and edge-encoders are both $2 \times n$ matrices, where n is the length of the eigenvector.

Let X denote the eigenvector of the pixel, W denote the encoder matrix, and R the encoded vector result. The encoding calculation is displayed in Equation (1).

$$
\begin{bmatrix} r_1 \\ r_2 \end{bmatrix} = \begin{bmatrix} w_{11} & w_{12} & \cdots & w_{1n} \\ w_{21} & w_{22} & \cdots & w_{2n} \end{bmatrix} \times \begin{bmatrix} x_1 & x_2 & \cdots & x_n \end{bmatrix}^T + \begin{bmatrix} b_1 \\ b_2 \end{bmatrix} \tag{1}
$$

where each row of matrix w represents a fitting function for a particular type of pixel, b_1 and b_2 are the respective biases, and the corresponding component of r is the encoded value of eigenvector x on that pixel type. The inner- and edge-encoders are trained separately.

2.2.3. Inner-Classifier and Edge-Classifier

For each pixel, the inner-classifier converts its vector of the encoded values given by the inner-encoder into a probability distribution over a set of classes, and classifies the pixel as an inner pixel or a non-inner pixel of the winter wheat plantation based on the location of the component with the highest probability. Similarly, the edge-classifier distinguishes between the edge and non-edge pixels of the winter wheat field using the vectors of the encoded values generated on the pixels by the edge-encoder.

In reference to the classic softmax classifier [60–65], Equation (2) is used here to convert vector r of the encoded values to vector p of the class probabilities for each pixel.

$$
\begin{bmatrix} p_1 \\ p_2 \end{bmatrix} = \begin{bmatrix} \dfrac{e^{r_1}}{e^{r_1} + e^{r_2}} \\ \dfrac{e^{r_2}}{e^{r_1} + e^{r_2}} \end{bmatrix} \tag{2}
$$

After the transformation, the index of $\max(p_1$ and $p_2)$ is taken as the predicted category of the pixel. For the inner-classifier, index numbers of 1 and 0 are assigned to the inner pixels of the winter wheat field and other pixels, respectively. Accordingly, index numbers of 1 and 0 are assigned to the edge pixels of the winter wheat field and other pixels, respectively.

2.2.4. Vote-Function

The vote-function determines the category number of a pixel given by the inner-classifier and edge-classifier and writes it to the output file. As described in the beginning of Section 2.2, the HSCNN indicates the winter wheat area using category number 100 and other land uses using category number 200. The category number of a pixel is calculated in Equation (3).

$$
o = \begin{cases} 100 & P_{\text{inner}} = 1 \text{ or } P_{\text{edge}} = 1 \\ 200 & P_{\text{inner}} = 0 \text{ and } P_{\text{edge}} = 0 \end{cases} \tag{3}
$$

where o represents the final category number of a pixel and P_{inner} and P_{edge} are the outputs of the inner-classifier and edge-classifier, respectively.

2.3. HSCNN Training

We manually labeled all images at the pixel level as ground truth (GT) label data. In other words, for each image, there exists a 7300×6900 label map, having a pixel-class (row-col indexed) correspondence with it. We used 36 images for training, and the remaining 3 images for testing. The GF-2 images and their corresponding artificial classification labels will be input to the HSCNN as training samples.

The training process includes error calculation, error back propagation and weight update. This process is iterated until the difference becomes smaller than the predetermined threshold.

We calculated the errors between the predicted classification label and manual classification label by the chain rule. The chain rule, the derivative rule in calculus, is used to find the derivative of a complex function, which is a common way to do the derivative calculation of calculus. The derivative of a composite function is the product of the derivatives of this finite number of functions at the corresponding point, as a chain. Then, the errors are back-propagated through the network. The backward propagation algorithm is a kind of training and learning method in deep learning, which can spread the error of the output layer backward to realize weight adjustment, adjust the weight between each node in the deep network, and achieve the goal that the sample tag output from the network is consistent with the actual tag. We use gradient descent method to update HSCNN parameters. Gradient descent method is the most commonly used optimization method. The idea is to use the negative gradient direction of the current position as the search direction, because that direction is the fastest descending direction of the current position.

2.3.1. Sample Labeling

We use the ENVI software for labeling and designing a preprocessor to build the labels. The process of artificial labeling is as follows:

1. The region-of-interest (RoI) tool in the ENVI software is used to select the winter wheat regions and other regions in the image. Then, the map locations of the pixels in each region are output to different files based on the category.
2. A band is added to the image file by the preprocessor as a mask band. The spatial resolution, size, and other parameters of the mask band are the same as the original image. Then, the category number of each pixel is written to the mask band according to the map location of the pixel previously output. We manually label all the images at the pixel level. Thus, for each image, there exists a 7300×6900 label map, with a row-column-indexed pixel-class correspondence.
3. The pixels marked as winter wheat are further categorized as edge pixels and inner pixels. Based on the parameters given above, the inner-layers comprise of eight convolutional layers, each with a 3×3 (length $\times$ width) convolution kernel. Therefore, the feature extraction from pixel s involves a 9×9 pixel block centered at s in the calculation. As defined in Section 2.2.1, the winter wheat pixels are divided sequentially into edge pixels and inner pixels. For training class by class, we use temporary code 160 to denote edge pixels and 170 to denote inner pixels in the mask band.

Figure 2 shows an example of an image-label pair.

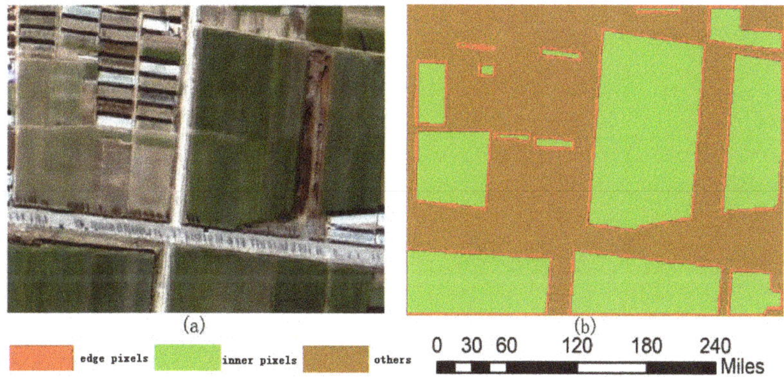

Figure 2. Image-label pair example: (**a**) original image; and (**b**) labels.

2.3.2. Loss Function

In our method, new loss functions are defined for the inner-CNN and edge-CNN, which still use the cross entropy as the basic element for the calculation, as expressed in Equation (4).

$$H(p,q) = -\sum_{i=1}^{2} q_i \log(p_i) \tag{4}$$

where p and q are, respectively the predicted and actual probability distribution, and i is index of a component in the probability distribution. On this basis, the loss function of the inner-CNN is defined as

$$loss = -\frac{1}{m} \sum_m \sum_{i=1}^{2} q_i \log(p_i) \tag{5}$$

When computing loss of inner-CNN, m is obtained by subtracting the number of edge pixels of the winter wheat field from the total number of samples, and when computing loss of edge-CNN, m is obtained by subtracting the number of inner pixels of the winter wheat field from the total number of samples.

2.3.3. Model Training

Images from two different periods were selected as the training data. We selected images from different periods for increasing the anti-interference abilities of the HSCNN and mitigating the complications such as the change in the seasons, and thus enhancing applicability. The training stage proceeded through the following steps:

1. Image-label pairs are input into the HSCNN as training samples. Network parameters are initialized.
2. Forward propagation is performed on the sample images.
3. The [loss]_inner is calculated and back propagated to the inner-CNN, whereas the [loss]_edge is calculated and back propagated to the edge-CNN.
4. The network parameters are updated using the stochastic gradient descent (SGD) [41,48] with momentum.
5. Steps (2)–(4) are iterated until both [loss]_inner and [loss]_edge are less than the predetermined threshold values.

The training yields two sub-networks, an inner-CNN and edge-CNN. The former can accurately extract the inner pixels of the winter wheat plantation from the sweet GF-2 remote sensing images, whereas the latter allows the best possible distinction between the edge pixels of the winter wheat planting region and other pixels.

In our training, the SGD method with momentum was used for parameter updates, which is illustrated in the following expression:

$$W^{(n+1)} = W^{(n)} - \Delta W^{(n+1)} \tag{6}$$

where $W^{(n)}$ denote the old parameters, $W^{(n+1)}$ denote the new parameters, and $\Delta W^{(n+1)}$ is the increment in the current iteration, which is a combination of the old parameters, gradient, and historical increment, i.e.,

$$\Delta W^{(n+1)} = \vartheta \left(d_w \cdot W^{(n)} + \frac{\partial J(W)}{\partial W^{(n)}} \right) + m \cdot \Delta W^{(n)} \tag{7}$$

where J(W) is the loss function, ϑ is the learning rate for step length control, d_w denotes the weight decay, and m denotes the momentum.

2.4. Segmentation Using the Trained Network

After successful training, the HSCNN can be used to segment an input imagery pixel-by-pixel. According to our design, the output is written in a new band. The benefit of this design avoids damaging the original file.

3. Experiments and Results

The data used in the experiment are presented in Section 2.1. In this section, the models used for comparison are described in Section 3.1, and the experimental results and assessment of accuracy are given in Section 3.2.

3.1. Comparison Model

Feature selection is the basis of remote sensing image segmentation. At present, there are mainly two methods based on artificial feature selection and machine learning. Haralick et al. (1973) put forward the method of gray level co-occurrence matrix, which is a classical artificial selection feature method. This method is mainly used to select image texture features. Since the texture is formed by repeated alternating changes of gray distribution in image space, so there is a certain gray-scale relationship between two separate pixels away certain distance, Haralick et al. described this correlation through a matrix [70]. Based on the artificial selection feature, only limited, shallow features can generally be selected. The feature selection based on machine learning can fully explore the deep feature and spatial semantic feature of the image. SegNet and DeepLab are classic semantic segmentation models of images, which have achieved very good results in the processing of images. Moreover, the working principles of these two models are very similar to our work, so we choose these two models as the comparison model, which can better reflect the advantages of our model in feature extraction. A comparative experiment was conducted using the methods established in the published literature.

3.1.1. SegNet

For the SegNet model, we directly employed the structure proposed by Badrinarayanan et al. [64], which consists of an encoder, a decoder, and a classifier. The encoder uses the first 13 convolutional layers of the VGG16 network, each having its corresponding decoder layer, totaling 13 decoder layers. The last decoder generates a multi-channel feature map as the input to the classifier, which outputs a probability vector of length K, where K is the number of classes. The final predicted category corresponds to the class having maximum probability at each pixel. In terms of training, SegNet can be trained end-to-end using SGD.

3.1.2. DeepLab

For DeepLab, we directly employed the DeepLab v3 model proposed by Liang-Chieh Chen et al. [66]. DeepLab was also developed based on the VGG network. To ensure that the output size would not be not too small without excessive padding, DeepLab changes the stride of the pool4 and pool5 layers of the VGG network from the original 2 to 1, plus 1 padding. To compensate for the effect of the stride change on the receptive field, DeepLab uses a convolution method called "atrous convolution" to ensure that the receptive field after pooling remains unchanged and the output is more refined. Finally, DeepLab incorporates a fully connected conditional random field (CRF) model to refine the segmentation boundary.

3.2. Results and Result Comparison

In the comparative experiment, we applied our trained model to three GF-2 images for segmentation. These images were only used for testing and not involved in training. Figure 3 illustrates the results obtained from the comparison methods and proposed method. In Figure 3, the first column illustrates the results of Experiment 1, the second column illustrates the results of Experiment 2 and the third column illustrates the results of Experiment 3.

Tables 1–3 are confusion matrices c for the segmentation results of SegNet model, DeepLab model, and HSCNN model, respectively. Each row of the confusion matrix represents the proportion taken by the actual category, and each column represents the proportion taken by the predicted category. As can be seen from the tables, our method achieves better classification results. In the example above, the proportion of "winter wheat" wrongly categorized as "background" is on average 0.069, and the proportion of "background" wrongly classified as "winter wheat" is on average 0.019, resulting in an overall accuracy of 91.2%.

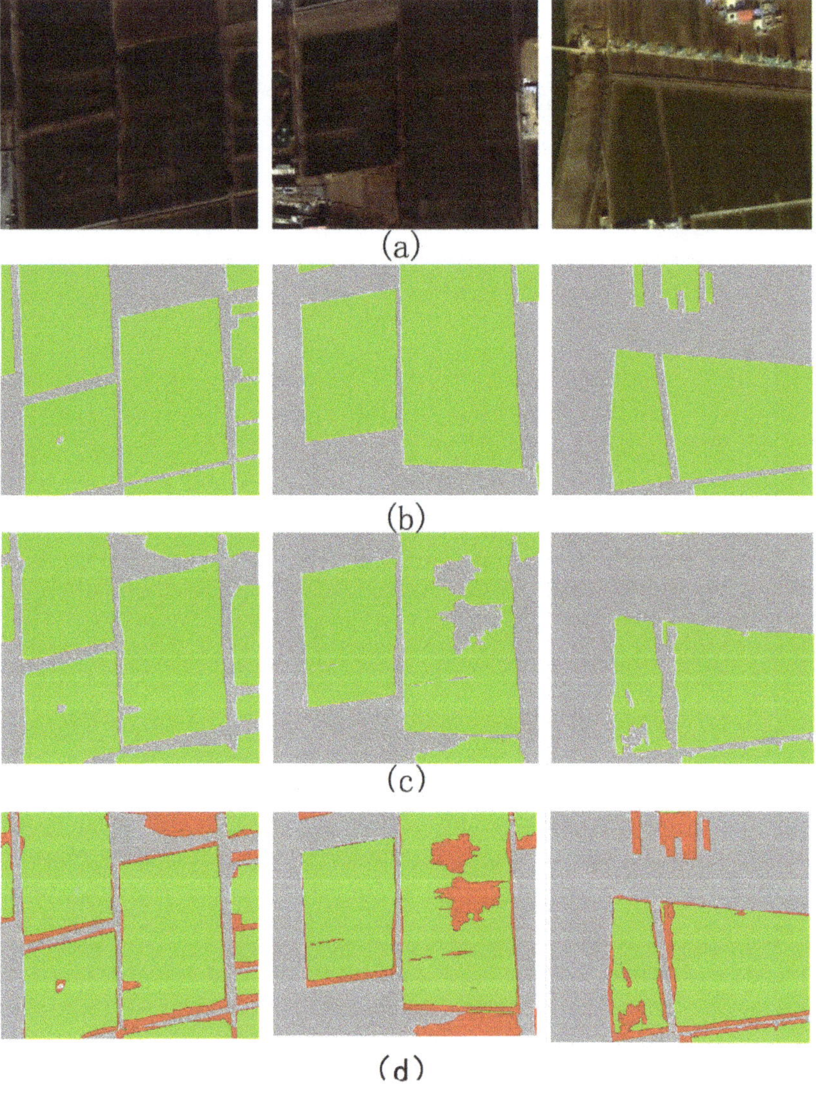

Figure 3. *Cont.*

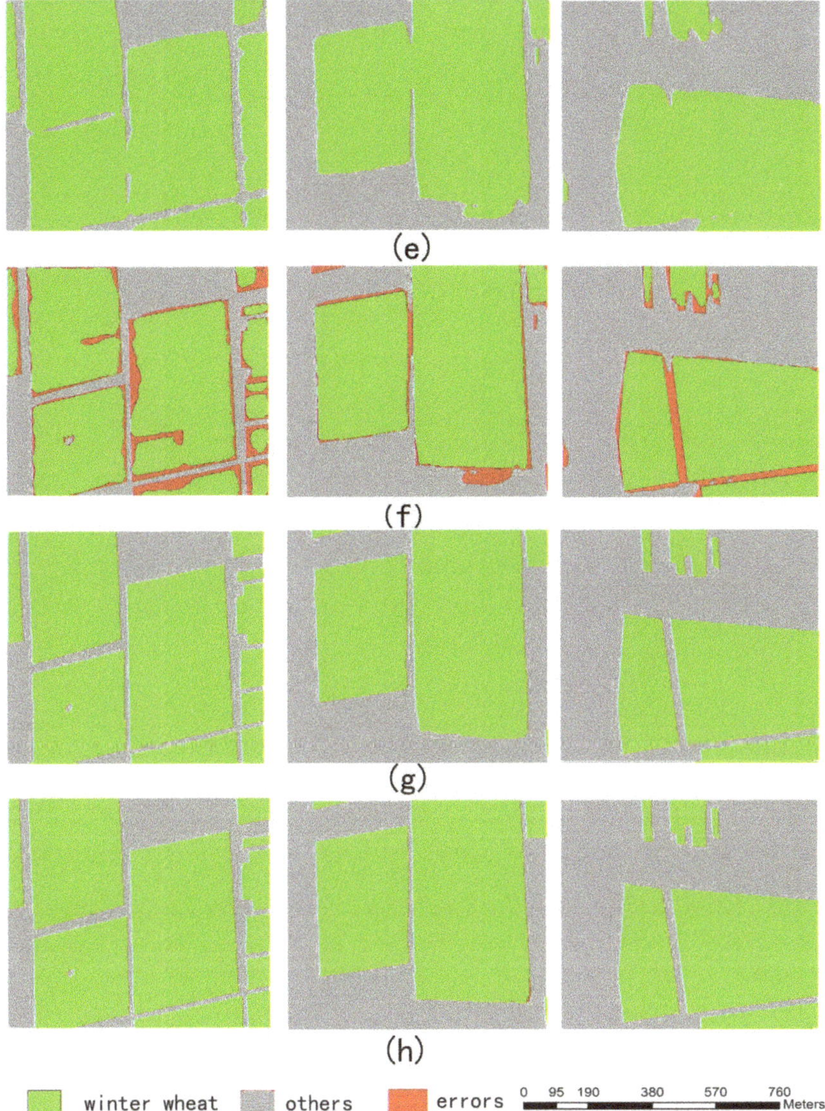

Figure 3. Segmentation results for Gaofen-2 (GF-2) images: (**a**) original images; (**b**) ground truth, (**c**) results of SegNet corresponding to the images in (**a**); (**d**) errors of SegNet; (**e**) results of DeepLab; (**f**) errors of DeepLab; (**g**) results of HSCNN; and (**h**) errors of HSCNN.

Table 1. Confusion matrix of the SegNet approach for Figure 4.

Experiment	GT/Predicted	Winter Wheat	Others
Experiment-1	winter wheat	0.621	0.129
	Others	0.087	0.163
Experiment-2	winter wheat	0.387	0.153
	Others	0.084	0.376
Experiment-3	winter wheat	0.217	0.123
	Others	0.129	0.531

Table 2. Confusion matrix of the DeepLab approach for Figure 4.

Experiment	GT/Predicted	Winter Wheat	Others
Experiment-1	winter wheat	0.653	0.107
	Others	0.055	0.185
Experiment-2	winter wheat	0.432	0.086
	Others	0.039	0.443
Experiment-3	winter wheat	0.301	0.108
	Others	0.045	0.546

Table 3. Confusion matrix of our HSCNN approach for Figure 4.

Experiment	GT/Predicted	Winter Wheat	Others
Experiment-1	winter wheat	0.681	0.075
	Others	0.027	0.217
Experiment-2	winter wheat	0.458	0.062
	Others	0.013	0.467
Experiment-3	winter wheat	0.329	0.071
	Others	0.017	0.583

Accuracy, precision, recall, and the Kappa coefficient were used to evaluate the models. These indices are calculated via mixing matrix c.

Accuracy is the ratio of the number of correctly classified samples to the total number of samples, and is given in this case by the following equation:

$$\text{Accuracy} = \frac{\sum_{i=1}^{2} C_{ii}}{\sum_{i=1}^{2} \sum_{j=1}^{2} C_{ij}} \tag{8}$$

Here, C_{ii} denotes the number of correctly classified samples, and C_{ij} denotes the number of samples of class i misidentified as class j.

Precision denotes the average proportion of pixels correctly classified to one class from the total retrieved pixels. Precision is calculated as:

$$\text{Precision} = \frac{1}{2} \sum_{i} C_{ii} / \sum_{j} C_{ij} \tag{9}$$

Recall represents the average proportion of pixels that are correctly classified in relation to the actual total pixels of a given class. Recall is computed as:

$$\text{Recall} = \frac{1}{2} \sum_{i} C_{ii} / \sum_{i} C_{ij} \tag{10}$$

The Kappa coefficient measures the consistency of the predicted classes with artificial labels. The Kappa coefficient is computed as:

$$Kappa = \frac{\frac{\sum_{i=1}^{2} C_{ii}}{\sum_{i=1}^{2}\sum_{j=1}^{2} C_{ij}} - \frac{\sum_{i=1}^{2} C_{ii}\sum_{j=1}^{2} C_{ij}}{(\sum_{i=1}^{2}\sum_{j=1}^{2} C_{ij})^2}}{1 - \frac{\sum_{i=1}^{2} C_{ii}\sum_{j=1}^{2} C_{ij}}{(\sum_{i=1}^{2}\sum_{j=1}^{2} C_{ij})^2}} \tag{11}$$

Equations (8)–(11) use the definitions given in Reference [18] and are modified according to our actual situation. The minimum accepted precision is 89% according to practical application. The indicator values are listed in Table 4.

Table 4. Comparison of the approaches using SegNet, DeepLab, and HSCNN.

Approach	Index	Experiment-1	Experiment-2	Experiment-3	Average
SegNet	Accuracy	0.784	0.763	0.748	0.765
	Precision	0.740	0.767	0.721	0.743
	Recall	0.718	0.766	0.720	0.734
	Kappa	0.579	0.617	0.564	0.586
DeepLab	Accuracy	0.838	0.875	0.847	0.853
	Precision	0.815	0.877	0.830	0.840
	Recall	0.778	0.877	0.852	0.836
	Kappa	0.665	0.778	0.716	0.720
HSCNN	Accuracy	0.898	0.925	0.912	0.912
	Precision	0.895	0.927	0.897	0.906
	Recall	0.853	0.927	0.921	0.900
	Kappa	0.776	0.860	0.826	0.821

4. Analysis and Discussion

From the experimental results in Section 3, it is clear that our method significantly improves the accuracy of winter wheat extraction. In this section, the advantages of our model are discussed in terms of the differences between the remote sensing images and camera images. This is followed by more specific comparisons with SegNet and DeepLab. Finally, the role of our model in the classification of land uses by remote sensing is discussed briefly.

4.1. Advantages of the HSCNN Model

CNNs have achieved significant success in camera image segmentation, which has motivated researchers to apply them to remote sensing images. The HSCNN model proposed here is developed based on a previous work followed by a further in-depth analysis of the fundamental difference between camera images and remote sensing images. Thus, it possesses clear advantages compared with the traditional practice of the straightforward application of camera image segmentation model to remote sensing images.

Camera images and remote sensing images essentially differ in information representation. Owing to their advantages in shooting distance and the pixel quality of the camera, camera images are superior in terms of the rich details they contain, such that one object is formed by multiple pixels. Thus, the color of a pixel reflects the information at a certain point on an object but not the spatial relation between the pixels, which is found and expressed only by convolution. The nature of convolution is to represent the spatial correlation between the pixels by constructing a complex fitting function by operating on the pixel value of a pixel block. Particularly because it makes good use of the essential characteristics of camera images, deep convolution is extremely successful in camera image processing.

In remote sensing images, particularly for crop fields, a pixel generally contains multiple objects. For example, generally in GF-2 images 1 m^2 of ground is covered by a pixel, which contains 600–700 winter wheat plants. A pixel embodies the color information of the plants and the spatial information between them. However, at the edge of a winter wheat field, the region covered by a pixel is often a mixture of the winter wheat and bare land or winter wheat and other geographical objects, with varying percentages of winter wheat in the space. In this view, the information contained in a pixel at the edge region is significantly different from that at the interior. These two types of pixels can even be regarded as two different types of objects.

Based on the above analyses, the HSCNN network architecture is designed with a complete consideration of the properties of the remote sensing image and extraction target, and it makes good use of the characteristics of the winter wheat field captured in the GF-2 remote sensing images. The strengths of this model are exhibited in the following three aspects:

1. Considering the significant difference between the pixels of the interior and edge region of the winter wheat plantation (during extraction), these two regions are treated as two subclasses. Accordingly, the features of the inner pixels are more focused, which facilitates the model training. Two sub-networks with different depths are then designed with respect to the characteristics of the two subclasses. The deep sub-network extracts the pixels at the interior, whereas the shallower sub-network extracts those at the edge. This scheme reduces the effect of the non-winter wheat pixels on the features and improves the stability of the model for edge pixel extraction.

2. Two types of kernels are used in the first convolutional layer. The $1 \times 1 \times 4$ kernels are used to extract the feature of the pixels, and the $3 \times 3 \times 4$ kernels are used to extract the spatial relation between pixels. This design takes advantage of the ability of convolution for extracting higher-level spatial semantics and for obtaining the rich pixel information contained in the remote sensing images.

3. Our model does not utilize pooling, instead the convolution result is taken as the eigenvalue of the central pixel of the pixel block. In the application of the convolutional network to image classification, the basic target (sample) for the classification is the entire image. Thus, pooling can produce the main features of the feature map and reduce the amount of subsequent computation. Although the information on the accurate position of the features is lost during this process, their relative positions are nevertheless retained, which ensures the normal operation of the subsequent computational steps. However, in image segmentation, the basic target (sample) for the classification is an individual pixel, whose exact location must be mapped by the eigenvalue. Therefore, the major advantage of our model is its ability to preserve the spatial location of the eigenvalue, which makes it possible to remove the deconvolution adopted by the traditional FCN. Accordingly, the amount of computation is reduced. Further, the loss in precision due to positioning error is reduced, as the accurate position of the eigenvalue is kept.

4.2. Comparison with SegNet and Analysis

SegNet is founded on the FCN model. Its main strength lies in the search and extraction of the rich details of an image by deep convolution, and it is very distinct when extracting target objects with relatively few pixels. If the target objects contain only a few pixels or even one pixel, the deep convolution does not generate more details and may introduce more noise owing to the expanded field of view, affecting the determination of the pixel type.

In the remote sensing images of GF-2, the edge and interior of the winter wheat plantation are very different in composition, which makes it more difficult for SegNet to locate the common features, because of its structure containing a single convolutional network. In comparison, the HSCNN is equipped with two sub-networks of different depths and is adaptable to the characteristics of the edge and interior. It also uses two different sizes of kernel, which are capable of uncovering the spatial relation between the pixels, and the information embedded in the pixels.

As shown in Figure 3, the segmentation results of HSCNN and SegNet are nearly identical for the interior of the winter wheat field. SegNet, however, produces prominent errors at the edge of the field, while HSCNN does not.

Both HSCNN and SegNet use classifiers to generate the probability distribution of the classes, and consider the class with the maximum probability (max) in the distribution as the type to which the pixel belongs. Clearly, a larger difference between the max and background implies a higher separability of the pixels and more reliable results. The probability differences given by the HSCNN and SegNet model for the inner wheat and edge wheat classes are presented in Figures 4 and 5, respectively. It is clear in Figure 4 that HSCNN and SegNet lead to significant probability differences for many pixels in the interior, which demonstrates the high separability of this region and the strength of CNN. In the probability distribution in Figure 5, fewer pixels are noted as having large probability differences in both the HSCNN and SegNet; nevertheless, the number is maintained at a quite high level for the HSCNN, whereas SegNet exhibits a reduced performance.

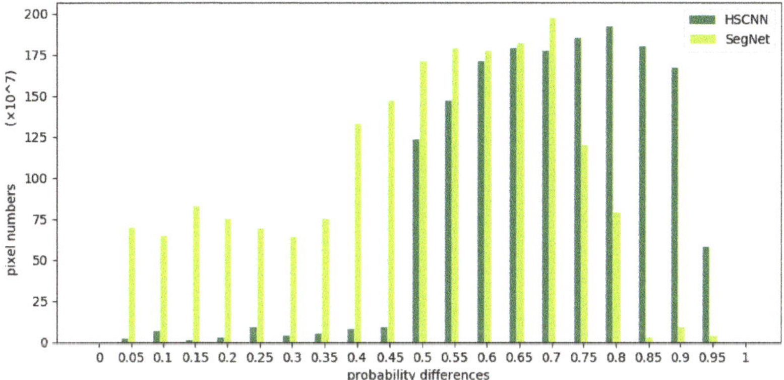

Figure 4. Distribution of the probability differences for the inner wheat pixels.

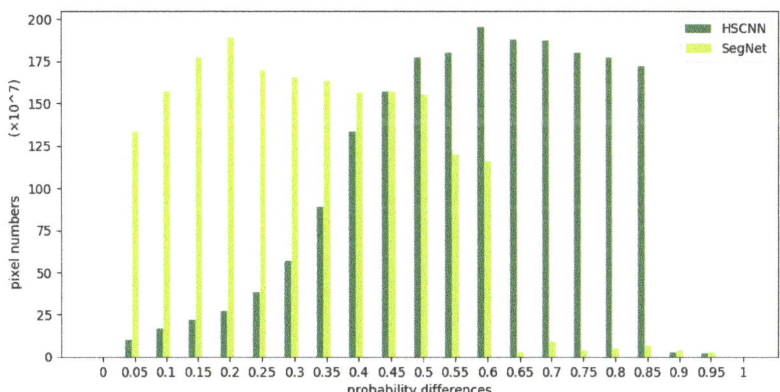

Figure 5. Distribution of the probability differences for the edge wheat pixels.

4.3. Comparison with DeepLab and Analysis

Compared with the FCN and SegNet, DeepLab has significant improvements in two aspects: (1) the deconvolution; and (2) the refinement of the boundary of the segmentation result by fully connected CRFs. These two improvements are beneficial for the segmentation of individual objects covering numerous pixels. Based on the published literature, DeepLab displays a higher segmentation accuracy at the boundary than the FCN and SegNet, because it better utilizes the detailed information contained in the image and the large-scale spatial correlation between the pixels. However, in its application to winter wheat identification, the strength of DeepLab is not fully realized, because the details within a pixel block of the winter wheat plantation do not change significantly. Therefore, less information is available to the model, and the spatial correlation within the farmlands and woods is not strong over large regions.

As mentioned in Section 4.2, the HSCNN completely utilizes the characteristics of the pixels and the spatial relation between them. Therefore, it is well adapted to the data characteristics of the winter wheat plantation. Further, it effectively avoids the deficiencies of DeepLab and ensures the accuracy of segmentation.

As in Section 4.2, the probability differences between the HSCNN and DeepLab models in the inner wheat and edge wheat class are displayed in Figures 6 and 7, respectively. It is clear in Figure 6 that both the models produce large probability differences for many pixels in the interior. In the probability distribution of Figure 7, a considerable number of pixels still display large probability differences after the HSCNN processing, whereas DeepLab shows a much poorer performance (even lower than SegNet), proving again the notion that the atrous convolution is not suitable for farmlands.

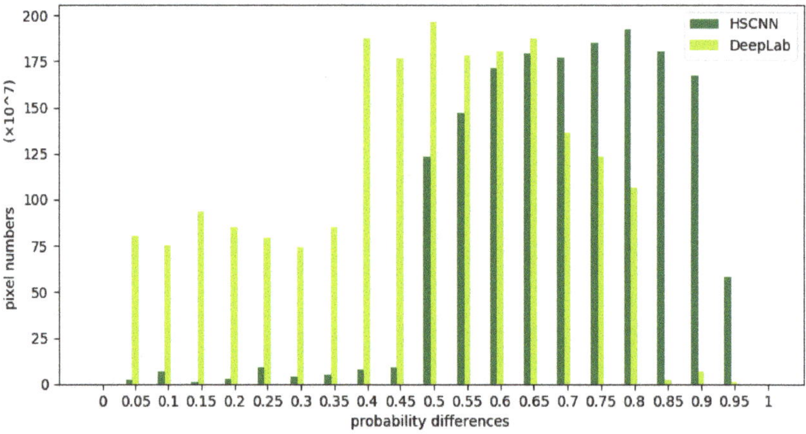

Figure 6. Distribution of the probability differences for the inner wheat pixels.

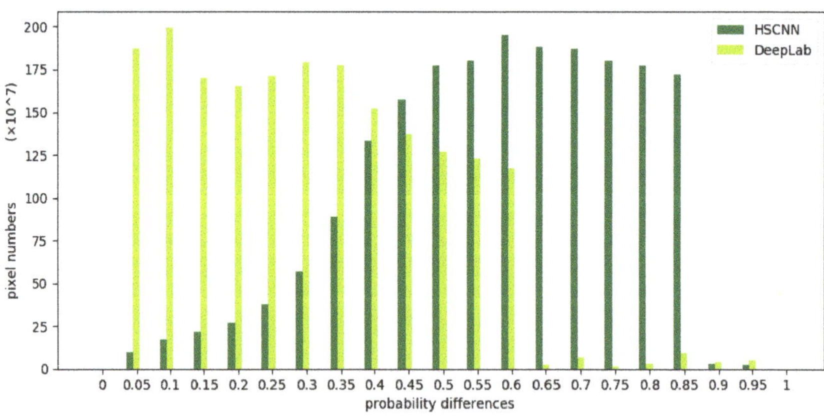

Figure 7. Distribution of the probability differences for the edge wheat pixels.

4.4. Benefits of Using the Proposed Approach to Classify Land Use

Accurate land use classification is of tremendous importance in scientific research and agriculture with the use of remote sensing data as an increasingly common practice for this purpose. Based on a CNN and taking complete advantage of the convolution in feature extraction, the design of the CNN architecture adapting to the features of the remote-sensing images is the key in land use classification by this method.

We have taken the feature difference between the edge pixel and the inner pixel in the white wheat planting area into full consideration, this significantly improve the extraction accuracy of the edge pixel. Compared with earlier research, the model presented in this paper has the following advantages.

Firstly, two types of kernels were used in the convolution of the model, which allowed the full utilization of the strength of the convolution in the extraction of spatial semantics and made appropriate use of the rich information contained in the pixels of the remote sensing images, thus achieving a more accurate segmentation.

Secondly, pooling layers were not used in the model. Although the speed of the feature aggregation was consequently reduced, the information of the exact location to which an eigenvalue corresponds was retained, thereby effectively mitigating the loss in the accuracy due to the positioning error of the deconvolution and improving the overall effect of the segmentation.

The model presented in this paper provides a solution for the edge extraction problem or the segmentation of the winter wheat plantation using GF-2 images. It has an important role to play and enhances the efficiency of the agricultural survey. This model has been utilized by the Department of Agriculture and the Meteorological Bureau of Shandong Province, China.

5. Conclusions

This paper presents a novel approach for the extraction of the winter wheat distribution from GF-2 remote-sensing images. Compared with the two typical deep learning-based approaches, the extraction accuracy is obviously improved. Our approach combines the segmentation and classification stages, taking the accuracy as the only constraint, and achieves high quality classification in an end-to-end way. The GT classes of ground objects are taken as the supervised information that guides both the feature extraction and the region generation. Taking into account the significant differences between pixel and edge pixel in the planting area, different convolution structures were used to extract the feature of edge and interior pixels, focusing on the common features in the two subclasses for more effective model training, and obtained a high resolution class prediction.

Our model is still limited in many aspects, and further improvements could be made in the following two areas: (1) The current encoder uses a relatively simple regression algorithm for encoding; thus, a regression that can express the complex relationship between the eigenvalues needs to be explored. (2) A new pooling method, which allows for expedited feature aggregation without the loss of the spatial information of the eigenvalues, should be established. We will continue our work in the future to improve the current model and obtain better classification performance.

Author Contributions: C.Z. wrote the manuscript; C.Z. and S.G. presented the direction of this study and designed the experiments; X.Y. and F.L. carried out the experiments and analyzed the results; M.Y. and Y.Z. carried out ground investigation and preprocessing; and Y.H., H.Z. and K.F. carried out sample labeling.

Funding: This work was funded by National Key R&D Program of China, grant number 2017YFA0603004; Science Foundation of Shandong, grant numbers ZR2017MD018 and ZR2016DP01; the National Science Foundation of China, grant number 41471299; and Open research project of Key Laboratory on meteorological disaster monitoring, early warning and risk management in characteristic agricultural areas of arid area, grant numbers CAMF-201701 and CAMF-201803.

Conflicts of Interest: The authors declare no conflict of interest.

References

1. Announcement of the National Statistics Bureau on Grain Output in 2017. Available online: http://www.gov.cn/xinwen/2017-12/08/content_5245284.htm (accessed on 8 December 2017).
2. Wang, L.M.; Liu, J.; Yao, B.M.; Ji, F.H.; Yang, F.G. Area change monitoring of winter wheat based on relationship analysis of GF-1 NDVI among different years. *Trans. CSAE* **2018**, *34*, 184–191. [CrossRef]
3. He, H.; Zhu, X.F.; Pan, Y.Z.; Zhu, W.Q.; Zhang, J.S.; Jia, B. Study on scale issues in measurement of winter wheat plant area by remote sensing. *J. Remote Sens.* **2008**, *1*, 168–176.
4. Zhang, J.H.; Feng, L.L.; Yao, F.M. Improved maize cultivated area estimation over a large scale combining MODIS-EVI time series data and crop phonological information. *ISPRS J. Photogramm. Remote Sens.* **2014**, *94*, 102–113. [CrossRef]
5. Wu, M.Q.; Wang, C.Y.; Niu, Z. Mapping paddy field in large areas, based on time series multi-sensors data. *Trans. CSAE* **2010**, *26*, 240–244.
6. Xu, Q.Y.; Yang, G.J.; Long, H.L.; Wang, C.C.; Li, X.C.; Huang, D.C. Crop information identification based on MODIS NDVI time-series data. *Trans. CSAE* **2014**, *30*, 134–144.
7. Becker-Reshef, I.; Vermote, E.; Lindeman, M.; Justice, C. A generalized regression-based model for forecasting winter wheat yields in Kansas and Ukraine using MODIS data. *Remote Sens. Environ.* **2010**, *114*, 1312–1323. [CrossRef]
8. Zhang, J.G.; Li, X.W.; Wu, Y.L. Object oriented estimation of winter wheat planting area using remote sensing data. *Trans. CSAE* **2008**, *24*. [CrossRef]
9. Zhu, C.M.; Luo, J.C.; Shen, Z.F.; Chen, X. Winter wheat planting area extraction using multi-temporal remote sensing data based on filed parcel characteristic. *Trans. CSAE* **2011**, *27*, 94–99.
10. Lu, L.L.; Guo, H.D. Extraction method of winter wheat phenology from time series of SPOT/VEGETATION data. *Trans. CSAE* **2009**, *25*, 174–179.
11. Jha, A.; Nain, A.S.; Ranjan, R. Wheat acreage estimation using remote sensing in tarai region of Uttarakhand. *Vegetos* **2013**, *26*, 105–111. [CrossRef]
12. Wu, M.Q.; Yang, L.C.; Yu, B.; Wang, Y.; Zhao, X.; Niu, Z.; Wang, C.Y. Mapping crops acreages based on remote sensing and sampling investigation by multivariate probability proportional to size. *Trans. CSAE* **2014**, *30*, 146–152.
13. You, J.; Pei, Z.Y.; Wang, F.; Wu, Q.; Guo, L. Area extraction of winter wheat at county scale based on modified multivariate texture and GF-1 satellite images. *Trans. CSAE* **2016**, *32*, 131–139. [CrossRef]
14. Wang, L.M.; Liu, J.; Yang, F.G.; Fu, C.H.; Teng, F.; Gao, J. Early recognition of winter wheat area based on GF-1 satellite. *Trans. CSAE* **2015**, *31*, 194–201. [CrossRef]
15. Ma, S.J.; Yi, X.S.; You, J.; Guo, L.; Lou, J. Winter wheat cultivated area estimation and implementation evaluation of grain direct subsidy policy based on GF-1 imagery. *Trans. CSAE* **2016**, *32*, 169–174. [CrossRef]
16. Wang, L.M.; Liu, J.; Yang, L.B.; Yang, F.G.; Teng, F.; Wang, X.L. Remote sensing monitoring winter wheat area based on weighted NDVI index. *Trans. CSAE* **2016**, *32*, 127–135. [CrossRef]
17. Wu, M.Q.; Huang, W.J.; Niu, Z.; Wang, Y.; Wang, C.Y.; Li, W.; Hao, P.Y.; Yu, B. Fine crop mapping by combining high spectral and high spatial resolution remote sensing data in complex heterogeneous areas. *Comput. Electron. Agric.* **2017**, *139*, 1–9. [CrossRef]
18. Fu, G.; Liu, C.J.; Zhou, R.; Sun, T.; Zhang, Q.J. Classification for high resolution remote sensing imagery using a fully convolutional network. *Remote Sens.* **2017**, *9*, 498. [CrossRef]
19. Liu, Y.D.; Cui, R.X. Segmentation of Winter Wheat Canopy Image Based on Visual Spectral and Random Forest Algorithm. *Spectrosc. Spect. Anal.* **2015**, *35*, 3480–3485.
20. Dong, Z.P.; Wang, M.; Li, D.R. A High Resolution Remote Sensing Image Segmentation Method by Combining Superpixels with Minimun Spanning Tree. *Acta Geod. Cartogr. Sin.* **2017**, *46*, 734–742. [CrossRef]
21. Basaeed, E.; Bhaskar, H.; Al-Mualla, M. Supervised remote sensing image segmentation using boosted convolutional neural networks. *Knowl.-Based Syst.* **2016**, *99*, 19–27. [CrossRef]
22. Liu, D.W.; Han, L.; Han, X.Y. High Spatial Resolution Remote Sensing Image Classification Based on Deep Learning. *Acta Opt. Sin.* **2016**, *36*, 0428001. [CrossRef]
23. Luo, B.; Zhang, L.P. Robust autodual morphological profiles for the classification of high-resolution satellite images. *IEEE Trans. Geosci. Remote* **2014**, *52*, 1451–1462. [CrossRef]

24. Li, D.R.; Zhang, L.P.; Xia, G.S. Automatic Analysis and Mining of Remote Sensing Big Data. *Acta Geod. Cartogr. Sin.* **2014**, *43*, 1211–1216. [CrossRef]

25. Chan, T.H.; Jia, K.; Guo, S.H.; Lu, J.; Zeng, Z.; Ma, Y. PCANet: A Simple Deep Learniing Baseline for Image Classification. *IEEE Trans. Image Process.* **2015**, *24*, 5017–5032. [CrossRef] [PubMed]

26. Mas, J.F.; Flores, J.J. The application of artificial neural networks to the analysis of remotely sensed data. *Int. J. Remote Sens.* **2008**, *29*, 617–663. [CrossRef]

27. Gustavo, C.V.; Bruzzone, L. Kernel-based methods for hyperspectral image classification. *IEEE Trans. Geosci. Remote Sens.* **2005**, *43*, 1351–1362.

28. Mountrakis, G.; Im, J.; Ogole, C. Support vector machines in remote sensing: A review. *ISPRS J. Photogramm. Remote Sens.* **2011**, *66*, 247–259. [CrossRef]

29. Pacifici, F.; Chini, M.; Emery, W.J. A neural network approach using multi-scale textural metrics from very high-resolution panchromatic imagery for urban land-use classification. *Remote Sens. Environ.* **2009**, *113*, 1276–1292. [CrossRef]

30. Huang, X.; Zhang, L. An SVM ensemble approach combining spectral, structural, and semantic features for the classification of high resolution remotely sensed imagery. *IEEE Trans. Geosci. Remote Sens.* **2013**, *51*, 257–272. [CrossRef]

31. Liu, C.; Hong, L.; Chen, J.; Chun, S.S.; Deng, M. Fusion of pixel-based and multi-scale region-based features for the classification of high-resolution remote sensing image. *J. Remote Sens.* **2015**, *19*, 228–239. [CrossRef]

32. Yuan, Y.; Lin, J.; Wang, Q. Hyperspectral Image Classification via Multitask Joint Sparse Representation and Stepwise MRF Optimization. *IEEE Trans. Cybern.* **2016**, *46*, 2966–2977. [CrossRef] [PubMed]

33. Xie, F.D.; Li, F.F.; Lei, C.K.; Ke, L.N. Representative Band Selection for Hyperspectral Image Classification. *ISPRS Int. J. Geo-Inf.* **2018**, *7*, 338. [CrossRef]

34. Bengio, Y. Learning deep architectures for AI. *Found. Trends Mach. Learn.* **2009**, *2*, 1–127. [CrossRef]

35. Larochelle, H.; Bengio, Y.; Louradour, J.; Lamblin, P. Exploring strategies for training deep neural networks. *J. Mach. Learn. Res.* **2009**, *10*, 1–40.

36. Jones, N. The learning machines. *Nature* **2014**, *505*, 146–148. [CrossRef] [PubMed]

37. Gao, Q.S.; Lim, S.S.; Jia, X.P. Hyperspectral Image Classification Using Convolutional Neural Networks and Multiple Feature Learning. *Remote Sens.* **2018**, *10*, 299. [CrossRef]

38. Dong, Y.; Liu, Y.N.; Lian, S.G. Automatic age estimation based on deep learning algorithm. *Neurocomputing* **2016**, *187*, 4–10. [CrossRef]

39. Taormina, R.; Chau, K.W. Data-driven input variable selection for rainfall–runoff modeling using binary-coded particle swarm optimization and Extreme Learning Machines. *J. Hydrol.* **2015**, *529*, 1617–1632. [CrossRef]

40. Liang, Z.; Shan, S.; Liu, X.; Wen, Y. Fuzzy prediction of AWJ turbulence characteristics by using typical multi-phase flow models. *Eng. Appl. Comput. Fluid Mech.* **2017**, *11*, 225–257. [CrossRef]

41. Bellary, S.A.I.; Adhav, R.; Siddique, M.H.; Chon, B.H.; Kenyery, F.; Samad, A. Application of computational fluid dynamics and surrogate-coupled evolutionary computing to enhance centrifugal-pump performance. *Eng. Appl. Comput. Fluid Mech.* **2016**, *10*, 171–181. [CrossRef]

42. Zhang, J.; Chau, K.W. Multilayer Ensemble Pruning via Novel Multi-sub-swarm Particle Swarm Optimization. *J. Univ. Comput. Sci.* **2009**, *15*, 840–858.

43. Wang, W.C.; Chau, K.W.; Xu, D.M.; Chen, X.Y. Improving forecasting accuracy of annual runoff time series using ARIMA based on EEMD decomposition. *Water Resour. Manag.* **2015**, *29*, 2655–2675. [CrossRef]

44. Zhang, S.; Chau, K.W. Dimension reduction using semi-supervised locally linear embedding for vegetation leaf classification. *Emerg. Intell. Comput. Technol. Appl.* **2009**, *5754*, 948–955.

45. Wu, C.; Chau, K.; Fan, C. Prediction of rainfall time series using modular artificial neural networks coupled with data-preprocessing techniques. *J. Hydrol.* **2010**, *389*, 146–167. [CrossRef]

46. Castelluccio, M.; Poggi, G.; Sansone, C.; Verdoliva, L. Land Use Classification in Remote Sensing Images by Convolutional Neural Networks. *arXiv*, 2015; arXiv:1508.00092.

47. Hu, F.; Xia, G.S.; Hu, J.; Zhang, L. Transferring deep convolutional neural networks for the scene classification of high-resolution remote sensing imagery. *Remote Sens.* **2015**, *7*, 14680–14707. [CrossRef]

48. Zhu, C.; Cheng, T. Research on geological hazard identification based on deep learning. In Proceedings of the 6th International Conference on Computer-Aided Design, Manufacturing, Modeling and Simulation, Busan, Korea, 14–15 April 2018. [CrossRef]

49. Wu, Z.; Zhang, Q. On combining spectral, textural and shape features for remote sensing image segmentation. *Acta Geod. Cartogr. Sin.* **2013**, *42*, 44–50.

50. Noh, H.; Hong, S.; Han, B. Learning deconvolution network for semantic segmentation. In Proceedings of the IEEE International Conference on Computer Vision, Santiago, Chile, 7–13 December 2015; pp. 1520–1528.

51. Paisitkriangkrai, S.; Sherrah, J.; Janney, P.; Hengel, V.D. Effective semantic pixel labelling with convolutional networks and conditional random fields. In Proceedings of the IEEE Conference on Computer Vision and Pattern Recognition Workshops, Boston, MA, USA, 7–12 June 2015; pp. 36–43.

52. Papandreou, G.; Kokkinos, I.; Savalle, P.A. Untangling local and global deformations in deep convolutional networks for image classification and sliding window detection. *arXiv*, 2014; arXiv:1412.0296.

53. Badrinarayanan, V.; Handa, A.; Cipolla, R. SegNet: A deep convolutional encoder-decoder architecture for robust semantic pixel-wise labelling. *arXiv*, 2015; arXiv:1505.07293.

54. Ioffe, S.; Szegedy, C. Batch normalization: Accelerating deep network training by reducing internal covariate shift. *arXiv*, 2015; arXiv:1502.03167.

55. Liu, J.; Liu, B.; Lu, H. Detection guided deconvolutional network for hierarchical feature learning. *Pattern Recognit.* **2015**, *48*, 2645–2655. [CrossRef]

56. Volpi, M.; Ferrari, V. Semantic segmentation of urban scenes by learning local class interactions. In Proceedings of the IEEE Conference on Computer Vision and Pattern Recognition Workshops, Boston, MA, USA, 7–12 June 2015; pp. 1–9.

57. Simonyan, K.; Zisserman, A. Very deep convolutional networks for large-scale image recognition. *arXiv*, 2014; arXiv:1409.1556.

58. Mittal, A.; Hooda, R.; Sofat, S. LF-SegNet: A Fully Convolutional Encoder–Decoder Network for Segmenting Lung Fields from Chest Radiographs. *Wirel. Pers. Commun.* **2018**, *101*, 511–529. [CrossRef]

59. Kendall, A.; Badrinarayanan, V.; Cipolla, R. Bayesian SegNet: Model uncertainty in deep convolutional encoder-decoder architectures for scene understanding. *arXiv*, 2015; arXiv:1511.02680.

60. Chen, L.C.; Papandreou, G.; Kokkinos, I.; Murphy, K.; Yuille, A.L. Semantic Image Segmentation with Deep Convolutional Nets and Fully Connected Crfs. *arXiv*, 2014; arXiv:1412.7062.

61. Long, J.; Shelhamer, E.; Darrell, T. Fully convolutional networks for semantic segmentation. In Proceedings of the IEEE Conference on Computer Vision and Pattern Recognition, Boston, MA, USA, 7–12 June 2015; pp. 3431–3440.

62. Längkvist, M.; Kiselev, A.; Alirezaie, M.; Loutfi, A. Classification and segmentation of satellite orthoimagery using convolutional neural networks. *Remote Sens.* **2016**, *8*, 239. [CrossRef]

63. Dolz, J.; Desrosiers, C.; Ben, A. 3D fully convolutional networks for subcortical segmentation in MRI: A large-scale study. *NeuroImage* **2018**, *170*, 456–470. [CrossRef] [PubMed]

64. Badrinarayanan, V.; Kendall, A.; Cipolla, R. SegNet: A deep convolutional encoderdecoderarchitecture for image segmentation. *IEEE Trans. Pattern Anal. Mach. Intell.* **2017**, *39*, 2481–2495. [CrossRef] [PubMed]

65. Ronneberger, O.; Fischer, P.; Brox, T. U-Net: Convolutional Networks for Biomedical Image Segmentation. *arXiv*, 2015; arXiv:1505.04597.

66. Chen, L.C.; Papandreou, G.; Kokkinos, I.; Murphy, K.; Yuille, A.L. Deeplab: Semantic image segmentation with deep convolutional nets, atrous convolution, and fully connected crfs. *arXiv*, 2016; arXiv:1606.00915.

67. Lin, H.N.; Shi, Z.W.; Zou, Z.X. Maritime Semantic Labeling of Optical Remote Sensing Images with Multi-Scale Fully Convolutional Network. *Remote Sens.* **2017**, *9*, 480–501. [CrossRef]

68. Visin, F.; Ciccone, M.; Romero, A.; Kastner, K.; Cho, K.; Bengio, Y.; Matteucci, M.; Courville, A. Reseg: A recurrent neural network-based model for semantic segmentation. *arXiv*, 2016; arXiv:1511.07053.

69. Statistical Yearbook of Shandong Province. Available online: http://www.stats-sd.gov.cn/col/col6279/index.html (accessed on 27 October 2017).

70. Haralick, R.M.; Shanmugam, K. Texture features for image classification. *IEEE Trans. Syst. Man Cybern.* **1973**, *SMC-3*, 610–621. [CrossRef]

 *applied sciences*

Article

A New Weighting Approach with Application to Ionospheric Delay Constraint for GPS/GALILEO Real-Time Precise Point Positioning

Tianjun Liu [1], Jian Wang [2],*, Hang Yu [1,3], Xinyun Cao [4,5] and Yulong Ge [6]

[1] NASG Key Laboratory of Land Environment and Disater Monitoring, China University of Mining and Technology, Xuzhou 221116, China; tjliu_cumt@126.com (T.L.); yhecit@163.com (H.Y.)

[2] School of Geomatics and Urban Spatial Information, Beijing University of Civil Engineering and Architecure, Beijing 100044, China

[3] School of Environment Science and Spatial Informatics, China University of Mining and Technology 1Daxue Road, Xuzhou 221116, China

[4] School of Geography, NanJing Normal University, No.1 Wenyuan Road, Xianlin University District, Nanjing 210023, China; xycao@whu.edu.cn

[5] School of Geodesy and Geomatics, Wuhan University, 129 Luoyu Road, Wuhan 430079, China

[6] University of Chinese Academy of Sciences, Beijing 100049, China; geyulong15@mails.ucas.ac.cn

*   Correspondence: wjiancumt@163.com

Received: 8 November 2018; Accepted: 5 December 2018; Published: 7 December 2018

Abstract: The real-time precise point positioning (RT PPP) technique has attracted increasing attention due to its high-accuracy and real-time performance. However, a considerable initialization time, normally a few hours, is required in order to achieve the proper convergence of the real-valued ambiguities and other estimate parameters. The RT PPP convergence time may be reduced by combining quad-constellation global navigation satellite system (GNSS), or by using RT ionospheric products to constrain the ionosphere delay. But to improve the performance of convergence and achieve the best positioning solutions in the whole data processing, proper and precise variances of the observations and ionospheric constraints are important, since they involve the processing of measurements of different types and with different accuracy. To address this issue, a weighting approach is proposed by a combination of the weight factors searching algorithm and a moving-window average filter. In this approach, the variances of ionospheric constraints are adjusted dynamically according to the principle that the sum of the quadratic forms of weighted residuals is the minimum, and the filter is applied to combine all epoch-by-epoch weight factors within a time window. To evaluate the proposed approach, datasets from 31 Multi-GNSS Experiment (MGEX) stations during the period of DOY (day of year) 023-054 in 2018 are analyzed with different positioning modes and different data processing methods. Experimental results show that the new weighting approach can significantly improve the convergence performance, and that the maximum improvement rate reaches 35.9% in comparison to the traditional method of priori variance in the static dual-frequency positioning mode. In terms of the RMS (Root Mean Square) statistics of positioning errors calculated by the new method after filter convergence, the same accuracy level as that of RT PPP without constraints can be achieved.

Keywords: real-time precise point positioning; convergence time; ionospheric delay constraints; precise weighting

---

1. Introduction

With the rapid development of Global Navigation Satellite System (GNSS), real-time precise point positioning (RT PPP) with integer ambiguities resolution is possible thanks to the real-time precise orbits, clocks, and code/phase biases products of satellites, provided freely by the Center National d'Etudes Spatiales (CNES) [1]. These real-time products in the CLK93/CLK92 stream have been broadcasted by the CNES real-time analysis center since 14/09/2014 [2]. Based on these products, the integer property of user ambiguities can be recovered to improve positioning accuracy and reliability. However, ambiguities and other parameters (e.g., tropospheric delay) need a few hours to achieve the proper convergence, even with good satellite geometries and observation quality [3]. Therefore, how to reduce the RT PPP convergence time has become one of the key issues for further improving RT PPP performance [4].

Three different methods have been proposed to reduce the convergence time in RT PPP. One is to fix ambiguity parameters to integer values. Many researchers have demonstrated that ambiguity resolution can improve the PPP in terms of both precision and convergence performance [5–7]. The second method is to utilize multi-frequency and/or other GNSS constellation observations [8–10]. With the application of observations from multi-frequency or other satellite positioning systems, the PPP convergence time can be reduced due to high measurement redundancy and improved degrees of freedom [11]. The third method is to apply RT ionospheric or tropospheric correction products [7,12,13]. Accuracy better than 10 cm can be achieved in a few minutes with dual-frequency signals by using precise ionospheric and tropospheric correction products [13]. Compared to the other two methods, ionosphreric products are the more effective in terms of reducing the convergence time [14]. Thanks to the standardized RT message of Vertical Total Electron Content (VTEC) models in CLK93/CLK92 stream from CNES [15], the convergence time of RT PPP can be reduced significantly by using an uncombined functional model with RT ionospheric correction products [3]. When RT ionospheric products are applied as an extra constraint in the RT PPP, proper and precise variances of the raw observations and ionospheric constraint are important as they involve the processing of measurements of different types whose quality is different in terms of residual errors. Currently, a priori variances are mainly used to determine the weights of observation and ionospheric constraint in the whole process of data processing. However, this method may not be precise, especially when the accuracy of ionospheric products is uncertain, and it will lead to unreliable positioning results. To address this issue, a weighting approach is proposed by combining a weight factor searching algorithm with a moving-window average filter. Weight factors are utilized to adjust the priori variances of ionospheric constraint and the moving-window average filter is introduced to improve the precision and reliability of the weight factors. In this paper, we adopt this method in the Precise Point Positioning With Integer and Zero-difference Ambiguity Resolution Demonstrator (PPP-WIZARD) by the CNES. Both static and kinematic experiments in single-frequency and dual-frequency cases are conducted to assess the performance of the new weighting approach. The results indicate that this new method can significantly reduce convergence time and improve reliability of positioning solutions in RT PPP.

The rest of the paper is organized as follows: Firstly, the uncombined functional model with ionospheric constraint for GPS/GALILEO RT PPP is presented. Secondly, the RT ionospheric products from CNES are compared with post-processing GIM (global ionospheric map) products from CODE (Center for Orbit Determination in Europe) agencies. Afterwards, the weight factors searching algorithm with a moving-window average filter is proposed. Finally, the converging performance and positioning accuracy of the proposed weighting approach are assessed in RT PPP by different methods and different positioning modes.

2. Approach of GPS/GALILEO RT PPP with Ionospheric Constraint

2.1. Function Models of GPS/GALILEO RT PPP

In the RT PPP model, satellite clock and position are calculated by broadcast ephemerides with RT precise orbit/clock corrections products from CNES. The uncombined raw observable model for GPS/GALILEO PPP can be written as [3]:

$$
\begin{aligned}
P_{r,i}^{sys,s} &= D_{r,i}^{sys,s} + \gamma_i^{sys,s} I_1^{sys,s} + b_{r,P_i}^{sys} - b_{P_i}^{sys,s} + cdt_r^{sys} - cdt^{sys,s} \\
\lambda_i^{sys,s} L_{r,i}^{sys,s} &= D_{r,i}^{sys,s} - \gamma_i^{sys,s} I_1^{sys,s} + b_{r,L_i}^{sys} - b_{L_i}^{sys,s} + cdt_r^{sys} - cdt^{sys,s} + \lambda_i^{sys,s} W^{sys,s} - \lambda_i^{sys,s} N_i^{sys,s}
\end{aligned}
\tag{1}
$$

where $P_{r,i}^{sys,s}$ and $L_{r,i}^{sys,s}$ are the raw pseudorange (or code) and phase measurements at frequency f_i ($i = 1,2$) to the system "*sys*" (for GPS and GALILEO) for the specific satellite s and receiver r. The pseudorange measurements are expressed in meters, while phase measurements are expressed in cycles. $\lambda_i^{sys,s}$ denotes the wavelength of phase observations at frequency f_i ($i = 1,2$). $D_{r,i}^{sys,s}$ is the geometrical propagation distances between the satellite and receiver phase centers at each frequency including troposphere elongation, relativistic effects, etc. $I_1^{sys,s}$ is the slant ionospheric delay at frequency f_1 in meters. $\gamma_i^{sys,s} = (f_1/f_i)^2$ is a frequency-dependent scale factor. This elongation varies with the inverse of the square of the frequency and with opposite signs between phase and pseudorange [3]. $b_{P_i}^{sys,s}$ and $b_{L_i}^{sys,s}$ are the frequency-dependent uncalibrated code delyas (UCD) and uncalibrated phase delyas (UPD) of the satllite. b_{r,P_i}^{sys} and b_{r,L_i}^{sys} are the UCD and UPD of receiver. $W_r^{sys,s}$ is the contribution of the wind-up effect in cycles. $N_i^{sys,s}$ stands for the undifferenced ambiguities for each frequency f_i in cycle. $dt^{sys,s}$ and dt_r^{sys} are satellite and receiver code clock offsets. In order to eliminate the UCD and UPD of satellite, the code biases $\overline{b}_{P_i}^{s}$ and phase biases $\overline{b}_{L_i}^{s}$ from CNES caster are applied to the raw observable model, the reparameterization of (1) can be written as [6]:

$$
\begin{aligned}
P_{r,1}^{sys,s} + \overline{b}_{P_1}^{sys,s} &= \overline{D}_r^{sys,s} + c\overline{d}t_r^{sys} + \gamma_1^{sys,s} \overline{I}_1^{sys,s} \\
P_{r,2}^{sys,s} + \overline{b}_{P_2}^{sys,s} &= \overline{D}_r^{sys,s} + c\overline{d}t_r^{sys} + \gamma_2^{sys,s} \overline{I}_1^{sys,s} \\
\lambda_1^{sys,s} \cdot \left(L_{r,1}^{sys,s} + \overline{b}_{L_1}^{s} \right) &= \overline{D}_r^{sys,s} + c\overline{d}t_r^{sys} - \gamma_1^{sys,s} \overline{I}_1^{sys,s} + \lambda_1^{sys,s} W^s + \lambda_1^{sys,s} \overline{N}_1^{sys,s} \\
\lambda_2^{sys,s} \cdot \left(L_{r,2}^{sys,s} + \overline{b}_{L_2}^{s} \right) &= \overline{D}_r^{sys,s} + c\overline{d}t_r^{sys} - \gamma_2^{sys,s} \overline{I}_1^{sys,s} + \lambda_2^{sys,s} W^s + \lambda_2^{sys,s} \overline{N}_W^{sys,s}
\end{aligned}
\tag{2}
$$

With

$$
\left\{
\begin{aligned}
\overline{I}_1^{sys,s} &= I_1^{sys,s} + \beta_{12}^{sys} DCB_{r,12}^{sys} \\
c \cdot \overline{d}t_r^{sys} &= c \cdot dt_r^{sys} + b_{r,P_{IF}}^{sys} \\
\lambda_1^{sys} \cdot \overline{N}_1^{sys,s} &= \lambda_1^{sys} \cdot \left(N_1^{sys,s} + b_{r,L_1}^{sys} \right) - b_{r,P_{IF}}^{sys} + \beta_{12}^{sys} DCB_{r,12}^{sys} \\
\lambda_2^{sys} \cdot \left(\overline{N}_W^{sys,s} + \overline{N}_1^{sys,s} \right) &= \lambda_2^{sys} \cdot \left(N_2^{sys,s} + b_{r,L_2}^{sys} \right) - b_{r,P_{IF}}^{sys} + \gamma_2^{sys} \beta_{12}^{sys} DCB_{r,12}^{sys} \\
DCB_{r,12}^{sys} &= b_{r,P_1}^{sys} - b_{r,P_2}^{sys} \qquad b_{r,P_{IF}}^{sys} = \alpha_{12}^{sys} \cdot b_{r,P_1}^{sys} + \beta_{12}^{sys} \cdot b_{r,P_1}^{sys} \\
\alpha_{12}^{sys} &= f_1^2 / \left(f_1^2 - f_2^2 \right) \qquad \beta_{12}^{sys} = -f_2^2 / \left(f_1^2 - f_2^2 \right)
\end{aligned}
\right.
\tag{3}
$$

where $\overline{D}_r^{sys,s}$ is the geometric distance with satellite orbit, satellite code clock and tropospheric delay T_w fixed. $DCB_{r,12}^{sys}$ is differential code bias (*DCB*) of receiver. Note that both the ionospheric delay $I_1^{sys,s}$ and $DCB_{r,12}^{sys}$ of receiver are perfectly correlated, and they are estimated as lumped terms $\overline{I}_1^{sys,s}$ in the traditional uncombination PPP functional model. The estimated parameter vector can be expressed as:

$$
X = \left[x, c \cdot \overline{d}t_r, T_w, \overline{I}_1^{sys,s}, \overline{N}_1^{sys,s}, \overline{N}_W^{sys,s} \right]
\tag{4}
$$

If the ionosphere products are introduced as pseudo-measurements to constraint slant the ionosphere delay, the *DCB* of the receiver estimated parameter needs to be added in (4) to keep consistency between the estimated ionospheric delay and ionospheric products. The estimated parameter vector in the ionospheric delay constraint PPP can be expressed as:

$$X = \left[x, c \cdot \bar{d}t_r, DCB_{r,12}^{sys}, T_w, I_1^{sys,s}, \overline{N}_1^{sys,s}, \overline{N}_W^{sys,s} \right] \tag{5}$$

2.2. RT Ionosphere Products and Post-Processing GIM Products

The RT ionospheric products using spherical harmonic expansions are broadcasted with an updating rate of 60 s in CLK92/CLK93 RT stream from the CNES caster [3]. A spherical harmonic expansion allows a global and continuous model of the ionosphere, but can also be applied to regional representation [3]. Based on these products and ionospheric mapping function, the slant TEC of each satellite can be used for positioning in real time. In contrast to RT ionospheric products, the post-processing GIM products from CODE agency are maps that contain a globally-distributed grid [16]. The spatial resolution of latitude and longitude is 2.5° and 5° in these maps, respectively, and the map is updated at an interval of one or two hours [16]. When the GNSS users obtain the vertical total electron content value $VTEC$ from the ionosphere products, the slant ionosphere delay can be computed as follows:

$$I_{product}^{sys,s} = m \cdot \left(40.3 \cdot \frac{VTEC}{f_1^2} \right) \tag{6}$$

$$m = \cos^{-1} \left(\arcsin \left(\frac{R_E}{R_E + H} \cdot \sin z \right) \right) \tag{7}$$

where $I_{product}^{sys,s}$ is the ionospheric delay of the pseudo-observable, m is the ionospheric mapping function as expressed in Equation (7), z is the zenith angle from the satellites to the receiver, R_E is the radius of the Earth (m) and H is the height of the ionosphere shell (m), where the value of H is 450 km for the products of CNES and CODE agency. Since the post-processing GIM products form CODE agency, that are computed by the stations distributed globally, has the highest accuracy (about 2~4 TECU) compared with those of other agencies [17], we use the post-processing GIM product as a reference to assess the RT ionosphere products. Figure 1 shows the slant TEC of GPS/GALILEO satellites for stations GMSD on day of year (DOY) 045, 2018. The slant TEC variation of RT ionosphere products and post-processing GIM products are quite consistent for a whole day, but there are obvious offsets with different products at some times, and the maximum offsets can be up to about 10 TECU, which is equivalent to a range error of about 1.6 m of the f_1 frequency of GPS. If the imprecise and unreliable variances of ionospheric constraint are used, these offsets have a negative impact on the accuracy of positioning.

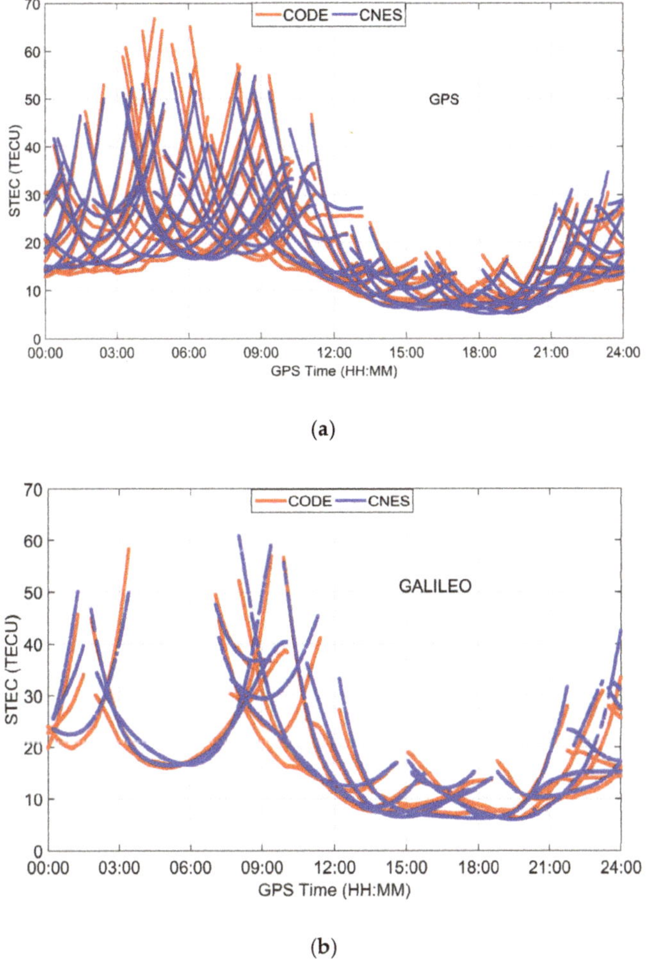

(a)

(b)

Figure 1. Slant TEC of GPS (**a**) and GALILEO (**b**) derived from different agencies at station GMSD.

2.3. Weight Factors Searching Algorithm with Moving-Window Average Filter

In order to achieve fast convergence and high-accuracy positioning solutions after convergence, a weighting approach is presented which combines a weight factors searching algorithm and a moving-window average filter. The weight factors searching algorithm is similar to the method of Helmert variance component estimation; it is based on the principle that the post-fit weighted sum residuals of squares is the minimum. A moving-window average filter is applied to improve the precision and reliability of this searching algorithm.

When RT ionospheric products are introduced, the variances matrix of measurement errors R and measurement error vector V in the Extended Kalman Filter (EKF) can be written as follows [18]:

$$R = \begin{bmatrix} R^2_{observa} & 0 \\ 0 & R^2_{constra} \end{bmatrix} \tag{8}$$

$$V = \begin{bmatrix} V_{observa} & V_{constra} \end{bmatrix}^T \tag{9}$$

where the subscript *"observa"* and *"constra"* denote the raw observation (pseudorange and phase measurements) and ionospheric constraints, respectively. The variance matrix of observation errors $R^2_{observa}$ depends on the elevation E^s of satellite s and can be expressed as follows [19]:

$$R^2_{observa} = \begin{bmatrix} \sigma_0^2/\sin^2 E^1 & 0 & 0 \\ 0 & \ddots & 0 \\ 0 & 0 & \sigma_0^2/\sin^2 E^s \end{bmatrix} \tag{10}$$

In the Equation (10), the value of the standard deviation σ_0 for GPS/GALILEO are set as 1 m and 0.01 m for the pseudorange and phase observations, respectively. Different from the variance matrix of observation, the variance of ionospheric constraints $R^2_{constra}$ would be expressed as a product of a weight factor λ and an initial variance matrix $\widetilde{R}^2_{constra}$, as follows:

$$R^2_{constra} = \lambda \cdot \widetilde{R}^2_{constra} = \lambda \cdot \begin{bmatrix} \sigma^2_{constra,1} & 0 & 0 \\ 0 & \ddots & 0 \\ 0 & 0 & \sigma^2_{constra,s} \end{bmatrix} \tag{11}$$

The initial ionospheric constraints $\sigma^2_{constra}$ are set as the same as σ_0^2 of pseudorange observations (this value of $\sigma^2_{constra}$ is suggested in PPP-WIZARD ducumentation). The computation procedure of the weight factors searching algorithm comprises the following steps:

(1) Assign an initial weight factor ($\lambda = 1$) to the variance of ionospheric constraints $R^2_{constra}$.
(2) Initialize the variance matrix of measurement errors R by using Equation (10) and (11).
(3) Compute post-fit measurement error vector V after performing the EKF.
(4) Compute the post-fit weighted sum residuals of squares $V^T R^{-1} V$.
(5) Update the weight factor $\lambda = \lambda + 1$ ($\lambda \leq T$), where T is a search space, which will be determined through the case studies later in the paper.

Repeat steps (2)–(5) to find the optimal weight factor $\lambda_{optimal}$ that satisfies the following equation:

$$V^T R^{-1} V = \min \tag{12}$$

After step (6) is fulfilled, the final variance matrix of measurement errors R can be determined by using the optimal weight factor $\lambda_{optimal}$.

In order to determine a suitable search space T of the weight factors searching algorithm, different positioning modes are conducted with ionospheric constraints. Figure 2 displays the relationships between the weight factors and the post-fit weighted sum-squared residuals. It is clearly seen that varying trend of post-fit sum-squared residuals are quite consistent in different positioning modes. Due to the imprecise $\widetilde{R}^2_{constra}$ used in RT PPP, the post-fit weighted sum residuals of squares is large when the variance matrix of ionospheric constraints is small. As the weight factor increases, the ionospheric constraints are weakened, and the post-fit weighted sum residuals of squares decreases gradually and then tends to be convergent. It was found that different positioning modes have the same characteristics, the post-fit weighted sum residuals of squares was close to stable when the distance between the point and the origin of the coordinate is the shortest. Therefore, to improve the efficiency of the searching algorithm, the Equation (12) can be replaced:

$$D_i = \sqrt{i^2 + \left(V_i^T R_i^{-1} V_i\right)^2} = \min \tag{13}$$

where subscript *"i"* denotes $i - th$ search of searching algorithm, D_i is the distance from the $i - th$ point to origin of the coordinate. If the distance D_{i+1} is greater than distance D_i, the search will be stopping and the λ_{i+1} of $(i + 1) - th$ searching will be taken as the optimal weight factor.

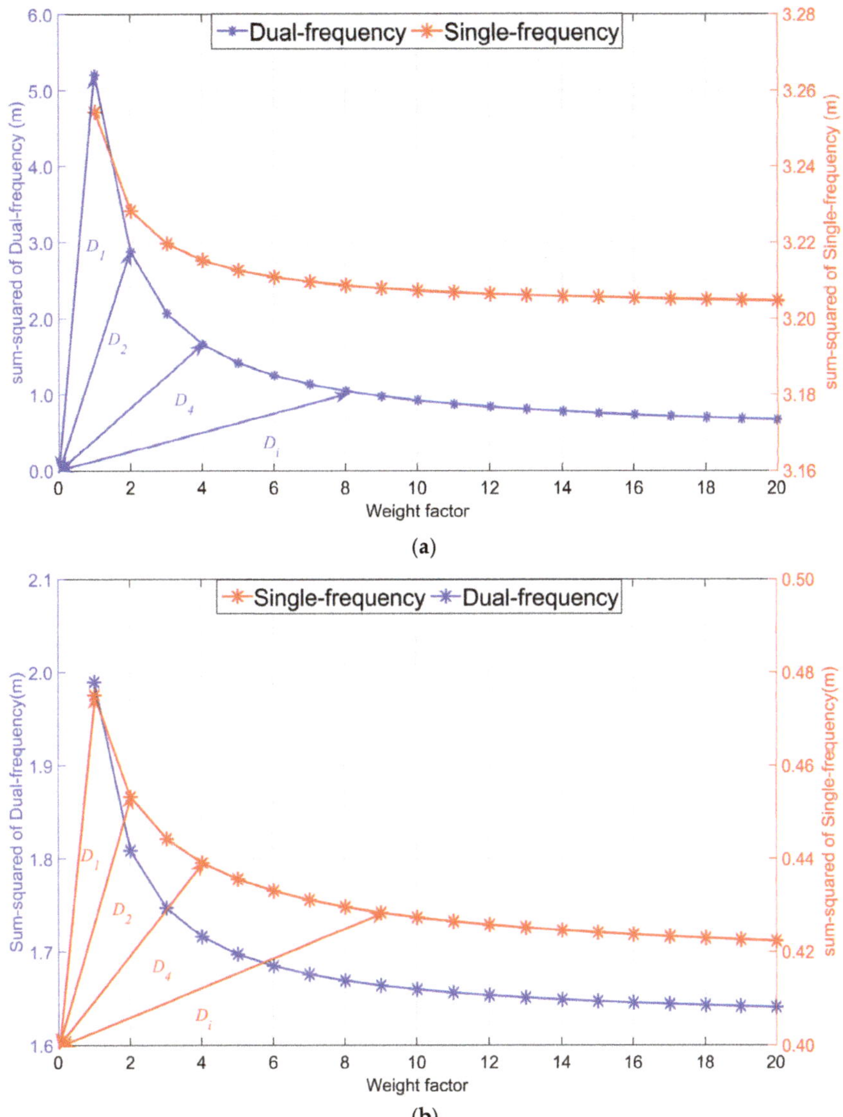

Figure 2. Relationships between the weight factors and the post-fit weighted sum-squared residuals for kinematic (**a**) and static (**b**).

Since the epoch-by-epoch determines the optimal weight factor, $\lambda_{optimal}$ is not always available due to the limited number of the visible satellites and low redundancy of single-epoch data. To further enhance the reliability of the solutions, a moving-window average filter is applied to determine the smoothed optimal weight factor as follows:

$$\lambda_{MW}(k) = \frac{1}{n} \sum_{i=k-n+1}^{k} \lambda(i) \tag{14}$$

where n is the size of the smoothing window in average filter, $\lambda_{MW}(k)$ is a smoothed optimal weight factor over n multiple epochs within a time window from epoch $(k - n + 1)$ to epoch k. A long window size would tend to blend these changes into the previous open sky conditions, while a short one would change quickly to the conditions but would be prone to larger errors [20]. Since the observation data of MGEX station is obtained under good operating conditions, a suitable window size is set as 10. The effectiveness of this window size will be verified through experiments later in the paper.

3. Results and Analysis

3.1. Data Description and Process Schemes

In order to test the proposed weighting approach for GPS/GALILEO RT PPP with ionospheric constraints, both static and kinematic experiments are conducted at single- and dual-frequencies. Three data processing methods, as listed in Table 1, were established to evaluate the impact on convergence time and positioning with different variances of ionospheric constraint. The first is using the GPS/GALILEO raw observations without ionospheric constraints, and the estimated parameter vector can be expressed as Equation (4). The second method is to determine the weight between observations and ionospheric constraints using priori variances by Equation (10). The third method is to determine the weight using the proposed approach. The estimated parameter vector of the second and the third method can be expressed as Equation (5). As shown in Figure 3, the observation data of 31 MGEX stations for 30-days (23 January 2018 to 23 February 2018) are selected. All these stations can track GPS/GALILEO satellites. The station coordinates were estimated every 30 s. The strategy of RT GPS/GALILEO PPP is summarized in Table 2. The RT precise orbit/clock correction and code/phase biases products in CLK92 stream from the CNES caster are used. Moreover, the correction of receiver PCO (phase center offset) and PCV (Phase Center Variations) for GALILEO are assumed to be the same as that of GPS.

Table 1. List of Data Processing Methods.

Modes	Details
Without constraint	GPS/GALILEO observations without ionospheric constraints
Priori variance	GPS/GALILEO observations+ionospheric constraints with a priori variance
New method	GPS/GALILEO observations+ionospheric constraints with proposed method

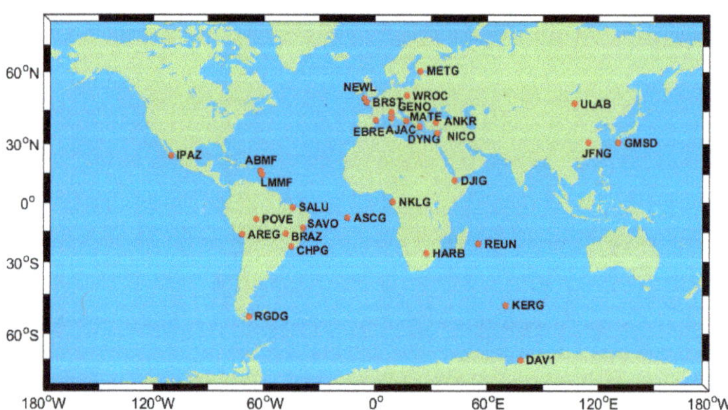

Figure 3. The distribution of MGEX stations in the experiment.

Table 2. Strategies for RT GPS/GALILEO PPP.

Item	Setting
Observations	Raw pseudo range and phase observations
Frequency	GPS: L1/L2; GALILEO: E1/E5
Estimator	Extended Kalman filter
Elevation cutoff	$10°$
Sampling offset	30 s
Observations weight	Elevation dependent weighting; 0.01 m and 1 m for GPS/GALILEO phase and pseudo range observables in zenith direction;
Phase windup	Phase polarization effects applied [21]
Attitude law	Nominal attitude for GPS and GALILEO
Station displacement	Solid Earth tides, ocean tide loading and pole tides [22]
A priori Troposphere delay	Saastamoinen model and Niell mapping function [23]
Zenith wet tropospheric delay	Estimated as random walk ($1 \times 10^{-8} \mathrm{m^2/s}$)
Ionosphere	Estimated as random walk processes ($1 \times 10^{-6} \mathrm{m^2/s}$);
Station coordinate	Estimated as constant/white noise ($60^2 \mathrm{\ m^2}$) in static/kinematic modes
Receiver clock	Estimated as white noise for each GNSS system
Satellite antenna PCO and PCV	PCV and PCO values for GPS/GALILEO were corrected with igs14.atx;
Receiver antenna PCO and PCV	Corrected by igs14.atx; Applied the same values as GPS to GALILEO;
Phase ambiguities	Float solution [24,25]

3.2. Data Processing and Analysis

In GPS/GALILEO RT PPP, the position filter is considered to have converged when the absolute values of the positioning errors in East and North directions reach 0.1 m and keep within 0.1 m for consecutive 20 epochs (ten minutes) in dual-frequency static and kinematic cases. Given that the RT PPP errors in the single-frequency case are larger than those of dual-frequency case, the definition of the convergence of the single-frequency cases is enlarged to 0.3 m in the East and North directions. The convergence time is the period from the first epoch to the converged epoch. We compare the station coordinates calculated from different methods with coordinates supplied by IGS weekly solutions and calculate the RMS in three directions of them.

3.2.1. The Static RT PPP with Different Data Processing Methods

Figure 4a shows the positioning errors of static RT PPP with dual-frequency for stations ANRK on DOY 045, 2018. We can see that the result of three data processing methods is very similar when the filter converges to stable values. But there is great difference in convergence time with different processing methods. As seen from the blue curves, the convergence time of GPS/GALILEO RT PPP is about 116 min. After the ionospheric products are introduced for positioning with priori variance, it is clear that the convergence time can be reduced distinctly, especially in the UP direction. Compared with the priori variance method, the weights calculated by using new method are more reliable, and thus, convergence time can be further reduced; it only takes 26 min to achieve vertical accuracy of better than 0.1 m. The number of visible GPS/GALILEO satellites is shown in Figure 4b, together with the PDOP (position dilution of precision). The average number of visible GPS and GALILEO satellites are 9.3 and 5.4, respectively, leading to an average PDOP of 1.7. Figure 4c shows the weight factors of ionospheric constraints for different processing methods. The weight factors solutions vary in a range of 2.0 to 8.0 by using the new method. Since the priori variance for the ionospheric constraints does not correspond to reality, the weight factors of the new method are larger than that of a priori variance when the filter is not convergent; this is the reason why the convergence performance can be improved by using the new method.

To test the effectiveness of the new method in static single-frequency RT PPP, an experiment was conducted for station METG on DOY 051, 2018. Figure 4d shows the time series of single-frequency RT PPP positioning errors. The convergence performance of the without-constraints method is the worst of the three methods, especially in the East and Up direction, and its convergence time is up to 106 min.

When ionospheric products are introduced as a pseudo-observable to constrain ionosphere delay in RT PPP, the convergence time can be reduced greatly. Using the new method, the convergence time is further reduced to 37 min and the positioning solutions are more precise than those of a priori variance. Figure 4f provides the weight factors for different processing methods. Due to the small weight factors calculated by the new method during the process of filter convergence, the contribution of ionospheric constraints is proper, and thus, convergence time can be reduced. The reason why positioning accuracy is high after filter convergence is that the weight factors calculated by the new method is larger than that of the priori variance method, and the contribution of low-accuracy (compared to carrier-phase measurements) RT ionospheric products is small. Table 3 summarizes the RMS of positioning error and convergence time for static single- and dual-frequency GPS/GALILEO RT PPP with three processing methods. The RMS computations are based on the position errors without considering the process of the filter convergence. From the Table 3, we can see that three methods have similar positioning accuracies after filter convergence in the dual-frequency case, but the positioning accuracy of the priori variance method becomes slightly worse than new methods in the single-frequency case. Compared to the other two methods, the proposed approach significantly improves convergence performance in static dual- and single-frequency cases. The convergence time improvement rate of the new method refers to the that of other two methods which are listed in the right column of "convergence time", in which the largest improvement rate is 77%.

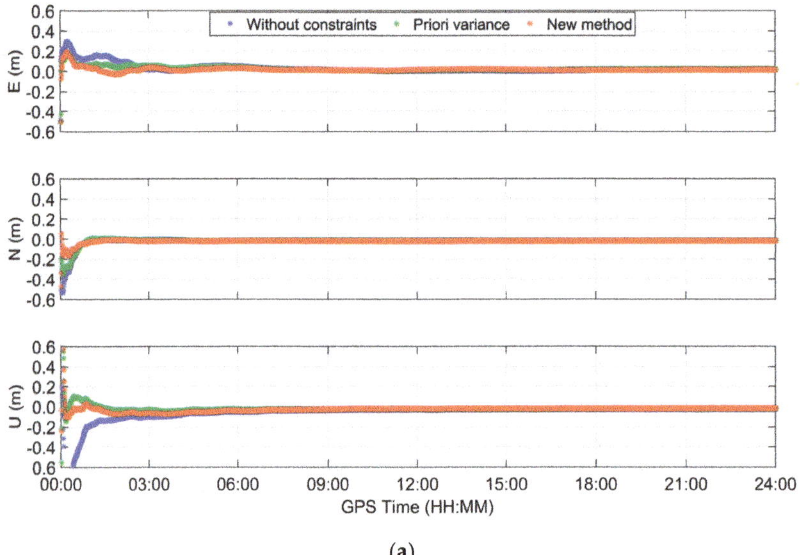

(a)

Figure 4. *Cont.*

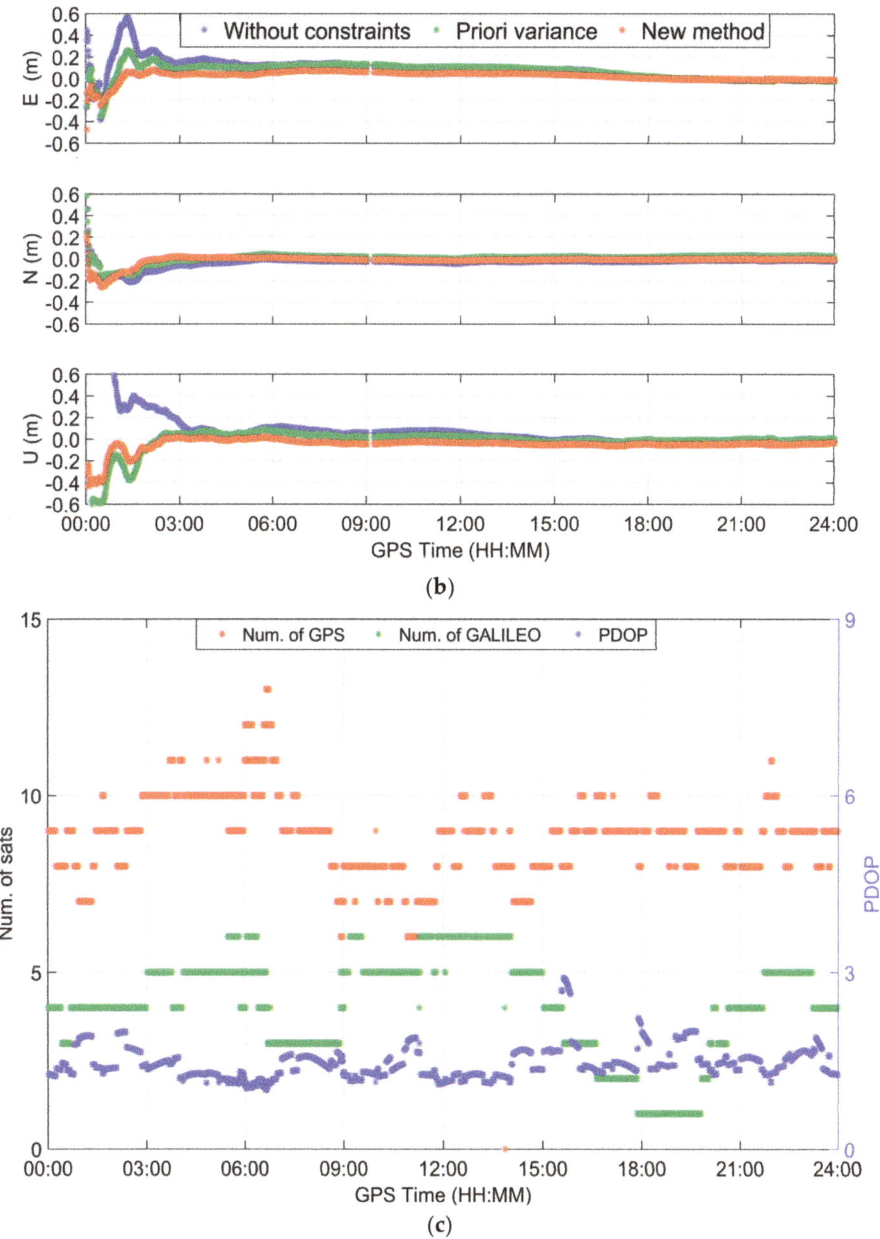

Figure 4. *Cont.*

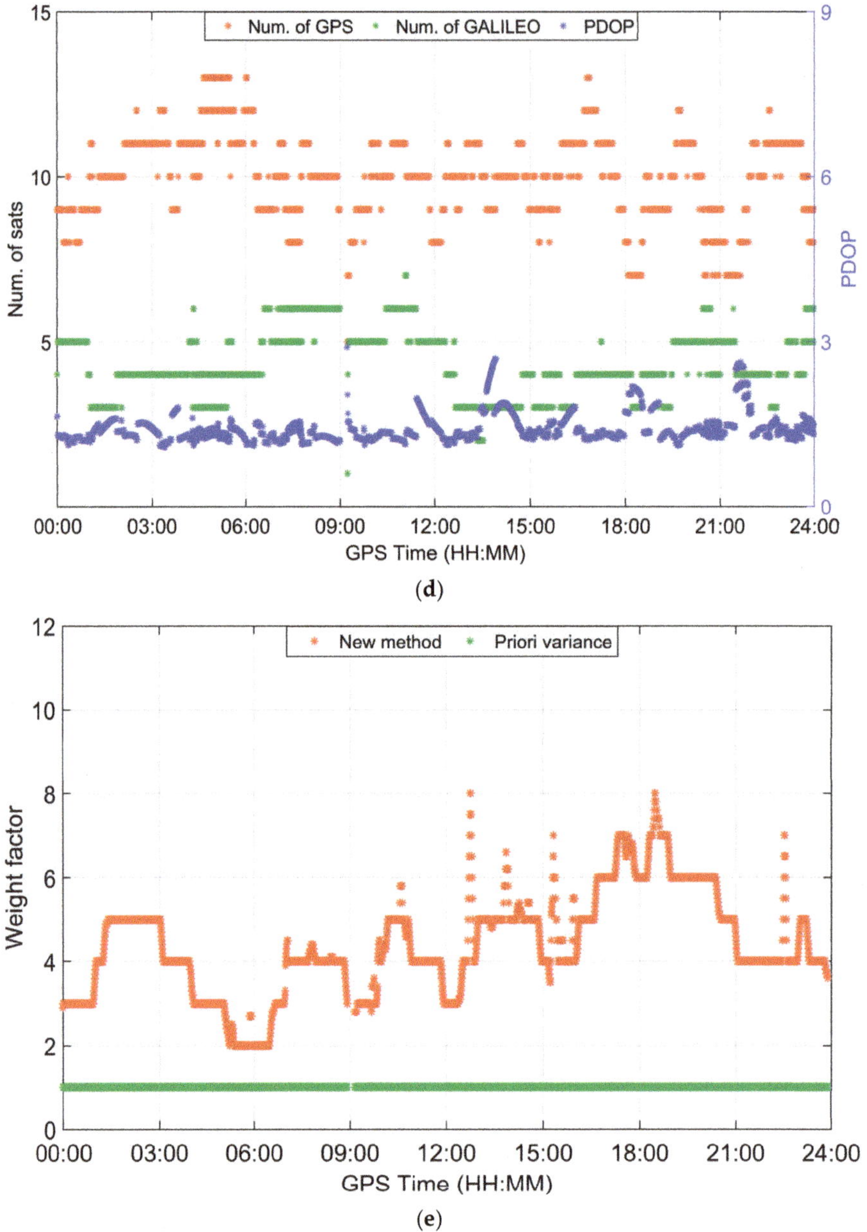

(**d**)

(**e**)

Figure 4. *Cont.*

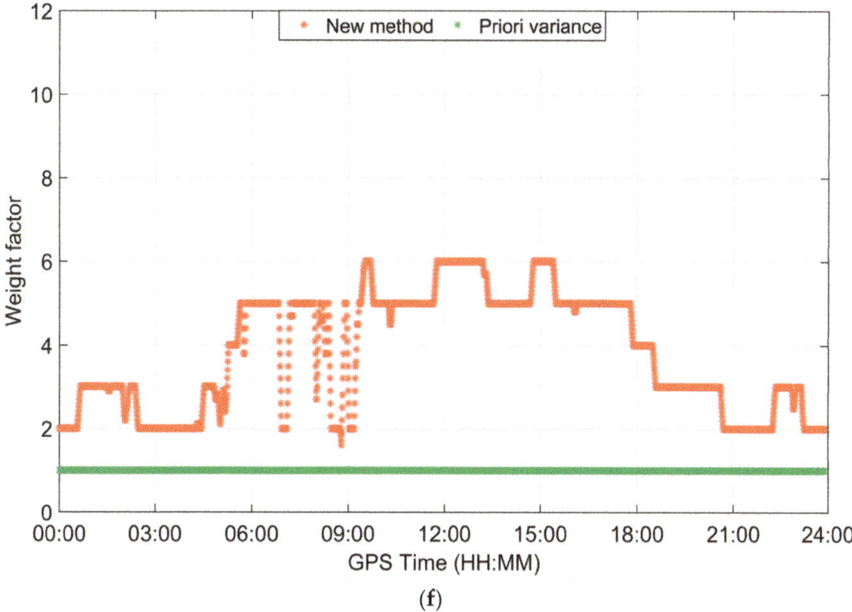

Figure 4. Static positioning errors (a,b), number of satellites, PDOP (c,d), weight factors (e,f) of dual-frequency RT PPP by station ANRK (**a,c,e**) and single-frequency RT PPP by station METG (**b,d,f**).

Table 3. Static RT PPP for three processing methods.

Frequency	Methods	Convergence Time (min)	E (m)	N (m)	U (m)
	Without constraint	116 (77%)	0.0033	0.0038	0.0346
Dual-frequency	A priori variance	50 (48%)	0.0030	0.0039	0.0349
	New method	26	0.0019	0.0038	0.0336
	Without constraint	106 (65%)	0.0102	0.0748	0.0832
Single-frequency	A priori variance	52 (28%)	0.0129	0.0848	0.0732
	New method	37	0.0092	0.0687	0.0514

In order to ensure a reliable value of the window size in the proposed weighting approach, Figure 5 provides the convergence time of dual- and single-frequencies using different window sizes of 1, 5, 10, 15, and 20. As a result, the varying trend of convergence times is quite consistent in dual- and single-frequency cases, the improvement of the convergence time is less significant when the window size is increased from 10 to 20. This suggests that the window size of 10 is suitable, which will be applied for the rest of our data analysis.

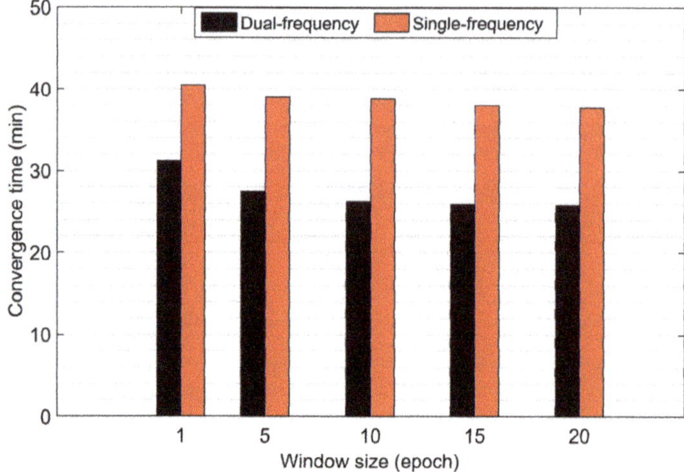

Figure 5. Convergence time using the proposed weighting approach with different smoothing window sizes for dual-frequency and single-frequency METG station.

3.2.2. The Kinematic RT PPP with Different Data Processing Methods

Different from static RT PPP positioning, the station coordinates will be estimated epoch-by-epoch in kinematic positioning. To assess the performance of the new weighting approach for single- and dual-frequency kinematic RT PPP, the datasets at MAT1 and WROC station on DOY 047, 2018 are conducted in the three processing methods.

The kinematic GPS/GALILEO RT PPP solutions with different methods are illustrated in Figure 6a,d for the dual- and single-frequency cases, respectively. The PDOP and number of satellites for MAT1 and KRGG station are provided in Figure 6b,e. Figure 6c,f shows the weight factors based on the new method and priori variance. We can find that the weight factors calculated by the new method are increased gradually when the positioning filter converges to a stable value, resulting in high-accuracy positioning solutions. Similar to the static RT PPP, we calculated the statistics of positioning accuracy and convergence times in different methods; these are given in Table 4. It is noted that both single- and dual-frequency convergence performance are improved after adding ionospheric constraints. Using our proposed approach, the RT PPP solution can converge within 15 min and 20 min for dual- and single-frequency cases, respectively, and the largest improvement rate can reach 73%. In terms of the RMS statistics of positioning errors, using the new method can achieve nearly the same positioning performance compared to the method without constraints. But using the priori variance method, obvious offsets exist in three directions after filter convergence, especially in the single-frequency case, which is caused by unreasonable weight in the ionospheric constraints.

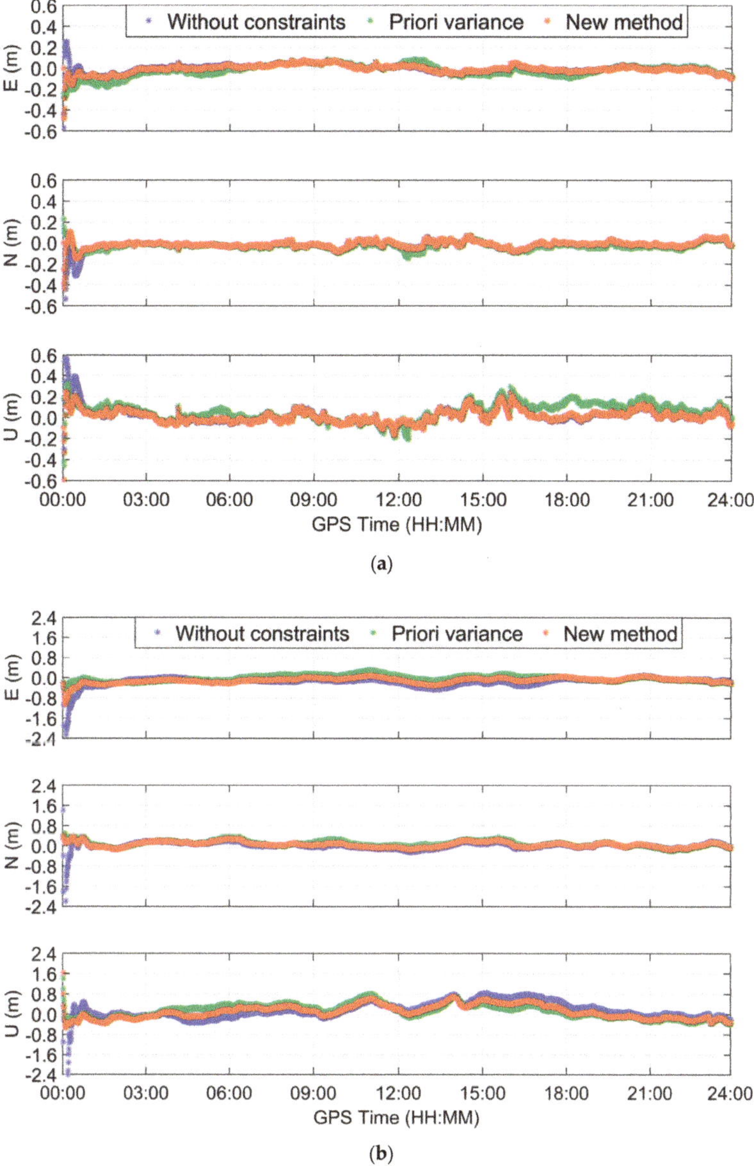

(**a**)

(**b**)

Figure 6. *Cont.*

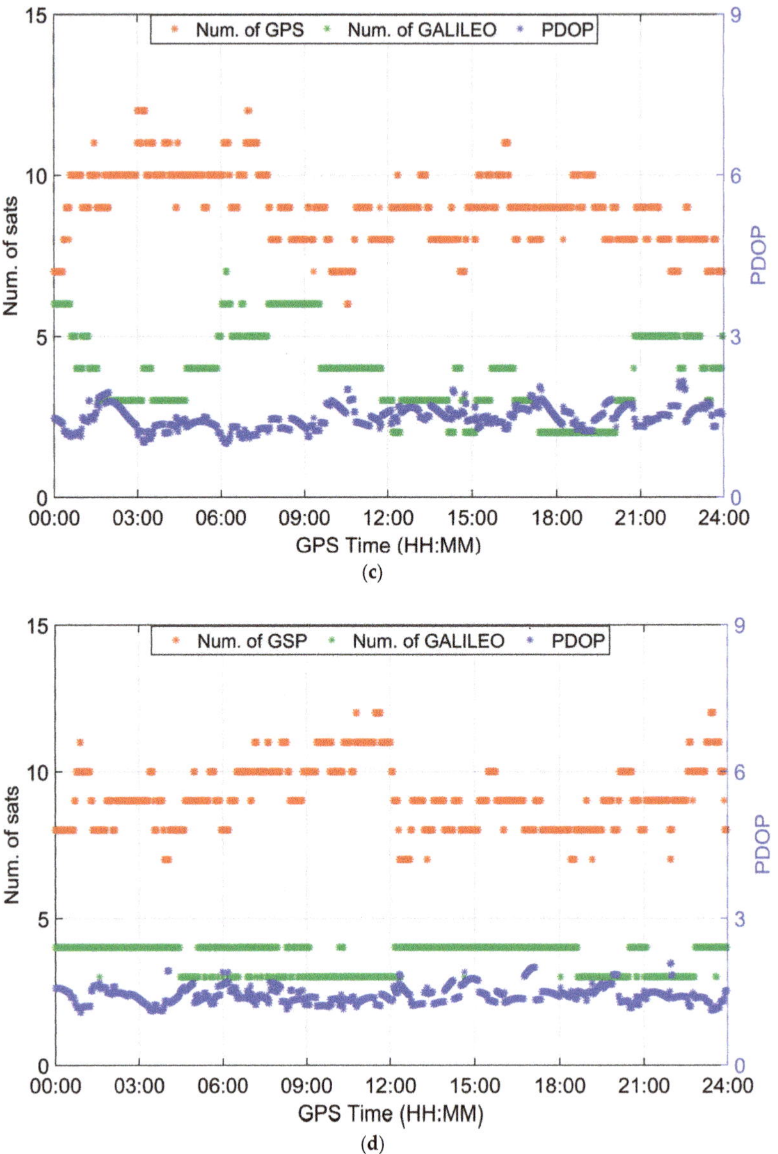

(c)

(d)

Figure 6. *Cont.*

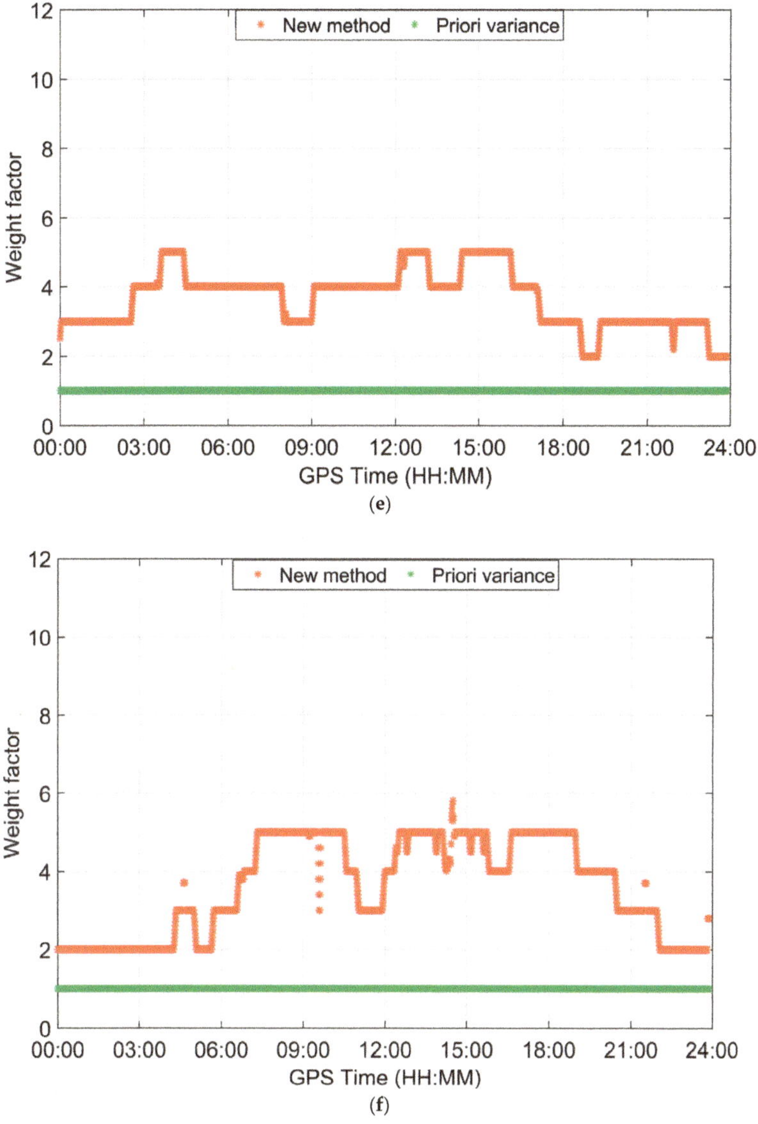

Figure 6. The kinematic positioning errors (a,b), number of satellites, PDOP (c,d), weight factors (e,f) of dual-frequency by station MAT1 (**a,c,e**) and single-frequency by station KRGG (**b,d,f**).

Table 4. Kinematic RT PPP for three processing methods.

Frequency	Methods	Convergence Time (min)	E (m)	N (m)	U (m)
	Without constraint	56 (73%)	0.0682	0.0821	0.1186
Dul-frequency	A priori variance	38 (32%)	0.0852	0.0798	0.1634
	New method	15	0.0568	0.0825	0.1183
	Without constraint	61 (67%)	0.1152	0.1007	0.3257
Single-frequency	A priori variance	31 (35%)	0.1655	0.1534	0.3760
	New method	20	0.1062	0.1186	0.3009

3.2.3. Convergence Performance and Positioning Accuracy Assessment

To further assess the convergence time and positioning accuracy in RT PPP using our proposed approach, the method of priori variance and without constraints, as listed in Table 1, are compared with a weight factors searching algorithm using datasets collected at 31 MGEX stations over 30 consecutive days from 23 January 2018 to 23 February 2018. A total of 11,160 sets of results are used to derive a statistical estimate on the convergence time as well as the positioning accuracy. The definition of the convergence time is the same as that described in Section 3.2. The distribution of the 11,160 sets of convergence times is plotted in minutes in Figure 7. It is observed that the performance of convergence can be improved in all positioning modes by using ionospheric constraints, but only to a slight degree. Using our proposed weighting approach, the convergence time of RT PPP is significantly decreased. The percentage of position solutions converging within 20 min is up to about 54%, 39%, 52%, and 50% in the dual-frequency static, dual-frequency kinematic, single-frequency static, and single-frequency kinematic positioning modes, respectively. The statistical results in terms of mean convergence time are also given in Figure 7. According to the mean values, the improvement of the new method on the convergence time is about 35.9%, 25.9%, 20.4%, and 25.2% over the method of priori variance in four positioning modes, respectively.

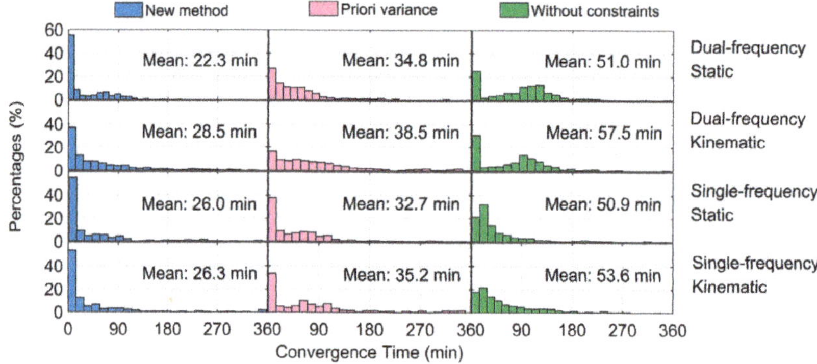

Figure 7. Distribution of convergence time for dual-frequency static, dual-frequency kinematic, single-frequency static, and single-frequency kinematic using datasets at 31 MGEX stations over thirty days.

During the converging procedure of the RT PPP position filter, the size of position errors over time can also reflect the converging speed of the position filter [17]. Figure 8 illustrates the RMS statistical values of vertical and horizontal for all stations at the beginning of 30 min (left) and all day (right). Since the position solutions in the first two hours are still in the converging stage, they are not used for the accuracy statistics of all day. From the statistical results of 30 min, it is noted that positioning errors in four positioning modes sharply decrease after using RT ionospheric products at the beginning of the filter processing, especially in the single-frequency case. There is no denying that RT ionospheric products play an important role in accelerating filter convergence. But the method of priori variance exhibits slightly worse performance than the other two methods in the RMS statistical values of all day, as shown in Figure 8 (right), which indicates the negative impact of unreliable weighting on the convergence position accuracy. In terms of the RMS statistical values of all day, our proposed weighting approach can achieve the same precision as the method without constraints.

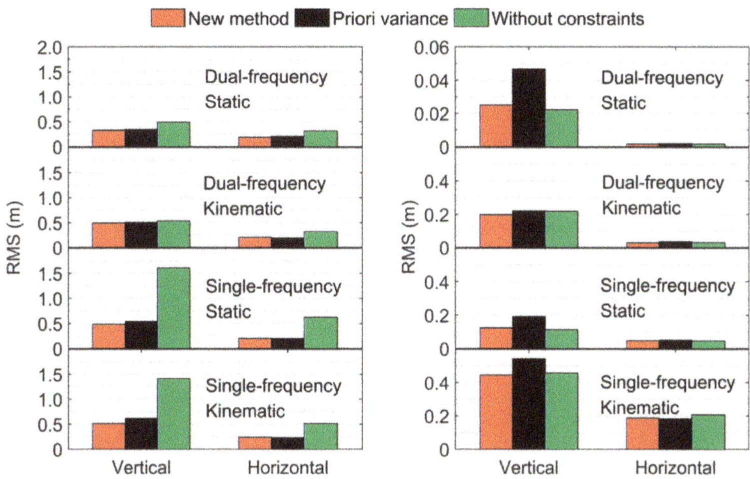

Figure 8. RMS statistics of positioning errors for different RT PPP processing methods at the beginning of 30 min (**left**) and all day (**right**).

4. Conclusions

RT PPP can provide centimeter accuracy-level solutions based on real-time precise products, for orbits, clocks, and code/phase biases, provided by CNES. Nevertheless, a significant challenge for the RT PPP to achieve high-accuracy position solutions is its long convergence time, i.e., up to a few hours. Thanks to the standardization of RT message of VTEC models in CLK92/CLK93 stream from CNES, the convergence time of RT PPP can be reduced by using an uncombined functional model with RT ionospheric correction products. In order to significantly reduce the convergence time and achieve the high-accuracy positioning solutions after filter convergence, the proper weight of ionospheric constraints are important.

To solve this issue, a weight factors searching algorithm with a moving-window average filter is proposed. This approach is similar to the method of Helmert variance component estimation; it searches for the optimal weight of ionospheric constraints according to the principle that the sum of the quadratic forms of weighted residuals is the minimum, and makes good use of the weight information at previous epochs. Datasets collected at 31 MGEX stations on 30 consecutive days are exploited to evaluate the proposed approach. Both static and kinematic experiments have been carried out in dual- and single-frequency, and the statistical results indicate that the new method significantly improves the performance of RT PPP convergence. The maximum improvement reaches 35.9% in comparison to the method of priori variance. By using the new method, the final positioning accuracy is not affected by the accuracy of RT ionosphere products, and the same accuracy as that of without constraints can be achieved. Overall, our proposed weighting approach can not only accelerate the convergence at the beginning of filter processing, but can also achieve high-accuracy position solutions after filter convergence. Future work will include the application of the proposed weighting approach to multi-GNSS combinations with tropospheric constraints.

Author Contributions: T.L., J.W. and H.Y. came up with the idea of the constrain method, T.L. and X.C. analyzed the data, T.L. and Y.G. conceived and designed the experiments. T.L. wrote the paper.

Funding: The Fundamental Research Funds for the Central Universities with a grant number as 2015XKMS051.

Acknowledgments: The authors gratefully acknowledge Center National d'Etudes Spatiales (CNES) and IGS Multi-GNSS Experiment (MGEX) for providing GNSS data and precise products.

Conflicts of Interest: The authors declare no conflict of interest.

References

1. Junior, O. *Definition and Implementation of a New Service for Precise GNSS Positioning*; UNESP: Sao Paulo, Brazil, 2018.
2. Abdi, N.; Ardalan, A.A.; Karimi, R.; Rezvani, M.-H. Performance assessment of multi-GNSS real-time PPP over Iran. *Adv. Space Res.* **2017**, *59*, 2870–2879. [CrossRef]
3. Laurichesse, D.; Privat, A. An open-source PPP client implementation for the CNES PPP-WIZARD demonstrator. In Proceedings of the ION GNSS+ 2015, Tampa, FL, USA, 14–18 September 2015; pp. 15–18.
4. Geng, J.; Meng, X.; Dodson, A.H.; Ge, M.; Teferle, F.N. Rapid re-convergences to ambiguity-fixed solutions in precise point positioning. *J. Geod.* **2010**, *84*, 705–714. [CrossRef]
5. Pan, L.; Xiaohong, Z.; Fei, G. Ambiguity resolved precise point positioning with GPS and BeiDou. *J. Geod.* **2016**, *91*, 25–40. [CrossRef]
6. Laurichesse, D.; Langley, R. Handling the Biases for Improved TripleFrequency PPP Convergence. *GPS World* **2015**, *26*, 49–54.
7. Li, X.; Ge, M.; Zhang, H.; Nischan, T.; Wickert, J. The GFZ real-time GNSS precise positioning service system and its adaption for COMPASS. *Adv. Space Res.* **2013**, *51*, 1008–1018. [CrossRef]
8. Cai, C.; Gao, Y. Modeling and assessment of combined GPS/GLONASS precise point positioning. *GPS Solut.* **2012**, *17*, 223–236. [CrossRef]
9. Lou, Y.; Zheng, F.; Gu, S.; Wang, C.; Guo, H.; Feng, Y. Multi-GNSS precise point positioning with raw single-frequency and dual-frequency measurement models. *GPS Solut.* **2015**, *20*, 849–862. [CrossRef]
10. Li, P.; Zhang, X. Integrating GPS and GLONASS to accelerate convergence and initialization times of precise point positioning. *GPS Solut.* **2014**, *18*, 461–471. [CrossRef]
11. Cai, C.; Gao, Y.; Pan, L.; Zhu, J. Precise point positioning with quad-constellations: GPS, BeiDou, GLONASS and Galileo. *Adv. Space Res.* **2015**, *56*, 133–143. [CrossRef]
12. Li, W.; Teunissen, P.; Zhang, B.; Verhagen, S. Precise Point Positioning Using GPS and Compass Observations. In Proceedings of the China Satellite Navigation Conference (CSNC) 2013 Proceedings, Wuhan, China, 15–17 May 2013; pp. 367–378. [CrossRef]
13. Zhang, H.; Hao, J.; Xie, J. The Weight Matrix Determination of Ionospheric Delay Constraintfor MultiGNSS Precise Point Positioning Using Raw Observations. *Acta Geod. Cartogr. Sin.* **2018**, *47*, 308–315. [CrossRef]
14. Juan, J.M.; Hernández-Pajares, M.; Sanz, J.; Ramos-Bosch, P.; Aragon-Angel, A.; Orus, R. Enhanced Precise Point Positioning for GNSS Users. *IEEE Trans. Geosci. Remote* **2012**, *50*, 4213–4222. [CrossRef]
15. RTCM Standard 10403.2. *Differential GNSS (Global Navigation Satellite Systems) Services-Version 3*; Radio Technical Commission for Maritime Services: Arlington, VA, USA, 2013.
16. Schaer, S. IONEX: The ionosphere Map EXchange Format Version 1. In Proceedings of the Igs Analysis Center Workshop, Darmstadt, Germany, 9–11 February 1998.
17. Cai, C.; Gong, Y.; Gao, Y.; Kuang, C. An Approach to Speed up Single-Frequency PPP Convergence with Quad-Constellation GNSS and GIM. *Sensors* **2017**, *17*, 1302. [CrossRef] [PubMed]
18. Hernandez-Pajares, M.; Juan, J.M.; Sanz, J.; Ramos-Bosch, P.; Rovira-Garcia, A.; Salazar, D. The ESA/UPC GNSS-Lab tool (gLAB): An advanced multipurpose package for GNSS data processing. In Proceedings of the Satellite Navigation Technologies and European Workshop on GNSS Signals and Signal Processing, Noordwijk, The Netherlands, 8–10 December 2011; pp. 1–8.
19. Gerdan, G.P. A comparison of four methods of weighting double difference pseudorange measurements. *Surveyor* **1995**, *40*, 60–66. [CrossRef]
20. Cai, C.; Pan, L.; Gao, Y. A Precise Weighting Approach with Application to Combined L1/B1 GPS/BeiDou Positioning. *J. Navig.* **2014**, *67*, 911–925. [CrossRef]
21. Wu, J.T.; Wu, S.C.; Hajj, G.A.; Bertiger, W.I.; Lichten, S.M. Effects of antenna orientation on GPS carrier phase. *Astrodynamics* **1991**, *1993*, 1647–1660.
22. Petit, G.; Luzum, B. IERS Conventions. Bureau International des Poids et Mesures Sevres (France). 2010. Available online: https://www.iers.org/IERS/EN/Publications/TechnicalNotes/tn36.html (accessed on 1 July 2017).
23. Niell, A.E. Global mapping functions for the atmosphere delay at radio wavelengths. *J. Geophys. Res.-Solid Earth* **1996**, *101*, 3227–3246. [CrossRef]

24. Teunissen, P. A Comparision of TCAR, CIR and LAMBDA GNSS Ambiguity Resolution. In Proceeding of the 15th International Technical Meeting of the Satellite Division of the Institute of Navigation, Portland, OR, USA, 24–27 September 2001; pp. 2799–2808.
25. Ge, Y.; Zhou, F.; Sun, B.; Wang, S.; Shi, B. The Impact of Satellite Time Group Delay and Inter-Frequency Differential Code Bias Corrections on Multi-GNSS Combined Positioning. *Sensors* **2017**, *17*, 602. [CrossRef] [PubMed]

Article

Landslide Susceptibility Modeling Using Integrated Ensemble Weights of Evidence with Logistic Regression and Random Forest Models

Wei Chen [1,2,3,*,†], **Zenghui Sun** [1,2,†] **and Jichang Han** [1,2,*]

[1] Key Laboratory of Degraded and Unused Land Consolidation Engineering, Ministry of Land and Resources of China, Xi'an 710075, China

[2] Shaanxi Provincial Land Engineering Construction Group Co. Ltd., Xi'an 710075, China; sunzenghui061@126.com

[3] College of Geology & Environment, Xi'an University of Science and Technology, Xi'an 710054, Shaanxi, China

* Correspondence: chenwei0930@xust.edu.cn (W.C.); hanjc_sxdj@126.com (J.H.)

† These authors contribute equally to this paper.

Received: 19 November 2018; Accepted: 27 December 2018; Published: 4 January 2019

Abstract: The main aim of this study was to compare the performances of the hybrid approaches of traditional bivariate weights of evidence (WoE) with multivariate logistic regression (WoE-LR) and machine learning-based random forest (WoE-RF) for landslide susceptibility mapping. The performance of the three landslide models was validated with receiver operating characteristic (ROC) curves and area under the curve (AUC). The results showed that the areas under the curve obtained using the WoE, WoE-LR, and WoE-RF methods were 0.720, 0.773, and 0.802 for the training dataset, and were 0.695, 0.763, and 0.782 for the validation dataset, respectively. The results demonstrate the superiority of hybrid models and that the resultant maps would be useful for land use planning in landslide-prone areas.

Keywords: landslide; weights of evidence; logistic regression; random forest; hybrid model

---

1. Introduction

Landslides are common geological hazards caused by multiple factors including landform [1,2], geological evolution [3], groundwater [4], land use type [5], precipitation [6,7], irrigation [8], earthquake [9], engineering construction [10], and climate change [11–13]. To avoid casualties caused by landslides and guarantee the stable development of mountainous areas, it is critical to determine a control and prevention scheme for landslides in a region. Generally, regional landslide susceptibility maps are beneficial to mitigate the effects of landslide hazards.

At present, various methods have been proposed and introduced into landslide susceptibility mapping. The existing modeling approaches can be put into two categories: qualitative approaches and quantitative approaches [14,15]. In recent years, conventional qualitative approaches have been gradually abandoned by many researchers due to the risk that expert opinion can make the results stray from objective reality [16]. Compared with qualitative approaches, quantitative approaches are mainly based on the hidden information of objective data instead of subjective experience. Additionally, quantitative approaches mainly include traditional mathematical statistic methods, deterministic models, and some state-of-the-art machine learning algorithms.

For traditional statistical methods, the probability-frequency ratio (FR) [17,18], weight of evidence (WoE) [19,20], statistical index (SI) [21,22], index of entropy (IoE) [23,24], certainty factors (CF) [25–27], evidential belief function (EBF) [28–30], and logistic regression (LR) [31,32] models have been extensively adopted in landslide susceptibility mapping. However, one limitation for all traditional statistical methods is that some hypotheses exist [33]. In deterministic models, the detail characteristics of slopes are necessary to construct the calculation model [34]. Although deterministic models conform to basic physical laws of landslide, these models are not very suitable for regional landslide susceptibility assessments due to the complex process of modeling and computing [34].

In the past decade, with the rise of machine learning and data mining, a number of relevant algorithms have been developed for landslide susceptibility zonation [35–39]. For instance, the logistic regression model (LRM), artificial neural network (ANN), support vector machine (SVM), and decision tree (DT) were the top four machine learning algorithms in landslide susceptibility mapping during the period of 2005–2016 [16]. It is clear that machine learning algorithms improve the prediction accuracy of regional landslide occurrence, but the generalization performance of single classifiers still needs to be promoted [40]. In this way, a series of ensemble approaches have recently become more and more popular in geo-hazard susceptibility mapping [37,41–43].

In terms of ensemble approaches, several single classifiers have been combined using ensemble frameworks including random subspace [44], random forest [45,46], Bagging [47], AdaBoost [48], MultiBoost [49], and so on [37,50–52]. Currently, some novel ensemble techniques have been proposed and applied in landslide susceptibility assessment, flood susceptibility mapping, and groundwater potential analysis [41,53,54]. Additionally, the excellent performance of ensemble algorithms on predictive ability and generalization capacity has also been proven. For example, Kadavi et al. [55] compared four ensemble-based machine learning models (AdaBoost, LogitBoost, Multiclass Classifier, and Bagging) with the traditional frequency ratio model (FRM) in the task of landslide susceptibility mapping. Furthermore, the results demonstrated that all of the AUC values of the four ensemble-based machine learning models were higher than that of FRM. In addition, many scholars preferred to construct ensemble learning models by integrating machine learning algorithms with bivariate statistical models because some of the hypotheses of the conventional models can be weakened through hybrid models [56]. Meanwhile, part of the merits of bivariate statistical models and machine learning models can remain by integrating together. Weights of evidence models, as a classic bivariate statistical approach, can calculate the weights of various categories of a conditioning factor based on sturdy mathematical theories [57]. Furthermore, the weights of evidence models can be integrated with other machine learning approaches to reveal the hidden correlations between different conditioning factors and landslide occurrence. Therefore, in the present study, based on GIS tools, the integrated ensemble weights of evidence with logistic regression and random forest models were employed to map landslide susceptibility, and the results were compared and analyzed quantitatively by receiver operating characteristic curves (ROC) and area under the curve (AUC).

2. Study Area

The study area was located in Shaanxi Province, China (Figure 1) where the average annual temperature is 14.2 °C, the average annual rainfall is 909.8 mm, and the evaporation is 1537.1 mm. Topographically, the study area is part of the Qinba Mountain. The general trend is high in the south and low in the north. Elevation ranges from 442 m to 2410 m above sea level, with an average elevation of 1171 m. Slope angles in the study area are in the range of 0 to 70°. Most of the slope angles are in classes of 10–20° (29.27%), followed by 20–30° (26.29%), 0–10° (23.64%), and 30–40° (14.99%). Only 5.81% of slope angles are higher than 40°.

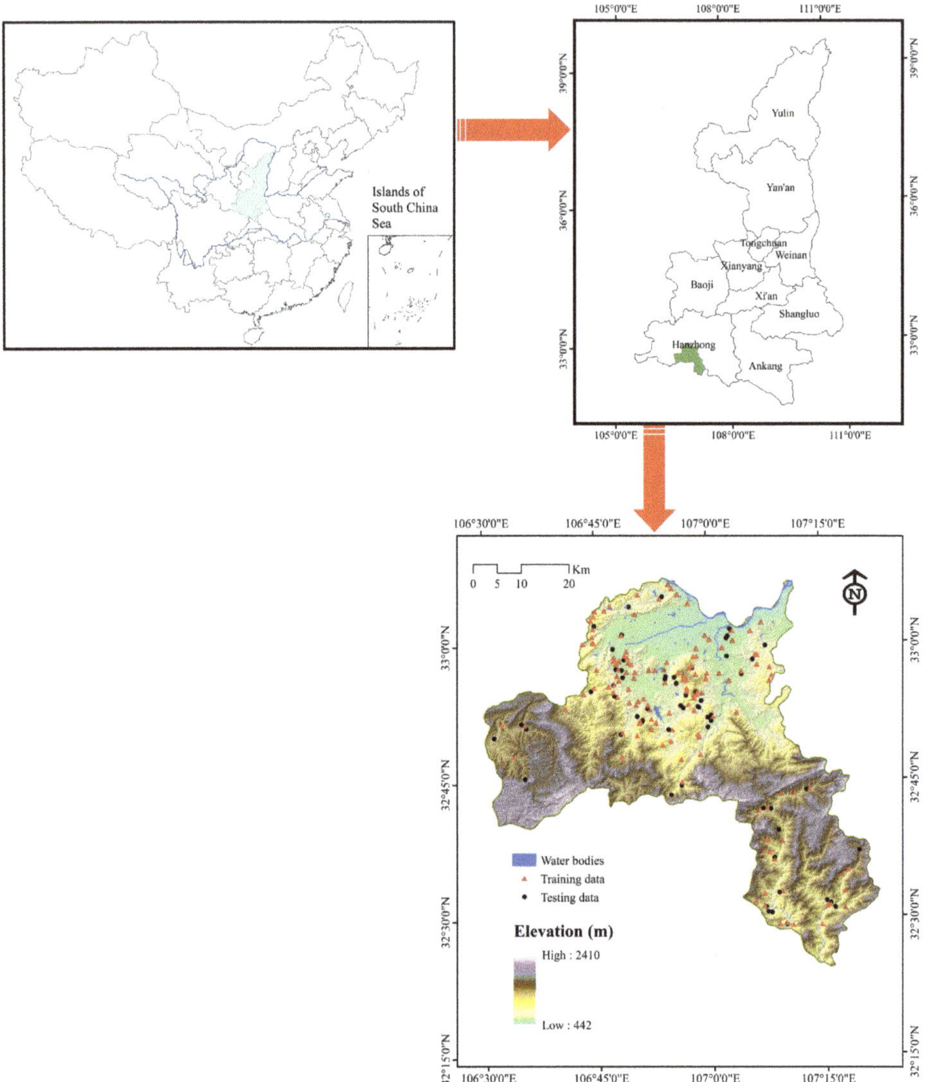

Figure 1. Location of the study area.

Geologically, the study area is located at the northern margin of the Yangtze plate. There are five major faults crossing the area including (1) the Gangchang fault (SW–NE direction), (2) the Xiaolengba–Qinjiaba fault (NW–SE direction), (3) the Xiaoba–Haitang fault (SW–NE direction), (4) the Moujiaba–Shuimohe fault (W–E direction), and (5) the Jiangjiawan–Zhujiaba–Tuqiangping fault (SW–NE direction) (Figure 2).

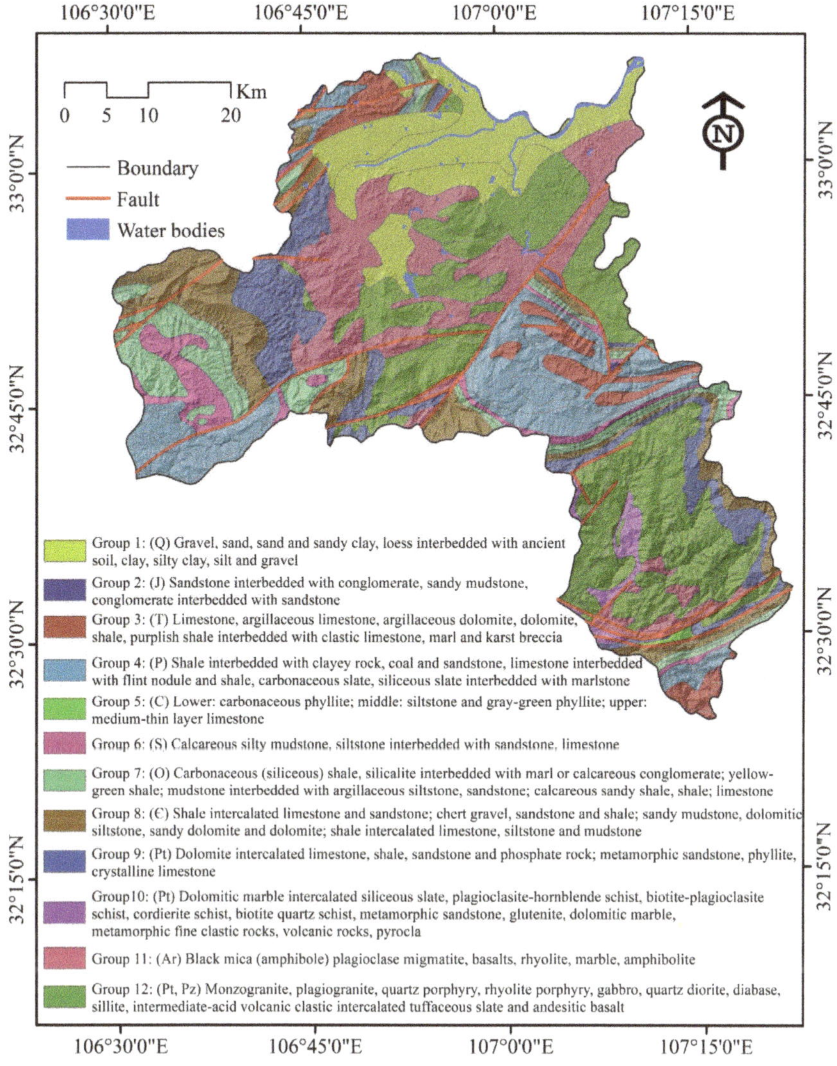

Figure 2. Geological map of the study area.

3. Materials and Methods

3.1. Data Preparation

A landslide inventory includes the locations of the past and recent landslides [21]. A landslide inventory can give insight into landslide location, dates, type, frequency of occurrence, state of activity, magnitude or size, failure mechanisms, causal factors, and damage caused [58,59]. In the present study, the landslide inventory map was prepared on the basis of satellite images (Google Earth and ZY03 images) and historical landslide records of the area, which were verified by GPS. A total of 202 landslides were identified to prepare the landslide susceptibility map, of which most of the landslides were slides (190), the others included 12 rock falls [60]. According to an analysis in the GIS environment, the smallest landslide was nearly 160 m^3, the largest landslide was more than 1,000,000 m^3, while the average was 33,000 m^3. Finally, 141 landslides were randomly selected as training data and rest of them were used for the verification of the landslide susceptibility map (Figure 1).

There are no universal guidelines for selecting landslide conditioning factors [33,61]. A total of 16 landslide conditioning factors were used for landslide susceptibility mapping including slope angle, slope aspect, elevation, plan curvature, profile curvature, topographic wetness index (TWI), stream power index (SPI), sediment transport index (STI), distance to rivers, distance to roads, distance to faults, soil, land use, normalized difference vegetation index (NDVI), lithology, and rainfall, which are considered as controlling factors in the occurrence of landslides in the study area.

Slope angle is an important factor that affects the stress state of slope mass, and these positions where stress exceeds failure strength may contribute to landslide hazards [62,63]. In this case, as shown in Figure 3a, the thematic data layer of the slope angle was reclassified into seven categories with an interval of $10°$, namely, $(0–10°)$, $(10–20°)$, $(20–30°)$, $(30–40°)$, $(40–50°)$, $(50–60°)$, and $(60–72.83°)$.

Slope aspect is another common conditioning factor for the task of landslide susceptibility mapping [64,65]. It has been proven that most landslides usually occur at a certain slope aspect for a given study area, but the mechanism has not been revealed clearly [66]. Therefore, slope aspect was also employed as a conditioning factor. Here, slope aspect categories include flat, north, northeast, east, southeast, south, southwest, west, and northwest (Figure 3b).

Generally, it is considered that elevation has a firm relationship with landslide occurrence [67]. There is no denying that elevation can influence the topography, vegetation, temperature, humidity, human activities, and many other conditions that have a connection with slope stability [30,68]. In Figure 3c, the elevation of the study area was divided into ten classes with an interval of 200 m, i.e., (442–600 m), (600–800 m), (800–1000 m), (1000–1200 m), (1200–1400 m), (1400–1600 m), (1600–1800 m), (1800–2000 m), (2000–2200 m), and (2200–2410 m).

Plan curvature and profile curvature are two quantitative indices that embody topographic characteristics and trend from different perspectives [69]. Various curvature values indicate different runoff and erosion conditions of water. For instance, the upwardly convex surfaces have positive curvature values while negative curvature values mean upwardly concave surfaces [30]. In this study, the plan curvature and profile curvature values were both reclassified into three groups (Figure 3d,e).

TWI was proposed to indicate the local groundwater potential by Moore [70] in 1991. Currently, TWI is regarded as an extensively-used causative factor in landslide susceptibility assessment [71]. It is expressed as $\mathrm{TWI} = \ln(\frac{\alpha}{\tan \beta})$, where β is the slope angle (radian), and α is the flow accumulation through a point [72]. The TWI values of the study area can be calculated by GIS software and reclassified as (<4), (4–5), (5–6), (6–7), and (>7) with an interval of 1 (Figure 3f).

SPI can directly measure the erosion capacity of the stream. A higher SPI value indicates that the stream has more powerful erosion on the slope surface [55]. The SPI values are mainly determined as $\mathrm{SPI} = \alpha \tan \beta$ [54,70]. In this study, the SPI values were identified as five categories with an interval of 20, namely, (<20), (20–40), (40–60), (60–80), and (>80) (Figure 3g).

As another topographic index, STI has also been considered to construct the landslide susceptibility model [73]. Similar to SPI, STI can quantitatively reflect the regional topographic features and erosion conditions [74]. For the present study, STI values contained five categories with an interval of 10: (<10), (10–20), (20–30), (30–40), (>40) (Figure 3h).

Rivers can not only affect the moisture distribution in slopes, but can also erode the toes of slopes, which cause slope deformation and failure [75]. Thus, it is necessary to consider the river effects when producing landslide susceptibility maps. In this study, based on the distance to rivers, five buffer zones with an interval of 200 m were generated for each river: (<200 m), (200–400 m), (400–600 m), (600–800 m), and (>800 m) (Figure 3i).

Generally speaking, road construction in mountainous areas, which always produce an engineering load and destroy the integrity of slope structure, have significant negative impacts on the slope stability [76]. Hence, the distance to roads is usually selected as a conditioning factor to embody the influence of road engineering activities on landslide occurrence [77]. Here, values of the distance to roads were divided into five groups with an interval of 300 m, i.e., (<300 m), (300–600 m), (600–900 m), (900–1200 m), and (>1200 m) (Figure 3j).

Fault structures affect the spatial distribution and characteristics of landslides in a certain region [50]. According to relevant studies [30,78], the integrity of rock and soil mass generally decrease as the distance to the faults shorten. In this way, landslide hazards are more likely to occur in the neighboring area of faults. Ultimately, buffers of various faults in the study area were obtained and reclassified into five categories with an interval of 1000 m: (<1000 m), (1000–2000 m), (2000–3000 m), (3000–4000 m), and (>4000 m) (Figure 3k).

In terms of soil, this is an essential factor that has a strong correlation with landslide occurrence [79]. To a great extent, the strength, root cohesion, permeability, and vegetation coverage of the soil mass depend on the soil type [80,81], which can impact the failure characteristics of slopes [82,83]. In this study area, a total of nine soil types were identified including cumulic anthrosol, dystric cambisol, eutric cambisol, calcaric fluvisol, haplic luvisol, chromic luvisol, eutric planosol, calcaric regosol, and eutric regosol (Figure 3l).

Land use is one of the most frequently used conditioning factors, and the correlation between landslides and land use has been confirmed [84]. For instance, in some farmland regions, landslides are frequent and common under long-term irrigation [85]. For the study area, the types of land use mainly consist of farmland, forestland, grassland, water, residential areas, and bareland (Figure 3m).

NDVI is a very popular index to measure the degree of vegetation in a region. NDVI values can be figured out by the formula NDVI $= (I - R)/(IR + R)$, where IR is the infrared band and R is the red band of the electromagnetic spectrum [86]. The range of NDVI values is from -1 to 1, and a positive value means that the local ground is covered by vegetation. Five categories of NDVI values were generated based on the natural break method [87], namely (–0.21–0.21), (0.21–0.36), (0.36–0.44), (0.44–0.52), and (0.52–0.65) (Figure 3n).

Like soil, lithology is one of the most important factors that directly determines slope stability. According to many existing studies, the physical and mechanical properties of rock mass usually change dramatically with lithological units [88]. Therefore, most landslides occur in the sliding-prone lithological units that have lower strength and a higher moisture content. For this study area, the strata were mainly reclassified into twelve lithological units based on the lithofacies and geological ages, and the specific distribution of various lithologies was illustrated in Figure 3o.

Rainfall is a crucial triggering factor that causes massive landslides by means of raising the groundwater level and increasing pore water pressure [89]. It can be observed that the probability of landslide occurrence indeed grows under the actions of long-term or heavy rainfall. Based on the meteorological data of the study area, the corresponding rainfall map with an interval of 100 mm/yr was produced, i.e., (<900 mm/yr), (900–1000 mm/yr), (1000–1100 mm/yr), (1100–1200 mm/yr), (1200–1300 mm/yr), (1300–1400 mm/yr), (1400–1500 mm/yr), (1500–1600 mm/yr), (1600–1700 mm/yr), and (>1700 mm/yr) (Figure 3p).

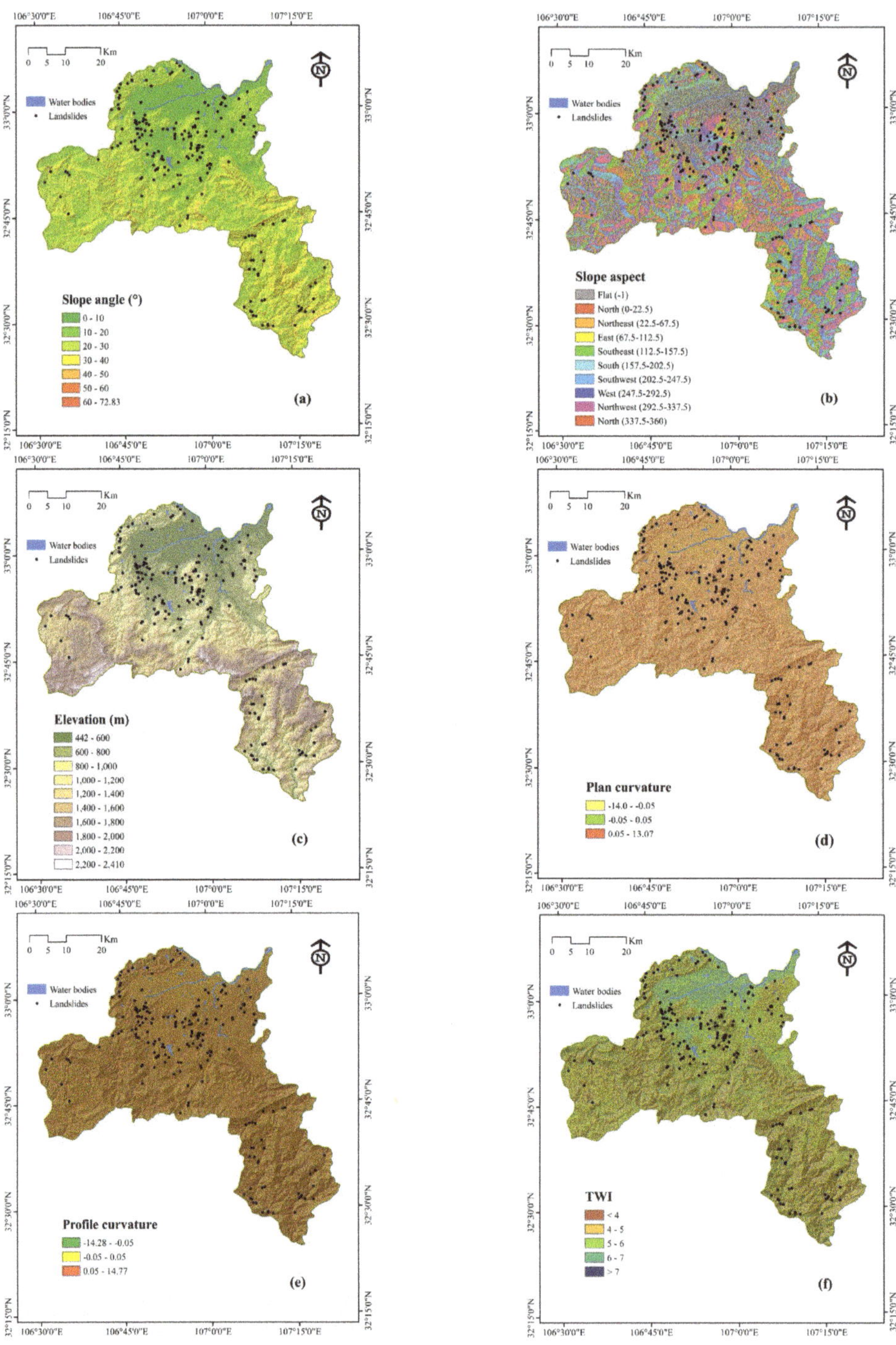

Figure 3. *Cont.*

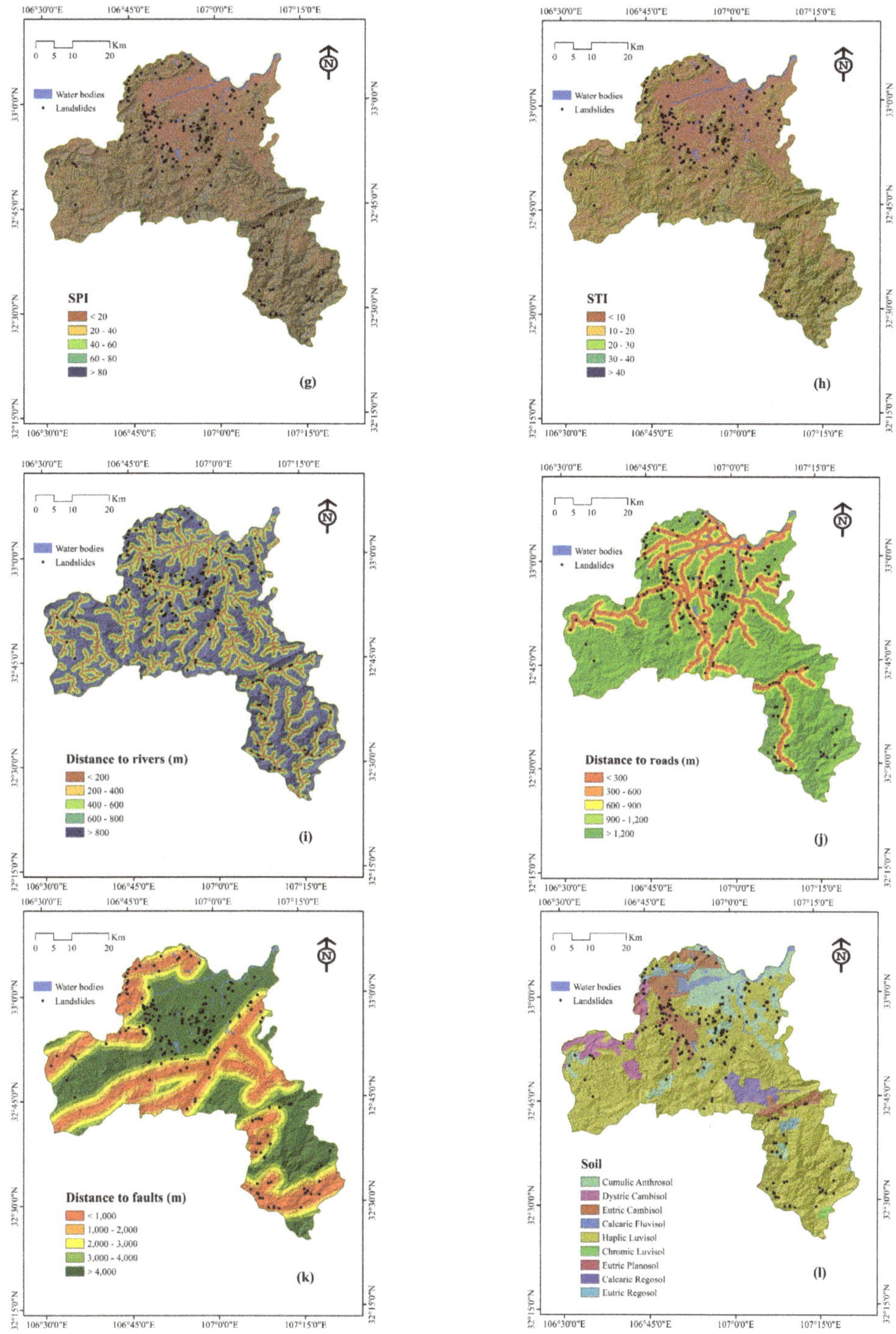

Figure 3. *Cont.*

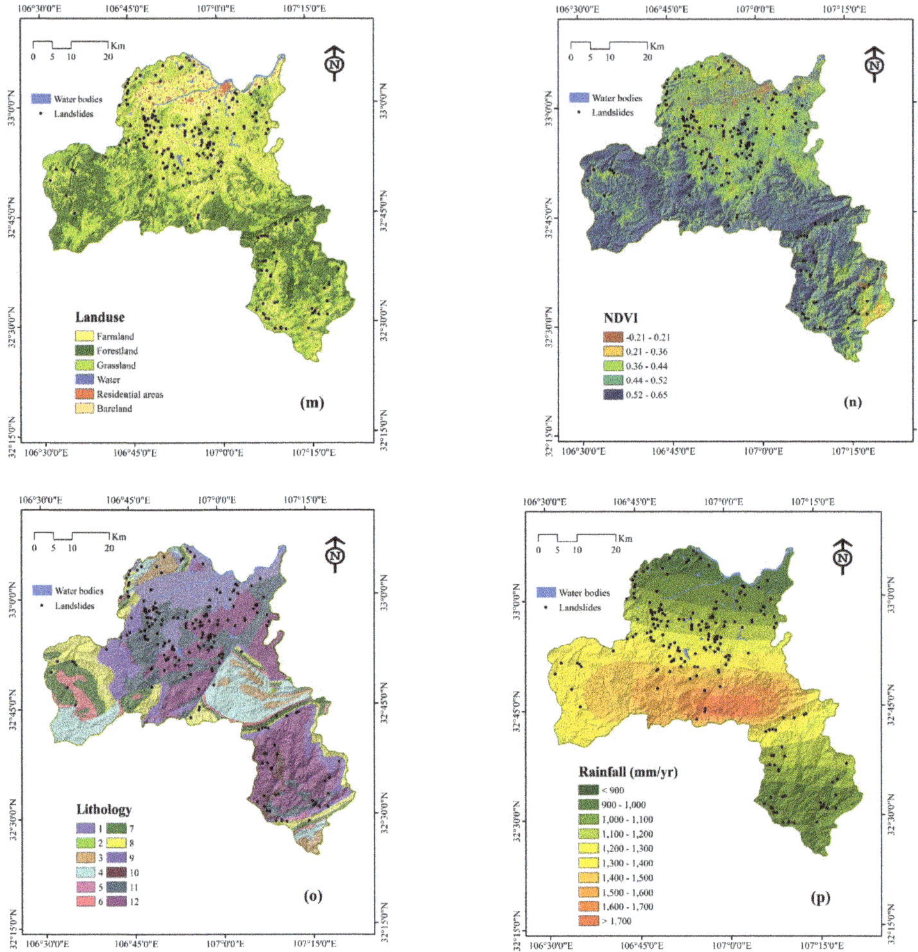

Figure 3. Thematic maps: (**a**) Slope angle; (**b**) Slope aspect; (**c**) Elevation; (**d**) Plan curvature; (**e**) Profile curvature; (**f**) TWI; (**g**) SPI; (**h**) STI; (**i**) Distance to rivers; (**j**) Distance to roads; (**k**) Distance to faults; (**l**) Soil; (**m**) Land use; (**n**) NDVI; (**o**) Lithology; and (**p**) Rainfall.

3.2. Weight of Evidence

Weight of evidence (WoE) is one of the most popular models that uses the Bayesian theory of conditional probability to quantify spatial associations between evidence layers and known mineral occurrences [90]. In the WoE method, conditional independence is the most important issue that should be considered. The WoE is based on the calculation of positive weight W^+ and negative weight W^- as follows:

$$W^+ = \ln \frac{p\{B|A\}}{p\{B|\overline{A}\}} \tag{1}$$

$$W^- = \ln \frac{p\{\overline{B}|A\}}{p\{\overline{B}|\overline{A}\}} \tag{2}$$

where B is the presence predictive factor; $\overline{B}$ is the absence of the predictive factor; A is the presence of landslide; and $\overline{A}$ is the absence of landslide. In landslide susceptibility prediction, the weight contrast $W_f = W^+ - W^-$ was used to measure and reflect the spatial association between the landslide conditioning factors and landslide occurrence [91].

3.3. Logistic Regression

Logistic regression (LR) is one type of regression analysis where categorical outcomes can be predicted based on a certain predictor [92]. By using the logistic functions, probabilities of the possible outcomes can be modeled [93].

The logistic regression model is useful for two-class classification. Assuming there are n samples of the pairs, (x_i, y_i), $i = 1, 2, \ldots, n$, $y_i \in \{-1, +1\}$ is a binary class label for each sample $i = 1, 2, \ldots, n$ and weights (w, b). In the logistic regression for binary classification, the occurrence probability of the class is modeled with the below function:

$$P(y = \pm 1|x, w) = \frac{1}{1 + \exp(-y(w^T x + b))} \tag{3}$$

where b is the intercept; T is the matrix transposition; and the k-dimensional coefficient vector, $w = (w_1, w_2, \ldots w_k)^T$ are parameters to be estimated.

3.4. Random Forest

Random forests (RF) are an ensemble of separately trained binary decision trees [94]. In the random forest algorithm, a random vector i_k is naturally produced, independent from the previous random vectors and distributed to all trees, and each tree is grown using the training dataset and random vector i_k, and outcomes are in the collection of tree-structured classifiers $h(x, i_k)$, $k = 1, 2, \ldots n$ at input vector x. In this study, i_k is the landslide conditioning factors. The random forest consisted of two trees, namely, landslide and non-landslide, each constructed while considering sixteen random features.

Generally, in a random forest algorithm, the generalization error is described as below [95]:

$$GE = P_{x,y}(mg(x, y) < 0) \tag{4}$$

where x and y are the landslide conditioning factors indicating the probability over the x, y space, and mg is the margin function, which is defined as below:

$$mg(x, y) = av_k I(h_k(x) = y) - \max_{j \neq y} av_k I(h_k(x) = j) \tag{5}$$

Which measures the extent to which the average number of votes at random vectors for the right output exceeds the average vote for any other output. The $I(*)$ is the indicator function [96].

4. Results

4.1. Correlation Analysis

The correlation between the conditioning factors and probability of landslides occurrence was measured by the weight contrast W_f, and the calculation results of the WoE model are listed in Table 1. The LR method was employed to produce the landslide susceptibility map, and one of the most critical applicable conditions of LR is that the landslide conditioning factors are mutually independent [97]. Therefore, it is necessary to diagnose the multicollinearity of various conditioning factors when evaluating landslide susceptibility [98]. Currently, the tolerance (TOL) (TOL $= 1 - R^2$, and R is the coefficient of determination of the regression equation) and variance inflation factor (VIF) (VIF $= 1/\text{TOL}$) have been applied in multicollinearity diagnosis [99–101].

Table 1. Correlation between landslides and conditioning factors using the WoE model.

Factors	Class	No. of Landslide	No. of Pixels in Domain	W^+	W^-	W_f
Slope angle (°)	0–10	36	738,360	0.077	−0.025	0.102
	10–20	63	914,163	0.423	−0.246	0.669
	20–30	29	821,142	−0.246	0.075	−0.320
	30–40	11	468,077	−0.653	0.081	−0.734
	40–50	2	155,377	−1.255	0.037	−1.292
	50–60	0	24,357	0.000	0.008	0.000
	60–72.83	0	1710	0.000	0.001	0.000
Slope aspect	Flat	0	874	0.000	0.000	0.000
	North	16	443,863	−0.225	0.033	−0.258
	Northeast	16	405,251	−0.134	0.019	−0.153
	East	17	376,207	0.001	0.000	0.001
	Southeast	23	390,547	0.266	−0.044	0.310
	South	32	374,222	0.639	−0.130	0.769
	Southwest	13	344,928	−0.181	0.020	−0.201
	West	9	354,647	−0.576	0.055	−0.631
	Northwest	15	432,647	−0.264	0.037	−0.301
Elevation (m)	442–600	28	413,571	0.405	−0.079	0.485
	600–800	48	512,157	0.730	−0.237	0.968
	800–1000	31	377,619	0.598	−0.119	0.717
	1000–1200	17	326,381	0.143	−0.018	0.161
	1200–1400	15	398,407	−0.182	0.024	−0.206
	1400–1600	2	385,439	−2.163	0.117	−2.281
	1600–1800	0	376,083	0.000	0.128	0.000
	1800–2000	0	247,350	0.000	0.083	0.000
	2000–2200	0	78,216	0.000	0.025	0.000
	2200–2410	0	7963	0.000	0.003	0.000
Plan curvature	−14.0−−0.05	58	144,0116	−0.114	0.088	−0.203
	−0.05–0.05	13	215,290	0.291	−0.025	0.316
	0.05–13.07	70	1,467,780	0.055	−0.051	0.106
Profile curvature	−14.28−−0.05	66	1,428,952	0.023	−0.020	0.042
	−0.05–0.05	16	177,891	0.689	−0.062	0.751
	0.05–14.77	59	1,516,343	−0.149	0.123	−0.271
TWI	<4	11	558,428	−0.829	0.116	−0.945
	4–5	50	1,000,955	0.101	−0.052	0.153
	5–6	48	746,522	0.354	−0.143	0.497
	6–7	20	393,490	0.119	−0.018	0.137
	>7	12	423,791	−0.467	0.057	−0.523
SPI	<20	88	1,740,663	0.113	−0.164	0.277
	20–40	20	497,521	−0.116	0.021	−0.137
	40–60	12	231,236	0.139	−0.012	0.151
	60–80	5	133,800	−0.189	0.008	−0.197
	>80	16	519,966	−0.383	0.062	−0.445
STI	<10	90	1,722,652	0.146	−0.215	0.361
	10–20	32	702,426	0.009	−0.003	0.012
	20–30	6	295,062	−0.798	0.056	−0.853
	30–40	5	141,300	−0.244	0.010	−0.254
	>40	8	261,746	−0.390	0.029	−0.419
Distance to rivers (m)	<200	27	521,129	0.138	−0.030	0.168
	200–400	22	463,390	0.050	−0.009	0.059
	400–600	18	427,717	−0.070	0.011	−0.081
	600–800	19	374,831	0.116	−0.017	0.133
	>800	55	1,336,119	−0.092	0.064	−0.156
Distance to roads (m)	<300	33	343,852	0.754	−0.150	0.904
	300–600	16	279,559	0.237	−0.027	0.264
	600–900	8	245,226	−0.325	0.023	−0.348
	900–1200	15	219,752	0.413	−0.040	0.453
	>1200	69	2,034,797	−0.286	0.382	−0.668

Table 1. *Cont.*

Factors	Class	No. of Landslide	No. of Pixels in Domain	W^+	W^-	W_f
Distance to faults (m)	<1000	32	671,796	0.054	−0.015	0.069
	1000–2000	18	503,008	−0.232	0.039	−0.271
	2000–3000	21	412,189	0.121	−0.020	0.141
	3000–4000	8	348,794	−0.677	0.060	−0.737
	>4000	62	1,187,399	0.145	−0.101	0.246
Soil	Cumulic Anthrosol	20	360,361	0.206	−0.030	0.237
	Dystric Cambisol	4	113,893	−0.251	0.008	−0.259
	Eutric Cambisol	31	249,592	1.012	−0.165	1.177
	Calcaric Fluvisol	0	37,035	0.000	0.012	0.000
	Haplic Luvisol	80	2,211,459	−0.222	0.393	−0.615
	Chromic Luvisol	0	10,045	0.000	0.003	0.000
	Eutric Planosol	3	14,836	1.500	−0.017	1.516
	Calcaric Regosol	1	82,141	−1.311	0.020	−1.330
	Eutric Regosol	2	43,824	0.011	0.000	0.011
Land use	Farmland	86	90,0284	0.750	−0.601	1.351
	Forestland	4	96,7369	−2.390	0.342	−2.732
	Grassland	51	1,202,442	−0.062	0.037	−0.100
	Water	0	18,838	0.000	0.006	0.000
	Residential areas	0	33,563	0.000	0.011	0.000
	Bareland	0	690	0.000	0.000	0.000
NDVI	−0.21–0.21	4	67,502	0.272	−0.007	0.279
	0.21–0.36	10	207,991	0.063	−0.005	0.068
	0.36–0.44	63	651,020	0.762	−0.358	1.121
	0.44–0.52	56	1,089,392	0.130	−0.077	0.207
	0.52–0.65	8	1,107,281	−1.832	0.379	−2.212
Lithology	1	27	363,139	0.499	−0.089	0.588
	2	0	1694	0.000	0.001	0.000
	3	2	136,901	−1.128	0.031	−1.159
	4	6	398,403	−1.098	0.093	−1.191
	5	0	7470	0.000	0.002	0.000
	6	0	107,848	0.000	0.035	0.000
	7	5	225,834	−0.713	0.039	−0.751
	8	10	319,450	−0.366	0.034	−0.401
	9	9	276,290	−0.326	0.027	−0.353
	10	1	39,158	−0.570	0.005	−0.575
	11	32	435,539	0.487	−0.107	0.594
	12	49	811,460	0.291	−0.126	0.417
Rainfall (mm/yr)	<900	8	189,533	−0.067	0.004	−0.071
	900–1000	29	582,217	0.098	−0.024	0.122
	1000–1100	23	282,006	0.591	−0.083	0.675
	1100–1200	35	329,319	0.856	−0.174	1.030
	1200–1300	16	271,086	0.268	−0.030	0.298
	1300–1400	18	629,601	−0.457	0.089	−0.545
	1400–1500	7	351,254	−0.818	0.068	−0.886
	1500–1600	3	270,784	−1.405	0.069	−1.474
	1600–1700	1	135,625	−1.812	0.037	−1.849
	>1700	1	81,761	−1.306	0.019	−1.325

Generally, a TOL value less than 0.1 or a VIF value larger than 10 is regarded as a symbol of multicollinearity [61]. In this study, the results of the WoE model were used as inputs to calculate the TOL and VIF values of all of the conditioning factors. In accordance with the calculated results, there was no multicollinearity among the landslide conditioning factors (Table 2).

Table 2. Multicollinearity analysis.

Landslide Conditioning Factors	Collinearity Statistics	
	Tolerance (TOL)	Variance inflation factors (VIF)
Slope angle	0.761	1.315
Slope aspect	0.883	1.133
Elevation	0.650	1.539
Plan curvature	0.714	1.400
Profile curvature	0.855	1.170
TWI	0.828	1.208
SPI	0.434	2.303
STI	0.402	2.489
Distance to rivers	0.946	1.057
Distance to roads	0.779	1.284
Distance to faults	0.908	1.101
NDVI	0.774	1.292
Soil	0.642	1.557
Land use	0.627	1.595
Lithology	0.765	1.308
Rainfall	0.664	1.507

4.2. Application of the WoE Model

In terms of slope angle, the slope angle between 10°–20° (0.669) is more prone to landslide occurrence. Additionally, the W_f values of the region where slope angles larger than 50° are 0. For the slope aspect factor, W_f was the highest for south-facing (0.769). Furthermore, southeast-facing (0.310) and east-facing (0.001) also had a positive correlation with landslide occurrence. In the case of elevation, most landslides were distributed in the classes of 442–600 m (0.485), 600–800 m (0.968), and 800–1000 m (0.717). When the elevation was larger than 1200 m, elevation had an inhibitory effect on landslides. In the case of plan curvature, flat areas had a more important impact on landslides, whereas the W_f values of convex areas and concave areas were 0.106 and −0.203, respectively. In the case of profile curvature, the W_f values of the concave class, flat class, and convex class ere 0.042, 0.751, and −0.271, respectively. For TWI, the highest W_f value was observed for the interval of 5–6 (0.497) while the class <4 (−0.945) had the lowest value. For SPI, the class <20 (0.277) had the highest W_f value, and the areas of SPI 20–40 and >60 were negative for landslides. For STI, the class <10 had the highest W_f value of 0.361, while the class 20–30 had the lowest value of −0.853. In the case of distance to rivers, the classes of <200 m (0.168) and 600–800 m (0.133) occupied higher W_f values when compared to the other classes. In the case of distance to roads, the class of <300 m (0.904) had a more intimate correlation with landslide occurrence. In the case of distance to faults, it can be seen that the class >4000 m had the highest W_f value of 0.246. For soil, eutric cambisol (1.177) and eutric planosol (1.516) were more likely to induce landslides due to the dramatic falling of soil strength under saturated conditions [85]. For land use, farmland (1.351) had the highest probability of landslide occurrence, which may be essentially caused by irrigation. According to the W_f values of NDVI, the class of 0.36–0.44 (1.121) mainly contributed to landslide occurrence, while the lowest value was for the class of 0.52–0.65 (−2.212), which indicates that high vegetation coverage can restrain landslides. In the case of lithology, the W_f values of group 1 (Q) (0.588), group 11 (Ar) (0.594), and group 12 (Pt, Pz) (0.417) were larger than 0, indicating that these lithological groups had the highest susceptibility to landslide. In the case of rainfall, the range between 1100–1200 mm/yr (1.030) showed high susceptibility for landslide occurrence.

The calculated W_f values for all landslide conditioning factors were summed using the following equation to construct the landslide susceptibility map (LSM):

$$
\begin{aligned}
LSM_{WoE} = \ & \text{Slope angle}_{Wf} + \text{Slope aspect}_{Wf} + \text{Elevation}_{Wf} + \text{Plan curvature}_{Wf} \\
& + \text{Profile curvature}_{Wf} + \text{TWI}_{Wf} + \text{SPI}_{Wf} + \text{STI}_{Wf} + \text{Distance to rivers}_{Wf} \\
& + \text{Distance to roads}_{Wf} + \text{Distance to faults}_{Wf} + \text{NDVI}_{Wf} + \text{Soil}_{Wf} + \text{Landuse}_{Wf} \\
& + \text{Lithology}_{Wf} + \text{Rainfall}_{Wf}
\end{aligned}
\tag{6}
$$

The integrated result of the WoE model is shown in Figure 4. The LSM was reclassified into five classes based on the natural break method: very low, low, moderate, high, and very high.

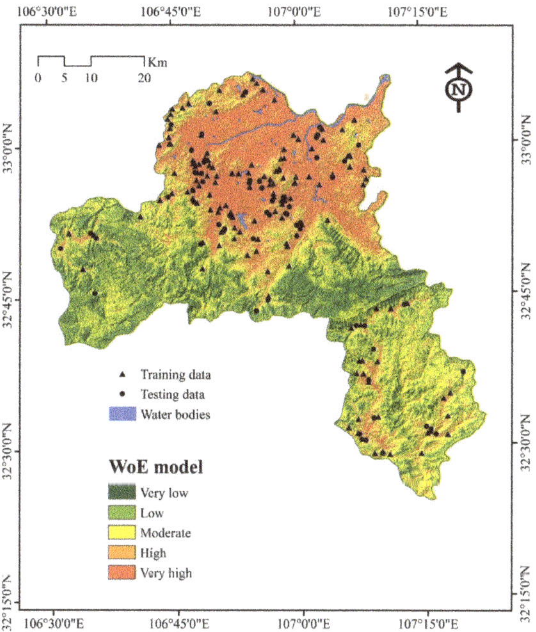

Figure 4. Landslide susceptibility map using the WoE model.

4.3. Application of the WoE-LR Model

In this case, SPSS 18.0 software was applied to build a landslide susceptibility model with the WoE-LR model. The input table of the LR model can be generated by the determined class values of variables based on the WoE model [54]. In the analysis process, a forward stepwise LR was adopted, and the analysis results are given in Tables 3 and 4. The Cox and Snell R Square (0.245) and Nagelkerke R Square (0.326) are two pseudo determined coefficients that are used to reflect the degree of independent variables explaining dependent variables [102,103]. According to Table 4, the LR equation and landslide occurrence probability P can be expressed as Equations (7) and (8), respectively.

$$
\begin{aligned}
y = \ & 1.122 \times \text{Slope angle} + 2.157 \times \text{Slope aspect} + 0.986 \times \text{Elevation} \\
& + 2.505 \times \text{Plan curvature} + 0.868 \times \text{Profile curvature} + 1.764 \times \text{TWI} \\
& + 1.427 \times \text{SPI} + 1.142 \times \text{STI} + 0.512 \times \text{Distance to rivers} \\
& + 1.445 \times \text{Distance to roads} + 0.972 \times \text{Distance to faults} \\
& + 0.859 \times \text{NDVI} + 1.392 \times \text{Soil} + 1.634 \times \text{Landuse} + 1.032 \times \text{Lithology} \\
& + 1.594 \times \text{Rainfall} + 0.806
\end{aligned}
\tag{7}
$$

$$
P = \frac{e^y}{1 + e^y}
\tag{8}
$$

Table 3. Maximum likelihood estimation and Cox and Snell's and Nagelkerke's R-square.

−2 Log Likelihood	Cox & Snell R Square	Nagelkerke R Square
311.780	0.245	0.326

Table 4. Coefficients of WoE-LR model.

Landslide Conditioning Factors	Coefficients
Slope angle	1.122
Slope aspect	2.157
Elevation	0.986
Plan curvature	2.505
Profile curvature	0.868
TWI	1.764
SPI	1.427
STI	1.142
Distance to rivers	0.512
Distance to roads	1.445
Distance to faults	0.972
NDVI	0.859
Soil	1.392
Land use	1.634
Lithology	1.032
Rainfall	1.594
Constant	0.806

Ultimately, the landslide susceptibility index (LSI) for the LR model were obtained based on Equation (8), moreover, the LSI values were reclassified into five categories by the natural break method: very low, low, moderate, high, and very high (Figure 5).

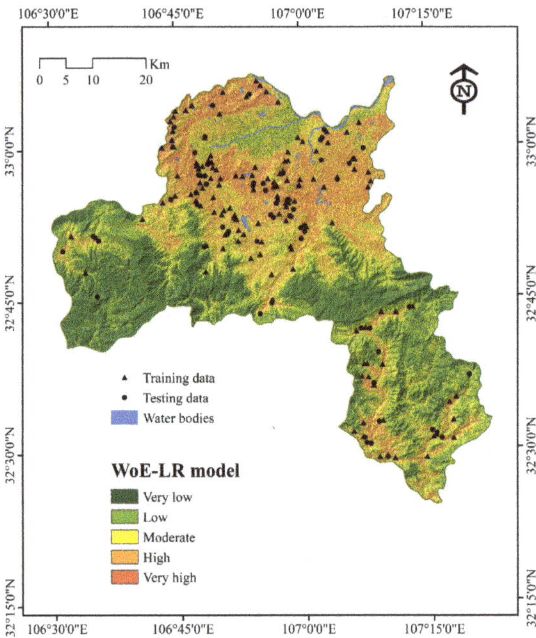

Figure 5. Landslide susceptibility map using the WoE-LR model.

4.4. Application of the WoE-RF Model

Similarly, the calculated results of the WoE model can be used as input for the RF model. In this study, training of the RF model was implemented by WEKA software. During the analyzing process, the importance of various conditioning factors can be measured quantitatively and ordered by MDA (mean decrease accuracy) and MDG (mean decrease Gini). Generally, MDA is determined during the Out-Of-Bag error calculation phase, while MDG is a measure of how each variable contributes to the homogeneity of the nodes and leaves [104]. The values of the above-mentioned two metrics of the conditioning factors are illustrated in Figure 6, and a larger value of MDA or MDG means a higher importance of the corresponding variable. Accordingly, in terms of MDA, land use is the most critical factor in the RF model, while soil is second in importance only to land use. For the MDG, the importance of elevation was first, followed by rainfall and land use. Finally, based on ArcGIS software, the landslide susceptibility map using the WoE-RF was generated and is shown in Figure 7.

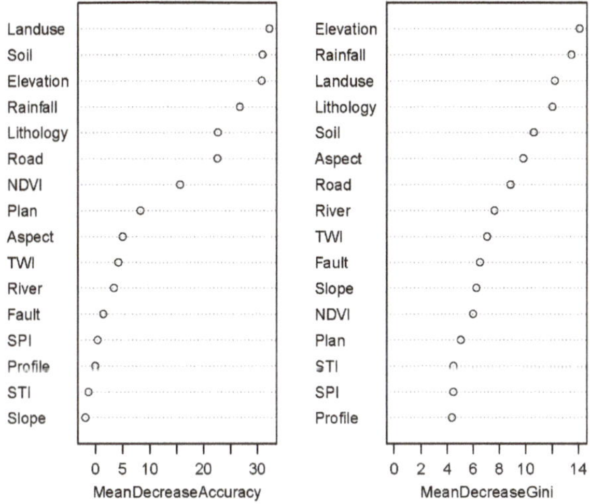

Figure 6. Mean decrease accuracy and mean decrease Gini.

4.5. Validation of Landslide Models

Currently, the ROC and AUC have been widely applied to validate the performance of determined landslide susceptibility models [64,69,105]. The ROC curve can be generated by plotting the false positive rate (100-specificity) in the x-axis versus the sensitivity in the y-axis [71]. The area under the ROC curve (AUC) is an indicator of the global summary measure of the performance of a model [106–108]. In the present study, to assess the validation of the WoE, WoE-LR, and WoE-RF models, the ROC curves of three models with training and validation datasets are described in Figures 8 and 9.

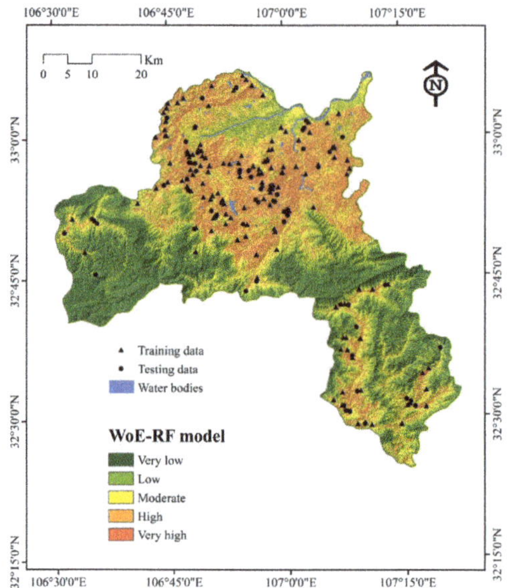

Figure 7. Landslide susceptibility map using the WoE-RF model.

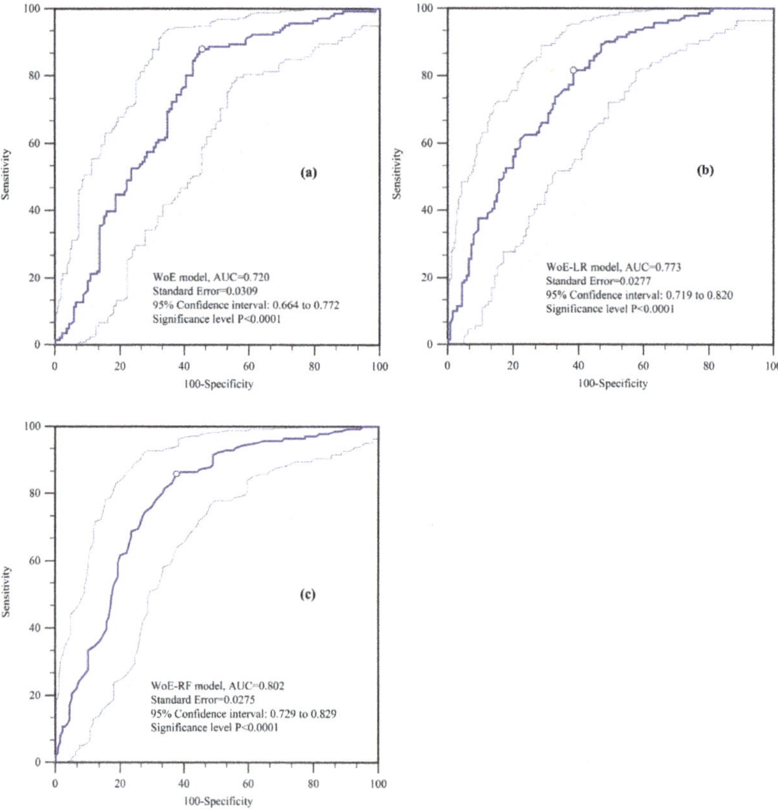

Figure 8. ROC curves using the training dataset.

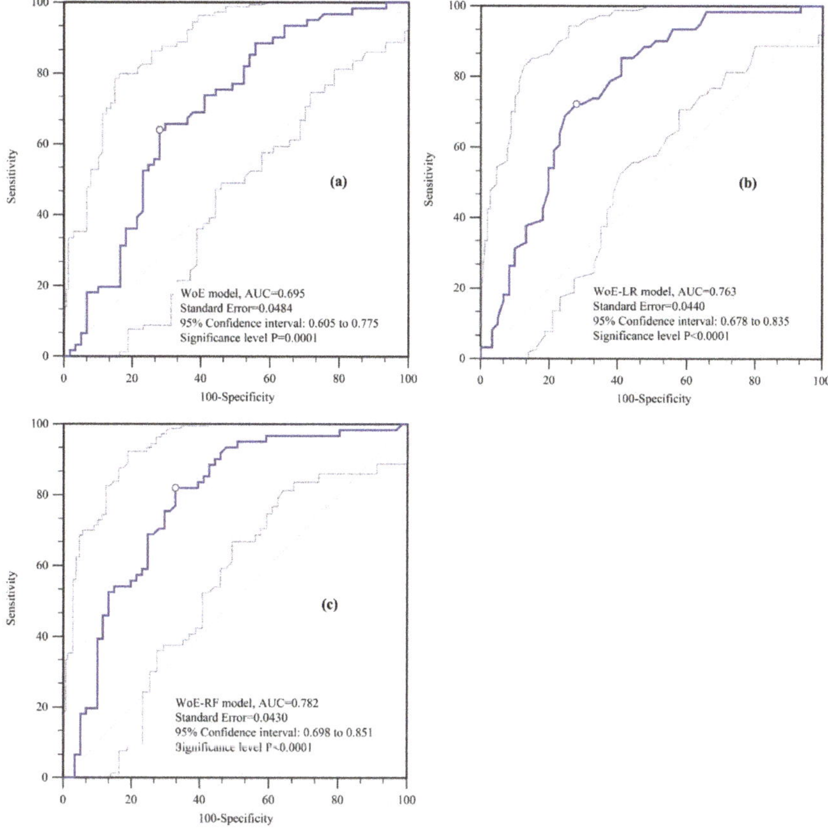

Figure 9. ROC curves using the validation dataset.

In the case of the training dataset, the WoE-RF model had the best performance with the highest AUC value of 0.802, while the AUC values of the WoE model and WoE-LR model were 0.720 and 0.773, respectively. Meanwhile, the WoE-RF model had the lowest standard error (0.0275) and a 95% confidence interval of 0.729–0.829. Thus, the WoE-RF and WoE-LR models can improve the accuracy of the traditional WoE model in this study, and the WoE-RF model showed a relatively better performance.

In the case of the validation data, it can be seen that the AUC values of the various models decreased slightly when compared with the training dataset. The AUC values were 0.695, 0.763, and 0.782 for the WoE model, WoE-LR model, and WoE-RF model, respectively. Similarly, the lowest standard error was 0.0430 for the WoE-RF model, followed by the WoE-LR model (0.0440), and the WoE model (0.0484). The detailed results demonstrated that the WoE-RF model had a prominent prediction capacity on landslide susceptibility mapping.

5. Discussions

Under the action of environmental factors and human activities, the frequency of landslide occurrence has been increasing in recent decades, which may result in catastrophic losses on lives, resources, and property [109,110]. Currently, numerous approaches have been used in landslide susceptibility mapping such as FR [23], WoE [19], IoE [111], machine learning [64,112], and ensemble learning models [53,54]. In the above-mentioned models, the probabilistic meaning and calculation procedure of the WoE model are relatively concise and specific, which makes the WoE a classical and widely used method in landslide susceptibility mapping. Nevertheless, due to the uncertainties and fuzziness in the data of the conditioning factors [113], for different datasets, the performance of the WoE models were significantly distinguished [114–116]. In the present study, the integrated ensemble WoE with LR and RF models were proposed and applied for landslide susceptibility modeling in order to improve the accuracy and generalization ability of the traditional WoE model.

Landslide inventory map is a preliminary step toward landslide susceptibility, hazard and risk assessment [59]. Generally, there are two classes of Landslide inventories: landslide-event inventories that are associated with a trigger and historical landslide inventories [59,117]. In the present study, we adopted the latter formation, which was the sum of many landslide events over a long time. However, the evidence of many smaller landslides might has been lost due to various degrees of modification by subsequent landslides, erosional processes, vegetation growth and anthropic influences [59]. Therefore, application of multi-temporal high-resolution satellites images for interpretation of smaller landslides may be an effective supplement to the current landslide inventory and efficient for improving the accuracies of landslide susceptibility maps.

According to the existing literature and multicollinearity analysis, sixteen conditioning factors were selected: slope angle, slope aspect, elevation, plan curvature, profile curvature, TWI, STI, SPI, distance to rivers, distance to roads, distance to faults, NDVI, soil, land use, lithology, and rainfall. Furthermore, based on the W_f values, the relationships between landslide occurrence and these factors were analyzed. It was demonstrated that all factors had nonlinear relationships with landslides. In addition, the RF model was employed to measure the importance of factors with two indices, the MDA and MDG. In terms of MDA, it could be observed that the most critical factor was land use, followed by soil and elevation. Slope angle had the lowest impact on landslide occurrence. However, for MDG, the importance of elevation, rainfall, and land use ranked first, second, and third, respectively, while the lowest MDG value was for profile curvature.

There are some classification techniques for a landslide susceptibility map in GIS software, such as manual, defined interval, natural break, equal interval, quantile, standard deviation, geometrical interval, and landslide percentage [118]. Generally, user-defined classification is more difficult for the reader to interpret and justify. Therefore, current automatic classification systems should be used instead of a user-defined classification [118]. Besides, when landslide susceptibility indexes have positive or negative skewness, the best classification methods are quantile or natural break [119]. In the present study, natural break method, which is the most commonly used models [120,121], is the most suitable method for modelling landslide susceptibility according to the histogram of data distribution.

In this paper, a comparison study of the WoE, WoE-LR, and WoE-RF models was implemented. LR is a widely used model for classification, particularly for binary classification problems [122]. Thus, we integrated the WoE with the LR model to acquire a better classifier. The WoE-RF model is a combination of the weight of evidence and random forest approach. It has been proven that RF is one of the most popular classification algorithms and can improve the performance of single classifiers [96,123]. Moreover, RF can decrease the dependence of the WoE model on independence among the conditioning factors. Accordingly, the results showed that both the LR model (AUC = 0.773 for training data; AUC = 0.763 for validation data) and RF model (AUC = 0.802 for training data; AUC = 0.782 for validation data) can increase the performance of the traditional WoE model (AUC = 0.720 for training data; AUC = 0.695 for validation data), and the WoE-RF model produced the best results.

Comparing the overall classification results of the three models, the results confirmed that the RF model had a better performance on improving the generalization ability of a weak classifier and raising the corresponding prediction accuracy. Therefore, the landslide susceptibility maps generated by the WoE-RF and WoE-LR models contain reference meaning for the study area to a certain extent. Furthermore, the procedure of factor selection and ensemble model construction is of some value to similar studies.

6. Conclusions

The results are indicative of the quality of the maps drawn by the hybrid approaches of traditional bivariate weights of evidence (WoE) with multivariate logistic regression (WoE-LR) and machine learning-based random forest (WoE-RF). In general, the following conclusions can be drawn:

(1) Geomorphological factors, geological factors, geo-environmental factors, and anthropogenic factors were used for the development of the landslide model. The preliminary selection of these 16 conditioning factors was based on the multicollinearity diagnosis. The TOL and VIF values of all the conditioning factors indicated no multicollinearity.

(2) According to the results of the WoE model, most occurred at slopes of 10–20° with the south aspect, elevations of 600–800 m, distance to rivers of <200 m, distance to roads of <300 m, and a farmland land cover category.

(3) WoE-RF possessed relatively good accuracy when compared to the WoE-LR and WoE models. By using the ROC curve, the AUC values of the training dataset produced by these three methods were 0.802, 0.773, and 0.720, respectively. For the validation dataset, the AUC values were 0.782, 0.763, and 0.695, respectively. It can be concluded that the proposed hybrid models are promising approaches for the spatial prediction of landslides and can also be applied in other landslide-prone areas.

Author Contributions: W.C. and Z.S. collected the field data and conducted the landslide mapping and analysis. W.C. and Z.S. wrote the manuscript. J.H. provided critical comments in planning this paper and edited the manuscript. All authors discussed the results and edited the manuscript.

Acknowledgments: This study was supported by the Opening Fund of Key Laboratory of Degraded and Unused Land Consolidation Engineering, the Ministry of Land and Resources (Grant No. SXDJ2018-04), the National Natural Science Foundation of China (Grant No. 41807192), the China Postdoctoral Science Foundation (Grant No. 2018T111084, 2017M613168), and the Project funded by the Shaanxi Province Postdoctoral Science Foundation (Grant No. 2017BSHYDZZ07).

Conflicts of Interest: No potential conflict of interest was reported by the authors.

References

1. Kim, M.S.; Onda, Y.; Kim, J.K.; Kim, S.W. Effect of topography and soil parameterisation representing soil thicknesses on shallow landslide modelling. *Quat. Int.* **2015**, *384*, 91–106. [CrossRef]
2. Liucci, L.; Melelli, L.; Suteanu, C.; Ponziani, F. The role of topography in the scaling distribution of landslide areas: A cellular automata modeling approach. *Geomorphology* **2017**, *290*, 236–249. [CrossRef]
3. Agostini, A.; Tofani, V.; Nolesini, T.; Gigli, G.; Tanteri, L.; Rosi, A.; Cardellini, S.; Casagli, N. A new appraisal of the ancona landslide based on geotechnical investigations and stability modelling. *Q. J. Eng. Geol. Hydrogeol.* **2014**, *47*, 29–43. [CrossRef]
4. Peng, D.; Xu, Q.; Liu, F.; He, Y.; Zhang, S.; Qi, X.; Zhao, K.; Zhang, X. Distribution and failure modes of the landslides in heitai terrace, china. *Eng. Geol.* **2018**, *236*, 97–110. [CrossRef]
5. Persichillo, M.G.; Bordoni, M.; Meisina, C. The role of land use changes in the distribution of shallow landslides. *Sci. Total Environ.* **2017**, *574*, 924–937. [CrossRef] [PubMed]
6. Chang, J.-M.; Chen, H.; Jou, B.J.-D.; Tsou, N.-C.; Lin, G.-W. Characteristics of rainfall intensity, duration, and kinetic energy for landslide triggering in taiwan. *Eng. Geol.* **2017**, *231*, 81–87. [CrossRef]
7. Segoni, S.; Rosi, A.; Lagomarsino, D.; Fanti, R.; Casagli, N. Brief communication: Using averaged soil moisture estimates to improve the performances of a regional-scale landslide early warning system. *Nat. Hazards Earth Syst. Sci.* **2018**, *18*, 807–812. [CrossRef]
8. Hou, X.; Vanapalli, S.K.; Li, T. Water infiltration characteristics in loess associated with irrigation activities and its influence on the slope stability in heifangtai loess highland, china. *Eng. Geol.* **2018**, *234*, 27–37. [CrossRef]
9. Wang, T.; Wu, S.R.; Shi, J.S.; Xin, P.; Wu, L.Z. Assessment of the effects of historical strong earthquakes on large-scale landslide groupings in the wei river midstream. *Eng. Geol.* **2018**, *235*, 11–19. [CrossRef]
10. Mohammadi, S.; Taiebat, H. Finite element simulation of an excavation-triggered landslide using large deformation theory. *Eng. Geol.* **2016**, *205*, 62–72. [CrossRef]

11. Alvioli, M.; Melillo, M.; Guzzetti, F.; Rossi, M.; Palazzi, E.; von Hardenberg, J.; Brunetti, M.T.; Peruccacci, S. Implications of climate change on landslide hazard in central italy. *Sci. Total Environ.* **2018**, *630*, 1528–1543. [CrossRef] [PubMed]

12. Peres, D.J.; Cancelliere, A. Modeling impacts of climate change on return period of landslide triggering. *J. Hydrol.* **2018**, *567*, 420–434. [CrossRef]

13. Gariano, S.L.; Guzzetti, F. Landslides in a changing climate. *Earth-Sci. Rev.* **2016**, *162*, 227–252. [CrossRef]

14. Fell, R.; Corominas, J.; Bonnard, C.; Cascini, L.; Leroi, E.; Savage, W.Z. Guidelines for landslide susceptibility, hazard and risk zoning for land use planning. *Eng. Geol.* **2008**, *102*, 85–98. [CrossRef]

15. Corominas, J.; van Westen, C.; Frattini, P.; Cascini, L.; Malet, J.P.; Fotopoulou, S.; Catani, F.; Van Den Eeckhaut, M.; Mavrouli, O.; Agliardi, F.; et al. Recommendations for the quantitative analysis of landslide risk. *Bull. Eng. Geol. Environ.* **2014**, *73*, 209–263. [CrossRef]

16. Pourghasemi, H.R.; Teimoori Yansari, Z.; Panagos, P.; Pradhan, B. Analysis and evaluation of landslide susceptibility: A review on articles published during 2005–2016 (periods of 2005–2012 and 2013–2016). *Arab. J. Geosci.* **2018**, *11*, 193. [CrossRef]

17. Lee, S.; Dan, N.T. Probabilistic landslide susceptibility mapping in the lai chau province of vietnam: Focus on the relationship between tectonic fractures and landslides. *Environ. Geol.* **2005**, *48*, 778–787. [CrossRef]

18. Chen, W.; Pourghasemi, H.R.; Panahi, M.; Kornejady, A.; Wang, J.; Xie, X.; Cao, S. Spatial prediction of landslide susceptibility using an adaptive neuro-fuzzy inference system combined with frequency ratio, generalized additive model, and support vector machine techniques. *Geomorphology* **2017**, *297*, 69–85. [CrossRef]

19. Xu, C.; Xu, X.; Lee, Y.H.; Tan, X.; Yu, G.; Dai, F. The 2010 yushu earthquake triggered landslide hazard mapping using gis and weight of evidence modeling. *Environ. Earth Sci.* **2012**, *66*, 1603–1616. [CrossRef]

20. Xie, Z.; Chen, G.; Meng, X.; Zhang, Y.; Qiao, L.; Tan, L. A comparative study of landslide susceptibility mapping using weight of evidence, logistic regression and support vector machine and evaluated by sbas-insar monitoring: Zhouqu to wudu segment in bailong river basin, china. *Environ. Earth Sci.* **2017**, *76*, 313. [CrossRef]

21. Mandal, S.; Mandal, K. Bivariate statistical index for landslide susceptibility mapping in the rorachu river basin of eastern sikkim himalaya, india. *Spat. Inf. Res.* **2018**, *26*, 59–75. [CrossRef]

22. Regmi, A.D.; Devkota, K.C.; Yoshida, K.; Pradhan, B.; Pourghasemi, H.R.; Kumamoto, T.; Akgun, A. Application of frequency ratio, statistical index, and weights-of-evidence models and their comparison in landslide susceptibility mapping in central nepal himalaya. *Arab. J. Geosci.* **2014**, *7*, 725–742. [CrossRef]

23. Jaafari, A.; Najafi, A.; Pourghasemi, H.R.; Rezaeian, J.; Sattarian, A. Gis-based frequency ratio and index of entropy models for landslide susceptibility assessment in the caspian forest, northern iran. *Int. J. Environ. Sci. Technol.* **2014**, *11*, 909–926. [CrossRef]

24. Tien Bui, D.; Shahabi, H.; Shirzadi, A.; Chapi, K.; Alizadeh, M.; Chen, W.; Mohammadi, A.; Ahmad, B.; Panahi, M.; Hong, H.; et al. Landslide detection and susceptibility mapping by airsar data using support vector machine and index of entropy models in cameron highlands, malaysia. *Remote Sens.* **2018**, *10*, 1527. [CrossRef]

25. Hong, H.; Chen, W.; Xu, C.; Youssef, A.M.; Pradhan, B.; Tien Bui, D. Rainfall-induced landslide susceptibility assessment at the chongren area (china) using frequency ratio, certainty factor, and index of entropy. *Geocarto Int.* **2017**, *32*, 139–154. [CrossRef]

26. Chen, W.; Li, W.; Chai, H.; Hou, E.; Li, X.; Ding, X. Gis-based landslide susceptibility mapping using analytical hierarchy process (ahp) and certainty factor (cf) models for the baozhong region of baoji city, china. *Environ. Earth Sci.* **2016**, *75*, 1–14. [CrossRef]

27. Dou, J.; Oguchi, T.; Hayakawa, Y.S.; Uchiyama, S.; Saito, H.; Paudel, U. Gis-based landslide susceptibility mapping using a certainty factor model and its validation in the chuetsu area, central japan. In *Landslide Science for a Aafer Geoenvironment*; Springer: New York, NY, USA, 2014; pp. 419–424.

28. Chen, W.; Shahabi, H.; Shirzadi, A.; Hong, H.; Akgun, A.; Tian, Y.; Liu, J.; Zhu, A.-X.; Li, S. Novel hybrid artificial intelligence approach of bivariate statistical-methods-based kernel logistic regression classifier for landslide susceptibility modeling. *Bull. Eng. Geol. Environ.* **2018**, 1–23. [CrossRef]

29. Pradhan, A.M.S.; Kim, Y.-T. Spatial data analysis and application of evidential belief functions to shallow landslide susceptibility mapping at mt. Umyeon, seoul, korea. *Bull. Eng. Geol. Environ.* **2017**, *76*, 1263–1279. [CrossRef]

30. Ding, Q.; Chen, W.; Hong, H. Application of frequency ratio, weights of evidence and evidential belief function models in landslide susceptibility mapping. *Geocarto Int.* **2017**, *32*, 619–639. [CrossRef]

31. Zhang, T.; Han, L.; Chen, W.; Shahabi, H. Hybrid integration approach of entropy with logistic regression and support vector machine for landslide susceptibility modeling. *Entropy* **2018**, *20*, 884. [CrossRef]

32. Mandal, S.; Mandal, K. Modeling and mapping landslide susceptibility zones using gis based multivariate binary logistic regression (lr) model in the rorachu river basin of eastern sikkim himalaya, india. *Modeling Earth Syst. Environ.* **2018**, *4*, 69–88. [CrossRef]

33. Chen, W.; Zhang, S.; Li, R.; Shahabi, H. Performance evaluation of the gis-based data mining techniques of best-first decision tree, random forest, and naïve bayes tree for landslide susceptibility modeling. *Sci. Total Environ.* **2018**, *644*, 1006–1018. [CrossRef]

34. Youssef, A.M.; Pourghasemi, H.R.; Pourtaghi, Z.S.; Al-Katheeri, M.M. Landslide susceptibility mapping using random forest, boosted regression tree, classification and regression tree, and general linear models and comparison of their performance at wadi tayyah basin, asir region, saudi arabia. *Landslides* **2016**, *13*, 839–856. [CrossRef]

35. Zhou, C.; Yin, K.; Cao, Y.; Ahmed, B.; Li, Y.; Catani, F.; Pourghasemi, H.R. Landslide susceptibility modeling applying machine learning methods: A case study from longju in the three gorges reservoir area, china. *Comput. Geosci.* **2018**, *112*, 23–37. [CrossRef]

36. Pham, B.T.; Jaafari, A.; Prakash, I.; Bui, D.T. A novel hybrid intelligent model of support vector machines and the multiboost ensemble for landslide susceptibility modeling. *Bull. Eng. Geol. Environ.* **2018**, 1–22. [CrossRef]

37. Hong, H.; Liu, J.; Bui, D.T.; Pradhan, B.; Acharya, T.D.; Pham, B.T.; Zhu, A.X.; Chen, W.; Ahmad, B.B. Landslide susceptibility mapping using j48 decision tree with adaboost, bagging and rotation forest ensembles in the guangchang area (china). *Catena* **2018**, *163*, 399–413. [CrossRef]

38. Chen, W.; Shahabi, H.; Shirzadi, A.; Li, T.; Guo, C.; Hong, H.; Li, W.; Pan, D.; Hui, J.; Ma, M.; et al. A novel ensemble approach of bivariate statistical-based logistic model tree classifier for landslide susceptibility assessment. *Geocarto Int.* **2018**, *33*, 1398–1420. [CrossRef]

39. Chen, W.; Shahabi, H.; Zhang, S.; Khosravi, K.; Shirzadi, A.; Chapi, K.; Pham, B.T.; Zhang, T.; Zhang, L.; Chai, H.; et al. Landslide susceptibility modeling based on gis and novel bagging-based kernel logistic regression. *Appl. Sci.* **2018**, *8*, 2540. [CrossRef]

40. Truong, X.; Mitamura, M.; Kono, Y.; Raghavan, V.; Yonezawa, G.; Truong, X.; Do, T.; Tien Bui, D.; Lee, S. Enhancing prediction performance of landslide susceptibility model using hybrid machine learning approach of bagging ensemble and logistic model tree. *Appl. Sci.* **2018**, *8*. [CrossRef]

41. Pham, B.T.; Tien Bui, D.; Prakash, I.; Dholakia, M.B. Hybrid integration of multilayer perceptron neural networks and machine learning ensembles for landslide susceptibility assessment at himalayan area (india) using gis. *CATENA* **2017**, *149*, 52–63. [CrossRef]

42. Pham, B.T.; Shirzadi, A.; Tien Bui, D.; Prakash, I.; Dholakia, M.B. A hybrid machine learning ensemble approach based on a radial basis function neural network and rotation forest for landslide susceptibility modeling: A case study in the himalayan area, india. *Int. J. Sediment. Res.* **2017**. [CrossRef]

43. Chen, W.; Pourghasemi, H.R.; Kornejady, A.; Zhang, N. Landslide spatial modeling: Introducing new ensembles of ann, maxent, and svm machine learning techniques. *Geoderma* **2017**, *305*, 314–327. [CrossRef]

44. Pham, B.T.; Prakash, I.; Tien Bui, D. Spatial prediction of landslides using a hybrid machine learning approach based on random subspace and classification and regression trees. *Geomorphology* **2018**, *303*, 256–270. [CrossRef]

45. Zabihi, M.; Pourghasemi, H.R.; Pourtaghi, Z.S.; Behzadfar, M. Gis-based multivariate adaptive regression spline and random forest models for groundwater potential mapping in iran. *Environ. Earth Sci.* **2016**, *75*, 665. [CrossRef]

46. Lagomarsino, D.; Tofani, V.; Segoni, S.; Catani, F.; Casagli, N. A tool for classification and regression using random forest methodology: Applications to landslide susceptibility mapping and soil thickness modeling. *Environ. Modeling Assess.* **2017**, *22*, 201–214. [CrossRef]

47. Breiman, L. Bagging predictors. *Mach. Learn.* **1996**, *24*, 123–140. [CrossRef]

48. Tien Bui, D.; Ho, T.-C.; Pradhan, B.; Pham, B.-T.; Nhu, V.-H.; Revhaug, I. Gis-based modeling of rainfall-induced landslides using data mining-based functional trees classifier with adaboost, bagging, and multiboost ensemble frameworks. *Environ. Earth Sci.* **2016**, *75*, 1101. [CrossRef]

49. Xia, C.-k.; Su, C.-l.; Cao, J.-t.; Li, P. Multiboost with enn-based ensemble fault diagnosis method and its application in complicated chemical process. *J. Cent. South. Univ.* **2016**, *23*, 1183–1197. [CrossRef]

50. Pham, B.T.; Shirzadi, A.; Tien Bui, D.; Prakash, I.; Dholakia, M.B. A hybrid machine learning ensemble approach based on a radial basis function neural network and rotation forest for landslide susceptibility modeling: A case study in the himalayan area, india. *Int. J. Sediment. Res.* **2018**, *33*, 157–170. [CrossRef]

51. Fanos, A.M.; Pradhan, B.; Mansor, S.; Yusoff, Z.M.; Abdullah, A.F.b. A hybrid model using machine learning methods and gis for potential rockfall source identification from airborne laser scanning data. *Landslides* **2018**, *15*, 1833–1850. [CrossRef]

52. Chen, W.; Panahi, M.; Tsangaratos, P.; Shahabi, H.; Ilia, I.; Panahi, S.; Li, S.; Jaafari, A.; Ahmad, B.B. Applying population-based evolutionary algorithms and a neuro-fuzzy system for modeling landslide susceptibility. *CATENA* **2019**, *172*, 212–231. [CrossRef]

53. Naghibi, S.A.; Moghaddam, D.D.; Kalantar, B.; Pradhan, B.; Kisi, O. A comparative assessment of gis-based data mining models and a novel ensemble model in groundwater well potential mapping. *J. Hydrol.* **2017**, *548*, 471–483. [CrossRef]

54. Chen, W.; Li, H.; Hou, E.; Wang, S.; Wang, G.; Panahi, M.; Li, T.; Peng, T.; Guo, C.; Niu, C.; et al. Gis-based groundwater potential analysis using novel ensemble weights-of-evidence with logistic regression and functional tree models. *Sci. Total Environ.* **2018**, *634*, 853–867. [CrossRef] [PubMed]

55. Kadavi, P.; Lee, C.-W.; Lee, S. Application of ensemble-based machine learning models to landslide susceptibility mapping. *Remote Sens.* **2018**, *10*, 1252. [CrossRef]

56. Chen, W.; Xie, X.; Peng, J.; Shahabi, H.; Hong, H.; Bui, D.T.; Duan, Z.; Li, S.; Zhu, A.X. Gis-based landslide susceptibility evaluation using a novel hybrid integration approach of bivariate statistical based random forest method. *CATENA* **2018**, *164*, 135–149. [CrossRef]

57. Vakhshoori, V.; Pourghasemi, H.R. A novel hybrid bivariate statistical method entitled froc for landslide susceptibility assessment. *Environ. Earth Sci.* **2018**, *77*, 686. [CrossRef]

58. Fell, R.; Glastonbury, J.; Hunter, G. Rapid landslides: The importance of understanding mechanisms and rupture surface mechanics. *Q. J. Eng. Geol. Hydrogeol.* **2007**, *40*, 9–27. [CrossRef]

59. Rosi, A.; Tofani, V.; Tanteri, L.; Stefanelli, C.T.; Agostini, A.; Catani, F.; Casagli, N. The new landslide inventory of tuscany (italy) updated with ps-insar: Geomorphological features and landslide distribution. *Landslides* **2018**, *15*, 5–19. [CrossRef]

60. Hungr, O.; Leroueil, S.; Picarelli, L. The varnes classification of landslide types, an update. *Landslides* **2014**, *11*, 167–194. [CrossRef]

61. Tien Bui, D.; Tuan, T.A.; Klempe, H.; Pradhan, B.; Revhaug, I. Spatial prediction models for shallow landslide hazards: A comparative assessment of the efficacy of support vector machines, artificial neural networks, kernel logistic regression, and logistic model tree. *Landslides* **2016**, *13*, 361–378. [CrossRef]

62. Dai, F.C.; Lee, C.F.; Li, J.; Xu, Z.W. Assessment of landslide susceptibility on the natural terrain of lantau island, hong kong. *Environ. Geol.* **2001**, *40*, 381–391.

63. Nefeslioglu, H.A.; Duman, T.Y.; Durmaz, S. Landslide susceptibility mapping for a part of tectonic kelkit valley (eastern black sea region of turkey). *Geomorphology* **2008**, *94*, 401–418. [CrossRef]

64. Pham, B.T.; Khosravi, K.; Prakash, I. Application and comparison of decision tree-based machine learning methods in landside susceptibility assessment at pauri garhwal area, uttarakhand, india. *Environ. Process.* **2017**, *4*, 711–730. [CrossRef]

65. Wu, Y.; Li, W.; Liu, P.; Bai, H.; Wang, Q.; He, J.; Liu, Y.; Sun, S. Application of analytic hierarchy process model for landslide susceptibility mapping in the gangu county, gansu province, china. *Environ. Earth Sci.* **2016**, *75*, 422. [CrossRef]

66. Saadatkhah, N.; Kassim, A.; Lee, L.M. Susceptibility assessment of shallow landslides in hulu kelang area, kuala lumpur, malaysia using analytical hierarchy process and frequency ratio. *Geotech. Geol. Eng.* **2015**, *33*, 43–57. [CrossRef]

67. Aditian, A.; Kubota, T.; Shinohara, Y. Comparison of gis-based landslide susceptibility models using frequency ratio, logistic regression, and artificial neural network in a tertiary region of ambon, indonesia. *Geomorphology* **2018**, *318*, 101–111. [CrossRef]

68. Riaz Muhammad, T.; Basharat, M.; Hameed, N.; Shafique, M.; Luo, J. A data-driven approach to landslide-susceptibility mapping in mountainous terrain: Case study from the northwest himalayas, pakistan. *Nat. Hazards Rev.* **2018**, *19*, 05018007. [CrossRef]

69. Chen, W.; Li, W.; Hou, E.; Bai, H.; Chai, H.; Wang, D.; Cui, X.; Wang, Q. Application of frequency ratio, statistical index, and index of entropy models and their comparison in landslide susceptibility mapping for the baozhong region of baoji, china. *Arab. J. Geosci.* **2015**, *8*, 1829–1841. [CrossRef]

70. Moore, I.D.; Grayson, R.B.; Ladson, A.R. Digital terrain modelling: A review of hydrological, geomorphological, and biological applications. *Hydrol. Process.* **1991**, *5*, 3–30. [CrossRef]

71. Chen, W.; Peng, J.; Hong, H.; Shahabi, H.; Pradhan, B.; Liu, J.; Zhu, A.-X.; Pei, X.; Duan, Z. Landslide susceptibility modelling using gis-based machine learning techniques for chongren county, jiangxi province, china. *Sci. Total Environ.* **2018**, *626*, 1121–1135. [CrossRef]

72. Beven, K.J.; Kirkby, M.J. A physically based, variable contributing area model of basin hydrology/un modèle à base physique de zone d'appel variable de l'hydrologie du bassin versant. *Int. Assoc. Sci. Hydrol. Bull.* **1979**, *24*, 43–69. [CrossRef]

73. Ge, Y.; Chen, H.; Zhao, B.; Tang, H.; Lin, Z.; Xie, Z.; Lv, L.; Zhong, P. A comparison of five methods in landslide susceptibility assessment: A case study from the 330-kv transmission line in gansu region, china. *Environ. Earth Sci.* **2018**, *77*, 662. [CrossRef]

74. Wu, Z.; Wu, Y.; Yang, Y.; Chen, F.; Zhang, N.; Ke, Y.; Li, W. A comparative study on the landslide susceptibility mapping using logistic regression and statistical index models. *Arab. J. Geosci.* **2017**, *10*, 187. [CrossRef]

75. Yalcin, A. Gis-based landslide susceptibility mapping using analytical hierarchy process and bivariate statistics in ardesen (turkey): Comparisons of results and confirmations. *CATENA* **2008**, *72*, 1–12. [CrossRef]

76. Akgun, A. A comparison of landslide susceptibility maps produced by logistic regression, multi-criteria decision, and likelihood ratio methods: A case study at İzmir, turkey. *Landslides* **2012**, *9*, 93–106. [CrossRef]

77. Chen, W.; Yan, X.; Zhao, Z.; Hong, H.; Bui, D.T.; Pradhan, B. Spatial prediction of landslide susceptibility using data mining-based kernel logistic regression, naive bayes and rbfnetwork models for the long county area (china). *Bull. Eng. Geol. Environ.* **2018**, 1–20. [CrossRef]

78. Xu, C.; Dai, F.; Xu, X.; Lee, Y.H. Gis-based support vector machine modeling of earthquake-triggered landslide susceptibility in the jianjiang river watershed, China. *Geomorphology* **2012**, *145–146*, 70–80. [CrossRef]

79. Ohlmacher, G.C.; Davis, J.C. Using multiple logistic regression and gis technology to predict landslide hazard in northeast kansas, USA. *Eng. Geol.* **2003**, *69*, 331–343. [CrossRef]

80. Al-Abadi, A.M.; Shahid, S. A comparison between index of entropy and catastrophe theory methods for mapping groundwater potential in an arid region. *Environ. Monit. Assess.* **2015**, *187*, 576. [CrossRef]

81. Salvatici, T.; Tofani, V.; Rossi, G.; D'Ambrosio, M.; Tacconi Stefanelli, C.; Masi, E.B.; Rosi, A.; Pazzi, V.; Vannocci, P.; Petrolo, M.; et al. Application of a physically based model to forecast shallow landslides at a regional scale. *Nat. Hazards Earth Syst. Sci.* **2018**, *18*, 1919–1935. [CrossRef]

82. Peng, J.; Tong, X.; Wang, S.; Ma, P. Three-dimensional geological structures and sliding factors and modes of loess landslides. *Environ. Earth Sci.* **2018**, *77*, 675. [CrossRef]

83. Santo, A.; Di Crescenzo, G.; Forte, G.; Papa, R.; Pirone, M.; Urciuoli, G. Flow-type landslides in pyroclastic soils on flysch bedrock in southern italy: The bosco de' preti case study. *Landslides* **2018**, *15*, 63–82. [CrossRef]

84. Glade, T. Landslide occurrence as a response to land use change: A review of evidence from new zealand. *CATENA* **2003**, *51*, 297–314. [CrossRef]

85. Cui, S.-H.; Pei, X.-J.; Wu, H.-Y.; Huang, R.-Q. Centrifuge model test of an irrigation-induced loess landslide in the heifangtai loess platform, northwest china. *J. Mt. Sci.* **2018**, *15*, 130–143. [CrossRef]

86. Justice, C.O.; Townshend, J.R.G.; Holben, B.N.; Tucker, C.J. Analysis of the phenology of global vegetation using meteorological satellite data. *Int. J. Remote Sens.* **1985**, *6*, 1271–1318. [CrossRef]

87. Chen, W.; Xie, X.; Wang, J.; Pradhan, B.; Hong, H.; Tien Bui, D.; Duan, Z.; Ma, J. A comparative study of logistic model tree, random forest, and classification and regression tree models for spatial prediction of landslide susceptibility. *CATENA* **2017**, *151*, 147–160. [CrossRef]

88. Peruccacci, S.; Brunetti, M.T.; Luciani, S.; Vennari, C.; Guzzetti, F. Lithological and seasonal control on rainfall thresholds for the possible initiation of landslides in central italy. *Geomorphology* **2012**, *139–140*, 79–90. [CrossRef]

89. Tsukamoto, Y.; Ohta, T. Runoff process on a steep forested slope. *J. Hydrol.* **1988**, *102*, 165–178. [CrossRef]

90. Agterberg, F.P. Systematic approach to dealing with uncertainty of geoscience information in mineral exploration. *APCO* **1989**, *89*, 165–178.

91. Dahal, R.K.; Hasegawa, S.; Nonomura, A.; Yamanaka, M.; Masuda, T.; Nishino, K. Gis-based weights-of-evidence modelling of rainfall-induced landslides in small catchments for landslide susceptibility mapping. *Environ. Geol.* **2008**, *54*, 311–324. [CrossRef]

92. Lachenbruch, P.A.; Mccullagh, P.; Nelder, J.A. Generalized linear models. *Biometrics* **1990**, *46*, 291–303. [CrossRef]

93. Agarwal, S.; Kachroo, P.; Regentova, E. A hybrid model using logistic regression and wavelet transformation to detect traffic incidents. *IATSS Res.* **2016**, *40*, 56–63. [CrossRef]

94. Ravì, D.; Bober, M.; Farinella, G.M.; Guarnera, M.; Battiato, S. Semantic segmentation of images exploiting dct based features and random forest. *Pattern Recognit.* **2016**, *52*, 260–273. [CrossRef]

95. Masetic, Z.; Subasi, A. Congestive heart failure detection using random forest classifier. *Comput. Methods Programs Biomed.* **2016**, *130*, 54–64. [CrossRef] [PubMed]

96. Breiman, L. Random forests. *Mach. Learn.* **2001**, *45*, 5–32. [CrossRef]

97. Ozdemir, A. Landslide susceptibility mapping using bayesian approach in the sultan mountains (akşehir, turkey). *Nat. Hazards* **2011**, *59*, 1573–1607. [CrossRef]

98. Chen, W.; Xie, X.; Peng, J.; Wang, J.; Duan, Z.; Hong, H. Gis-based landslide susceptibility modelling: A comparative assessment of kernel logistic regression, naïve-bayes tree, and alternating decision tree models. *Geomat. Nat. Hazards Risk* **2017**, *8*, 950–973. [CrossRef]

99. Lin, G.-F.; Chang, M.-J.; Huang, Y.-C.; Ho, J.-Y. Assessment of susceptibility to rainfall-induced landslides using improved self-organizing linear output map, support vector machine, and logistic regression. *Eng. Geol.* **2017**, *224*, 62–74. [CrossRef]

100. Lee, J.-H.; Sameen, M.I.; Pradhan, B.; Park, H.-J. Modeling landslide susceptibility in data-scarce environments using optimized data mining and statistical methods. *Geomorphology* **2018**, *303*, 284–298. [CrossRef]

101. Yu, H.; Jiang, S.; Land, K.C. Multicollinearity in hierarchical linear models. *Soc. Sci. Res.* **2015**, *53*, 118–136. [CrossRef]

102. Ozdemir, A.; Altural, T. A comparative study of frequency ratio, weights of evidence and logistic regression methods for landslide susceptibility mapping: Sultan mountains, sw turkey. *J. Asian Earth Sci.* **2013**, *64*, 180–197. [CrossRef]

103. Ozdemir, A. Gis-based groundwater spring potential mapping in the sultan mountains (konya, turkey) using frequency ratio, weights of evidence and logistic regression methods and their comparison. *J. Hydrol.* **2011**, *411*, 290–308. [CrossRef]

104. Hong, H.; Tsangaratos, P.; Ilia, I.; Chen, W.; Xu, C. Comparing the Performance of a Logistic Regression and a Random Forest Model in Landslide Susceptibility Assessments. the Case of Wuyaun Area, China. In *Workshop on World Landslide Forum*; Mikos, M., Tiwari, B., Yin, Y., Eds.; Springer: Cham, Switzerland, 2017; pp. 1043–1050.

105. Pham, B.T.; Pradhan, B.; Tien Bui, D.; Prakash, I.; Dholakia, M.B. A comparative study of different machine learning methods for landslide susceptibility assessment: A case study of uttarakhand area (india). *Environ. Model. Softw.* **2016**, *84*, 240–250. [CrossRef]

106. Tien Bui, D.; Tuan, T.A.; Hoang, N.-D.; Thanh, N.Q.; Nguyen, D.B.; Van Liem, N.; Pradhan, B. Spatial prediction of rainfall-induced landslides for the lao cai area (vietnam) using a hybrid intelligent approach of least squares support vector machines inference model and artificial bee colony optimization. *Landslides* **2017**, *14*, 447–458. [CrossRef]

107. Chen, W.; Pourghasemi, H.R.; Naghibi, S.A. Prioritization of landslide conditioning factors and its spatial modeling in shangnan county, china using gis-based data mining algorithms. *Bull. Eng. Geol. Environ.* **2018**, *77*, 611–629. [CrossRef]

108. Chen, W.; Pourghasemi, H.R.; Naghibi, S.A. A comparative study of landslide susceptibility maps produced using support vector machine with different kernel functions and entropy data mining models in china. *Bull. Eng. Geol. Environ.* **2018**, *77*, 647–664. [CrossRef]

109. He, S.; Pan, P.; Dai, L.; Wang, H.; Liu, J. Application of kernel-based fisher discriminant analysis to map landslide susceptibility in the qinggan river delta, three gorges, china. *Geomorphology* **2012**, *171*, 30–41. [CrossRef]

110. Chen, W.; Panahi, M.; Pourghasemi, H.R. Performance evaluation of gis-based new ensemble data mining techniques of adaptive neuro-fuzzy inference system (anfis) with genetic algorithm (ga), differential evolution

(de), and particle swarm optimization (pso) for landslide spatial modelling. *CATENA* **2017**, *157*, 310–324. [CrossRef]

111. Wang, Q.; Li, W.; Wu, Y.; Pei, Y.; Xie, P. Application of statistical index and index of entropy methods to landslide susceptibility assessment in gongliu (xinjiang, china). *Environ. Earth Sci.* **2016**, *75*, 599. [CrossRef]

112. Naghibi, S.A.; Ahmadi, K.; Daneshi, A. Application of support vector machine, random forest, and genetic algorithm optimized random forest models in groundwater potential mapping. *Water Resour. Manag.* **2017**, *31*, 2761–2775. [CrossRef]

113. Chen, W.; Shirzadi, A.; Shahabi, H.; Ahmad, B.B.; Zhang, S.; Hong, H.; Zhang, N. A novel hybrid artificial intelligence approach based on the rotation forest ensemble and naïve bayes tree classifiers for a landslide susceptibility assessment in langao county, china. *Geomat. Nat. Hazards Risk* **2017**, *8*, 1955–1977. [CrossRef]

114. Ilia, I.; Tsangaratos, P. Applying weight of evidence method and sensitivity analysis to produce a landslide susceptibility map. *Landslides* **2016**, *13*, 379–397. [CrossRef]

115. Polykretis, C.; Chalkias, C. Comparison and evaluation of landslide susceptibility maps obtained from weight of evidence, logistic regression, and artificial neural network models. *Nat. Hazards* **2018**. [CrossRef]

116. Lee, S. Landslide detection and susceptibility mapping in the sagimakri area, korea using kompsat-1 and weight of evidence technique. *Environ. Earth Sci.* **2013**, *70*, 3197–3215. [CrossRef]

117. Malamud, B.D.; Turcotte, D.L.; Guzzetti, F.; Reichenbach, P. Landslide inventories and their statistical properties. *Earth Surf. Process. Landf.* **2004**, *29*, 687–711. [CrossRef]

118. Baeza, C.; Lantada, N.; Amorim, S. Statistical and spatial analysis of landslide susceptibility maps with different classification systems. *Environ. Earth Sci.* **2016**, *75*, 1318. [CrossRef]

119. Akgun, A.; Sezer, E.A.; Nefeslioglu, H.A.; Gokceoglu, C.; Pradhan, B. An easy-to-use matlab program (mamland) for the assessment of landslide susceptibility using a mamdani fuzzy algorithm. *Comput. Geosci.* **2012**, *38*, 23–34. [CrossRef]

120. Kumar, R.; Anbalagan, R. Landslide susceptibility mapping using analytical hierarchy process (ahp) in tehri reservoir rim region, uttarakhand. *J. Geol. Soc. India* **2016**, *87*, 271–286. [CrossRef]

121. Khosravi, K.; Pourghasemi, H.R.; Chapi, K.; Bahri, M. Flash flood susceptibility analysis and its mapping using different bivariate models in iran: A comparison between shannon(')s entropy, statistical index, and weighting factor models. *Environ. Monit. Assess.* **2016**, *188*, 656. [CrossRef] [PubMed]

122. Kleinbaum, D.G.; Klein, M. Introduction to logistic regression. In *Logistic Regression: A Self-Learning Text*; Kleinbaum, D.G., Ed.; Springer: New York, NY, USA, 2010.

123. Genuer, R.; Poggi, J.-M.; Tuleau-Malot, C.; Villa-Vialaneix, N. Random forests for big data. *Big Data Res.* **2017**, *9*, 28–46. [CrossRef]

Article

Modeling of CO Emissions from Traffic Vehicles Using Artificial Neural Networks

Omer Saud Azeez [1], Biswajeet Pradhan [2,*], Helmi Z. M. Shafri [1], Nagesh Shukla [2], Chang-Wook Lee [3,*] and Hossein Mojaddadi Rizeei [2]

1 Department of Civil Engineering, Faculty of Engineering, University Putra Malaysia (UPM), Serdang 43400, Malaysia; baghdad.eagle2016@gmail.com (O.S.A.); helmi@upm.edu.my (H.Z.M.S.)
2 Centre for Advanced Modelling and Geospatial Information Systems (CAMGIS), Faculty of Engineering and Information Technology, University of Technology Sydney, Sydney 2007, NSW, Australia; nagesh.shukla@uts.edu.au (N.S.); h.mojaddadi@gmail.com (H.M.R.)
3 Division of Science Education, Kangwon National University, 1 Kangwondaehak-gil, Chuncheon-si 24341, Gangwon-do, Korea
* Correspondence: Biswajet.Pradhan@uts.edu.au (B.P.); cwlee@kangwon.ac.kr (C.-W.L.)

Received: 21 November 2018; Accepted: 11 January 2019; Published: 16 January 2019

Abstract: Traffic emissions are considered one of the leading causes of environmental impact in megacities and their dangerous effects on human health. This paper presents a hybrid model based on data mining and GIS models designed to predict vehicular Carbon Monoxide (CO) emitted from traffic on the New Klang Valley Expressway, Malaysia. The hybrid model was developed based on the integration of GIS and the optimized Artificial Neural Network algorithm that combined with the Correlation based Feature Selection (CFS) algorithm to predict the daily vehicular CO emissions and generate prediction maps at a microscale level in a small urban area by using a field survey and open source data, which are the main contributions to this paper. The other contribution is related to the case study, which represents the spatial and quantitative variations in the vehicular CO emissions between toll plaza areas and road networks. The proposed hybrid model consists of three steps: the first step is the implementation of the correlation-based Feature Selection model to select the best model's predictors; the second step is the prediction of vehicular CO by using a multilayer perceptron neural network model; and the third step is the creation of micro scale prediction maps. The model was developed using six traffic CO predictors: number of vehicles, number of heavy vehicles, number of motorbikes, temperature, wind speed and a digital surface model. The network architecture and its hyperparameters were optimized through a grid search approach. The traffic CO concentrations were observed at 15-min intervals on weekends and weekdays, four times per day. The results showed that the developed model had achieved validation accuracy of 80.6 %. Overall, the developed models are found to be promising tools for vehicular CO simulations in highly congested areas.

Keywords: traffic CO; traffic CO prediction; neural networks; GIS; land use/land cover (LULC)

---

1. Introduction

The transport infrastructure like expressways and roads has a significant importance in the development of any country's economy by linking cities. These infrastructures are rapidly developing due to the changing in the traffic modes, leading to congested roads. Hence, road traffic emissions are increasing, creating many negative impacts on air quality on roadways, intersections and toll roads. Traffic emissions, such as carbon monoxide (CO), are the primary contributor to overall air pollution from this infrastructure, and the primary source of traffic emissions is vehicular exhausts.

Spatial prediction models are effectively used as a decision-making support tool for prediction and simulation of traffic emissions on road networks [1–3]. There are various negative impacts that can result from inappropriate traffic levels, including high levels of noise and high concentrations of gaseous pollutants [4,5]. Several diseases e.g., cancers, heart diseases, respiratory problems and preterm births, can occur when human beings are exposed to high concentrations of CO [6–8].

The measurement of vehicular emissions on roadways and toll gates may be costly, risky and requires a lot of time and effort. Moreover, the designers do not have the opportunity to determine the vehicular emissions through the design process. In the most recent planning techniques for design of highways and road networks, traffic emission models are often required to support sustainable transportation planning and the reduction of traffic emissions from sources such as congestion and tollgate areas. Thus, the GIS-based modeling of traffic emission and intra-urban air pollution exposure can be an effective tool in the environmental assessment for sustainable road planning. This tool can distinguish the areas affected by different types of pollutants and the related ecological and social factors. This would be able to determine the best strategy to support the decision makers [9]. On the other hand, GIS can save costs and time in the traffic emission modeling and can therefore be used in sustainable planning.

Different types of traffic CO prediction approaches are mentioned in the literature [10–12]. Early methods of traffic CO modelling were based on traditional techniques using data sampling and global position system (GPS) techniques. Several thematic maps and vehicle emission equations are combined to model the traffic emission distribution in a region and produce informative maps that could help in effective decision making [10]. Recent methods are mostly based on land-use regression analysis using statistical and soft computing algorithms [10,13]. These statistical and computing techniques allow the input of various traffic and road geometry factors. Almost all these models are designed by using experimental samples; consequently, these models are highly influenced by the traffic flow condition and the measurement style and the geographic locations [14]. The main drawback of these models is that they can not be generalized because of the local environment like vehicle model and type and the weather [15,16]. Ref. [17] presented an approach of recognizing the road geometric features from positioning information surveyed by collecting vehicle data.

2. Previous Works

Many models have been developed to predict CO emissions and other traffic emissions, such as NOx, NO_2, CO_2 and SO_2. In a paper, Ref. [18] presented a methodology by integrating the spatial analysis techniques and the neighborhood statistic function algorithm to evaluate the spatial diffusion of the gaseous pollutant in north of Italy by using the air pollutant records obtained from monitoring stations and GIS data (i.e., administrative borders, built-up areas, emission sources and road networks). Their results were illustrated on grids with a cell size of (4 × 4) km. Although this method showed a significant spatial representation of air pollution, the methodology was constrained by the limited spatial resolution. Therefore, it cannot be used for high-resolution data. Ref. [19] developed a GIS-based tool by combining the operational street pollution model (OSPM) and a multi-agent-based transportation model (MATSIM) to estimate the air pollutant concentrations in Munich, Germany. Their results showed hourly prediction of NOx from traffic. This approach can be used as an effective tool for air quality studies in urban areas. Nevertheless, its disadvantages appear in the complexity of a system that comprises different models where the non-expert users are not able to use it. Ref. [20] developed a model based on land-use regression algorithm and land-use types, meteorological variables and vehicle-kilometers-travelled (VKTs) and linear regression algorithm to estimate the concentrations of Nitrogen Dioxide (NO_2) in Seoul, Korea. The results showed the significant impacts of the residential, commercial land use, wind speed, temperature and humidity on the concentrations of NO_2. The air pollutants recorded by the fixed air quality monitoring stations can be affected by several factors such as terrain and buildings altitude. Moreover, the weather factors are not suitable to model and produce high-resolution products such as roadmaps. Ref. [21] presented a statistical model based on the fuzzy logic system to predict CO concentrations in Tehran, Iran. This model mainly relied on historical data, which were obtained from monitoring stations. Fuzzy logic algorithms were applied to combine the parameters. Their results showed that lowest Room Mean Square Error (RMSE) was recorded at 2.13. Another study related to statistical modeling was conducted by [22] to forecast air pollutants in Hong Kong based on the integration of two statistical models, i.e., the generalized additive models and the Global Forecast System, which linked the air pollution with meteorological data. Results showed a contrast in the air pollutant levels between urban and suburban areas. This model is useful for predicting air quality in complex terrain areas. These models lack the spatial aspect and could not be used to produce prediction maps. Ref. [23] developed a methodology by using two commercial programs to estimate the traffic emissions in small area in Madrid, Spain. The VISSIM program was used for traffic simulation to calculate a velocity-time profile. Then, the related emissions at the vehicle level were completed using the VERSIT + micro program. Results showed the spatial variation in NOx and PM_{10} concentrations are based on microscale maps with high resolution, cell size (5 × 5) m. This model depends on the estimated emissions data based on prediction simulations without using actual samples based on sampling equipment.

Recently, machine-learning technologies have attracted researchers. The neural network (NN) models are the most popular models in the Artificial Intelligence models. Ref. [24] developed a model by integrating the artificial neural network (ANN) algorithm and evolutionary polynomial regression (EPR) to estimate the CO concentrations in Tabriz City, Iran. The EPR is one of the data mining algorithms developed based on evolutionary computing and the integration of numerical regression and genetic algorithm. The EPR model involves two stages: a genetic algorithm is used in the first stage based on the numerical regression to search for symbolic structures, whereas in the second stage, symbolic structure parameters are determined based on the linear least squares techniques. Their results showed that the ANN model is more reliable than the EPR model. The highest value of the correlation coefficient was measured at 0.85 based on NN and 0.41 using EPR. This study indicated that NN modeling can be efficiently utilized for air quality forecasting. On the other hand, ref. [25] developed a model based on the NN algorithm and data obtained from field survey to estimate the hourly traffic emissions near roads. The authors used different parameters such as traffic data, meteorology, proximity to roads and road direction. This model is considered as an efficient approach for predicting pollutant near a road. Although they used geographic information as a parameter, their results did not contain spatial prediction results such as maps. They only presented a statistical analysis. Results showed that the highest correlation coefficient for the CO prediction was 0.879. Ref. [26] conducted the most relevant studies that combined the NN model and the spatial prediction model. They presented an approach that combined the linear-chain conditional random field algorithm and ANN model to generate real-time air pollution maps. They utilized the data recorded from monitoring stations and the traffic data collected from the field while geographic parameters like land use and road network were derived using GIS data. Their results showed the air pollutants prediction on maps with (1 × 1) km spatial resolution. However, the developed model did not consider many issues like uncertainty, modeling multifactor and nonlinearity. Although several authors have attempted to overcome these issues, they principally focused on the integration of big data and the large scale modeling. On the other hand, most of these models deal with large quantities of data, expensive equipment, and complex data processing models, which require substantial time, cost and other resources.

In this paper, we presented a hybrid model to produce microscale prediction maps considering toll gate locations, as well as the other parameters listed in the literature. The model is developed by combining the metaheuristic optimization technique and ANN algorithm to predict traffic emissions based on a small number of training data and avoiding transferability issues. The metaheuristic optimization algorithms like correlation-based feature selection models which have the ability to find best model's predictors in a short time were compared to other optimization techniques. Also, ANN algorithms are suitable for prediction based on few training data. The major contributions of this work lie in producing highly accurate predictive maps and providing a description of the high variation of traffic emissions on roads and tollgate areas. Other significant advantages may include easy implementation of the proposed model in open source GIS software where the non-expert users can utilize the model for rapid simulations and assessments of vehicular emissions based on microscale prediction maps. Also, the users can design GIS models based on their needs. We proposed that the combination of metaheuristic optimization and machine learning algorithms could help improving the forecasting of CO emissions on roads, highways and in tollgate areas.

3. Materials and Methods

3.1. Study Area

This study was conducted near of Subang jaya toll plaza which links the New Klang Valley Expressway (NKVE) and the federal expressway in Peninsular Malaysia. Subang toll plaza is located within a highly dense populated area in Petaling Jaya, Selangor, Malaysia (Figure 1). The total length of NKVE is 35 km, which connects urban and industrial areas in the capital city of Kuala Lumpur. It is a major highway for citizens who are living in the main cities like Kuala Lumpur, Subang, Shah Alam, Damansara, Sungai Buloh, Klang and Petaling Jaya. The vehicular speed limits are standardized to 110 and 90 km/h on Bukit Raja to Bukit Lanjan stretch and Bukit Lanjan to Jalan Duta stretch, respectively. The study area contains different types of land use such as tollgate area, commercial, industrial and residential areas, making it well suited for vehicular emissions studies.

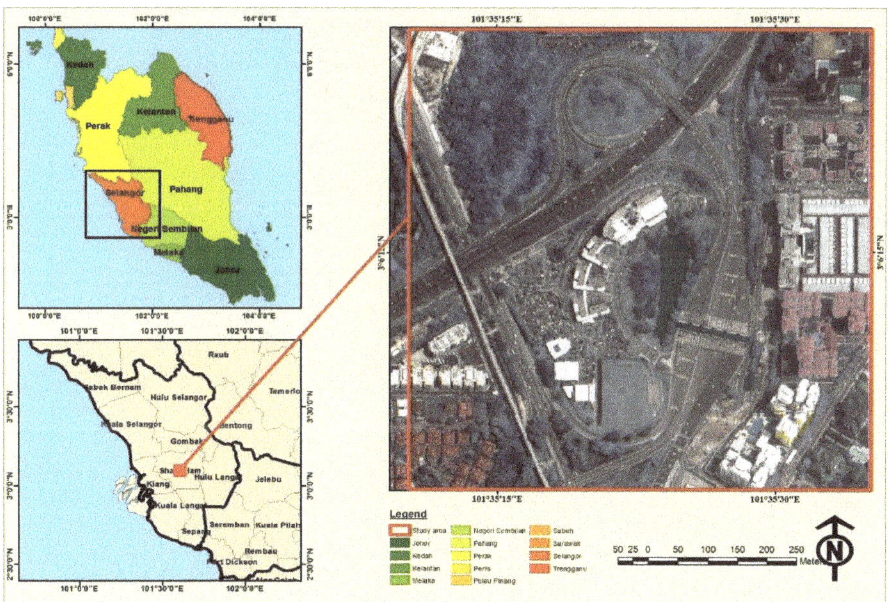

Figure 1. Location map of the study area.

3.2. Data and Method

Several data (i.e., vehicular CO samples, meteorology and traffic flow data) were collected from the field during April 2017. Light Detection and Ranging (LiDAR) data were collected in March 2017, and the Worldview3 satellite image was captured in May 2017. Figure 2 shows the overall methodology, which consists of several steps. The hybrid model is designed to predict CO emissions at a specific time and location, for example, prediction maps based on different times of a day. The first step is data collection, which was achieved based on a gas analyzer and data loggers to simultaneously collect CO, temperature, humidity, and traffic information. The LULC map was extracted from the Worldview-3 image with spatial resolution of 0.3 m. The digital surface model (DSM) was derived from the Airborne

LiDAR point clouds by using Environment for Visualizing Images (ENVI) software. The second step is the statistical modelling, which was applied by combining two models i.e., the correlation based feature selection (CFS) and Multilayer Perceptron (MLP) by using Weka software. The final step is the spatial modelling based on the regression equations derived from regression analysis and GIS techniques to generate microscale prediction maps for the traffic CO emissions during different times of the day.

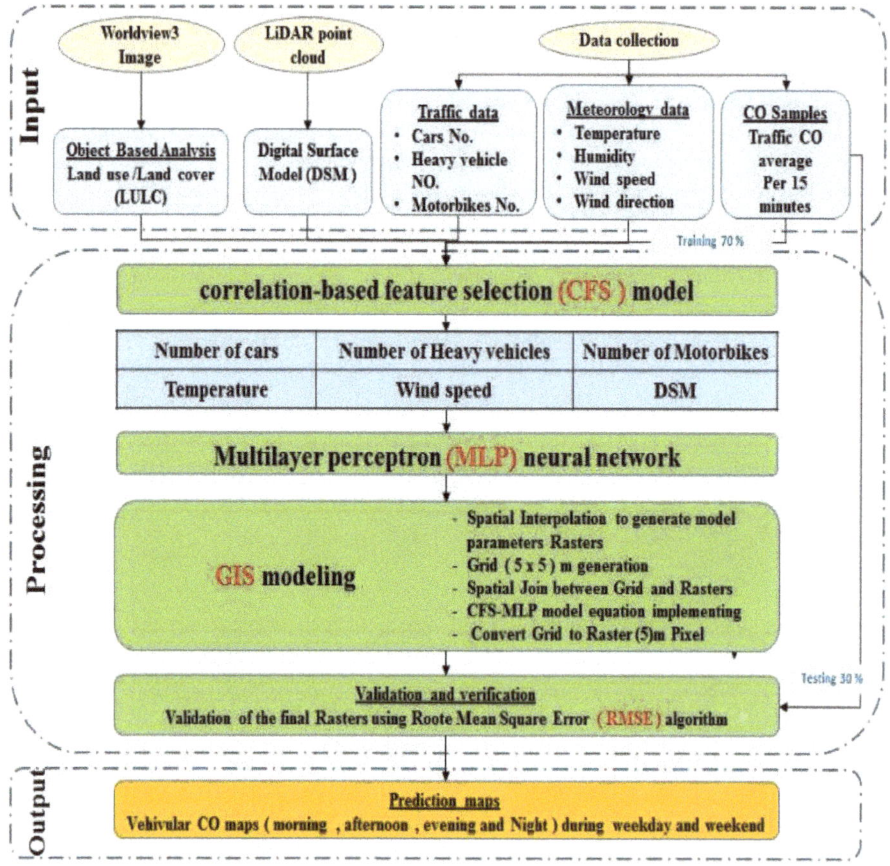

Figure 2. The overall methodological flow chart employed in this study.

3.3. Field Surveying

3.3.1. Sampling Selection

The vehicular CO, traffic condition and meteorology data are important for the development of vehicular CO emissions models [27]. Many studies have described approaches of collecting traffic flow data from the field, and the most important aspect has been determined to be the distribution of air pollutants samples and their suitability [28,29]. The accuracy of the spatial interpolation is highly affected by the sampling design and the variations between traffic CO samples [30]. Moreover, the density of points must be good enough to achieve high accuracy of interpolated data. Conversely, a large number of samples should be avoided to decrease the processing time. Most importantly, the samples density should be adjusted by considering the vehicular CO diffusion characteristics. In this paper, traffic CO data were collected according to a procedure given by [31]. Their method was implemented by creating sampling locations based on a random selection method by using spatial analysis techniques. This approach generates the optimum number of samples compared to the study area [31].

First, three layers (residential, commercial and industrial) from the land-use map were extracted. These layers were converted into points with a spatial constraint to force them inside the land-use polygons. Next, the density of points was estimated using a 150 m search radius and the resolution of the output density raster was set to 25 m. The density rasters were then integrated based on different coefficients (industrial = 1, commercial = 2, and residential = 3). The next step is the rescaling of the final density raster from 0 to 1 depending on linear approach led to the creation of the probability raster which was used to select the samples of the traffic CO. Next, the geospatial balanced points were then created within the study area by using the probability raster. The total number of points were generated based on the length of the road network, the cost of the project and the capability of traffic emissions detection equipment used in the data gathering. As a result, the locations of the traffic CO samples were distributed along the study area. Therefore, extra steps should be implemented to improve the created points in order to select the final sample locations depending on the transportation characteristics. Tessellated grids with a spatial resolution 25 m were generated. Then, these grids were intersected with the created points and the road network layer and the remaining tessellated grids were removed. Subsequently, the final traffic CO samples were selected within the intersected tessellated grids and road network.

3.3.2. Data Collection

Traffic CO concentrations were collected from the field by using a low cost Gas Analyzer device model Micro-clip5 (1 ppm resolution). The data were collected continuously in 15-min intervals (recording the 15-min minimum, maximum, and average). The traffic flow data and meteorology information were simultaneously collected using a data logger and GPS device (Garmin GPS etrex 10, Olathe, KS, USA; available at University Putra Malaysia). Figure 3 shows the sampling procedure. The traffic CO analyzer was installed in the location of samples by using a Global Navigation System (GPS), at least 2 m from the road edge. The GPS was used to determine the geographic location of samples and to manually verify the locations using land-use maps. The traffic CO was measured four times a day on weekends and weekdays, in the morning (6.30 a.m. to 8.30 a.m.), afternoon (11.30 a.m. to 1.30 p.m.), evening (6.30 p.m. to 8.30 p.m.), and at night (11 p.m. to 12 midnight). In addition, the traffic data were collected by using digital cameras installed in the road's side in the sample's location. The traffic flow data were classified into several types (the number of cars, the number of heavy vehicles and the number of motorbikes), where the cars were private cars and taxi cars. On the other hand, the heavy vehicles refer to the following: medium truck, heavy truck, super-heavy/special duty truck and buses, while the motorbikes refer to any type of motorbikes. The meteorological data (temperature, humidity, wind speed and wind direction) were collected for two days, including on weekdays and weekends, to examine various scenarios in the study area for accurate and effective study of traffic CO modelling and mapping (e.g., hazard maps, risk maps, and further analysis). Figure 4 shows the data collection procedure adopted in this study.

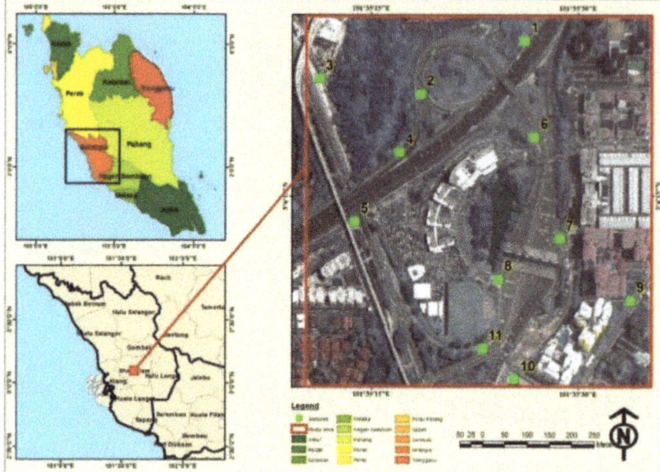

Figure 3. The sampling locations.

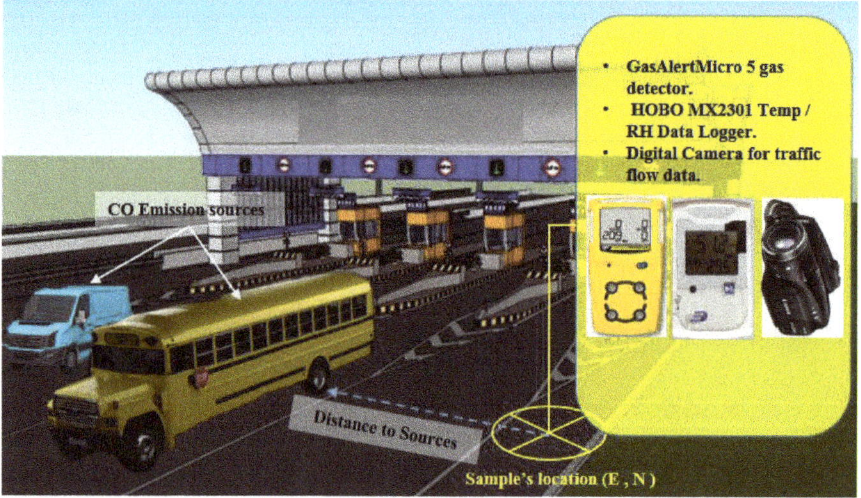

Figure 4. Location of traffic CO and meteorological parameter measurement on a highway section.

3.4. Vehicular CO Prediction Model

3.4.1. Vehicular CO Model Parameters

The presented model aims to predict the daily traffic CO emissions and create prediction maps at different times of a day by using GIS techniques. The traffic CO descriptor (i.e., the dependent parameter) in the current study is the vehicular CO emissions measured every 15 min. The contributing parameters to vehicular CO emission were first selected depending on the previous studies and with consideration of traffic condition, weather characteristics, the surrounding (LULC), topography, and the building heights in the study area. These parameters are the number of vehicles, number of heavy vehicles, the number of motorbikes, temperature, humidity, wind speed, wind direction, LULC and digital surface model (DSM). Many studies have been conducted based on these parameters [32–36]. There are many factors that could affect the vehicular emissions such as engine condition and the fuel type (gasoline, diesel); however, these data cannot be detected through field survey. On the other hand, the data collection did not contain vehicle speed information because the study area is very small and the variation between the vehicles speed is difficult to differentiate. Consequently, the vehicle speed profile is stable during the day. There are many vehicular emissions studies conducted without adopting vehicle speed, such as [25]. The parameters' statistics are shown in Table 1. The NN model was used to find out the degree of contribution of each mentioned parameter for estimating the vehicular CO emission. The NN has the ability to model the complex simulations of non-linear problems such as vehicular CO emission. However, contributing parameters to vehicular CO emission may not be directly used as inputs in the NN model, because of high levels of correlation between the factors resulting in multicollinearity, which can reduce the precision of estimating the vehicular CO emission. Moreover, using a large number of contributing parameters to vehicular CO emission (traffic CO predictors) as input layers of NN can generate over-fitting problems and increase the complexity ratio to run NN model. In the proposed NN model, only relevant and low-correlated parameters were used as inputs. The relevant and significant factors were selected using the CFS model. Table 2 shows the traffic CO measurements results.

Table 1. Summary statistics of traffic CO predictors.

Parameter	Average	Minimum	Maximum
Number of vehicles (per 15 min)	1172	126	2762
Number of heavy vehicles (per 15 min)	78	16	325
Number of motorbikes (per 15 min)	112	9	489
Temperature (°C)	29.9	25.6	37.7
Humidity (%)	73.5	54.3	94.5
Wind Speed (mph)	16.87	16	18.20
Wind Direction (angle)	247.1	0	350
DSM (m)	25.7	10.03	129.5

Table 2. Summary statistics of traffic CO measurements.

Time	Weekend			Weekday		
	Average CO Concentration (per 15 min) (ppm)			Average CO Concentration (per 15 min) (ppm)		
	Min	Max	Mean	Min	Max	Mean
Morning	0	8	2.36	0	30.5	8.5
Afternoon	0	14.5	3.5	0	12.8	4.5
Evening	0	9.3	3.92	0	27.3	5.84
Night	0	3.6	1.47	0	5.6	1.9

3.4.2. Correlation-Based Feature Selection (CFS) Model

The CFS algorithm is one of the machine learning algorithms which is considered as a filter algorithm that choose the features based on correlation concepts [37]. A major characteristic of the correlation-based function is the ability to choose sub-groups that include features that are unusually correlated to the targeted class but unassociated with each other. On the other hand, this algorithm neglects the features with low correlation with the targeted class, and this algorithm is used to delete the duplicated features because they will be correlated with one of the rest of the features at least. The acknowledgement of a feature will rely on the degree to which it predicts classes in territories of the instance space not currently anticipated by different features.

The CFS's feature subset assessment function is presented in Equation (1):

$$Ms = k\overline{r_{cf}} / \sqrt{k + k(k-1)\,\overline{r_{ff}}} \tag{1}$$

where Ms is the heuristic "merit" of a feature subset S containing k features, $\overline{r_{cf}}$ is the mean of the feature-class correlation ($f \in$ S), and $\overline{r_{ff}}$ is the average of the feature-feature intercorrelation.

3.4.3. Multilayer Perceptron (MLP) Neural Network

The MLP algorithm is one of the ANN algorithms which results from adding hidden layers to the simple perceptron. In this algorithm, the structure of the NN is generally trained based on backpropagation algorithm and some related variants. Therefore, the models are designed based on the integration of the MLP algorithm and a backpropagation algorithm called backpropagation neural network [38–43]. The multilayer perceptron algorithm was developed due to the computational limitations that resulted from single-layered perceptron models. According to the experiments, the multilayer perceptron algorithm has the ability to represent complex simulations and mapping and process high level non-linear problems. Also, the MLP algorithm has the ability to process the nonlinear features, thus allowing the representation of a continuous function of non-linear activation functions such as sigmoid functions [44], which has a clear analogy with the conventional representation of a periodic function such as a Fourier series (i.e., as the sum of simple sine waves). Therefore, the MLP can be considered as a universal functional approximation. Figure 3 shows the architecture of an MLP with several layers of neurons and nonlinear activation functions.

The MLP is considered as a feed-forward system with single or multiple layer of segments among network output and input layers [38–45]. With the assumption of L-layer MPL, the system can signified by $N_{n_0,n_1,\ldots,n_L}^{L}$, where n_l, $l = 0, 1, \ldots, L$ indicate the number of segments in the input layer $(l = 0)$, the l is number of hidden layers that can range $(l = 0, 2, \ldots, L-1)$, L is the number of layers, and the output layer is $(l = L)$. $X^{(0)} = \left[x_0^{(0)} u_1 u_2 \ldots u_{n0} \right]^T$ is the input vector and $X^{(l)} = \left[x_0^{(l)} x_1^{(l)} \ldots x_{ni}^{(l)} \right]^T$ denotes the lth layer output vector in the interval [0, T]. At this time, $\{u_i\}, j = 1, 2, \ldots n_0$ represents the input attribute pattern while $x_j^{(l)}$ indicates the output of the jth segments of lth network layer. The threshold input is represented by $x_0^{(l)}$ with a fixed value at one. The neuron weight of the jth segment of lth layer from the ith segment of $(l-1)$ can be represented by $\omega_{ji}^{(l)}$. The activation function, which is connected with all the segments of the system except the input layer, is the tanh function specified by $\rho(S) = \tan h(S) = \left(\frac{1-e^{-2S}}{1+e^{-2S}} \right)$. The restricted derivative of the $\rho(S)$ based on S is signified by $\rho\prime(S)$ that is known as $\rho\prime(S) = \left(1 - \rho^2(S)\right)$. Also, the linear sum of the jth segment of lth layer is symbolized by $S_j^{(l)}$.

In the forward part, at the kth time direct, the input attribute pattern vector of $X^{(0)}$ is implemented in the system, while the corresponding preferred output is $\{y_j\}$, for $j = 1, 2, \ldots, n_L$. Since no calculation is applied, the input layer of the MLP is known by $x_j^{(0)} = u_j$ for $j = 1, 2, \ldots, n_0$. For other layers, $l = 1, 2, \ldots, L$, and $j = 1, 2, \ldots, n_1$, the outputs are computed as:

$$x_j^{(0)} = \rho(S_j^{(l)}) \text{ and } S_j^{(l)} = \sum_{i=0}^{ni-1} \omega_{ji}^{(1)} x_i^{(l-1)}.$$

The probable output is identified by $\{\hat{y}_J\}$ and is assumed as $\{\hat{y}_J\} = \left\{ x_j^{(L)} \right\}$ for all $j = 1, 2, \ldots, n_L$. The mean square error for the system can be formulated as $\frac{e^2}{nL}$ where $e_j = y_j - \hat{y}_J$ is the error signal for the jth output. Furthermore, the instantons squared error can be computed by $e^2 = \sum_{j=1}^{n\,L} e_j^2$.

However, in the learning part, the BP procedure reduces the squared error by varying $\{\omega_{ji}^{(l)}\}$ according to the gradient search method, recursively. The squared error derivatives connected with the jth segment in layer l are described as Equation (2):

$$\delta_j^{(1)} = -\frac{1}{2} \frac{\delta e^2}{\delta S_i^{(1)}} \tag{2}$$

Then, these derivatives can be formulated as in Equation (3):

$$\delta_j^{(1)} = \begin{cases} \rho'\left(S_j^{(l)}\right).e_j \ for \ l = L \\ \rho'\left(S_j^{(l)}\right).\sum_{i=1}^{n\,l+1} \omega_i^{(l+1)}.\delta_i^{(l+1)} \ for \ l = L-1, L-2, \ldots, 1. \end{cases} \tag{3}$$

Eventually, the weights of the MLP are reorganized at the *k*th instant as in Equation (4):

$$\omega_{ji}^{(l)}(k+1) = \omega_{ji}^{(l)}(k) + \Delta_{ji}^{(l)}(k) \ and \ \Delta_{ji}^{(l)}(k) = \mu\delta_i^{(l)}x_i^{(l-1)} + \gamma\Delta_{ji}^{(l)}(k-1) \tag{4}$$

where μ is the learning rate and γ signifies the momentum rate hyper-parameters.

The final proposed network architecture for traffic CO prediction is illustrated in Figure 5. The proposed network is designed based on the results of the best traffic predictors that resulted from the CFS model, optimization and hyper-parameter, which are used to select the best predictors. By using the open source machine learning software (Weka), different MLP structures have been used to select the optimal MLP neural network model for the traffic CO prediction model. The proposed methodology was designed by combining the correlation-based feature selection model and multilayer perceptron.

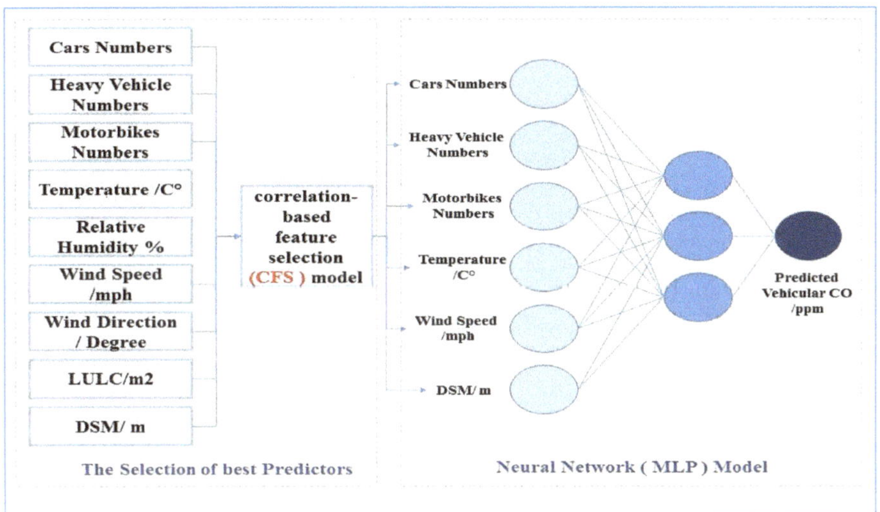

Figure 5. The architecture of the proposed neural network for traffic CO prediction (6-3-1).

3.4.4. Optimization Method

The capacity of the prediction in MLP model relies on its hyperparameter and structure. In this paper, several network structure hyperparameter combinations were tested to determine a sub-optimal network model for modeling vehicular CO. Table 3 shows the structures and hyperparameters evaluated in the current study and their search space domain. In general, there are two main categories of NN, MLP and radial basis function (RBF). The former uses dot products between inputs and weights and monotonic activation functions such as sigmoid. The MLP uses dot products between inputs and weights and monotonic activation functions such as sigmoid while the RBF uses Euclidean distances between inputs and weights and usually Gaussian activation functions. Both of the networks can be trained with the back-propagation algorithm. In the RBF model, it is not necessary to use multiple hidden layers whereas with MLP, multiple hidden layers are used. In addition, the RBF model is less sensitive to noise than the MLP model. Other parameters of the network are the number of hidden units, training algorithm, error function, activation function, learning rate, and momentum. The number of hidden layers controls the complexity of the designed network. A small number of hidden units may result in low prediction capacity due to insufficient learning whereas a large number of hidden layers can reduce the ability of the model to be generalized and can also create overfitting problems. The training algorithm is the optimization method for calculating the weight for each node in the network. There are many training algorithms for NN based on back-propagation; the Broyden–Fletcher–Goldfarb–Shanno (BFGS) and radial basis function training algorithm (RBFT) are the most recommended back-propagation algorithms used for optimization of NN architecture [46]. During training of the network, an objective function or optimization score function was minimized according to the labeled training dataset. The optimizer usually has a learning rate and gradient momentum parameters. In addition, many activation functions such as identity, logistic, and Gaussian could be used in the hidden layers and the output layers of the neural network.

Table 3. Hyper parameters of the proposed model for traffic CO prediction and their search spaces used for fine-tuning.

Parameter	Search Domain
Type of network	MLP, RBF
Number of hidden units	(3–40)
Training Algorithm	BFGS, RBFT
Hidden Activation	Identity, Logistic, Tanh, Exponential, Gaussian
Output Activation	Identity, Logistic, Tanh, Exponential, Gaussian
Learning rate	(0.1, 0.5)
Momentum	(0.1–0.9)

The total number of instances was 352, which was separated into training 70% (246) and testing 30% (106). This sample size is still small compared to other studies. However, it needs a good approach to handle overfitting problems. One approach is to design a cross-validation evaluation procedure. Other methods are data augmentation or collecting new samples. There are also methods such as transfer learning which requires retrained networks. Overall, training of the neural networks with small datasets requires careful analysis and evaluation before using them in practice. There are also some tricks to improve the performance of neural networks for small datasets. Those include batch normalization, rectified linear unit (relu) activations and regularization methods such as l1 and l2. On the other hand, the hyperparameters of the NN for predicting vehicular CO were chosen based on the implementation of systematic grid-based search that can be applied with the Scikit-Learn algorithm using 100 epochs. Although this method demands high quality of computation, more accurate outputs could result by fine tuning the hyperparameter values. Many models based on various integrations of parameters were generated. Cross validation (10-fold) was applied to validate each model. Therefore, the parameters that resulted in higher accuracy are the best parameters.

3.4.5. GIS Modelling

Collecting traffic CO data, especially on the highways, is dangerous and expensive. Therefore, predicting traffic CO concentrations on highways helps to generate traffic CO data that can be used for further studies. In this research, the measured and predicted traffic CO concentrations are used for GIS modelling based on a grid analysis. GIS modelling and mapping are mainly applied to assess affected people and environments due to inappropriate traffic emissions from traffic activities. The observed traffic CO samples are an important factor in the model, for computing the relationship between the predictive factors and carbon concentrations which were applied in the training data, and for the validation process. After training the NN model with the CO measurements from the field, the NN model produced a regression equation based on weighted values for each predictor factor, in order to calculate the predicted values based on the predictor factors that can be easily applied in a GIS platform; this is the main contribution of the proposed model. This equation has been applied using GIS to produce a spatial prediction of traffic CO concentration in the study area. This step was applied by implementing the final prediction equations on the 5×5 m GRID, using ArcGIS tools which are very efficient for spatial representation [23].

4. Results and Discussion

4.1. Contribution of Traffic CO Predictors

As shown in Table 4, the traffic CO predictors make different contributions to the traffic CO values that resulted from this study. The statistical analysis based on the Chi-square method shows that the parameters that contributed the most are the number of heavy vehicles (F = 32.784) and the number of vehicles (F = 18.277). Conversely, the findings indicated that the other traffic CO predictors (number of motorbikes, DSM, wind speed and temperature) did not make a significant contribution (Table 4).

Table 4. Results of assessing the contribution of traffic CO predictors using the Chi-square method.

Road Traffic CO Predictors	R-Squared	F-Statistic
Number of heavy vehicles	0.7546	32.784
Number of vehicles	0.5322	18.277
Number of motorbikes	0.0472	1.951
DSM (m)	0.0168	1.231
Wind speed (mph)	0.0016	0.124
Temperature (°C)	0.0014	0.1178

4.2. Traffic CO Prediction Results

Two MLP models were trained and tested. The first model was trained based on MLP algorithm using nine parameters (those listed in Table 1 plus LULC) to predict the traffic CO concentration. The second model was trained based on the combination with the correlation-based feature selection (CFS-MLP) and six parameters that resulted from the CFS model. These parameters were: number of vehicles, number of heavy vehicles, number of motorbikes, temperature, wind speed, and DSM. The CFS algorithm was implemented to select the highly correlated parameters and best parameters to predict the traffic CO that led to increased accuracy of the prediction process.

Table 5 shows the proposed model's results when the input parameters were filtered and reduced from nine parameters to six based on the CFS algorithm which finds features that have higher correlation with the class but are uncorrelated with each other. Therefore, the highest correlated parameters were used for the prediction analysis, which resulted in improving the prediction accuracy. The relative absolute error decreased from 30.94% to 21.99% and the root relative square error also decreased from 23.48% to 19.40%. On the other hand, the correlation coefficient increased from 0.866 ppm to 0.98 ppm. The mean absolute error (MAE) was reduced from 0.99 ppm to 0.89 ppm. The root mean square error (RMSE) also decreased from 1.29 ppm to 1.27 ppm. The prediction results showed that the prediction improvement occurred after the implementation of the CFS algorithm which is able to reduce the high dimensionality, remove the low correlated data and improve the learning accuracy.

Table 5. Results of predictions with MLP model and the proposed model (CFS-MLP) model.

MLP Model		CFS-MLP Model	
Best structure	9-4-1	Best structure	6-3-1
Correlation coefficient	0.8657	Correlation coefficient	0.980
Mean absolute error (ppm)	0.991	Mean absolute error (ppm)	0.8925
Root mean squared error (ppm)	1.2862	Root mean squared error (ppm)	1.2736
Relative absolute error %	30.94%	Relative absolute error %	21.99%
Root relative squared error %	23.48%	Root relative squared error %	19.40%
Total number of instances	247	Total number of instances	247

4.3. Traffic CO Prediction at Different Times of Day

Regression equations were created, based on the results from the CFS-MLP model; the coefficients of vehicular emissions predictors are calculated to formulate regression models to easily predict vehicular emission in the study area using a set of predictors that can be gathered from the field and existing databases in order to facilitate connection with the GIS-based model by applying parameter coefficients in the spatial model. The decision makers can notice the effect of causative parameters on vehicular emissions occurrence, which was assessed by the corresponding coefficient that appears in the regression function [47]. Regression equations were simulated at different times during the day (morning, afternoon, evening and night) to predict traffic CO at these times. The highest RMSE was 2.9817 ppm during evening observations, while the lowest RMSE was 0.387 ppm at night. Table 6 shows the results of the regression analysis at different times.

Table 6. Regression models for traffic CO prediction based on the CFS-MLP model.

Traffic CO Predictors	Estimated Coefficient			
	Morning	**Afternoon**	**Evening**	**Night**
Number of vehicles	−0.0016	0.0142	0.0108	0.0147
Number of heavy vehicles	0.0622	0.01	0.0319	−0.0216
Number of motorbikes	0.0135	−0.0378	−0.0376	−0.0093
Temperature °C	−0.4501	0.5512	0.4888	−0.0333
Wind speed mph	0.0752	−0.194	−0.4084	0.0135
DSM m	−0.2085	0.213	0.0812	0.1116
Intercept	16.8559	−22.2525	−15.8113	−2.1367
RMSE	2.914 ppm	2.0347 ppm	2.9817 ppm	0.387 ppm

4.4. GIS Modelling Results

A GIS model was applied to generate prediction maps at different times a day (Figures 6 and 7). The resultant spatial prediction maps showed that the concentrations of traffic CO increased more on weekdays than on weekends. These maps also showed that there was significant variation between traffic CO concentrations according to the time of day, with the traffic CO ranging from a high of 35.23 ppm per 15 min during the weekday morning to a low of 4.76 ppm per 15 min during the weekend night. The prediction maps showed that the highest values of traffic CO are located near the toll areas compared to other areas such as residential green areas, because of the traffic congestion at toll gates. The lowest values of traffic CO are concentrated near residential areas which reached zero values.

4.5. Comparison with Other Models

The developed model was compared with other popular models such as the support vector machine for regression (SVR) and the linear regression (LR) models. These models generated two statistical equations based on model parameters. Testing data were used for model validation; these equations are shown below:

$$
\begin{aligned}
\text{Traffic CO} = \ & 0.0022 \times \text{Number of vehicles} + 0.0403 \times \text{Number of heavy vehicles} \\
& - 0.0187 \times \text{Number of Motorbikes} + 0.1957 \times \text{Temperature} \\
& - 0.0984 \times \text{Wind speed} - 0.0102 \times \text{DSM} - 4.4382
\end{aligned}
\tag{4}
$$

For the LR model:

$$
\text{Traffic CO} = 0.05 \times \text{Number of heavy vehicles} + 0.21
\tag{5}
$$

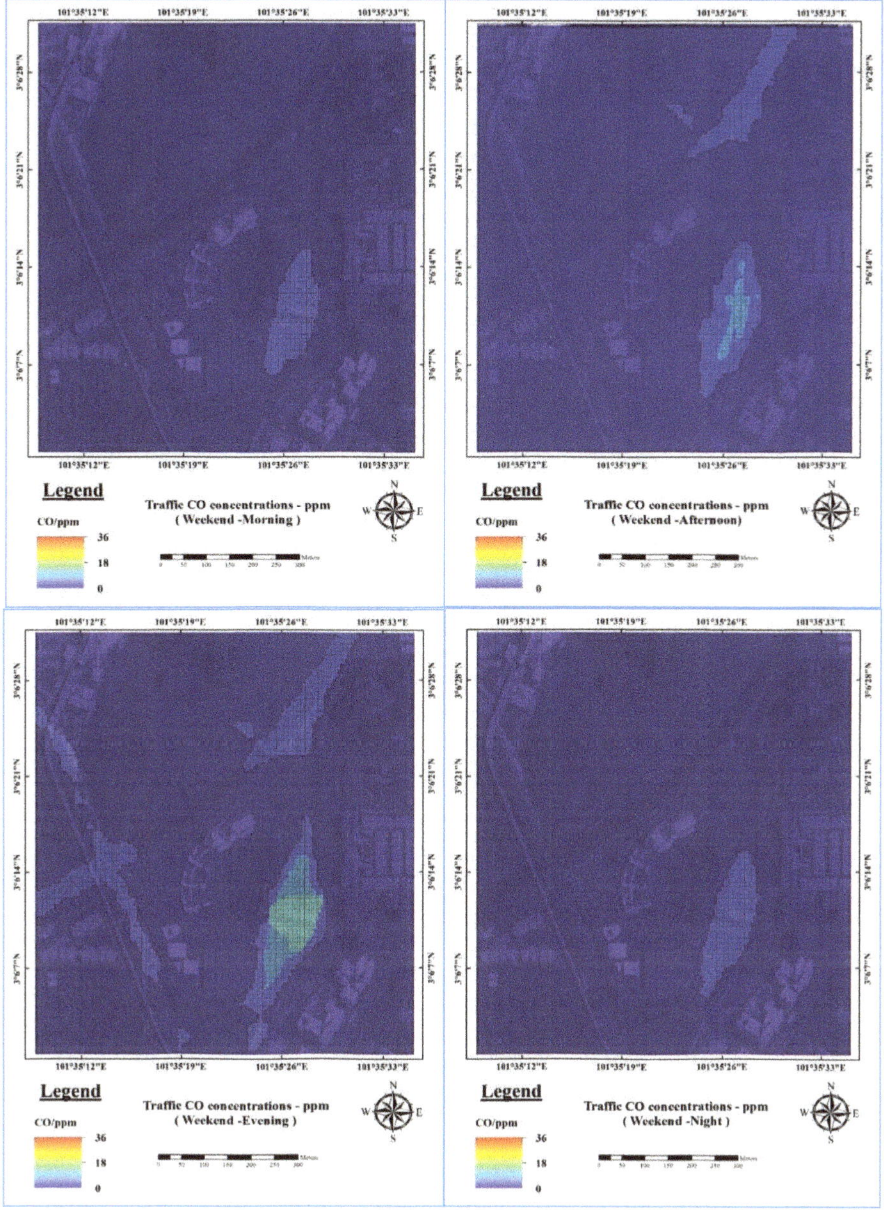

Figure 6. Prediction maps during the weekend at different times (morning, afternoon, evening and night).

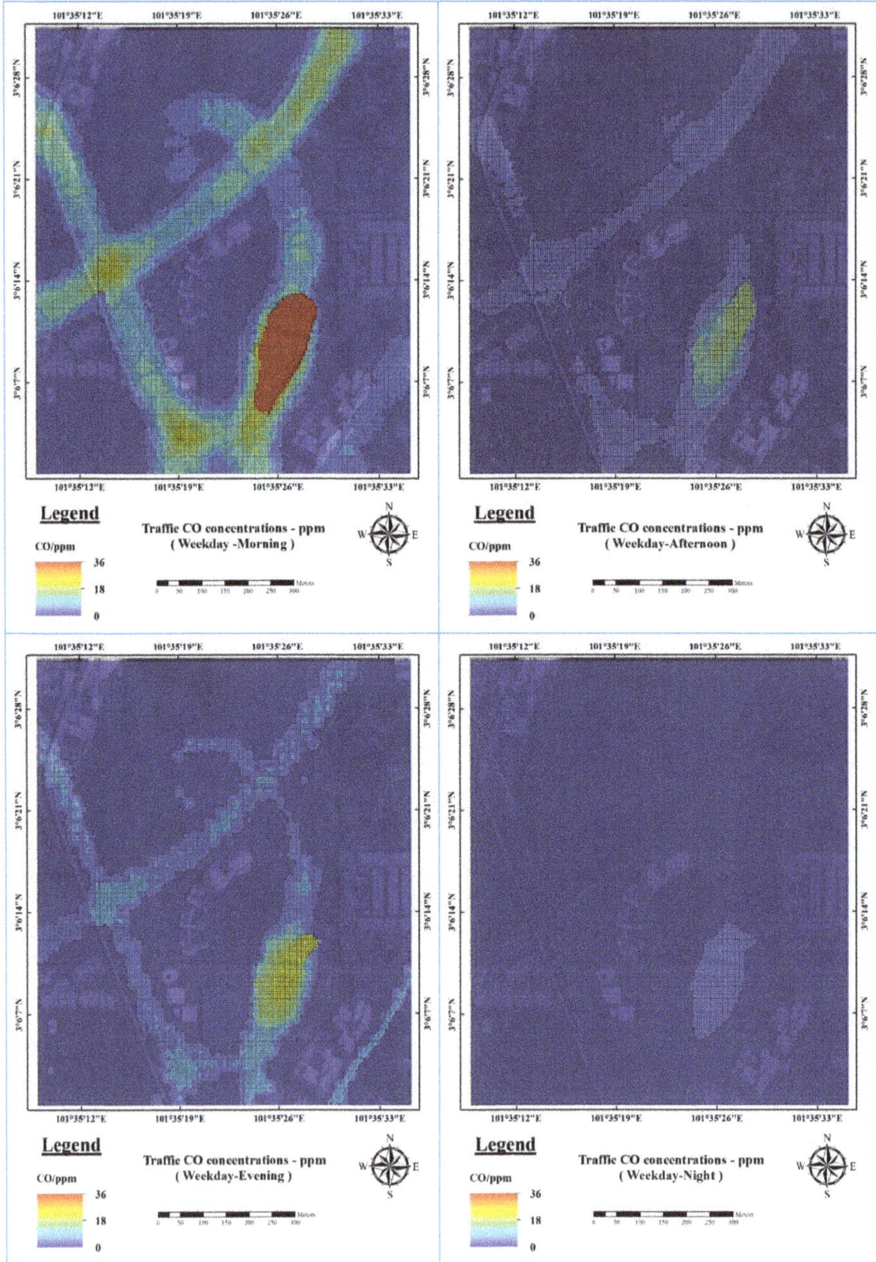

Figure 7. Prediction maps during weekdays at different times (morning, afternoon, evening and night).

4.6. Validation of Traffic CO Prediction Maps

The traffic CO spatial prediction maps were verified using the test sites of traffic CO samples, and the verification method was then performed by comparing the traffic CO test data and the traffic

CO spatial prediction maps. The lowest accuracy of the validation was at night time (72.48%) and the highest accuracy was during the evening (92.75%).

Table 7 shows the comparison between CFS-MLP, SVR and LR model. The comparative analysis was conducted by using the training data and the proposed model, support vector regression model, linear regression, land use regression model and dispersion model i.e., CALINE4 model based on many criteria not only Root mean square RMSE, but we also compared the proposed model with other baseline models based on mean absolute error (MAE), relative absolute error, root relative squared error and correlation coefficient. Results showed that the proposed model is superior to the compared models. The correlation coefficient based on our proposed model was 0.980, which is higher than the SVR (0.8668) and LR (0.851). The MAE of the proposed model was 0.896 ppm which is lower than SVR (1.640 ppm) and LR (1.851 ppm). On the other hand, the RMSE results indicated that the proposed model has the lowest RMSE (1.286 ppm) compared to SVR (2.752 ppm) and LR models (2.849 ppm). The RAE ratio for the proposed model was calculated to be 21.99%, which was lower than the RAE ratio of SVR (51.646%) and LR (55.048%). The root relative squared error indicated that the proposed model has the lowest value (19.40%) among the other models (49.784% and 48.292%).

Table 7. The comparison between CFS-MLP, SVR and LR models.

CFS-MLP Model		SVR Model		LR Model	
Correlation coefficient	0.980	Correlation coefficient	0.8668	Correlation coefficient	0.851
Mean absolute error (ppm)	0.896	Mean absolute error (ppm)	1.640	Mean absolute error (ppm)	1.851
Root mean squared error (ppm)	1.286	Root mean squared error (ppm)	2.752	Root mean squared error (ppm)	2.849
Relative absolute error (%)	21.99	Relative absolute error (%)	51.646	Relative absolute error (%)	55.048
Root relative squared error (%)	19.40	Root relative squared error (%)	49.784	Root relative squared error (%)	48.292
Total number of instances	247	Total number of instances	247	Total number of instances	247

Table 7 shows the comparison between CFS-MLP, SVR and LR model. The comparative analysis was conducted by using the training data and the proposed models, i.e., SVR, LR, land use regression model and dispersion model, i.e., CALINE4 model based on many criteria, not only RMSE. We also compared the proposed model with other baseline models based on mean absolute error (MAE), relative absolute error, root relative squared error and correlation coefficient. Results showed that the proposed model is superior to the other models. The CO emission prediction rate can be justified as highly accurate, where the accuracy is more than 90%; 90% to 80% is a good forecast; 80% to 50% is a reasonable forecast; and more than 50% is an inaccurate forecast [48–50].

The correlation coefficient based on our proposed model was 0.980, which is higher than the SVR (0.8668) and LR (0.851). The MAE of the proposed model was 0.896 ppm which is lower than SVR (1.640 ppm) and LR (1.851 ppm). On the other hand, the RMSE results indicated that the proposed model has the lowest RMSE (1.286 ppm) compared to SVR (2.752 ppm) and LR models (2.849 ppm). The RAE ratio for the proposed model was calculated to be 21.99%, which was lower than the RAE ratio of SVR (51.646%) and LR (55.048%). The root relative squared error indicated that the proposed model has the lowest value (19.40%) among the other models (49.784% and 48.292%). Table 8 shows the results of the prediction models when combining the correlation-based feature selection; the correlation coefficients of the CFS-SVR and CFS-LR are 0. 0.7578 and 0.82, respectively. These values indicated that there is decrease in the correlation coefficient. The MAE increased for the CFS-SVR and CFS-LR and reached 1.972 ppm and 1.9713 ppm, respectively. The EMSE also increased to 3.7109 ppm and 3.1057 ppm. Figure 8 illustrates the variations of the traffic CO concentrations in the testing data.

Table 8. The comparison between CFS-MLP, CFS-SVR and CFS-LR models.

CFS-MLP Model		CFS-SVR Model		CFS-LR Model	
Correlation coefficient	0.980	Correlation coefficient	0.7578	Correlation coefficient	0.82
Mean absolute error (ppm)	0.896	Mean absolute error (ppm)	1.972	Mean absolute error (ppm)	1.9713
Root mean squared error (ppm)	1.286	Root mean squared error (ppm)	3.7109	Root mean squared error (ppm)	3.1057
Relative absolute error (%)	21.99	Relative absolute error (%)	64.3605	Relative absolute error (%)	64.333
Root relative squared error (%)	19.40	Root relative squared error (%)	67.2464	Root relative squared error (%)	56.2795
Total number of instances	247	Total number of instances	247	Total number of instances	247

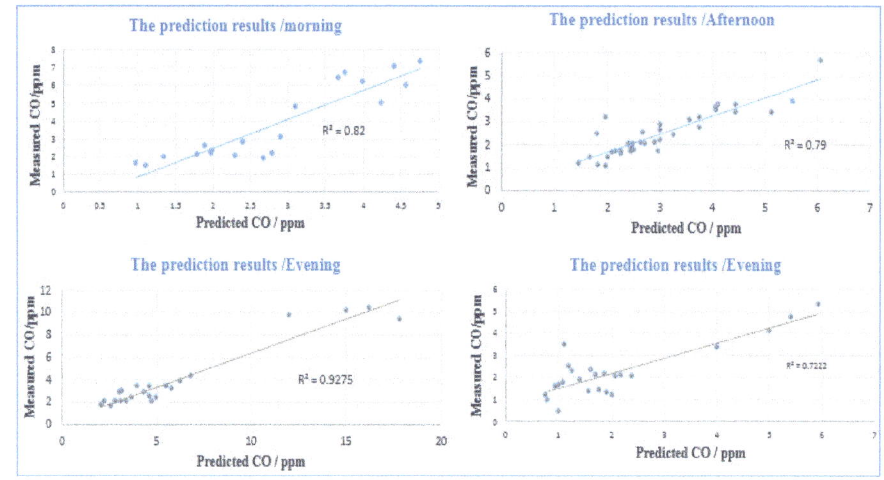

Figure 8. Accuracy assessment of the predicted maps (morning, afternoon, evening and night).

The first model that we compared with one of the baseline models is the CALINE4 model. The ALINE4 model is a dispersion model, which depends on a plume dispersion model used to predict the vehicular CO on roadways [48]. The CALINE4 model simulates the data based on a Gaussian diffusion algorithm and characterizes the pollutants dispersion on roads. We defined the proposed road network, weather data, traffic flow information, and receptor locations, and the prediction of traffic CO emissions was obtained. The MAE and RMSE values were 2.376 ppm and 4.2254 ppm, respectively, whereas and the correlation coefficient value was 0.6504. The prediction results appeared worse than our proposed model. This may due to the fact that the Gaussian diffusion, which was assumed in CL4, is not very realistic.

We also compared our work with the Land Use Regression (LUR) Model. This model is generally applied to predict air pollutants depending on the land use, traffic flow information, meteorology data and combined them based on a linear regression algorithm. The LUR model showed the following values: MAE 2.21 ppm, RMSE 4.50 ppm and a correlation coefficient of 0.5989 (Figure 9).

The final LUR equation used is given below:

$$\text{Predicted CO} = 0.0018 \times \text{Car} + 0.0423 \times \text{HV} - 0.0219 \times \text{Motorbike} + 0.2211 \times$$
$$\text{Temp/C} + 0.0312 \times \text{Relative Humidity} - 0.1315 \times \text{Wind speed} + 0.0018 \times \text{Wind}$$
$$\text{Angle Degree} - 0.0232 * \text{DSM} + 0.0006 \times \text{Builtup area} + 0.0064 \times \text{Highway} -$$
$$8.6627.$$

There are many models developed based on GIS and machine learning, for example [26] designed a model based on the integration of the ANN algorithm and the linear-chain conditional random field algorithm to produce real-time and fine-grained air pollution prediction maps.

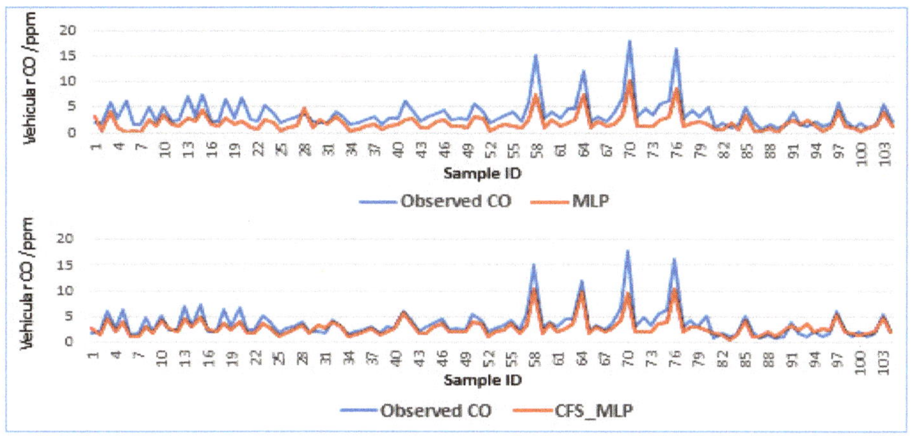

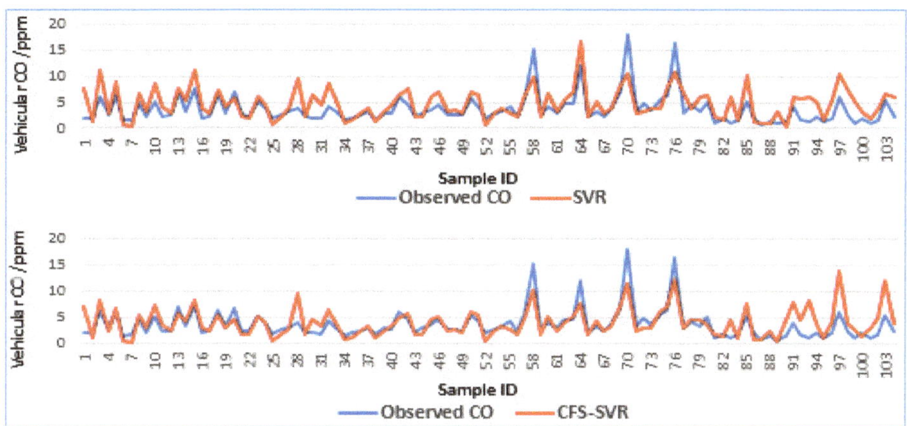

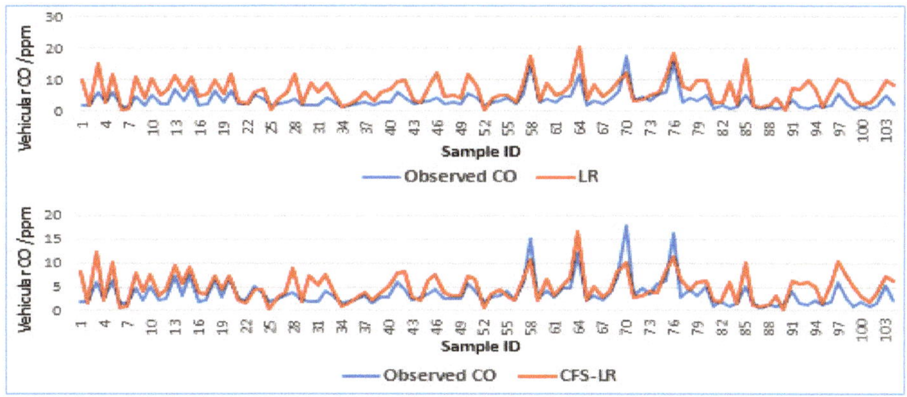

Figure 9. Comparison between the proposed and other models.

The air quality data were obtained from fixed air quality stations, the traffic data were collected from vehicle trajectory, and the meteorology data were collected from monitoring stations. Other data used were land use, road geometry and social information. Their results presented the prediction maps with low resolution, cell size (1×1) km^2 in local scale. The limitation of this model can be summarized in some points. The data collection from the standard fixed air quality monitoring stations may not be able to measure the air quality that people are exposed on the ground level due to the limitation of monitoring location and height. On the other hand, the data obtained from the fixed stations are not suitable for high-resolution prediction maps such as microscale maps. Moreover, this study did not contain information about the terrain and buildings.

The proposed model in this research is different from the aforementioned study based on some points. We developed a GIS-based NN and data were obtained from a field survey. Moreover, the land use and DSM were extracted from a very high resolution LiDAR data clouds. The proposed methodology is designed to predict the vehicular CO and produce high-resolution maps at a microscale level (5×5) m^2 whereas the aforementioned paper estimated air pollutants based on a low-resolution grid (1×1) km^2.

Regarding the meteorology parameters shown in Table 4, it is evident that the correlation between wind speed and vehicular emissions is 0.0016, which is considered weak in the short-term prediction and in a small area compared to other factors. In addition, the temperature has the lowest correlation with vehicular emissions 0.0014; therefore, the variation in the vehicular emissions (i.e., CO) may not be significantly affected by the weather condition in the prediction in small areas. On the other hand, the DSM, which was used to extract the terrain and building's altitude in the study area, has a good correlation with vehicular emissions (more than meteorology parameters). Therefore, the geographic factors are important in prediction studies. This study adopted high-resolution elevation data in order to extract results that are more accurate. In urban areas, building altitude is an important parameter because it can resist vehicular emissions and prevent the distribution of pollutants in urban areas.

5. Conclusions

Traffic emissions (e.g., traffic CO) are considered the major source of air pollution in congested urban areas, including road corridors in toll plaza areas. Traffic emission prediction models are utilized to evaluate the impacts of traffic CO emissions on the population and environment and some models are used to illustrate the spatial prediction of these emissions. In this paper, a hybrid prediction model was proposed by combining three models (CFS, MLP and GIS) to predict the traffic CO emissions and create micro-scale prediction maps in a small area at different times during the day. The final findings have shown that the proposed model scored an accuracy of 80.6% and the correlation coefficient of 0.980, RMSE of 1.2736 ppm and mean absolute error of 0.8925 ppm. We used CFS to identify and remove highly correlated parameters so that redundancy was reduced to choose the optimum parameters used in the prediction model through MLP. The simulation results showed that nine predictors were reduced to six, which contributed to an increase in the prediction accuracy.

The data were collected from the field and remote sensing data (i.e., LiDAR and very high-resolution WorldView-3 satellite image), and modelling was performed in a GIS environment.

In this study, we produced microscale maps for vehicular emissions. The simulated traffic CO emissions ranged from 35 ppm inside the toll plaza area to 0 ppm for areas that were located far away from the toll area. As per the microscale-prediction maps, high spatial variation in the traffic CO emissions was identified. The highest value of CO concentrations is found in traffic jam areas. Conversely, the lowest values of traffic CO emissions are distributed far from traffic activities. The highest concentrations of traffic CO were located inside the toll plaza because of the traffic jam that occurs daily in the toll areas these results give a clear indication about the relationship between the traffic activities and the traffic CO emissions. The traffic CO emissions may have a significant impact on the health of toll plaza workers, drivers and the passengers. Therefore, such a prediction model can aid decision makers to implement plans to mitigate the traffic emissions that can protect people who are working, passing through or living near toll plaza areas; this will be the main advantage of the proposed forecasting system.

Traffic CO emission prediction models and GIS modeling are both efficient tools for transportation planning and traffic emission assessment. The prediction maps produced by the proposed model can be used as an effective tool in the decision-making process to identify optimum solutions which can be used to mitigate traffic jams in toll plaza areas as well as on highways and road networks. As traffic emission pollution assessment by decision makers is very complicated and usually expensive because of the high level of requirements for expert knowledge and the developed support systems, the presented traffic CO assessment model is not expensive and can easily implemented. Moreover, the traffic CO pollution concentrations vary based on the traffic condition and the number of vehicles, which requires a periodic monitoring of traffic emissions by government agencies or relevant departments. The best parameter selection analysis could be used to reduce the data collection requirements, which can lead to reduced time required, resources utilized and processing time needed. GIS modeling is a useful tool for non-expert users to implement the traffic CO impact assessments in various applications. Finally, these models can be improved by using more advanced algorithms such as deep learning algorithms and a large number of samples that can be used to increase the accuracy of the prediction process.

Author Contributions: B.P. conceptualized, supervised, and obtained the grant for the study. B.P. and O.S.A. collected and analyzed the data, performed the analyses and validation, wrote the manuscript and contributed to the re-structuring and editing of the manuscript. B.P., O.S.A., N.S., H.Z.M.S., H.M.R. and C.-W.L. professionally optimized the manuscript.

Funding: This research is supported by the UTS under grant numbers 321740.2232335 and 321740.2232357.

Acknowledgments: The authors acknowledge and appreciate the provision of airborne laser scanning data (LiDAR), satellite images and logistic support by the PLUS Berhad. In addition, the second author, Biswajeet Pradhan, gratefully acknowledges the financial support from the UPM-PLUS industry project grant.

Conflicts of Interest: The authors declare no conflict of interest.

Abbreviations

GIS	A Geospatial Information System is a system designed to collect, manage, analyze, store and produce different types of spatial data.
CO	Carbon Monoxide is a toxic gas and it has no has no color, taste, or smell, resulting from the incomplete combustion of fuel.
RMSE	The algorithm of the root mean square is used to calculate the differences between values estimated by a model and the observed values.
VISSIM	Software designed for traffic flow simulation at a micro-scale level, which is designed by Planning Transport Verkehr (PTV), Germany.
EPR	The evolutionary polynomial regression, EPR, is one of the data-mining algorithms developed based on evolutionary computing and the integration of numerical regression and genetic algorithm.
CFS	A correlation-based feature selection algorithm, which is a type of filter algorithm that selects features based on a heuristic (correlation-based) function.
LiDAR	Light Detection and Ranging is an advanced surveying technology usually used to create 3D models by measure the distance between targets and the Laser Sensor.
ENVI	Environment for Visualizing Images: professional software used for image analysis and remote sensing applications.
MLP	A multilayer perceptron (MLP) is a class of feedforward artificial neural networks. An MLP consists of, at least, three layers of nodes: an input layer, a hidden layer and an output layer. Except for the input nodes, each node is a neuron that uses a nonlinear activation function.
LULC	Land Use and Land Cover are data files that describe the land surfaces such as water, vegetation and cultural features.
CFS-MLP	The proposed model that is the combination of two models, the correlation based feature selection algorithm and multilayer perceptron Neural Network algorithm.

CALINE4	California Line Source Dispersion is one of the dispersion models used to estimate carbon monoxide emissions near roads based on various parameters related to geographic locations.
MAE	Mean Absolute Error, MAE, measures the average magnitude of the errors in a set of predictions, without considering their direction. It is the average over the test sample of the absolute differences between prediction and actual observation where all individual differences have equal weight.
RAE	Relative Absolute Error is defined as the absolute error relative to the size of the measurement, and it depends on both the absolute error and the measured value. The relative error is large when the measured value is small, or when the absolute error is large.
ANN	An Artificial Neural Network is a computational model based on the structure and functions of biological neural networks. Information that flows through the network affects the structure of the ANN because a neural network changes—or learns, in a sense—based on that input and output.

References

1. Bastien, L.A.; McDonald, B.C.; Brown, N.J.; Harley, R.A. High-resolution mapping of sources contributing to urban air pollution using adjoint sensitivity analysis: Benzene and diesel black carbon. *Environ. Sci. Technol.* **2015**, *49*, 7276–7284. [CrossRef] [PubMed]
2. Fameli, K.M.; Assimakopoulos, V.D. Development of a road transport emission inventory for Greece and the Greater Athens Area: Effects of important parameters. *Sci. Total Environ.* **2015**, *505*, 770–786. [CrossRef]
3. Borge, R.; Narros, A.; Artinano, B.; Yagüe, C.; Gomez-Moreno, F.J.; de la Paz, D.; Quaassdorff, C. Assessment of microscale spatio-temporal variation of air pollution at an urban hotspot in Madrid (Spain) through an extensive field campaign. *Atmos. Environ.* **2016**, *140*, 432–445. [CrossRef]
4. Oftedal, B.; Krog, N.H.; Pyko, A.; Eriksson, C.; Graff-Iversen, S.; Haugen, M.; Aasvang, G.M. Road traffic noise and markers of obesity–a population-based study. *Environ. Res.* **2015**, *138*, 144–153. [CrossRef]
5. Ancona, C.; Badaloni, C.; Mattei, F.; Cesaroni, G.; Stafoggia, M.; Forastiere, F. Health Impact Assessment of Air Pollution, Noise, and Lack of Green in Rome. *J. Transp. Health* **2017**, *5*, S42–S43. [CrossRef]
6. Garshick, E.; Laden, F.; Hart, J.E.; Caron, A. Residence near a major road and respiratory symptoms in US veterans. *Epidemiology* **2003**, *14*, 728. [CrossRef] [PubMed]
7. Delfino, R.J.; Tjoa, T.; Gillen, D.L.; Staimer, N.; Polidori, A.; Arhami, M.; Longhurst, J. Traffic-related air pollution and blood pressure in elderly subjects with coronary artery disease. *Epidemiology* **2010**, *21*, 396–404. [CrossRef]
8. Crouse, D.L.; Goldberg, M.S.; Ross, N.A.; Chen, H.; Labrèche, F. Postmenopausal breast cancer is associated with exposure to traffic-related air pollution in Montreal, Canada: A case–control study. *Environ. Health Perspect.* **2010**, *118*, 1578. [CrossRef]
9. Domene, E.; Lopez, R.; Fauro, B.; Rojas-Rueda, D.; Conill, C.; Alsina, G.; Marull, J. Modelling Impacts of Mobility on Urban Air Quality and Health: Scenario Analysis for the Barcelona Metropolitan Area (Metropolitan Mobility Plan). *J. Transp. Health* **2017**, *5*, S60–S61. [CrossRef]
10. Singh, D.; Kumar, A.; Kumar, K.; Singh, B.; Mina, U.; Singh, B.B.; Jain, V.K. Statistical modeling of O3, NOx, CO, PM2. 5, VOCs and noise levels in commercial complex and associated health risk assessment in an academic institution. *Sci. Total Environ.* **2016**, *572*, 586–594. [CrossRef] [PubMed]
11. Behera, S.N.; Sharma, M.; Mishra, P.K.; Nayak, P.; Damez-Fontaine, B.; Tahon, R. Passive measurement of NO_2 and application of GIS to generate spatially-distributed air monitoring network in urban environment. *Urban Clim.* **2015**, *14*, 396–413. [CrossRef]
12. Johnson, M.; Isakov, V.; Touma, J.S.; Mukerjee, S.; Özkaynak, H. Evaluation of land-use regression models used to predict air quality concentrations in an urban area. *Atmos. Environ.* **2010**, *44*, 3660–3668. [CrossRef]
13. Kanaroglou, P.S.; Adams, M.D.; De Luca, P.F.; Corr, D.; Sohel, N. Estimation of sulfur dioxide air pollution concentrations with a spatial autoregressive model. *Atmos. Environ.* **2013**, *79*, 421–427. [CrossRef]
14. Vandaele, N.; Van Woensel, T.; Verbruggen, A. A queueing based traffic flow model. *Transp. Res. Part D Transp. Environ.* **2000**, *5*, 121–135. [CrossRef]
15. Tomić, J.; Bogojević, N.; Pljakić, M.; Šumarac-Pavlović, D. Assessment of traffic noise levels in urban areas using different soft computing techniques. *J. Acoust. Soc. Am.* **2016**, *140*, EL340–EL345. [CrossRef]

16. Hamad, K.; Khalil, M.A.; Shanableh, A. Modeling roadway traffic noise in a hot climate using artificial neural networks. *Transp. Res. Part D. Transp. Environ.* **2017**, *53*, 161–177. [CrossRef]
17. Di Mascio, P.; Di Vito, M.; Loprencipe, G.; Ragnoli, A. Procedure to determine the geometry of road alignment using GPS data. *Procedia-Soc. Behav. Sci.* **2012**, *53*, 1202–1215. [CrossRef]
18. Righini, G.; Cappelletti, A.; Ciucci, A.; Cremona, G.; Piersanti, A.; Vitali, L.; Ciancarella, L. GIS based assessment of the spatial representativeness of air quality monitoring stations using pollutant emissions data. *Atmos. Environ.* **2014**, *97*, 121–129. [CrossRef]
19. Hülsmann, F.; Gerike, R.; Ketzel, M. Modelling traffic and air pollution in an integrated approach–the case of Munich. *Urban Clim.* **2014**, *10*, 732–744. [CrossRef]
20. Kim, Y.; Guldmann, J.M. Land-use regression panel models of NO2 concentrations in Seoul, Korea. *Atmos. Environ.* **2015**, *107*, 364–373. [CrossRef]
21. Zarandi, M.F.; Faraji, M.R.; Karbasian, M. Interval type-2 fuzzy expert system for prediction of carbon monoxide concentration in mega-cities. *Appl. Soft Comput.* **2012**, *12*, 291–301. [CrossRef]
22. Kwok, L.K.; Lam, Y.F.; Tam, C.Y. Developing a statistical based approach for predicting local air quality in complex terrain area. *Atmos. Pollut. Res.* **2017**, *8*, 114–126. [CrossRef]
23. Quaassdorff, C.; Borge, R.; Pérez, J.; Lumbreras, J.; de la Paz, D.; de Andrés, J.M. Microscale traffic simulation and emission estimation in a heavily trafficked roundabout in Madrid (Spain). *Sci. Total Environ.* **2016**, *566*, 416–427. [CrossRef]
24. Shakerkhatibi, M.; Mohammadi, N.; Zoroufchi Benis, K.; Behrooz Sarand, A.; Fatehifar, E.; Asl Hashemi, A. Using ANN and EPR models to predict carbon monoxide concentrations in urban area of Tabriz. *Environ. Health Eng. Manag. J.* **2015**, *2*, 117–122.
25. Cai, M.; Yin, Y.; Xie, M. Prediction of hourly air pollutant concentrations near urban arterials using artificial neural network approach. *Transp. Res. Part D Transp. Environ.* **2009**, *14*, 32–41. [CrossRef]
26. Zheng, Y.; Liu, F.; Hsieh, H.P. U-Air: When urban air quality inference meets big data. In Proceedings of the 19th ACM SIGKDD International Conference on Knowledge Discovery and Data Mining, Chicago, IL, USA, 11–14 August 2013; pp. 1436–1444.
27. Namdeo, A.; Mitchell, G.; Dixon, R. TEMMS: An integrated package for modelling and mapping urban traffic emissions and air quality. *Environ. Model Softw.* **2002**, *17*, 177–188. [CrossRef]
28. Kho, F.W.L.; Law, P.L.; Ibrahim, S.H.; Sentian, J. Carbon monoxide levels along roadway. *Int. J. Environ. Sci. Technol.* **2007**, *4*, 27–34. [CrossRef]
29. Ranjbar, H.R.; Gharagozlou, A.R.; Nejad, A.R.V. 3D analysis and investigation of traffic noise impact from Hemmat highway located in Tehran on buildings and surrounding areas. *J. Geogr. Inf. Syst.* **2012**, *4*, 322. [CrossRef]
30. Li, F.; Liao, S.S.; Cai, M. A new probability statistical model for traffic noise prediction on free flow roads and control flow roads. *Transp. Res. Part D Transp. Environ.* **2016**, *49*, 313–322. [CrossRef]
31. Ragettli, M.S.; Goudreau, S.; Plante, C.; Fournier, M.; Hatzopoulou, M.; Perron, S.; Smargiassi, A. Statistical modeling of the spatial variability of environmental noise levels in Montreal, Canada, using noise measurements and land use characteristics. *J. Expos. Sci. Environ. Epidemiol.* **2016**, *26*, 597. [CrossRef]
32. Kirchstetter, T.W.; Singer, B.C.; Harley, R.A.; Kendall, G.R.; Chan, W. Impact of oxygenated gasoline use on California light-duty vehicle emissions. *Environ. Sci. Technol.* **1996**, *30*, 661–670. [CrossRef]
33. Goyal, P. Present scenario of air quality in Delhi: A case study of CNG implementation. *Atmos. Environ.* **2003**, *37*, 5423–5431. [CrossRef]
34. Chen, K.S.; Wang, W.C.; Chen, H.M.; Lin, C.F.; Hsu, H.C.; Kao, J.H.; Hu, M.T. Motorcycle emissions and fuel consumption in urban and rural driving conditions. *Sci. Total Environ.* **2003**, *312*, 113–122. [CrossRef]
35. Henderson, S.B.; Beckerman, B.; Jerrett, M.; Brauer, M. Application of land use regression to estimate long-term concentrations of traffic-related nitrogen oxides and fine particulate matter. *Environ. Sci. Technol.* **2007**, *41*, 2422–2428. [CrossRef]
36. Shi, Y.; Lau, K.K.L.; Ng, E. Developing street-level PM2.5 and PM10 land use regression models in high-density Hong Kong with urban morphological factors. *Environ. Sci. Technol.* **2016**, *50*, 8178–8187. [CrossRef]
37. Hall, M.A. Correlation-Based Feature Selection for Machine Learning. Ph.D. Thesis, The University of Waikato, Hamilton, New Zealand, 1999.

38. Boznar, M.; Lesjak, M.; Mlakar, P. A neural network-based method for short-term predictions of ambient SO2 concentrations in highly polluted industrial areas of complex terrain. *Atmos. Environ. Part B Urban Atmos.* **1993**, *27*, 221–230. [CrossRef]

39. Chaloulakou, A.; Saisana, M.; Spyrellis, N. Comparative assessment of neural networks and regression models for forecasting summertime ozone in Athens. *Sci. Total Environ.* **2003**, *313*, 1–13. [CrossRef]

40. De Cos Juez, F.J.; Nieto, P.G.; Torres, J.M.; Castro, J.T. Analysis of lead times of metallic components in the aerospace industry through a supported vector machine model. *Math. Comput. Model.* **2010**, *52*, 1177–1184. [CrossRef]

41. Gardner, M.W.; Dorling, S.R. Neural network modelling and prediction of hourly NOx and NO2 concentrations in urban air in London. *Atmos. Environ.* **1999**, *33*, 709–719. [CrossRef]

42. Hooyberghs, J.; Mensink, C.; Dumont, G.; Fierens, F.; Brasseur, O. A neural network forecast for daily average PM10 concentrations in Belgium. *Atmos. Environ.* **2005**, *39*, 3279–3289. [CrossRef]

43. Kukkonen, J.; Partanen, L.; Karppinen, A.; Ruuskanen, J.; Junninen, H.; Kolehmainen, M.; Cawley, G. Extensive evaluation of neural network models for the prediction of NO2 and PM10 concentrations, compared with a deterministic modelling system and measurements in central Helsinki. *Atmos. Environ.* **2003**, *37*, 4539–4550. [CrossRef]

44. Nieto, P.G.; Lasheras, F.S.; García-Gonzalo, E.; de Cos Juez, F.J. PM 10 concentration forecasting in the metropolitan area of Oviedo (Northern Spain) using models based on SVM, MLP, VARMA and ARIMA: A case study. *Sci. Total Environ.* **2018**, *621*, 753–761. [CrossRef]

45. Friedman, J.; Hastie, T.; Tibshirani, R. *The Elements of Statistical Learning*; Springer Series in Statistics: New York, NY, USA, 2001; Volume 1.

46. Rojek, I. Technological process planning by the use of neural networks. *AI EDAM* **2017**, *31*, 1–5. [CrossRef]

47. Mousavi, S.Z.; Kavian, A.; Soleimani, K.; Mousavi, S.R.; Shirzadi, A. GIS-based spatial prediction of landslide susceptibility using logistic regression model. *Geomat. Nat. Hazards Risk* **2011**, *2*, 33–50. [CrossRef]

48. Sharma, N.; Gulia, S.; Dhyani, R.; Singh, A. Performance evaluation of CALINE 4 dispersion model for an urban highway corridor in Delhi. *J. Sci. Ind. Res.* **2013**, *72*, 521–530.

49. Pao, Y.-H.; Phillips, S.M.; Sobajic, D.J. Neural-net computing and the intelligent control of systems. *Int. J. Control* **1992**, *56*, 263–289. [CrossRef]

50. Pao, H.-T.; Fu, H.-C.; Tseng, C.-L. Forecasting of CO_2 emissions, energy consumption and economic growth in China using an improved grey model. *Energy* **2012**, *40*, 400–409.

 *applied sciences*

Article

Impact of Texture Information on Crop Classification with Machine Learning and UAV Images

Geun-Ho Kwak and No-Wook Park *

Department of Geoinformatic Engineering, Inha University, Incheon 22212, Korea; root0109@inha.edu
* Correspondence: nwpark@inha.ac.kr; Tel.: +82-32-860-7607

Received: 28 January 2019; Accepted: 12 February 2019; Published: 14 February 2019

Abstract: Unmanned aerial vehicle (UAV) images that can provide thematic information at much higher spatial and temporal resolutions than satellite images have great potential in crop classification. Due to the ultra-high spatial resolution of UAV images, spatial contextual information such as texture is often used for crop classification. From a data availability viewpoint, it is not always possible to acquire time-series UAV images due to limited accessibility to the study area. Thus, it is necessary to improve classification performance for situations when a single or minimum number of UAV images are available for crop classification. In this study, we investigate the potential of gray-level co-occurrence matrix (GLCM)-based texture information for crop classification with time-series UAV images and machine learning classifiers including random forest and support vector machine. In particular, the impact of combining texture and spectral information on the classification performance is evaluated for cases that use only one UAV image or multi-temporal images as input. A case study of crop classification in Anbandegi of Korea was conducted for the above comparisons. The best classification accuracy was achieved when multi-temporal UAV images which can fully account for the growth cycles of crops were combined with GLCM-based texture features. However, the impact of the utilization of texture information was not significant. In contrast, when one August UAV image was used for crop classification, the utilization of texture information significantly affected the classification performance. Classification using texture features extracted from GLCM with larger kernel size significantly improved classification accuracy, an improvement of 7.72%p in overall accuracy for the support vector machine classifier, compared with classification based solely on spectral information. These results indicate the usefulness of texture information for classification of ultra-high-spatial-resolution UAV images, particularly when acquisition of time-series UAV images is difficult and only one UAV image is used for crop classification.

Keywords: unmanned aerial vehicle; texture; gray-level co-occurrence matrix; machine learning; crop

---

1. Introduction

Agricultural environments are known to be sensitive to abnormal weather conditions and climatic disasters such as drought and flood [1,2], thus rendering essential systematic monitoring of crop conditions and crop yield forecasting [3,4]. Remote sensing technology received attention in the agriculture community due to its ability to provide periodic and regional information for crop monitoring and thematic mapping [5,6].

Crop type maps derived from classification of remote sensing images are important resources for crop yield estimation and forecasting. Since any error in the crop type maps affects outputs of crop yield and forecasting models, it is critical to generate reliable crop type maps [6]. The most important elements of input remote sensing images for crop classification are their spatial and temporal resolutions. Since each individual crop has its own growth cycle, time-series images are necessary to fully account for variations of physical characteristics that accompany crop growth [7,8]. According to the scale of the target area of interest, satellite images with proper spatial resolution should be used as input for crop classification. If coarse-resolution satellite images are used, mixed pixel problems are likely and classification performance decreases [9,10]. This is a common issue in Korea, where various types of crops are cultivated in small areas. The use of high-resolution satellite images and aerial photos can contribute to resolving the mixed pixel issues [11,12]. Despite the increased discrimination capability of high-resolution images, it is difficult to collect time-series datasets over the full growth cycles of crops. Acquisition of optical satellite images depends heavily on atmospheric conditions; thus, the images are often contaminated and masked by clouds. In addition, it is difficult to acquire time-series aerial photos at desired times due to cost issues.

In recent years, there was a growing interest in imaging of unmanned aerial vehicles (UAV) [11–15]. The advantage of UAV images over satellite images is their ability to provide local thematic information with much higher spatial and temporal resolutions [15]. UAV images with ultra-high spatial resolution [16,17] can improve the discrimination capability of various surface objects, leading to an increase in the number of detectable targets. Compared with satellite images, low-cost flexible control of unmanned aerial systems (UAS) enables easier acquisition of images at the desired times between sowing and harvesting of crops [12,15,16].

Despite the great potential of UAV imaging, the technique has several practical issues. Firstly, the ultra-high spatial resolution of UAV images usually causes noise effects due to increased detectable targets when conventional pixel-based approaches are applied for classification [11,18,19]. The common approach to mitigate noise effects is to either use spatial contextual information or apply an object-oriented classification approach. For the spatial contextual information approach, texture information is firstly extracted from a gray-level co-occurrence matrix (GLCM) [20] and then combined with spectral information for classification [21–23]. The utilization of such texture information can reduce the impacts of isolated pixels within the pixel-based approach. The object-oriented approach first extracts meaningful objects via multi-resolution segmentation [24] and classification is then carried out on object units [25–27]. These two approaches are known to achieve better classification accuracy than the pixel-based approach based purely on spectral information [19,22]. The second issue is heavy computational load related to data preprocessing and processing [11]. Most UAV images are acquired using a narrow field-of-view, which requires mosaicking of many sub-images to obtain a complete image set. If the sub-images are taken at different solar conditions and flight altitudes, radiometric calibration should be employed during mosaicking. The ultra-high spatial resolution of UAV images makes preprocessing complex and requires much processing time for classification [11].

Another important issue is that it is not always possible to construct a time-series UAV image set for crop classification. Although the acquisition of UAV images is less affected by atmospheric conditions than satellite images, it may be difficult to take UAV images in some season [12], particularly the rainy season which coincides with the growing season of crops in Korea. From an operational viewpoint, the acquisition of time-series UAV images for crop classification essentially has a prerequisite that operators make several visits to the area of interest. From a practical point of view, it is necessary to acquire optimal images at certain times, achieving classification accuracy comparable to the use of a complete time-series image set. Crop classification using UAV images is primarily conducted using a single UAV image [21,28], but accuracy comparisons with the case using a time-series image set are yet to be considered fully.

In addition to data acquisition issues, selection of proper classification methodology is important in order to generate reliable crop classification results. Since the 2000s, machine learning algorithms such as random forest (RF) and support vector machine (SVM) were widely applied to crop classification with remote sensing data [29–34].

Along with the aforementioned issues related to crop classification with UAV images and selection of appropriate classification methodology, this paper focuses on the evaluation of the effectiveness of texture information for crop classification with UAV images. In particular, the classification performance using a single-date UAV image is compared with that of a time-series image set.

In this study, two machine learning algorithms, RF and SVM, are applied as classification models, and the GLCM-based texture features [20] are used as additional features to reduce noise effects. From a practical viewpoint, we also investigate how much the utilization of texture information can improve classification accuracy when only a single-time UAV image is available. A case study of crop classification with UAV images in Anbandegi, a highland Kimchi cabbage cultivation area in Korea, was conducted to illustrate and discuss the two issues including the limited use of the UAV image and the impact of GLCM-based texture features on classification performance.

2. Materials and Methods

2.1. Study Area

Anbandegi, in the Gangwon Province of Korea, a major highland Kimchi cabbage cultivation area, was selected as the case study area (Figure 1). Summer Kimchi cabbage is usually cultivated in highlands in Korea because high temperature and humidity causes physiological disorders, insect pests, and diseases [35]. The altitude of the study area is about 1000 meters above mean sea level and is relatively higher than the surrounding terrain, which is suitable for highland Kimchi cabbage cultivation [35]. In the study area, cabbage and potatoes are also grown along with highland Kimchi cabbage. The total area of all crop parcels in the study area is 42.5 ha and the average size of each crop parcel is about 0.6 ha. The land-cover type of non-crop areas is mainly forest.

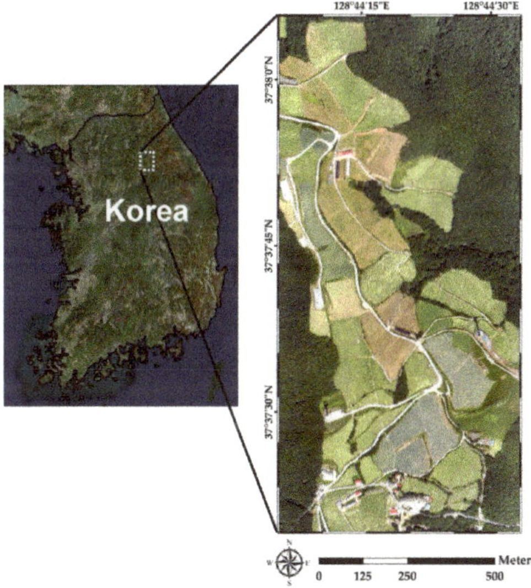

Figure 1. Location of the study area and the unmanned aerial vehicle (UAV) image mosaic acquired in the study area.

2.2. Datasets

2.2.1. UAV Images

We used six UAV image mosaics taken from June to September 2017, by considering the growth cycle of highland Kimchi cabbage (Table 1). The preprocessed UAV image mosaics provided by the National Institute of Agricultural Sciences (NAAS) were acquired from a fixed-wing unmanned aerial system (UAS; eBee, Sensefly, Swiss) equipped with a Cannon S110 camera that includes green (550 nm), red (625 nm), and near-infrared (NIR; 850 nm) spectral bands (hereafter referred to as VNIR). The UAV image mosaics with a ground sampling distance of 12 cm were upscaled to 25 cm resolution to facilitate data processing without loss of information. Upscaling may result in loss of textural image information. However, a significant change in the generation of texture features and classification results was not observed in our preliminary experiment at subareas in the study area, which was attributed to the size of crop parcels in the study area. Hence, the image mosaics with a 25 cm resolution were used as inputs for classification. To examine the applicability of a single-date image with texture information, the UAV image mosaic acquired on 25 August was selected due to the peak in vitality of highland Kimchi cabbage. This selection is explained in detail in Section 3.

Table 1. List of unmanned aerial vehicle (UAV) image mosaics acquired in the study area in 2017.

No.	Acquisition Date
1	29 June
2	12 July
3	27 July
4	25 August
5	13 September
6	21 September

2.2.2. Ground-Truth Data and Land-Cover Map

Ground-truth crop types were obtained by field surveys, which were also provided by NAAS. These data were used to both extract training data and to evaluate the classification performance. Table 2 presents crop classes for supervised classification and area information of each crop type. To mimic a case with limited available training data, 20,000 pixels (about 0.3% of ground-truth data) were randomly selected and used for training data for supervised learning. The remaining 6,710,210 pixels (99.7% of ground-truth data) were used as reference data. Note that a relatively small training dataset and a large reference dataset are used for classification and evaluation, respectively. Since the main targets of classification were crops in the study area, non-crop areas, including forests, were masked out prior to classification using land-cover maps from the Ministry of Environment [36].

Table 2. Crop classes and their respective area information in the study area.

Classes	Total Area (ha)	Average Area per Parcel (ha)
Highland Kimchi Cabbage	22.38	0.59
Cabbage	8.35	0.59
Potato	8.65	0.86
Fallow	3.08	0.31

2.3. Classification Methods and Feature Extraction

2.3.1. Random Forest

The RF classifier developed by Breiman [37] performs classification by extending decision trees to multiple trees rather than a single tree. Its classification performance is superior to a single decision tree due to its ability to maximize diversity through tree ensembles. It also demonstrates greater stability due to the synthesis of classification results from a large number of trees and the determination of final class labels through majority voting. In addition, RF requires a few parameters (i.e., the number of variables for node partitioning and the number of trees to be grown) to be set, unlike other machine learning algorithms.

The RF classifier applies bootstrap aggregating (bagging) to tree learners. Bagging repeatedly selects a random sample to replace the training data and fits trees to these samples. The remaining training data, the out-of-bag (OOB) data, are used to validate trees [37]. The OOB error that is the error rate of the OOB classifiers is often used as a measure of the generalization error on the training data [37]. To avoid overfitting the training data, each node of the trees determines the partitioning condition, and each tree chooses the random predictor variable and divide node using a genie index, as a measure of heterogeneity. An additional function of the RF classifier is to compute quantitative measures for variable importance using mean decrease impurity (MDI) and mean decrease accuracy (MDA) [28]. When constructing a large number of trees, MDI and MDA can be calculated by averaging the weighted impurity of each tree and the degree of accuracy improvement, respectively, by randomly changing the variable. In this study, the variable importance was used to quantify how useful texture information is for crop classification.

2.3.2. Support Vector Machine

SVM is a machine learning algorithm for finding the optimal decision boundary of training data located at the boundary of classes [38]. The SVM classifier is known to be effective for classification with a limited amount of training data [39]. The main concept of SVM is to solve the optimization problem which maximizes the margin between decision boundaries [40]. To solve non-linear optimization problems, kernel functions such as radial basis function (RBF) are commonly used [39]. When the RBF kernel is used, the parameters of cost and gamma should be optimally determined. Large values of cost and gamma result in overfitting to the training data, yielding poor generalization ability of the classifier [41]. In this study, these two hyper-parameters were determined using a grid search based on 10-fold cross-validation of training data [42].

2.3.3. Texture Information

To reduce the noise effects of isolated pixels in classification results, texture information is considered as an auxiliary feature for classification. Image texture analysis methods can be divided into four categories: statistical, geometric, model-based, and signal processing [43]. GLCM, developed by Haralick et al. [20], is a widely applied statistical method for remote sensing data processing such as vegetation structure modeling [44] and land-cover classification [45]. The original image is first converted to gray-scale. Then, the spatial features of the gray-scale image are extracted using the relationship of the brightness values between the center pixel and its neighborhood within the predefined kernel. The relationship of the brightness values is represented by a matrix which consists of the occurrence frequency of sequential pairs of pixel values existing simultaneously along the defined direction. By using this relationship, the GLCM can generate different texture information according to gray-scale level, kernel size, and direction. Fourteen texture features defined by Haralick et al. [20] are correlated, indicating that using all possible texture features provides redundant spatial contextual information which is not useful for classification. In this study, six texture features [46] were considered: (1) mean (ME), (2) standard deviation (STD), (3) homogeneity (HOM), (4) dissimilarity (DIS), (5) entropy (ENT), and (6) angular second moment (ASM), presented in Equations (1) to (6):

$$\text{ME} = \sum_{i=0}^{N-1} \sum_{j=0}^{N-1} i \times P(i,j), \tag{1}$$

$$\text{STD} = \sqrt{\sum_{i=0}^{N-1} \sum_{j=0}^{N-1} P(i,j) \times (i - \text{ME})^2}, \tag{2}$$

$$\text{HOM} = \sum_{i=0}^{N-1} \sum_{j=0}^{N-1} \frac{P(i,j)}{1 + (i-j)^2}, \tag{3}$$

$$\text{DIS} = \sum_{i=0}^{N-1} \sum_{j=0}^{N-1} P(i,j) \times |i-j|, \tag{4}$$

$$\text{ENT} = \sum_{i=0}^{N-1} \sum_{j=0}^{N-1} -P(i,j) \times \ln(p(i,j)), \tag{5}$$

$$\text{ASM} = \sum_{i=0}^{N-1} \sum_{j=0}^{N-1} P(i,j)^2, \tag{6}$$

where N denotes gray-scale level, while $P(i,j)$ is the normalized gray-scale value at positions i and j within the kernel, and its sum is 1.

The above texture features were generated from omnidirectional 64-shade gray-scale images. To test the impacts of the kernel size, we used three different kernel sizes: 3×3 (GK3), 15×15 (GK15), and 31×31 (GK31).

2.4. Classification Procedures

The entire procedure for crop classification with UAV images is presented in Figure 2. For each classifier, optimal parameters were first sought during a training phase. To investigate the impacts of both the number of input images and texture features, we tested eight combination cases for each classifier: UAV images (two cases: with the August image and with six multi-temporal images), and texture features (four cases: with texture features from three different kernel sizes (GK3, GK15, and GK31), and without texture features). These combinations were considered for comparison purposes since the main objective of this study was to evaluate the effectiveness of using texture information when a single-date UAV image is used for crop classification. The classification accuracy was assessed using quantitative measures based on a confusion matrix such as overall accuracy (OA), producer's accuracy (PA), and user's accuracy (UA).

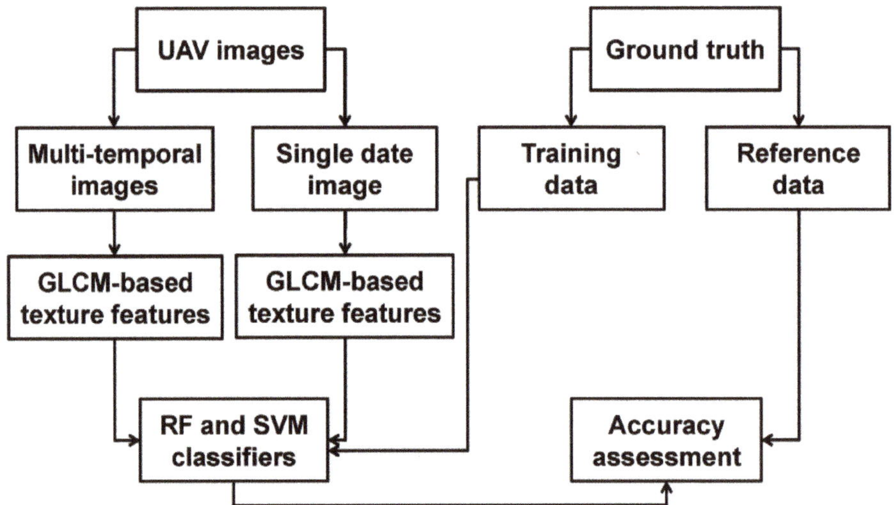

Figure 2. Schematic diagrams of all crop classification procedures applied in this study. GLCM: gray-level co-occurrence matrix; RF: random forest; SVM: support vector machine.

2.5. Implementation

ENVI software version 4.8 was used for generation of GLCM-based features and visualization of classification results. All procedures for classification and evaluation were done within the R software environment [47]. SVM and RF models were built using the R packages e1071 [42] and randomForest [48], respectively.

3. Results and Discussion

3.1. Parameterization of RF and SVM Classifiers

For the RF classifier, two parameters, the number of variables required for node partitioning and the number of trees to be grown, have to be selected. Firstly, the number of variables for node partitioning was set to $\sqrt{n}$ of the total number of variables. For example, for the case using the August image with texture information, there were nine variables (three spectral bands and six texture features); thus, the number of variables for node partitioning was set to 3. To determine the number of trees to be grown, variations of OOB errors with respect to the number of trees were investigated. From the variations of OOB errors, one can judge whether a sufficient number of trees were used for the RF modeling. In general, the OOB errors tend to decrease as the number of trees increases, and then converge to a certain value at the specific number of trees. When six multi-temporal UAV images were used as inputs, no distinctive differences in OOB errors were observed, and the error values were also very low for different texture feature cases. Figure 3 shows the variations of OOB errors when using the August image without and with texture features. The four combination cases showed different convergence values, but the variation patterns were very similar. As the number of trees increased to about 50, the OOB errors of all four combination cases decreased sharply. Then, the OOB errors reached the convergence values when the number of trees was about 150. By considering the convergence of OOB errors and processing time, the number of trees to be grown was set to 150.

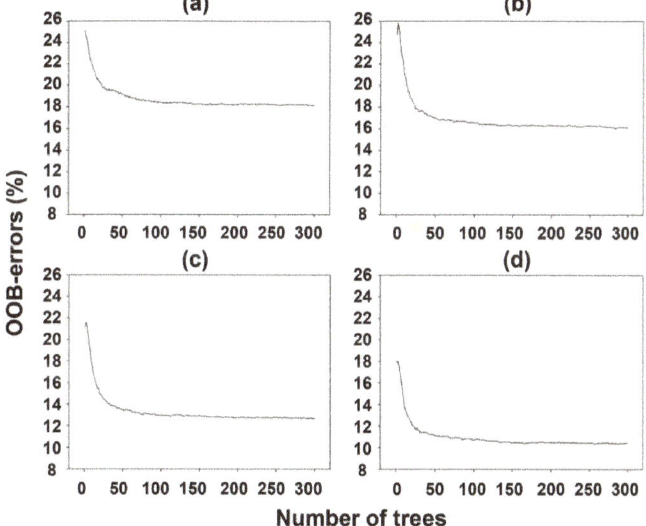

Figure 3. Variations of out-of-bag (OOB) errors of RF models with respect to the number of trees for the case using the August UAV image without and with texture features: (**a**) visible and near infrared (VNIR) spectral bands only; (**b**) VNIR + 3 × 3 kernel (GK3); (**c**) VNIR + 15 × 15 kernel (GK15); and (**d**) VNIR + 31 × 31 kernel (GK31).

Two parameters (cost and gamma) of the RBF kernel for the SVM classifier were tuned using a grid search. The optimal combination of the two parameters was determined through 10-fold cross-validation of training data. The optimal cost and gamma values were similar for combination cases of different kernel sizes of GLCM and input UAV images. Figure 4 presents the grid search results for the cases using the August image and six multi-temporal UAV images with texture feature GK31, showing the different training accuracy values. The training accuracy obtained by the grid search ranged between 52 and 82.4% for the case using the August image with texture features, while the maximum training accuracy for the case using six UAV images increased to 94%. It should be noted that this accuracy was obtained during the training phase; hence, higher training accuracy may fail to achieve higher prediction performance. It was found that the performance difference with respect to variations of the model parameters for the SVM classifier was also great, compared to the RF classifier, which indicates the importance of optimal parameter search for the SVM classifier.

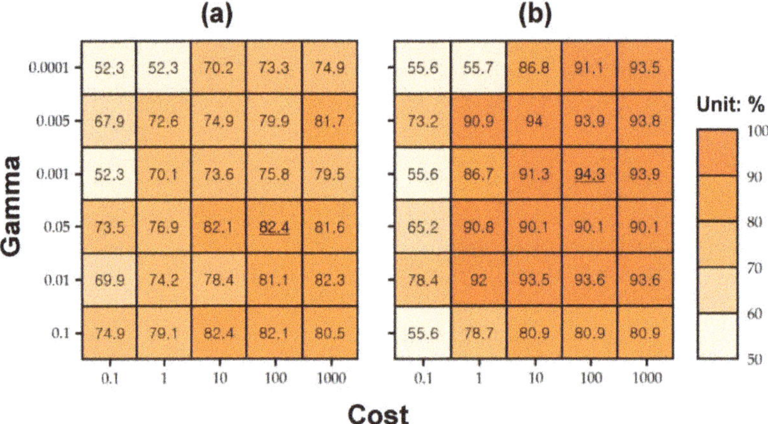

Figure 4. Cross-validation accuracy of SVM classifiers through a grid search. The case with the best accuracy is underlined: (a) using a single August image with VNIR and GK31 texture features; and (b) using six UAV images with VNIR and GK31 texture features.

3.2. Visual Assessment of Classification Results

Once optimal parameters were determined, the RF and SVM classifiers were applied to the different case combinations. Prior to quantitative accuracy assessment, the visual assessment of classification results was first conducted. When the RF and SVM classification results were compared for different combinations of input images and kernel sizes, the RF classifier showed misclassifications at some parcels in the southeastern parts of the study area, but significant differences in classification patterns were not observed. Figure 5 shows some classification results using the SVM classifier. When three spectral bands of the August image were used for classification, misclassification and noise effects by isolated pixels were the greatest in visual inspection of classification results. Confusion between highland Kimchi cabbage and cabbage was most common, as shown in Figure 5b, mainly due to their similar spectral characteristics in August (this is further discussed in Section 3.5). When texture features were combined with spectral information for the case using the August image only, the number of misclassified and isolated pixels decreased, but some misclassified pixels were still shown (Figure 5c). Using multi-temporal images greatly reduced misclassified pixels within each parcel, except for some around the parcel boundaries (Figure 5d). As expected, the use of texture features as additional information with multi-temporal spectral information showed the best agreement with the ground-truth data from visual inspection (Figure 5e), indicating the necessity of time-series images and texture features for crop classification.

The impacts of texture features generated by different kernel sizes on the classification results were also visually compared. The classified patterns were significantly affected by kernel size. When a very small kernel size, such as GK3, was used to extract texture features, the classification result was very similar to the case with spectral information only. As the kernel size increased, the noise effect was greatly alleviated. When multi-temporal images were used for classification, however, the combination of texture features with multi-temporal spectral information was less affected by the change in kernel size. The increase in kernel size resulted in the reduction of isolated noise patterns, but the difference was subtle compared to the case using the August UAV image.

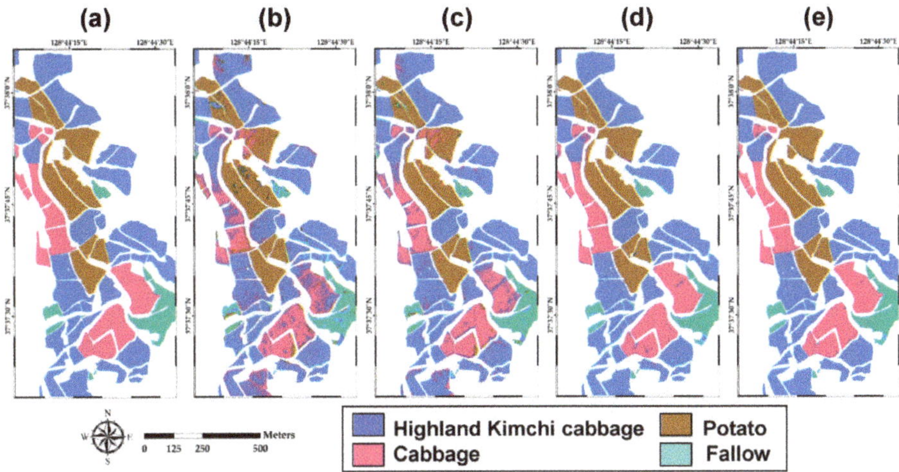

Figure 5. Comparison of SVM-based classification results with ground-truth data: (**a**) ground-truth data; (**b**) August image with VNIR; (**c**) August image with VNIR and GK31 texture features; (**d**) six multi-temporal images with VNIR; and (**e**) six multi-temporal with VNIR and GK31 texture features.

3.3. Quantitative Accuracy Assessment

The aforementioned visual and qualitative comparison results were further evaluated quantitatively by computing and comparing accuracy statistics. Confusion matrices were first prepared for all combination cases of each classifier, and related accuracy statistics were calculated by comparing classification results with reference data that were not used for training.

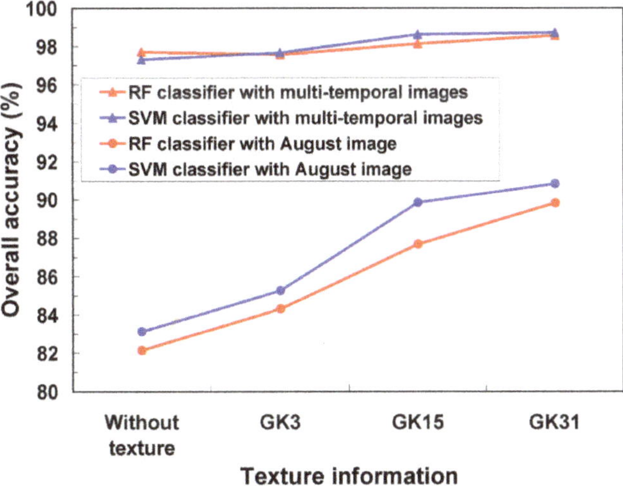

Figure 6. Overall accuracy of classification results without texture features and with texture features generated from different kernel sizes for the cases using the August image and six multi-temporal images.

Figure 6 shows variations in OA of classification results without texture features (VNIR) and with texture features generated from different kernel sizes (GK3, GK15, and GK31) using the August image and multi-temporal images. Although a very small portion of ground-truth data were used as training data, the OA values for the two classifiers were notably high (i.e. over 97%) when multi-temporal images were used for classification. Regardless of the number of input images and the classifier type, the combination of texture features and spectral information led to an increase in OA. The OA also increased with kernel size; however, only a slight improvement of OA was achieved for classification with multi-temporal images as kernel size increased. This result can be explained by the fact that most useful information for the discrimination of crops was already provided by time-series spectral information; hence, the contribution of texture features was minimal. In contrast, the improvement in OA by accounting for texture features was much more significant in the classification result using the August image only than using the multi-temporal images. Furthermore, the kernel size of GLCM greatly affected the OA using the August image. As kernel size increased, OA increased for both SVM and RF classifiers, and the use of GK31 texture features showed the best classification accuracy.

When comparing classification performance of both classifiers, the SVM classifier exhibited better OA than the RF classifier for the classification with the August image, indicating the superiority of the SVM classifier for the classification of crops in this study area. The difference in OA between SVM and RF classifiers was significant at the 5% significance level from the McNemar test [49], regardless of kernel sizes. It is noteworthy that the small difference in OA between two classifiers was significant at the 5% significance level even for all classification results based on multi-temporal images. Despite the similar OA values between two classifiers in the classification of multi-temporal images, this statistically significant difference was mainly due to evaluation with a very large amount of reference data (6,710,210 pixels). Even though parameter tuning is more demanding in the SVM classifier than the RF classifier, the optimal two parameters of the SVM classifier which were determined during a training stage with a relatively small training dataset could avoid overfitting the training data, leading to generalization ability for the large amount of reference data in this study.

Some confusion matrices for typical combination cases of the SVM classifier (one image versus multi-temporal images and with or without texture features) are listed in Table 3. Considering only the August image, combining texture features (GK31) with spectral information led to an increase of 7.72%p in OA, compared with the classification result with spectral information only (from 83.13% to 90.85%). The increase of class-wise accuracies was also achieved, as well as the improvement in OA. As discussed in the visual analysis of classification results, the confusion among four classes in Table 3 (particularly between highland Kimchi cabbage and cabbage) was significantly reduced, yielding increases in both PA and UA for all classes. When the August image with VNIR only was considered for classification, the similar vegetation vitality of highland Kimchi cabbage, cabbage, and weeds within the fallow class resulted in severe confusion. By accounting for texture features with spectral information, the confusion could be reduced. However, PA and UA of cabbage were relatively lower than that of other crops, indicating a persistent misclassification of cabbage to highland Kimchi cabbage. When multi-temporal images were used for classification, the accuracy values of all classes increased, particularly with cabbage. Texture features with multi-temporal spectral information proved most useful in the cabbage class because it alleviated the misclassification of cabbage to highland Kimchi cabbage.

Based on all evaluation results in Figure 6 and Table 3, it can be concluded that texture information extracted by the proper kernel size can improve classification performance, and the impact of using texture features is most significant when using a single image for crop classification. The latter finding implies the usefulness of texture information when only one UAV image is available for crop classification, due to difficulty acquiring time-series UAV images in the area of interest.

Table 3. Confusion matrices and accuracy statistics of some combination cases for the support vector machine (SVM) classifier. VNIR: visible and near infrared; UA: user's accuracy; PA: producer's accuracy; OA: overall accuracy; GK31: kernel size of 31 × 31.

August Image: VNIR Spectral Information					
Reference / Classification	Highland Kimchi Cabbage	Cabbage	Potato	Fallow	UA (%)
Highland Kimchi cabbage	3,074,131	342,355	49,838	84,726	86.57
Cabbage	230,661	869,250	65,288	6627	74.18
Potato	107,897	124,020	1,259,483	31,883	82.68
Fallow	67,045	12,343	9497	375,166	80.85
PA (%)	88.34	64.49	91.00	75.27	
OA (%)	83.13				

August Image: VNIR Spectral Information and GK31 Texture Features					
Reference / Classification	Highland Kimchi Cabbage	Cabbage	Potato	Fallow	UA (%)
Highland Kimchi cabbage	3,317,807	237,669	22,404	44,455	91.59
Cabbage	128,847	1,050,991	57,648	2636	84.75
Potato	16,522	54,544	1,294,756	18,465	93.53
Fallow	16,558	4764	9298	432,846	93.39
PA (%)	95.35	77.97	93.54	86.85	
OA (%)	90.85				

Multi-Temporal Images: VNIR Spectral Information					
Reference / Classification	Highland Kimchi Cabbage	Cabbage	Potato	Fallow	UA (%)
Highland Kimchi cabbage	3,421,871	46,150	11,031	41,618	97.19
Cabbage	15,143	1,294,009	10,189	4185	97.77
Potato	2092	4562	1,360,566	200	99.50
Fallow	40,628	3247	2320	452,399	90.73
PA (%)	98.34	96.00	98.30	90.77	97.30
OA (%)	97.30				

Multi-Temporal Images: VNIR Spectral Information and GK31 Texture Features					
Reference / Classification	Highland Kimchi Cabbage	Cabbage	Potato	Fallow	UA (%)
Highland Kimchi cabbage	3,461,811	35,558	5349	16,199	98.38
Cabbage	7159	1,309,847	5323	2337	98.88
Potato	1346	792	1,372,686	186	99.83
Fallow	9418	1771	748	479,680	97.57
PA (%)	99.48	97.17	99.17	96.24	98.72
OA (%)	98.72				

3.4. Comparison of Spectral and Texture Information

To examine which variable was most influential for classification performance, quantitative measures for variable importance were computed using the MDA in the RF classifier. MDA values of input variables with respect to different kernel sizes of GLCM are shown in Figure 7. Since 54 input variables were used for the classification of six multi-temporal images, only the top nine variables with the highest MDA values are presented for illustration purposes. Regardless of input images and the kernel size of GLCM, NIR and green bands were the most influential variables of the RF classifier. In particular, the NIR bands from July to September were included as important variables for the classification of multi-temporal images. Note that spectral information was more useful than texture information, and only one texture feature, such as ME, was helpful for multi-temporal images. ME, which is an estimate of the intensity of all pixels in spatial relationships that contribute to the GLCM, was the most important variable among the six texture features, irrespective of input images.

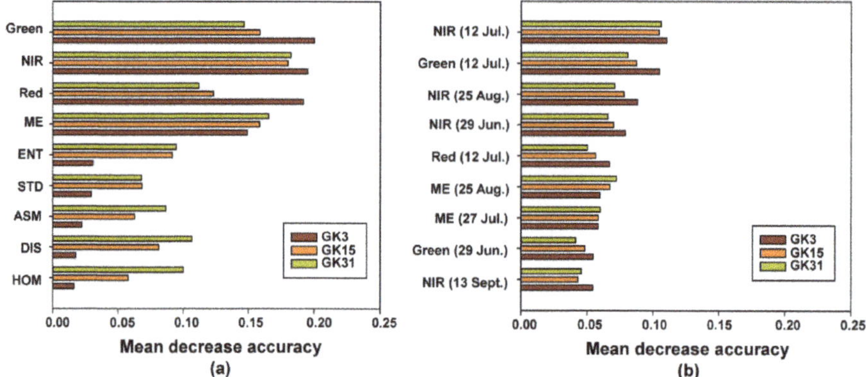

Figure 7. Mean decrease accuracy (MDA) values of input spectral and texture variables with respect to kernel size: (**a**) August image; and (**b**) multi-temporal images. ME: mean; ENT: entropy; ASM: angular second moment; STD: standard deviation; HOM: homogeneity; DIS: dissimilarity.

The MDA values of input variables were quite different according to the input images. When six multi-temporal images were used for classification, the MDA value for each variable was relatively small due to contributions of many input variables, but information content provided by many input variables led to very high classification accuracy, as shown in Table 3. Although multi-temporal spectral bands were considered the most informative, the influence of ME increased with kernel size (see the MDA value of ME for GK31 in Figure 7). With classification using only the August image, ME was the second most important variable for GK15 and GK31, indicating that the ME feature is very useful for the classification of crops in the study area. The contribution of other texture features increased with kernel size. For GK3, MDA values of texture features were much smaller than those of spectral bands. With increasing kernel size, gains in MDA values were most significant for texture features, including DIS and ENT. Texture information extracted from the GLCM with the proper kernel size can fill gaps in multi-spectral information, leading to an improvement in classification accuracy, as shown in Figure 6 and Table 3.

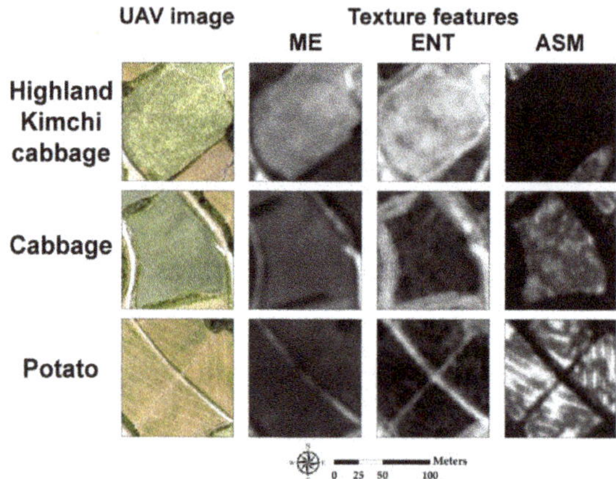

Figure 8. Some texture features (GK31) in subareas of the August image.

For further qualitative inspection of texture features, some texture features in four subareas of GK31 are provided in Figure 8. Brighter colors represent larger values in each texture feature. ME, which is regarded as the GLCM mean, provides low-pass filtered spatial information that is useful to mitigate noise effects in the ultra-high-resolution UAV image. As an index for measuring the randomness of contrast distributions, ENT increased with greater change of brightness values between the center pixel and its neighboring pixels. ENT values for different classes appear in Figure 8. ASM, which measures uniformity of contrast, also changed with the four classes. This visual inspection of texture features further confirmed the usefulness of texture information.

When considering the spatial resolution of the UAV image used for crop classification (i.e., 25 cm), GK3 and GK31 texture information represents 0.75 m and 7.75 m on the ground, respectively. The GK31 texture features are likely to represent the serial line patterns of crop cultivation well, consequently leading to superior OA. However, this is the particular result in the study area. If the spatial resolution of input images and the crop types change, the optimal kernel size of GLCM should be determined by considering spatial resolution, as well as cultivation patterns and crop characteristics such as size and shape.

3.5. Time-Series Analysis of Normalized Difference Vegetation Index for Selection of Optimal UAV Image

Spectral characteristics of crops depend on crop type and health conditions, but different crops may exhibit similar spectral response [35,50]. Accordingly, time-series images acquired during growth cycles of crops are often used to examine how well these images account for temporal variations of spectral response. For example, if temporal patterns in spectral responses of crops in the study area are significantly different, classification based on multi-temporal images can achieve satisfactory classification accuracy. Conversely, discrimination of crops with similar temporal variations of spectral responses may be difficult, even when multi-temporal images are used.

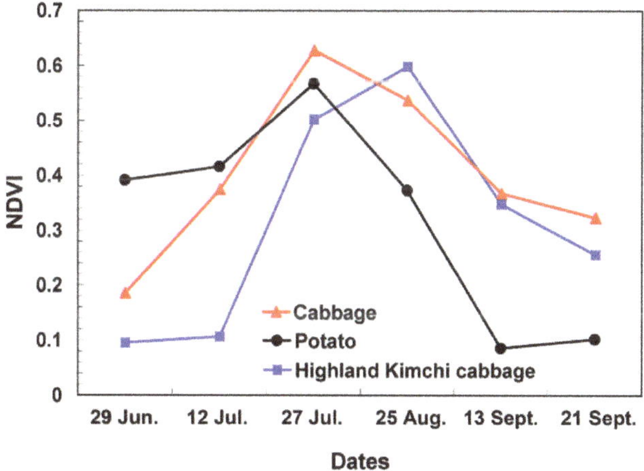

Figure 9. Temporal profiles of average normalized difference vegetation index (NDVI) values for three crops.

Figure 9 shows temporal variations in the average of normalized difference vegetation index (NDVI) values at pixels belonging to each crop. NDVI is a standardized index that quantifies greenness by using the difference in reflectance between NIR and red bands [51]. The average NDVI value of highland Kimchi cabbage was significantly lower than other crops on 12 July, and peaked in late August. In late July, cabbage had the highest NDVI value, followed by potatoes. The difference in average NDVI values between highland Kimchi cabbage and cabbage was not great in the August image (Figure 9), which led to difficulty in discerning the two crops. Although the difference was greater on 27 July, as shown in Figure 9, the lowest NDVI value of highland Kimchi may have resulted in the confusion with fallow and other small vegetation in the classification result using the 27 July image. If only one UAV image should be acquired, the image needs to be acquired when the vegetation vitality of the crop of interest reaches its maximum. Since highland Kimchi cabbage reached its maximum NDVI value in the 25 August image, we selected that image as the optimal single image. Actually, the classification accuracy using either the 12 July or the 27 July image was either similar to or lower than that using the August image. Despite the risk of misclassification using only the August image, similar spectral responses of different crops highlight the necessity of using additional information such as texture features, as applied in this study. Since the time to reach the maximum peak in NDVI may differ every year depending on weather conditions, however, the selection of the most appropriate acquisition date should be made by considering conditions and types of crops. Therefore, more extensive experiments should be carried out in other areas with different crop types. In addition, if phenological characteristics can be estimated from the entire time-series image set [52,53], a single-image acquisition date can be determined more optimally.

3.6. Classification Methods

In this study, two machine learning algorithms including RF and SVM were applied to crop classification. Recently, deep learning algorithms including convolutional neural network (CNN) were widely applied to remote sensing data classification [54–56]. Despite the promising performance of CNN, Kim et al. [57] reported that the training sample size has greater effects on the accuracy of CNN than that of SVM in crop classification, indicating a need for numerous training samples for improved CNN classification performance. Furthermore, Yu et al. [58] also reported that SVM with adjacent region features showed better accuracy than CNN for moderate-resolution land-cover classification. Therefore, deep learning is not always superior for all cases, and conventional machine learning algorithms can achieve classification performance comparable to, or even better than deep learning algorithms if proper spatial contextual features are combined with spectral information. To further evaluate the usefulness of texture features for crop classification, comparison with a patch-based CNN classifier will be conducted.

4. Conclusions

This study investigated the potential of GLCM-based texture information for crop classification with time-series UAV images and machine learning algorithms. The main focus was on the evaluation of the benefit of utilization of texture features along with spectral information when using a single UAV image. A case study of crop classification in the highland Kimchi cabbage cultivation area demonstrated the most accurate classification of multi-temporal UAV images with GLCM-based texture features. However, the utilization of texture features with spectral information from multi-temporal images did not lead to a significant improvement in classification accuracy. In contrast, when only a single UAV image was used, the utilization of texture features could significantly improve the classification accuracy. Therefore, when only one UAV image should be used for crop classification due to a difficulty in constructing a time-series UAV dataset, the information deficiency in spectral information can be complemented by structural information from texture features. Furthermore, the impact of texture information on classification accuracy was dependent on the kernel size of GLCM. Texture information extracted from the GLCM with larger kernel size improved classification performance in the study area. Therefore, proper kernel size selection is critical for the extraction of GLCM-based texture features. This indicates that both spatial resolution of input UAV images and shape characteristics of individual crops of interest should be considered in selection of optimal kernel size. However, these findings may be specific to this study area with particular crop types and not applicable to other areas. Therefore, more experiments on other areas with different combinations of crops should be carried out to strengthen the potential benefit of texture information from UAV images for crop classification. Experiments regarding determination of the minimum number of UAV images in crop classification with texture features, and comparison with deep learning algorithms will also be carried out in the future to extend key findings and recommendations presented herein.

Author Contributions: Conceptualization, G.-H.K. and N.-W.P.; methodology development and experiments, G.-H.K.; manuscript writing, G.-H.K. and N.-W.P.; supervision, N.-W.P.

Funding: This work was supported by INHA UNIVERSITY Research Grant.

Acknowledgments: The authors thank Kyung-Do Lee, Sang-Il Na, and Chan-Won Park at the National Institute of Agricultural Sciences for providing UAV images and ground-truth data used in this study. Constructive comments from the three anonymous reviewers are gratefully acknowledged.

Conflicts of Interest: The authors declare no conflicts of interest.

References

1. Rosenzweig, C.; Elliott, J.; Deryng, D.; Ruane, A.C.; Müller, C.; Arneth, A.; Boote, K.J.; Folberth, C.; Glotter, M.; Khabarov, N.; et al. Assessing agricultural risks of climate change in the 21st century in a global gridded crop model intercomparison. *Proc. Natl. Acad. Sci. USA* **2014**, *111*, 3268–3273. [CrossRef] [PubMed]

2. Zhong, L.; Gong, P.; Biging, G.S. Efficient corn and soybean mapping with temporal extendability: A multiyear experiment using Landsat imagery. *Remote Sens. Environ.* **2014**, *140*, 1–13. [CrossRef]

3. Na, S.I.; Park, C.W.; So, K.H.; Ahn, H.Y.; Lee, K.D. Development of biomass evaluation model of winter crop using RGB imagery based on unmanned aerial vehicle. *Korean J. Remote Sens.* **2018**, *34*, 709–720, (In Korean with English Abstract).

4. Tilman, D.; Balzer, C.; Hill, J.; Befort, B.L. Global food demand and the sustainable intensification of agriculture. *Proc. Natl. Acad. Sci. USA* **2011**, *108*, 20260–20264. [CrossRef] [PubMed]

5. Lee, J.; Seo, B.; Kang, S. Development of a biophysical rice yield model using all-weather climate data. *Korean J. Remote Sens.* **2018**, *33*, 721–732, (In Korean with English Abstract).

6. Kim, Y.; Park, N.-W.; Lee, K.-D. Self-learning based land-cover classification using sequential class patterns from past land-cover maps. *Remote Sens.* **2018**, *9*, 921. [CrossRef]

7. Clark, M.L. Comparison of simulated hyperspectral HyspIRI and multispectral Landsat 8 and Sentinel-2 imagery for multi-seasonal, regional land-cover mapping. *Remote Sens. Environ.* **2017**, *200*, 311–325. [CrossRef]

8. Sonobe, R.; Yamaya, Y.; Tani, H.; Wang, X.; Kobayashi, N.; Mochizuki, K.-I. Mapping crop cover using multi-temporal Landsat 8 OLI imagery. *Int. J. Remote Sens.* **2017**, *38*, 4348–4361. [CrossRef]

9. Friedl, M.A.; McIver, D.K.; Hodges, J.C.F.; Zhang, X.Y.; Muchoney, D.; Strahler, A.H.; Woodcock, C.E.; Gopal, S.; Schneider, A.; Cooper, A.; et al. Global land cover mapping from MODIS: Algorithms and early results. *Remote Sens. Environ.* **2002**, *83*, 287–302. [CrossRef]

10. Yang, C.; Wu, G.; Ding, K.; Shi, T.; Li, Q.; Wang, J. Improving land use/land cover classification by integrating pixel unmixing and decision tree methods. *Remote Sens.* **2017**, *9*, 122. [CrossRef]

11. Hall, O.; Dahlin, S.; Marstorp, H.; Archila Bustos, M.; Öborn, I.; Jirström, M. Classification of maize in complex smallholder farming systems using UAV imagery. *Drones* **2018**, *2*, 22. [CrossRef]

12. Böhler, J.; Schaepman, M.; Kneubühler, M. Crop classification in a heterogeneous arable landscape using uncalibrated UAV data. *Remote Sens.* **2018**, *10*, 1282. [CrossRef]

13. Pajares, G. Overview and current status of remote sensing applications based on unmanned aerial vehicles (UAVs). *Photogramm. Eng. Remote Sens.* **2015**, *81*, 281–330. [CrossRef]

14. Shahbazi, M.; Théau, J.; Ménard, P. Recent applications of unmanned aerial imagery in natural resource management. *GISci. Remote Sens.* **2014**, *51*, 339–365. [CrossRef]

15. Latif, M.A. An agricultural perspective on flying sensors. *IEEE Geosci. Remote Sens. Mag.* **2018**, *6*, 10–22. [CrossRef]

16. Poblete-Echeverría, C.; Olmedo, G.F.; Ingram, B.; Bardeen, M. Detection and segmentation of vine canopy in ultra-high spatial resolution RGB imagery obtained from unmanned aerial vehicle (UAV): A case study in a commercial vineyard. *Remote Sens.* **2017**, *9*, 268. [CrossRef]

17. Melville, B.; Fisher, A.; Lucieer, A. Ultra-high spatial resolution fractional vegetation cover from unmanned aerial multispectral imagery. *Int. J. Appl. Earth Obs. Geoinf.* **2019**, *78*, 14–24. [CrossRef]

18. Im, J.; Jensen, J.R.; Tullis, J.A. Object-based change detection using correlation image analysis and image segmentation. *Int. J. Remote Sens.* **2008**, *29*, 399–423. [CrossRef]

19. Li, M.; Zang, S.; Zhang, B.; Li, S.; Wu, C. A review of remote sensing image classification techniques: The role of spatio-contextual information. *Eur. J. Remote Sens.* **2014**, *47*, 389–411. [CrossRef]

20. Haralick, R.M.; Shanmugam, K.; Dinstein, I. Textural features for image classification. *IEEE Trans. Syst. Man Cybern.* **1973**, *SMC-3*, 610–621. [CrossRef]

21. Feng, Q.; Liu, J.; Gong, J. UAV remote sensing for urban vegetation mapping using random forest and texture analysis. *Remote Sens.* **2015**, *7*, 1074–1094. [CrossRef]

22. Yang, M.-D.; Huang, K.-S.; Kuo, Y.-H.; Tsai, H.P.; Lin, L.-M. Spatial and spectral hybrid image classification for rice lodging assessment through UAV imagery. *Remote Sen.* **2017**, *9*, 583. [CrossRef]

23. Zhang, X.; Cui, J.; Wang, W.; Lin, C. A study for texture feature extraction of high-resolution satellite images based on a direction measure and gray level co-occurrence matrix fusion algorithm. *Sensors* **2017**, *17*, 1474. [CrossRef] [PubMed]

24. Benz, U.C.; Hofmann, P.; Willhauck, G.; Lingenfelder, I.; Heynen, M. Multi-resolution, object-oriented fuzzy analysis of remote sensing data for GIS-ready information. *ISPRS J. Photogramm. Remote Sens.* **2004**, *58*, 239–258. [CrossRef]

25. Laliberte, A.S.; Rango, A.; Havstad, K.M.; Paris, J.F.; Beck, R.F.; McNeely, R.; Gonzalez, A.L. Object-oriented image analysis for mapping shrub encroachment from 1937 to 2003 in southern New Mexico. *Remote Sens. Environ.* **2004**, *93*, 198–210. [CrossRef]

26. Liu, H.; Zhang, J.; Pan, Y.; Shuai, G.; Zhu, X.; Zhu, S. An efficient approach based on UAV orthographic imagery to map paddy with support of field-level canopy height from point cloud data. *IEEE J. Sel. Top. Appl. Earth Obs. Remote Sens.* **2018**, *11*, 2034–2046. [CrossRef]

27. de Castro, A.I.; Torres-Sánchez, J.; Peña, J.M.; Jiménez-Brenes, F.M.; Csillik, O.; López-Granados, F. An automatic random forest-OBIA algorithm for early weed mapping between and within crop rows using UAV imagery. *Remote Sens.* **2018**, *10*, 285. [CrossRef]

28. Ahmed, O.S.; Shemrock, A.; Chabot, D.; Dillon, C.; Williams, G.; Wasson, R.; Franklin, S.E. Hierarchical land cover and vegetation classification using multispectral data acquired from an unmanned aerial vehicle. *Int. J. Remote Sens.* **2017**, *38*, 2037–2052. [CrossRef]

29. Tatsumi, K.; Yamashiki, Y.; Morante, A.K.M.; Fernández, L.R.; Nalvarte, R.A. Pixel-based crop classification in Peru from Landsat 7 ETM+ images using a random forest model. *J. Agric. Meteorol.* **2016**, *72*, 1–11. [CrossRef]

30. Moeckel, T.; Dayananda, S.; Nidamanuri, R.R.; Nautiyal, S.; Hanumaiah, N.; Buerkert, A.; Wachendorf, M. Estimation of vegetable crop parameter by multi-temporal UAV-borne images. *Remote Sens.* **2018**, *10*, 805. [CrossRef]

31. Song, Q.; Xiang, M.; Hovis, C.; Zhou, Q.; Lu, M.; Tang, H.; Wu, W. Object-based feature selection for crop classification using multi-temporal high-resolution imagery. *Int. J. Remote Sens.* **2018**, 1–16. [CrossRef]

32. Ma, L.; Fu, T.; Blaschke, T.; Li, M.; Tiede, D.; Zhou, Z.; Ma, X.; Chen, D. Evaluation of feature selection methods for object-based land cover mapping of unmanned aerial vehicle imagery using random forest and support vector machine classifiers. *ISPRS Int. J. Geo-Inf.* **2017**, *6*, 51. [CrossRef]

33. Yuan, Y.; Lin, J.; Wang, Q. Hyperspectral image classification via multitask joint sparse representation and stepwise MRF optimization. *IEEE Trans. Cybern.* **2016**, *46*, 2966–2977. [CrossRef]

34. Xie, F.; Li, F.; Lei, C.; Ke, L. Representative band selection for hyperspectral image classification. *ISPRS Int. J. Geo-Inf.* **2018**, *7*, 338. [CrossRef]

35. Lee, K.-D.; Park, C.-W.; So, K.-H.; Kim, K.-D.; Na, S.-I. Characteristics of UAV aerial images for monitoring of highland Kimchi cabbage. *Korean J. Soil Sci. Fertil.* **2017**, *50*, 162–178, (In Korean with English Abstract). [CrossRef]

36. EGIS (Environmental Geographic Information Service). Available online: https://egis.me.go.kr (accessed on 15 October 2018).

37. Breiman, L. Random forests. *Mach. Learn.* **2001**, *45*, 5–32. [CrossRef]

38. Gehler, P.V.; Schölkopf, B. An introduction to kernel learning algorithms. In *Kernel Methods for Remote Sensing Data Analysis*; Camps-Valls, G., Bruzzone, L., Eds.; Wiley: Chichester, UK, 2009; pp. 25–48.

39. Foody, G.M.; Mathur, A. A relative evaluation of multiclass image classification by support vector machines. *IEEE Trans. Geosci. Remote Sens.* **2004**, *42*, 1335–1343. [CrossRef]

40. Brereton, R.G.; Lloyd, G.R. Support vector machines for classification and regression. *Analyst* **2010**, *135*, 230–267. [CrossRef]

41. Foody, G.M.; Mather, A. Toward intelligent training of supervised image classifications: Directing training data acquisition for SVM classification. *Remote Sens. Environ.* **2004**, *93*, 107–117. [CrossRef]

42. Meyer, D.; Dimitriadou, E.; Hornik, K.; Weingessel, A.; Leisch, F. e1071: Misc Functions of the Department of Statistics, Probability Theory Group (Formerly: E1071), TU Wien. Available online: https://CRAN.R-project.org/package=e1071 (accessed on 28 December 2018).

43. Tuceryan, M.; Jain, A.K. Texture analysis. In *Handbook of Pattern Recognition & Computer Vision*, 2nd ed.; Chen, C.H., Pau, L.F., Wang, P.S.P., Eds.; World Scientific Publishing: Singapore, 1999; pp. 207–248.

44. Castillo-Santiago, M.A.; Ricker, M.; de Jong, B.H.J. Estimation of tropical forest structure from SPOT-5 satellite images. *Int. J. Remote Sens.* **2010**, *31*, 2767–2782. [CrossRef]

45. Johansen, K.; Coops, N.C.; Gergel, S.E.; Stange, Y. Application of high spatial resolution satellite imagery for riparian and forest ecosystem classification. *Remote Sens. Environ.* **2007**, *110*, 29–44. [CrossRef]

46. Szantoi, Z.; Escobedo, F.; Abd-Elrahman, A.; Smith, S.; Pearlstine, L. Analyzing fine-scale wetland composition using high resolution imagery and texture features. *Int. J. Appl. Earth Obs.* **2013**, *23*, 204–212. [CrossRef]

47. R Core Team. R: A Language and Environment for Statistical Computing. Available online: https://www.R-project.org (accessed on 28 December 2018).

48. Liaw, A.; Wiener, M. Classification and regression by randomForest. *R News* **2002**, *2*, 18–22.

49. Foody, G.M. Thematic map comparison: Evaluating the statistical significance of differences in classification accuracy. *Photogramm. Eng. Remote Sens.* **2004**, *70*, 627–633. [CrossRef]

50. Na, S.I.; Park, C.W.; So, K.H.; Ahn, H.Y.; Lee, K.D. Application method of unmanned aerial vehicle for crop monitoring in Korea. *Korean J. Remote Sens.* **2018**, *34*, 829–846, (In Korean with English Abstract).

51. Lillesand, T.M.; Kiefer, R.W.; Chipman, J.W. *Remote Sensing and Image Interpretation*, 6th ed.; Wiley: Hoboken, NJ, USA, 2008.

52. Atzberger, C.; Klisch, A.; Mattiuzzi, M.; Vuolo, F. Phenological metrics derived over the European continent from NDVI3g data and MODIS time series. *Remote Sens.* **2014**, *6*, 257–284. [CrossRef]

53. Lee, K.D.; Park, C.W.; So, K.H.; Kim, K.D.; Na, S.I. Estimating of transplanting period of highland Kimchi cabbage using UAV imagery. *J. Korean Soc. Agric. Eng.* **2017**, *59*, 39–50, (In Korean with English Abstract).

54. Ji, S.; Zhang, C.; Xu, A.; Shi, Y.; Duan, Y. 3D convolutional neural networks for crop classification with multi-temporal remote sensing images. *Remote Sens.* **2018**, *10*, 75. [CrossRef]

55. Kampffmeyer, M.; Salberg, A.-B.; Jenssen, R. Urban land cover classification with missing data modalities using deep convolutional neural networks. *IEEE J. Sel. Top. Appl. Earth Obs. Remote Sens.* **2018**, *11*, 1758–1768. [CrossRef]

56. Tan, K.; Wang, X.; Zhu, J.; Hu, J.; Li, J. A novel active learning approach for the classification of hyperspectral imagery using quasi-Newton multinomial logistic regression. *Int. J. Remote Sens.* **2018**, *39*, 3029–3054. [CrossRef]

57. Kim, Y.; Kwak, G.-H.; Lee, K.-D.; Na, S.-I.; Park, C.-W.; Park, N.-W. Performance evaluation of machine learning and deep learning algorithms in crop classification: Impact of hyper-parameters and training sample size. *Korean J. Remote Sens.* **2018**, *34*, 811–827, (In Korean with English Abstract).

58. Yu, L.; Su, J.; Li, C.; Wang, L.; Luo, Z.; Yan, B. Improvement of moderate resolution land use and land cover classification by introducing adjacent region features. *Remote Sens.* **2018**, *10*, 414. [CrossRef]

applied sciences

Article

Landslide Susceptibility Mapping Based on Random Forest and Boosted Regression Tree Models, and a Comparison of Their Performance

Soyoung Park [1] and Jinsoo Kim [2],*

[1] BK21 Plus Project of the Graduate School of Earth Environmental Hazard System,
    Pukyong National University, Busan 48513, Korea; yac100@pknu.ac.kr
[2] Department of Spatial Information Engineering, Pukyong National University, Busan 48513, Korea
* Correspondence: jinsookim@pknu.ac.kr; Tel.: +82-51-629-6658

Received: 2 February 2019; Accepted: 1 March 2019; Published: 6 March 2019

Abstract: This study aims to analyze and compare landslide susceptibility at Woomyeon Mountain, South Korea, based on the random forest (RF) model and the boosted regression tree (BRT) model. Through the construction of a landslide inventory map, 140 landslide locations were found. Among these, 42 (30%) were reserved to validate the model after 98 (70%) had been selected at random for model training. Fourteen landslide explanatory variables related to topography, hydrology, and forestry factors were considered and selected, based on the results of information gain for the modeling. The results were evaluated and compared using the receiver operating characteristic curve and statistical indices. The analysis showed that the RF model was better than the BRT model. The RF model yielded higher specificity, overall accuracy, and kappa index than the BRT model. In addition, the RF model, with a prediction rate of 0.865, performed slightly better than the BRT model, which had a prediction rate of 0.851. These results indicate that the landslide susceptibility maps (LSMs) produced in this study had good performance for predicting the spatial landslide distribution in the study area. These LSMs could be helpful for establishing mitigation strategies and for land use planning.

Keywords: landslide susceptibility; random forest; boosted regression tree; information gain; landslide susceptibility map

---

1. Introduction

A landslide is defined as a natural disaster that occurs when gravity causes a mass of debris, soil, or rock to move on a downward slope [1]. The majority of landslides occur as a result of hydroclimatic events, such as prolonged or intensive rain. Furthermore, mechanisms such as seismic triggers, wind, and freeze–thaw cycles are known to initiate landslides [2].

Mountains with shallow layers of soil that have formed in place from weathered gneiss and granite make up roughly 70% of the Korean peninsula [3]. Such terrain is vulnerable to weakening during heavy rainfall. Most of the annual precipitation occurs during the summer, when heavy rain and typhoons frequently occur. In particular, the heavy rain associated with typhoons has the potential to cause landslides in South Korea [4]. The year 2011 was a particularly devastating year, with 43 landslide-related casualties in Chuncheon and at Woomyeon Mountain in the area surrounding Seoul City. This is the largest number of landslide-related casualties since 2000.

South Korea has not been alone in experiencing an increase in such natural disasters. Other regions around the world have also experienced more frequent landslides on a larger scale and with more severe damage. In future decades, this trend will probably continue because of ongoing deforestation, increased urbanization, and an increase in regional precipitation in landslide-prone areas due to climate change [5]. It is essential that both susceptible and stable areas be identified to mitigate property damage, environmental degradation, and loss of life. Consequently, landslide susceptibility assessments, i.e., assessments of the spatial probability of a landslide occurring, are a huge step forward in the comprehensive hazard management of landslides [6,7]. The landslide susceptibility map (LSM) produced by a landslide susceptibility assessment can be a useful tool for authorities with decision-making capabilities.

Many methods and techniques have been proposed to evaluate landslide susceptibility. In the past few decades, statistical approaches have become popular in the use of remote sensing (RS) with a geographic information system (GIS). There are many statistical approaches used in landslide susceptibility assessment, including a frequency ratio (FR) [8,9], certainty factor (CF) [10], statistical index (SI) [11,12], as well as weight of evidence (WoE) [7,13,14] and logistic regression (LR) [15,16] approaches.

Recently, machine learning techniques have become popular in various fields. Machine learning, a branch of artificial intelligence, uses computer algorithms to analyze and predict information based on learning from training data [17,18]. Due to its robustness and high generalization capability, the use of machine learning has increased in landslide susceptibility analysis. Among the machine learning methods, artificial neural network [19,20], fuzzy logic [21,22], neuro-fuzzy [23], support vector machine [24,25], random forest [26,27], and naïve Bayes tree [17,28] methods have been popularly applied.

More recently, ensemble machine learning techniques have been used to enhance the prediction power and robustness of landslide susceptibility assessment. The ensemble methods, formed by a combination of variously based classifiers, have typically demonstrated significant improvement [17,24,29,30]. Ensemble techniques, which are relatively new approaches for producing a landslide susceptibility map, have been rarely used in the field. Therefore, the main objective of this research was to analyze and compare the performance of different ensemble models—namely, the random forest (RF) and boosted regression tree (BRT) models—for landslide susceptibility analysis. The RF and BRT models are very popular ensemble methods. Both are tree-based algorithms that predict the results by combining individual trees. However, the RF and BRT models build trees in different ways. Considering these characteristics, these models are appropriate for producing LSMs and for comparing LSM results. The results of the models were compared using the receiver operating characteristic (ROC) curve and statistical indices to determine the more robust model.

2. Study Area and Data Used

2.1. Study Area

The study area, Woomyeon Mountain, is located in the Seocho district of Seoul City, South Korea. This area lies within 37°27′00″–37°28′55″ N and 126°59′02″–127°01′41″ E (Figure 1). The average elevation is 293 m above sea level, and the slope is approximately 30°–35°. The bedrock is Precambrian banded biotite gneiss, which is believed to be highly susceptible to landslides because of severe weathering and abundant faults (Figure A1). In addition, granite gneiss with relatively poor compositional differentiation is excavated en masse, and there is partial distribution of an embedded dike. The gneiss outcrop is poor, as a result of severe weathering in the overall area, and its foliation structure is irregular, due to several folding events [31].

This area experienced concentrated precipitation from 26–29 July 2011. The maximum precipitation, which occurred during 2 h one morning, was 164 mm. This exceeded the 156-mm, 100-year return period. This heavy precipitation led to a debris flow landslide in the area near Woomyeon Mountain, and 1–1.5 m of stratum flowed over areas near the mountain. Seven locations in the study area, including two locations in the valley area damaged primarily from flooding and five locations damaged by debris flow, were affected by the landslide. The total area damaged by the debris was approximately 276,683 m², and the maximum length of damage from the upper part of the steep-slope disaster area to diffuse areas was approximately 764 m. The event caused 16 deaths and 10 building collapses [31].

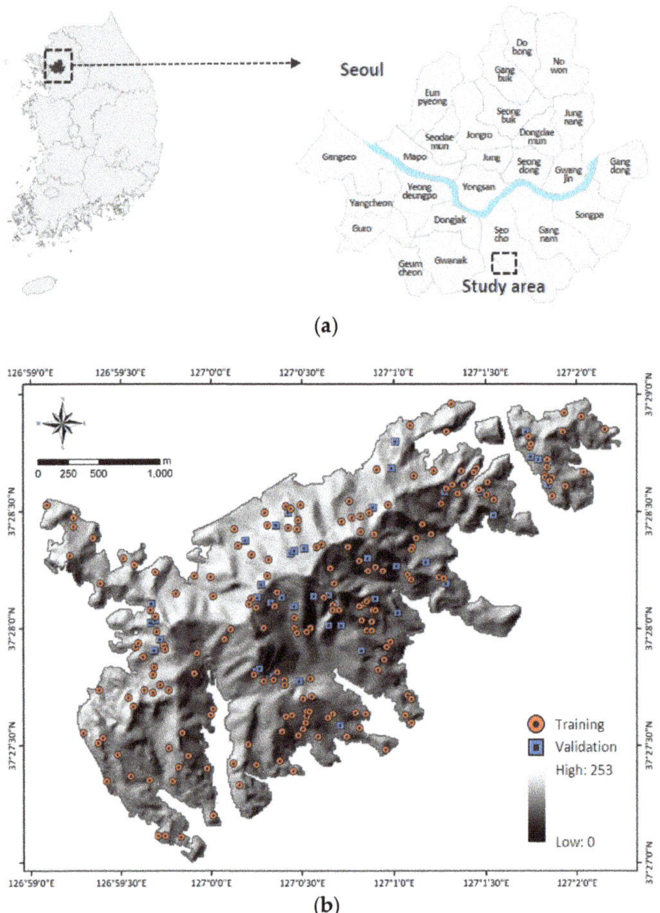

(a)

(b)

Figure 1. Location of study area (**a**) and landslide inventory map with hill shading (**b**).

2.2. Landslide Inventory

Landslide locations were identified using 32 aerial photographs of the study area, taken after the occurrence of the landslides. These aerial photographs were taken by a digital mapping camera with a spatial resolution of 10 cm. The orthorectified photographs were produced using the Leica Photogrammetry Suite (LPS) mounted on ERDAS Imagine 2011 (Erdas, Inc., Norcross, GA, United States). Landslide locations were digitized by visual interpretation using ArcGIS 10.2 (ESRI, Inc., Redlands, CA, USA). Among the digitized landslide locations, landslide locations belonging to rupture zones were converted to point data using a centroid technique. The point data representing the landslide locations were converted to a pixel format, with resolution of 10 m. From the 140 identified landslides, 42 (30%) were reserved to validate the model, after 98 (70%) had been chosen at random for model training. Additionally, non-landslide pixels were selected randomly from the non-landslide area: 98 non-landslide pixels were used for the training dataset, and 42 non-landslide pixels were used to build the validation dataset. This generating and splitting process was performed repeatedly more than 10 times. Finally, the combination utilized was found through the area under the receiver operating characteristic (ROC) curve (AUC) method.

2.3. Landslide Explanatory Variables

Landslides usually occur by complex interactions among various explanatory variables, and there is no consensus about which landslide explanatory variables to use. In this study, 14 explanatory variables were selected, based on a literature review and data availability. These factors were divided into the following three categories: topography, hydrology, and forestry (Table 1, Figure A2). These factors were produced in raster format with a cell size of 10×10 m, considering the scale of the input data, using ArcGIS 10.2 and ERDAS Imagine 2011; the total number of cells in the study area was 67,005. For the next process, the continuous variables among the explanatory variables were reclassified into seven classes, using ArcGIS 10.2. ArcGIS 10.2 provides various classification schemes, such as equal interval, standard deviation, natural break, quantile, etc. Natural break classification groups the classes based on break points that are relatively large jumps in data values. This classification method can be used to maximize the variance between classes. In addition, Cao et al. (2016) [32] indicated that natural break classification is more appropriate for the classification of variables, because their results showed that the LSM produced had higher accuracy compared to that using a different classification method. Therefore, natural break classification was used in this study.

Table 1. Information and sources of data used for the landslide susceptibility assessment at Woomyeon Mountain.

Category	Factor	Source	Scale (Resolution)	GIS and Data Type
-	Landslide inventory	Aerial photographs	1:5000	Raster
Topography	Altitude	Topographic maps	1:5000	Vector
	Slope degree	Digital elevation map	10×10 m	Raster
	Slope aspect			
	Profile curvature			
	Plan curvature			
Hydrology	Distance to streams	Digital elevation map	10×10 m	Raster
	Topographic wetness index			
	Stream power index			
	Sediment transport index			
	Terrain roughness index			
Forestry	Timber type	Forest map	1:5000	Vector
	Timber diameter			
	Timber age			
	Timber density			

2.3.1. Topography Factors

Topography factors include altitude, slope degree, slope aspect, profile curvature, and plan curvature. Altitude is an influential factor among the various landslide explanatory variables, because it is affected by several geomorphologic and geological processes. Slope, which can be described as the form between any section of the surface and a horizontal datum, has considerable influence on slope stability [33]. The degree of vulnerability to landslides may differ based on slope direction, because the water content of the surface, vegetation type, and soil strength may be different. In addition, both the profile and plan curvatures can be classified as flat, concave, or convex. During the rainy season, concave slopes may contain more moisture than convex slopes or flat slopes, so the concave slopes may be more vulnerable to landslides. All of these variables were extracted from the 10-m digital elevation model (DEM), using the spatial analyst tool of ArcGIS. The DEM was produced from 1:5000 topographic maps provided by the Korean National Geographic Information Institute.

2.3.2. Hydrology Factors

The hydrology factors were distance to streams, topographic wetness index (TWI), stream power index (SPI), sediment transport index (STI), and terrain roughness index (TRI). The streams were delineated by flow accumulation and converted to a vector format. The distance to streams was calculated using the Euclidean distance function in ArcGIS. Beven and Kirby (1979) [34] developed a TWI that reflects water's tendency to accumulate anywhere within the catchment area, accumulations that will then tend to move downslope as a result of gravity [35]. The water flow's power to erode is measured by the SPI, based on the assumption of proportionality of discharge to a catchment's specific area [36]. The STI is also often used to reflect the overland flow's power to erode [37]. The TWI, SPI, and STI were calculated with their base in specific catchment areas (A_s) and slope maps, using the following:

$$\text{TWI} = \ln\left(\frac{A_s}{tan\beta}\right) \tag{1}$$

$$\text{SPI} = A_s \times tan\beta \tag{2}$$

$$\text{STI} = \left(\frac{A_s}{22.13}\right)^{0.6}\left(\frac{sin\beta}{0.0896}\right)^{1.3} \tag{3}$$

where A_s represents the specific catchment area (m^2/m), and β represents the local slope gradient (degrees).

In addition, the TRI, which represents the concave and convex upward slopes [38], was calculated as

$$\text{TRI} = \sqrt{|x|(max^2 - min^2)}, \tag{4}$$

where *max* and *min* represent the maximum and minimum values of altitude among the nine rectangular neighbor pixels, respectively.

2.3.3. Forestry Factors

Vegetation prevents erosion on a slope by buffering the impact of rain falling on the slope, and vegetation roots increase the shear strength of the slope by increasing the shear strength of the soil. The forestry factors include timber type, timber diameter, timber age, and timber density. Here, timber type and timber age mean the species and average age of planted trees, respectively. In addition, timber diameter represents the size of the diameter at chest height. Timber density refers to the degree of closure of the crown canopy. These values were obtained from a 1:5000 scale forest map produced by the Korea Forest Research Institute.

3. Methodology

This study was performed using the following main steps: (1) collection and construction of database of landslides and landslide explanatory variables, (2) preparation of the training and test datasets through repeated random sampling, (3) feature selection using information gain (IG), (4) landslide susceptibility mapping using RF and BRT models, and (5) validation and comparison of performance among landslide susceptibility maps (LSMs) (Figure 2). The IG, RF, and BRT models were implemented in R (Foundation for Statistical Computing, Vienna, Austria) using the "FSelector," "randomForest," and "gbm" packages, respectively. These algorithms were performed employing a 10-fold cross-validation approach, to reduce the variability of the model results.

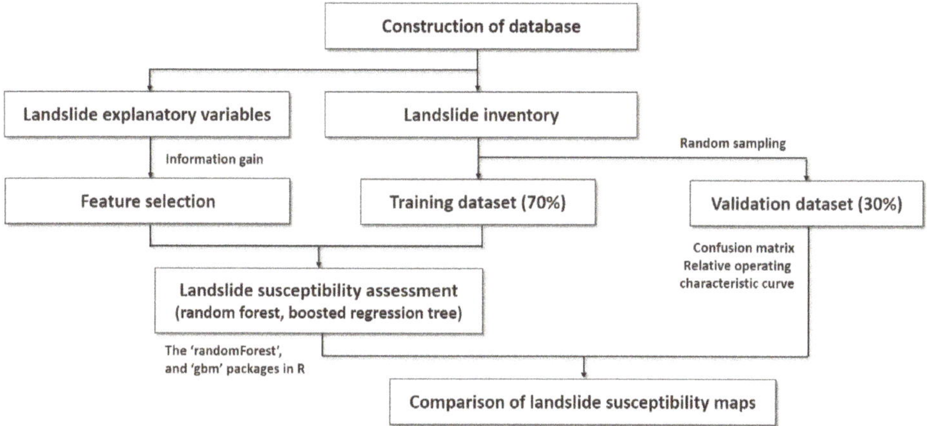

Figure 2. Flow chart of the overall methodology.

3.1. Landslide Dataset Preparation

In this study, correlations between landslide and landslide explanatory variables were analyzed using the FR (Table A1). The FR is the ratio of the area where landslides occurred to the total study area. The FR was calculated by dividing the ratio of landslide occurrence, for the class or type of each factor, into an area ratio, for the class or type of each factor to the total area. The calculated FRs pertaining to each landslide explanatory variable were normalized from 0 to 1. The normalized FRs were extracted for the landslide and non-landslide dataset. Subsequently, these data were used for the training and validation datasets, to run the models and evaluate the prediction capabilities of the models, respectively.

3.2. Information Gain

The landslide explanatory variables have a crucial role in producing LSM. Some landslide explanatory variables might be associated with reductions in model performance, overfitting, model training time, and predictive capability [39]. Therefore, it is necessary to recognize and choose proper landslide explanatory variables.

Various methods such as IG [17], chi-square statistics [30], and Relief-F [29] have been proposed for feature selection in landslide modeling. In this study, the IG, proposed by Quinlan (1993) [40], was used to determine irrelevant and unimportant variables. The IG evaluates an attribute by determining the overall information gain in terms of the class. Consequently, the result can determine the ranking of importance, based on the normalized average merit contributed by each attribute [41,42].

The IG value of landslide explanatory variable C_i belonging to class L (landslide and non-landslide) is calculated as [24,30]

$$IG(L, C_i) = IF(L) - IF(L|C_i),\qquad(5)$$

where $IF(L)$ is the entropy value of L, and $IF(L|C_i)$ is the entropy of L after integrating the values of landslide explanatory variable C_i. These values are calculated using Equation (6) and Equation (7), respectively:

$$IF(L) = -\sum_i P(L_i) \log_2(P(L_i)),\qquad(6)$$

$$IF(L|C_i) = -\sum_i P(L_i) \sum_j P(L_i|C_i) \log_2(P(L_i|C_i)),\qquad(7)$$

where $P(L_i)$ is the prior probability of the class L, and $P(L_i|C_i)$ is the posterior probabilities of L given the values of explanatory variable C_i. An explanatory variable with a higher IG value has higher rank, meaning that it is more important to landslide models. By contrast, an explanatory variable with an IG value of zero must be removed from the dataset, because that factor does not make a contribution.

3.3. Landslide Susceptibility Analysis

3.3.1. Random Forest

RF, developed by Breiman (2001) [43], is a popular ensemble learning method that has been used widely for classification, regression, clustering, and interaction detection. A single decision tree is a weak classification, because of its high variance and bias. However, RF tends to produce robust models, because it can mitigate these problems by using ensemble trees [44].

RF generates thousands of random binary trees to form a forest. Each tree is grown based on a bootstrap sample, using a classification and regression trees (CART) procedure with a random subset of variables selected at each node [26,45]. For each tree grown on a bootstrap sample, the "out-of-bag" (OOB) error rate is calculated using observations left out of the bootstrap sample. The final decisions of class membership and model construction (output) are determined by the majority vote among all trees [46].

Two types of error rate—the mean decrease in accuracy and the mean decrease in the Gini coefficient—were calculated. These measures have been widely used to rank and select variables [26,47]. To run the RF model, the user should optimize two priori parameters, the number of trees in the forest (*ntree*) and the number of variables tested at each node (*mtry*), to minimize the OOB error and obtain good model performance [44,45].

3.3.2. Boosted Regression Tree Model

The BRT model is a combination of statistical and machine learning techniques. The BRT model fits different techniques and combines them to improve the performance of a single model [48,49]. Two different algorithms, namely boosting and regression, are used in the model, and the strengths of these algorithms are combined to improve model accuracy and decrease model variance [45,50]. Boosting is one of the most powerful learning methods for improving model accuracy, by iteratively fitting new trees to the residual errors (RE) of the existing tree assemblage [45,51]. In addition to boosting, the BRT model uses regression trees in the modeling process. Regression trees are categorized from the classification and regression tree approaches from the decision tree group of models [52].

In the model, among the various parameters, the number of trees is automatically set through internal cross-validation. In addition, the learning rate, the number of nodes in a single tree, and bag fraction were determined through a trial-and-error approach [53]. The complexity of the model and the contribution of each tree to the model are controlled by a shrinkage parameter and the learning rate, respectively. The bag fraction and shrinkage parameter determine the number of trees required to reach the optimal solution [54].

3.4. Model Performance Assessment and Comparison

3.4.1. Confusion Matrix

The confusion matrix includes true positive (TP), false positive (FP), true negative (TN), and false negative (FN) categories. Using these values, various statistical indices, such as accuracy, sensitivity, specificity, threat score, equitable threat score, Pierce's skill score, odds ratio, and odds ratio skill score can be calculated [55]. The value calculated from the confusion matrix provides useful information on model performance and classification accuracy.

In this study, the sensitivity, specificity, overall accuracy, and kappa statistic were used to validate the performance of the LSMs. The percentages of landslide and non-landslide pixels classified correctly into those two categories enable the calculation of sensitivity and specificity, and the overall percentage classified correctly (in both categories together) indicates the accuracy of the LSMs [56]. In addition, the kappa statistic is used to evaluate the reliability of the landslide models. Its value ranges from −1 (non-reliable) to 1 (reliable) [57].

3.4.2. Receiver Operating Characteristic

The receiver operating characteristic (ROC) curve has been commonly used to validate the quality of a probabilistic model. The ROC curve is plotted by statistical index value pairs, with the false positive rate (sensitivity) on the x-axis and the "100−false negative rate" (100−specificity) on the y-axis. The ROC curve can be classified as a success rate curve or prediction rate curve, depending on the dataset used. The success rate curve, calculated using the training dataset, represents how well the LSMs fit the data. The prediction rate curve, calculated using the validation dataset, represents how well the model and landslide explanatory variables predict a landslide [11]. The ROC curve can be verified quantitatively when the area under the ROC curve (AUC) is calculated. AUC values range from 0.5 to 1.0. AUC values closer to 1 indicate a more accurate model.

4. Results

4.1. Selection of Landslide Explanatory Variables

The average information gain (AIG) value, and its standard deviation for each landslide explanatory variable, were calculated and ranked (Table 2). All landslide explanatory variables used in this study contributed to the landslide models, because the AIG values of these variables were more than 0. According to the results, the TRI had the highest AIG value (0.086), which means that this factor made the greatest contribution to the landslide models in this study area. By contrast, timber diameter made the smallest contribution to the landslide models, as indicated by the lowest AIG value (0.005).

Table 2. Information gain values for the landslide explanatory variables used in this study.

No.	Landslide Explanatory Variable	Average Merit	Standard Deviation
1	Terrain roughness index	0.086	±0.010
2	Slope aspect	0.071	±0.012
3	Distance to streams	0.06	±0.010
4	Altitude	0.049	±0.008
5	Timber type	0.049	±0.008
6	Stream power index	0.041	±0.008
7	Slope degree	0.038	±0.008
8	Sediment transport index	0.037	±0.006
9	Topographic wetness index	0.033	±0.011
10	Plan curvature	0.025	±0.006
11	Profile curvature	0.013	±0.005
12	Timber age	0.012	±0.002
13	Timber density	0.008	±0.002
14	Timber diameter	0.005	±0.002

4.2. Training the Random Forest and Boosted Regression Tree Models

The training dataset was used to train the RF and BRT models for landslide susceptibility assessment. During the training process, the optimum values of the parameters for the models were applied to obtain high model predictive capability. The optimized values for the RF model were 300 for *ntree* and 2 for *mtry*. In the case of the BRT model, the optimized values for *n.trees*, *interaction.depth*, *shrinkage*, and *n.minobsinnode* were 500, 1, 0.01, and 10, respectively. Subsequently, the RF and BRT models were constructed using the optimized parameters, based on the training dataset.

After their construction, the RF and BRT models were applied throughout the whole study area to produce LSMs.

4.3. Model Validation and Comparison

The performance of each model was analyzed using the training dataset. The RF model showed a higher sensitivity value (98.00%) than did the BRT model (79.57%). This result showed that the RF model classified more correctly than the BRT model in the landslide class. The specificity results also indicated that the RF model had higher specificity (100.00%) in the non-landslide class, indicating that the non-landslide pixels were more correctly classified. The specificity value of the BRT model was 76.70%. Because of the lower sensitivity and specificity values of the BRT model, the overall accuracy and kappa index values were lower, with values of 78.16% and 0.561, respectively. In the case of the RF model, the overall accuracy and kappa index were 98.98% and 0.980, respectively.

In addition, the success rate and the prediction rate were analyzed using the training dataset and the validation dataset, respectively (Figure 3). In the case of success rate, the RF and BRT models had values of 0.999 and 0.887, respectively. The prediction rate curve also showed that the RF model had a higher AUC (0.865) than the BRT model (0.851). Overall, the AUC values of all models were greater than 0.8. These results show that the LSMs constructed in this study have good accuracy in the spatial prediction of landslide susceptibility.

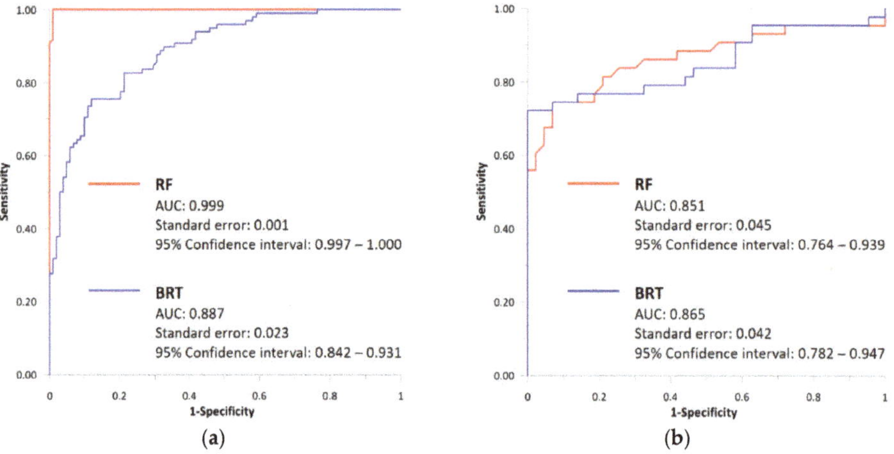

Figure 3. Analysis of the receiver operating characteristic (ROC) curve for the two landslide susceptibility maps: (**a**) success rate curve using the training dataset and (**b**) prediction rate curve using the validation dataset.

4.4. Generating Landslide Susceptibility Maps

The RF and BRT models were used to develop LSMs in the study area. The LSMs were prepared by generating landslide susceptibility indices (LSIs) and reclassifying the class. The LSIs were calculated based on the trained RF and BRT models. Using the natural breaks method, the LSMs were reclassified into five susceptibility classes: very high, high, moderate, low, and very low (Figure 4). Overall, the distribution of LSI for each susceptibility class was similar between the LSM produced by RF (RF LSM) and that produced by BRT (BRT LSM). The "high" and "very high" susceptibility classes covered about 30% of the total area. The RF model had a value of 34.69%, and the BRT model had the lower value of 31.11%.

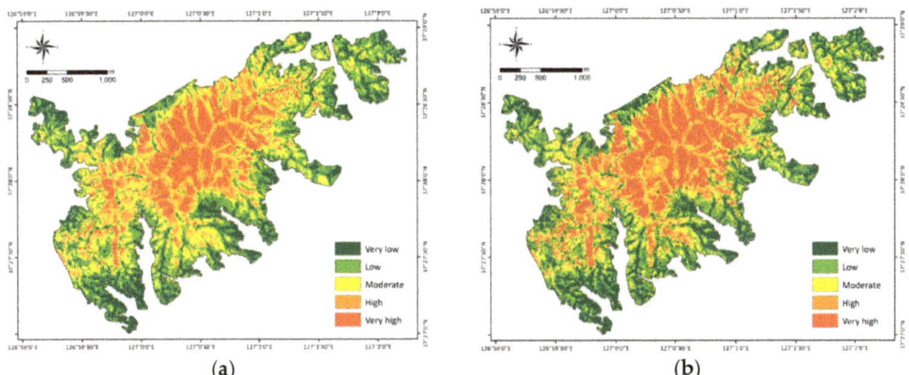

Figure 4. Landslide susceptibility maps produced by random forest (RF) (**a**) and boosted regression tree (BRT) (**b**) models.

The LSMs produced from the two models were validated based on the landslide density (LD) of each susceptibility class on the LSMs. The LD is the ratio of the percentage of landslide pixels to the percentage of all pixels for each susceptibility class shown on the map [56]. LD was calculated by overlaying the five LSMs and the landslide inventory map. Generally, for the study area, the value of LD increased gradually, from very low to very high susceptibility (Table 3). At the "very high" class, the RF and BRT models had LD values of 3.799 and 2.721, respectively. Overall, the models used in this study are suitable for LSM.

Table 3. Landslide density on landslide susceptibility maps produced from the different models.

	Random Forest			Boosted Regression Tree		
	Pixels of Class	Pixels of Landslide	Landslide Density	Pixels of Class	Pixels of Landslide	Landslide Density
Very low	13,034	2	0.073	10,671	3	0.135
Low	16,871	4	0.113	18,629	16	0.411
Moderate	13,854	16	0.553	16,860	35	0.994
High	12,917	36	1.334	12,930	41	1.518
Very high	10,329	82	3.799	7915	45	2.721
Total	67,005	140		67,005	140	

4.5. Discussion

The LSMs produced using the models were evaluated by statistical indices and ROC curves. The RF model had better sensitivity, specificity, overall accuracy, and kappa values. The AUC values of the LSMs used in this study were about 80%, indicating reasonable accuracy. The RF model had higher AUC values for the success rate and prediction rate curves than the BRT model. Thus, these models had very high predictive performance. Furthermore, the LSMs would be produced differently depending on the methods used and the landslide explanatory variables selected. The landslide explanatory variables may not make equal contributions, which can affect prediction ability. In this study, the landslide explanatory variables used made different contributions to the models. Table 4 illustrates the importance of each explanatory variable, calculated and normalized in the RF and BRT models. In general, TRI had the highest importance to the models, whereas timber diameter, timber age, and timber density had lower predictive capability.

Table 4. Relative importance of each landslide explanatory variable calculated in the random forest and boosted regression tree model.

	Random Forest		Boosted Regression Tree	
	Importance	Rank	Importance	Rank
Terrain roughness index	1.000	1	1.000	1
Distance to streams	0.857	2	0.556	3
Altitude	0.766	3	0.277	5
Sediment transport index	0.654	4	0.562	2
Timber type	0.484	5	0.158	7
Slope degree	0.469	6	0.000	-
Stream power index	0.449	7	0.084	9
Topographic wetness index	0.440	8	0.378	4
Slope aspect	0.408	9	0.242	6
Plan curvature	0.214	10	0.099	8
Profile curvature	0.118	11	0.016	10
Timber diameter	0.026	12	0.000	-
Timber age	0.009	13	0.000	-
Timber density	0.000	14	0.000	-

From the results, ensemble classification, such as that done by the model used in this study, can improve the performance of single (weak) classifiers and the prediction accuracy of LSM [56]. However, the models had overfitting problems, as indicated by the AUC values calculated using the training and validation datasets. The AUC values of the success rate curve were very high, almost reaching a value of 1, but the AUC values of the prediction rate curve were lower. Especially in the case of the RF model, the AUC value of the prediction rate was decreased by about 20%. This result showed that the RF model was trained excessively by the training data. This can be associated with poor generalization from training data and increased error for real data. Overfitting is a common problem affecting researchers performing machine learning and data mining. There can be many reasons for overfitting. However, in this study, the landslide explanatory variables used still included noise, despite the feature selection process. In addition, because the landslide area is very small compared to the non-landslide area, the model could not learn and predict the non-landslide area.

5. Conclusions

This study compared and analyzed landslide susceptibility at Woomyeon Mountain using different models. For this purpose, landslide-related spatial data consisting of a landslide inventory, and landslide explanatory variables were collected and prepared. The landslide inventory map was built using aerial photographs. The 14 landslide explanatory variables were constructed from spatial data collected by government organizations. These factors included altitude, slope degree, slope aspect, profile curvature, plan curvature, distance to streams, TWI, SPI, STI, TRI, timber type, timber diameter, timber age, and timber density.

The contribution of each landslide explanatory variable was evaluated using the average IG value with a 10-fold cross-validation approach. All of the landslide explanatory variables contributed to the models, because the IG values of all factors were greater than zero. Therefore, the landslide susceptibility analysis and mapping were performed with all landslide explanatory variables using the RF and BRT models. The RF and BRT models were implemented in R. A popular open-source software, R is helpful for statistical computing and data visualization [58]. The models were constructed using optimized parameters, and LSI was predicted over the study area.

The LSMs produced in this study may prove useful for decision makers, planners, and engineers in disaster planning to minimize economic losses and casualties. In a future study, the accuracy of the LSMs of this study could be enhanced by selecting more optimal landslide explanatory variables and solving the problem of overfitting.

Author Contributions: S.P. analyzed the data and wrote the paper. J.K. suggested the idea.

Funding: This research was financially supported by the BK21 Plus Project of the Graduate School of Earth Environmental Hazard System. In addition, this work was supported by a National Research Foundation of Korea (NRF) grant funded by the Korea government (MSIP) (No. NRF-2017R1A2B2009033).

Conflicts of Interest: The authors declare no conflicts of interest.

Appendix A

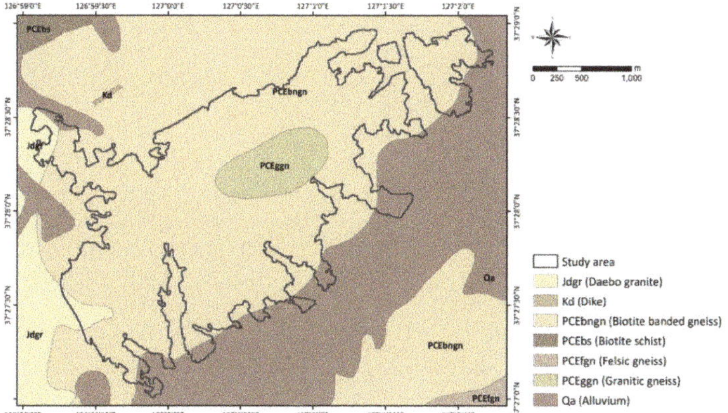

Figure A1. Geological features of the study area produced from 1:50,000 geological maps provided by the Korea Institute of Geoscience and Mineral Resources.

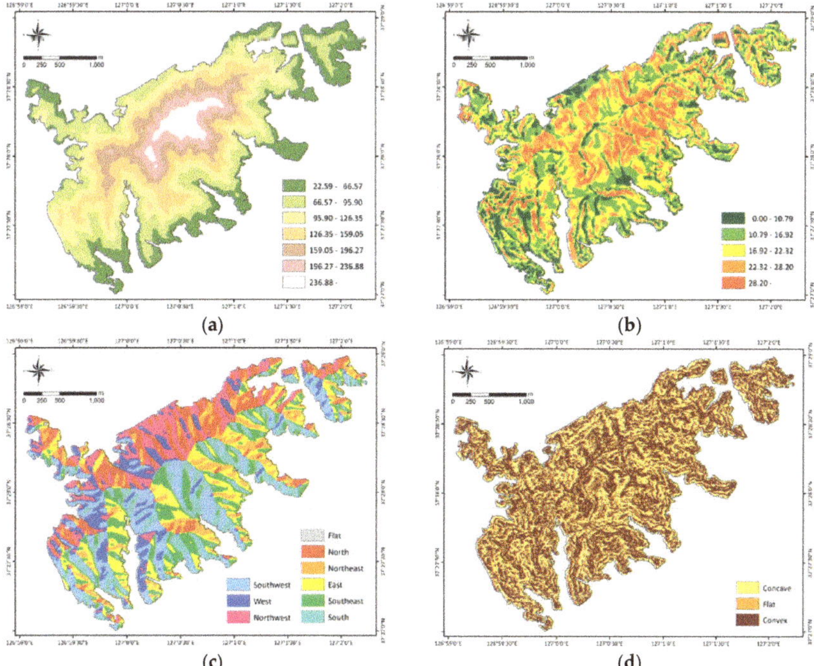

Figure A2. *Cont.*

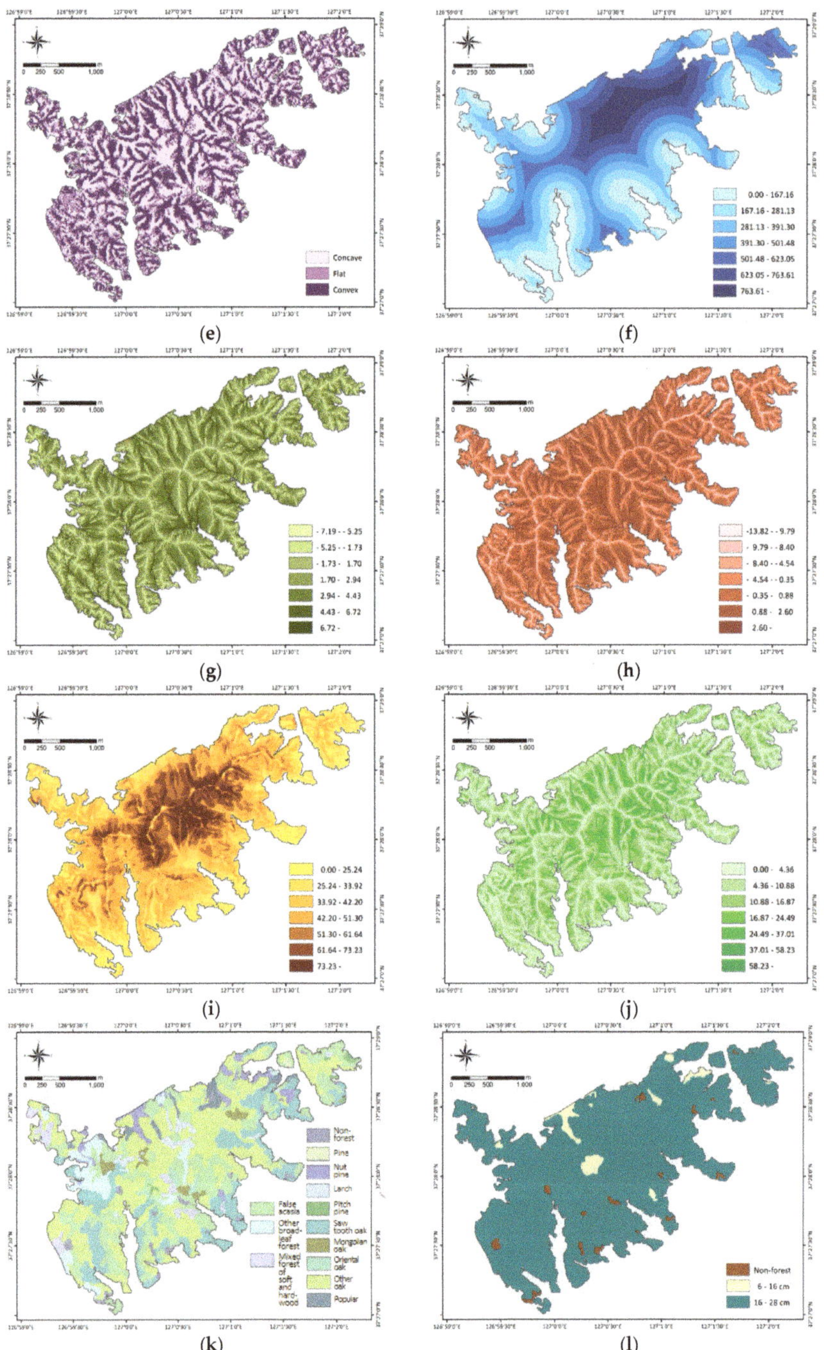

Figure A2. *Cont.*

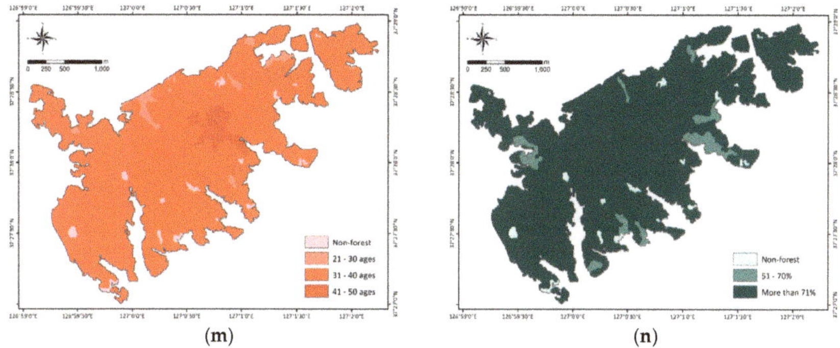

Figure A2. Landslide explanatory variables used to analyze landslide susceptibility: (**a**) altitude, (**b**) slope degree, (**c**) slope aspect, (**d**) profile curvature, (**e**) plan curvature, (**f**) distance to streams, (**g**) topographic wetness index, (**h**) stream power index, (**i**) sediment transport index, (**j**) terrain roughness index, (**k**) timber type, (**l**) timber diameter, (**m**) timber age, and (**n**) timber density.

Table A1. Correlations between landslide and landslide explanatory variables using the frequency ratio.

Factor	Class	No. of Pixels in Domains	No. of Landslide Pixels	Frequency Ratio	Normalized Frequency Ratio
	[22.59, 66.57]	12,477	10	0.38	0.00
	[66.57, 95.90]	17,042	20	0.56	0.11
	[95.90, 126.35]	13,614	31	1.09	0.42
Altitude (m)	[126.35, 159.05]	10,060	33	1.57	0.71
	[159.05, 196.27]	6766	17	1.20	0.49
	[196.27, 236.88]	4409	19	2.06	1.00
	[>236.88]	2637	10	1.81	0.85
	[0.00, 8.83]	4704	3	0.31	0.00
	[8.83, 13.98]	11,554	11	0.46	0.11
	[13.98, 18.15]	15,716	33	1.00	0.52
Slope degree (°)	[18.15, 22.07]	14,692	27	0.88	0.43
	[22.07, 26.24]	11,362	36	1.52	0.91
	[26.24, 31.39]	6708	23	1.64	1.00
	[>31.39]	2269	7	1.48	0.88
	Flat	69	0	0.00	0.00
	North	8483	12	0.68	0.50
	Northeast	7928	9	0.54	0.40
	East	9395	18	0.92	0.67
Slope aspect	Southeast	8837	20	1.08	0.80
	South	9522	26	1.31	0.96
	Southwest	8970	19	1.01	0.75
	West	6333	18	1.36	1.00
	Northwest	7468	18	1.15	0.85
Profile curvature	Concave	33,763	79	1.12	1.00
	Flat	247	0	0.00	0.00
	Convex	32,995	61	0.88	0.79
Plan curvature	Concave	31,592	82	1.24	1.00
	Flat	772	0	0.00	0.00
	Convex	34,641	58	0.80	0.65

Table A1. *Cont.*

Factor	Class	No. of Pixels in Domains	No. of Landslide Pixels	Frequency Ratio	Normalized Frequency Ratio
Distance to streams (m)	[0.00, 167.16]	8947	3	0.61	0.28
	[167.16, 281.13]	12,553	16	1.46	0.81
	[281.13, 391.30]	13,430	41	0.77	0.38
	[391.30, 501.48]	12,372	20	1.28	0.69
	[504.48, 623.05]	9369	25	1.77	1.00
	[623.05, 763.61]	6205	23	1.39	0.76
	[>763.61]	4129	12	0.63	0.50
Topographic wetness index	[−7.19, −5.25]	8350	11	0.63	0.50
	[−5.25, −1.73]	3148	2	0.30	0.24
	[−1.73, 1.70]	13,316	24	0.86	0.69
	[1.70, 2.94]	21,415	56	1.25	1.00
	[2.94, 4.43]	14,843	38	1.23	0.98
	[4.43, 6.72]	4652	9	0.93	0.74
	[>6.72]	1281	0	0.00	0.00
Stream power index	[−13.82, −9.79]	487	0	0.00	0.00
	[−9.79, −8.40]	3850	2	0.25	0.17
	[−8.40, −4.54]	7251	11	0.73	0.48
	[−4.54, −0.35]	13,013	16	0.59	0.39
Stream power index	[−0.35, 0.88]	22,507	51	1.08	0.72
	[0.88, 2.60]	16,613	52	1.50	1.00
	[>2.60]	3284	8	1.17	0.78
Sediment transport index	[0.00, 4.35]	15,486	17	0.53	0.20
	[4.35, 10.88]	21,160	31	0.70	0.27
	[10.88, 16.87]	16,659	44	1.26	0.48
	[16.87, 24.49]	9054	29	1.53	0.58
	[24.49, 37.01]	3267	18	2.64	1.00
	[37.01, 58.23]	1065	0	0.00	0.00
	[>58.23]	314	1	1.52	0.58
Terrain roughness index	[0.00, 25.24]	6302	0	0.00	0.00
	[25.24, 33.92]	13,887	14	0.48	0.20
	[33.92, 42.20]	14,614	23	0.75	0.31
	[42.20, 51.30]	12,675	33	1.25	0.52
	[51.30, 61.64]	9415	27	1.37	0.57
	[61.64, 73.23]	6747	34	2.41	1.00
	[>73.23]	3365	9	1.28	0.53
Timber type	Non-forest	1027	0	0.00	0.00
	Pine	143	0	0.00	0.00
	Nut pine	2319	3	0.62	0.24
	Larch	1389	2	0.69	0.27
	Pitch pine	431	0	0.00	0.00
	Sawtooth oak	11,482	18	0.75	0.30
	Mongolian oak	1183	2	0.81	0.32
	Oriental oak	565	3	2.54	1.00
	Other oak	25,845	63	1.17	0.46
	Poplar	1419	6	2.02	0.80
	False acasia	15,809	30	0.91	0.36
	Other broadleaf forest	3707	12	1.55	0.61
	Mixed forest of soft and hardwood	1686	1	0.28	0.11
Timber diameter (cm)	Non-forest	1027	0	0.00	0.00
	[6, 16]	2124	3	0.68	0.66
	[16, 28]	63,854	137	1.03	1.00

Table A1. *Cont.*

Factor	Class	No. of Pixels in Domains	No. of Landslide Pixels	Frequency Ratio	Normalized Frequency Ratio
Timber age (ages)	Non-forest	1027	0	0.00	0.00
	[21, 30]	1508	1	0.32	0.16
	[31, 40]	62,234	130	1.00	0.52
	[41, 50]	2236	9	1.93	1.00
Timber density (%)	Non-forest	1027	0	0.00	0.00
	[51, 70]	2994	5	0.80	0.78
	[>71]	62,984	135	1.03	1.00

References

1. Das, I.; Stein, A.; Kerle, N.; Dadhwal, V.K. Landslide susceptibility mapping along road corridors in the Indian Himalayas using Bayesian logistic regression models. *Geomorphology* **2012**, *179*, 116–125. [CrossRef]
2. Jakob, M.; Lambert, S. Climate change effects on landslides along the southwest coast of British Columbia. *Geomorphology* **2009**, *107*, 275–284. [CrossRef]
3. Vasu, N.N.; Lee, S.R. A hybrid feature selection algorithm integrating an extreme learning machine for landslide susceptibility modeling of Mt Woomyeon, South Korea. *Geomorphology* **2016**, *263*, 50–70. [CrossRef]
4. Park, S.; Choi, C.; Kim, B.; Kim, J. Landslide susceptibility mapping using frequency ratio, analytic hierarchy process, logistic regression, and artificial neural network methods at the Inje area, Korea. *Environ. Earth Sci.* **2013**, *68*, 1443–1464. [CrossRef]
5. Schuster, R.L. *Socioeconomic Significance of Landslides: Investigation and Mitigation*; National Academy Press Transportation Research Board Special Report; National Academy Press: Washington, DC, USA, 1996; Volume 247, pp. 12–35.
6. Guzzetti, F.; Reichenbach, P.; Ardizzone, F.; Cardinali, M.; Galli, M. Estimating the quality of landslide susceptibility models. *Geomorphology* **2006**, *81*, 166–184. [CrossRef]
7. Mohammady, M.; Pourghasemi, H.R.; Pradhan, B. Landslide susceptibility mapping at Golestan Province, Iran: A comparison between frequency ratio, Dempster–Shafer, and weights-of-evidence models. *J. Asian Earth Sci.* **2012**, *61*, 221–236. [CrossRef]
8. Akgun, A.; Dag, S.; Bulut, F. Landslide susceptibility mapping for a landslide-prone area (Findikli, NE of Turkey) by likelihood-frequency ratio and weighted linear combination models. *Environ. Geol.* **2008**, *54*, 1127–1143. [CrossRef]
9. Yilmaz, I. Landslide susceptibility mapping using frequency ratio, logistic regression, artificial neural networks and their comparison: A case study from Kat landslides (Tokat—Turkey). *Comput. Geosci.* **2009**, *35*, 1125–1138. [CrossRef]
10. Kanungo, D.P.; Sarkar, S.; Sharma, S. Combining neural network with fuzzy, certainty factor and likelihood ratio concepts for spatial prediction of landslides. *Nat. Hazards* **2011**, *59*, 1491–1512. [CrossRef]
11. Bui, D.T.; Lofman, O.; Revhaug, I.; Dick, O. Landslide susceptibility analysis in the Hoa Binh Province of Vietnam using statistical index and logistic regression. *Nat. Hazards* **2011**, *59*, 1413–1444. [CrossRef]
12. Zhang, G.; Cai, Y.; Zheng, Z.; Zhen, J.; Liu, Y.; Huang, K. Integration of the statistical index method and the analytic hierarchy process technique for the assessment of landslide susceptibility in Huizhou, China. *Catena* **2016**, *142*, 233–244. [CrossRef]
13. Ozdemir, A.; Altural, T. A comparative study of frequency ratio, weights of evidence and logistic regression methods for landslide susceptibility mapping: Sultan Mountains, SW Turkey. *J. Asian Earth Sci.* **2013**, *64*, 180–197. [CrossRef]
14. Xu, C.; Xu, X.; Lee, Y.H.; Tan, X.; Yu, G.; Dai, F. The 2010 Yushu earthquake triggered landslide hazard mapping using GIS and weight of evidence modeling. *Environ. Earth Sci.* **2012**, *66*, 1603–1616. [CrossRef]
15. Pourghasemi, H.R.; Moradi, H.R.; Aghda, S.F. Landslide susceptibility mapping by binary logistic regression, analytical hierarchy process, and statistical index models and assessment of their performances. *Nat. Hazards* **2013**, *69*, 749–779. [CrossRef]

16. Yalcin, A.; Reis, S.; Aydinoglu, A.C.; Yomralioglu, T. A GIS-based comparative study of frequency ratio, analytical hierarchy process, bivariate statistics and logistics regression methods for landslide susceptibility mapping in Trabzon, NE Turkey. *Catena* **2011**, *85*, 274–287. [CrossRef]

17. Chen, W.; Shirzadi, A.; Shahabi, H.; Ahmad, B.B.; Zhang, S.; Hong, H.; Zhang, N. A novel hybrid artificial intelligence approach based on the rotation forest ensemble and naïve Bayes tree classifiers for a landslide susceptibility assessment in Langao County, China. *Geomat. Nat. Hazards* **2017**, *8*, 1955–1977. [CrossRef]

18. Jordan, M.; Mitchell, T. Machine learning: Trend, perspectives, and prospects. *Science* **2015**, *349*, 255–260. [CrossRef] [PubMed]

19. Gomez, H.; Kavzoglu, T. Assessment of shallow landslide susceptibility using artificial neural networks in Jabonosa River Basin, Venezuela. *Eng. Geol.* **2005**, *78*, 11–27. [CrossRef]

20. Nefeslioglu, H.A.; Gokceoglu, C.; Sonmez, H. An assessment on the use of logistic regression and artificial neural networks with different sampling strategies for the preparation of landslide susceptibility maps. *Eng. Geol.* **2008**, *97*, 171–191. [CrossRef]

21. Shahabi, H.; Hashim, M.; Ahmad, B.B. Remote sensing and GIS-based landslide susceptibility mapping using frequency ratio, logistic regression, and fuzzy logic methods at the central Zab basin, Iran. *Environ. Earth Sci.* **2015**, *73*, 8647–8668. [CrossRef]

22. Zhu, A.X.; Wang, R.; Qiao, J.; Qin, C.Z.; Chen, Y.; Liu, J.; Zhu, T. An expert knowledge-based approach to landslide susceptibility mapping using GIS and fuzzy logic. *Geomorphology* **2014**, *214*, 128–138. [CrossRef]

23. Vahidnia, M.H.; Alesheikh, A.A.; Alimohammadi, A.; Hosseinali, F. A GIS-based neuro-fuzzy procedure for integrating knowledge and data in landslide susceptibility mapping. *Comput. Geosci.* **2010**, *36*, 1101–1114. [CrossRef]

24. Bui, D.T.; Ho, T.C.; Pradhan, B.; Pham, B.T.; Nhu, V.H.; Revhaug, I. GIS-based modeling of rainfall-induced landslides using data mining-based functional trees classifier with AdaBoost, Bagging, and MultiBoost ensemble frameworks. *Environ. Earth Sci.* **2016**, *75*, 1101.

25. Kavzoglu, T.; Sahin, E.K.; Colkesen, I. An assessment of multivariate and bivariate approaches in landslide susceptibility mapping: A case study of Duzkoy district. *Nat. Hazards* **2015**, *76*, 471–496. [CrossRef]

26. Pourghasemi, H.R.; Kerle, N. Random forests and evidential belief function-based landslide susceptibility assessment in Western Mazandaran Province, Iran. *Environ. Earth Sci.* **2016**, *75*, 1–17. [CrossRef]

27. Youssef, A.M.; Pourghasemi, H.R.; Pourtaghi, Z.S.; Al-Katheeri, M.M. Landslide susceptibility mapping using random forest, boosted regression tree, classification and regression tree, and general linear models and comparison of their performance at Wadi Tayyah Basin, Asir Region, Saudi Arabia. *Landslides* **2015**, *13*, 1–18.

28. Tsangaratos, P.; Ilia, I. Comparison of a logistic regression and Naïve Bayes classifier in landslide susceptibility assessments: The influence of models complexity and training dataset size. *Catena* **2016**, *145*, 164–179. [CrossRef]

29. Hong, H.; Liu, J.; Bui, D.T.; Pradhan, B.; Acharya, T.D.; Pham, B.T.; Ahmad, B.B. Landslide susceptibility mapping using J48 Decision Tree with AdaBoost, Bagging and Rotation Forest ensembles in the Guangchang area (China). *Catena* **2018**, *163*, 399–413. [CrossRef]

30. Pham, B.T.; Bui, D.T.; Dholakia, M.B.; Prakash, I.; Pham, H.V.; Mehmood, K.; Le, H.Q. A novel ensemble classifier of rotation forest and Naïve Bayer for landslide susceptibility assessment at the Luc Yen district, Yen Bai Province (Viet Nam) using GIS. *Geomat. Nat. Hazards Risk* **2017**, *8*, 649–671. [CrossRef]

31. Korean Geotechnical Society (KGS). *The Study on Investigation of Cause and Development of Restoration Policy about Landslide in Wumyon Area*; Korean Geotechnical Society: Seoul, Korea, 2011. (In Korean)

32. Cao, C.; Xu, P.; Wang, Y.; Chen, J.; Zheng, L.; Niu, C. Flash flood hazard susceptibility mapping using frequency ratio and statistical index methods in coalmine subsidence areas. *Sustainability* **2016**, *8*, 948. [CrossRef]

33. Kalantar, B.; Pradhan, B.; Naghibi, S.A.; Motevalli, A.; Mansor, S. Assessment of the effects of training data selection on the landslide susceptibility mapping: A comparison between support vector machine (SVM), logistic regression (LR) and artificial neural networks (ANN). *Geomat. Nat. Hazards Risk* **2018**, *9*, 49–69. [CrossRef]

34. Beven, K.J.; Kirkby, M.J. A physically based, variable contributing area model of basin hydrology/Un modèle à base physique de zone d'appel variable de l'hydrologie du bassin versant. *Hydrol. Sci. J.* **1979**, *24*, 43–69. [CrossRef]

35. Poudyal, C.P.; Chang, C.; Oh, H.J.; Lee, S. Landslide susceptibility maps comparing frequency ratio and artificial neural networks: A case study from the Nepal Himalaya. *Environ. Earth Sci.* **2010**, *61*, 1049–1064. [CrossRef]

36. Moore, I.D.; Grayson, R.B.; Ladson, A.R. Digital terrain modelling: A review of hydrological, geomorphological, and biological applications. *Hydrol. Process.* **1991**, *5*, 3–30. [CrossRef]

37. Moore, I.D.; Burch, G.J. Sediment transport capacity of sheet and rill flow: Application of unit stream power theory. *Water Resour. Res.* **1986**, *22*, 1350–1360. [CrossRef]

38. Althuwaynee, O.F.; Pradhan, B.; Park, H.J.; Lee, J.H. A novel ensemble bivariate statistical evidential belief function with knowledge-based analytical hierarchy process and multivariate statistical logistic regression for landslide susceptibility mapping. *Catena* **2014**, *114*, 21–36. [CrossRef]

39. Liu, H.; Motoda, H. Ensemble-based variable selection using independent probes. In *Computational Methods of Feature Selection*; Chapman and Hall/CRC: London, UK, 2007; pp. 147–162.

40. Quinlan, J.T. *C4.5: Programs for Machine Learning*; Morgan Kaufmann Publishers: San Francisco, CA, USA, 1993.

41. Oommen, T.; Cobin, P.F.; Gierke, J.S.; Sajinkumar, K.S. Significance of variable selection and scaling issues for probabilistic modeling of rainfall-induced landslide susceptibility. *Spat. Inf. Res.* **2018**, *26*, 21–31. [CrossRef]

42. Witten, I.G.; Frank, E. *Data Mining: Practical Machine Learning Tools and Techniques*, 2nd ed.; Morgan Kaufmann Publishers: San Francisco, CA, USA, 2005.

43. Breiman, L. Random forests. *Mach. Learn.* **2001**, *45*, 5–32. [CrossRef]

44. Taalab, K.; Cheng, T.; Zhang, Y. Mapping landslide susceptibility and types using Random Forest. *Big Earth Data* **2018**, *2*, 1–20. [CrossRef]

45. Rahmati, O.; Naghibi, S.A.; Shahabi, H.; Bui, D.T.; Pradhan, B.; Azareh, A.; Melesse, A.M. Groundwater spring potential modelling: Comprising the capability and robustness of three different modeling approaches. *J. Hydrol.* **2018**, *565*, 248–261. [CrossRef]

46. Micheletti, N.; Foresti, L.; Robert, S.; Leuenberger, M.; Pedrazzini, A.; Jaboyedoff, M.; Kanevski, M. Machine learning feature selection methods for landslide susceptibility mapping. *Math. Geosci.* **2014**, *46*, 33–57. [CrossRef]

47. Calle, M.L.; Urrea, V. Letter to the editor: Stability of random forest importance measures. *Brief Bioinform.* **2010**, *12*, 86–89. [CrossRef] [PubMed]

48. Schapire, R.E. The boosting approach to machine learning: An overview. In *Nonlinear Estimation and Classification*; Springer: New York, NY, USA, 2003; pp. 149–171.

49. Elith, J.; Leathwick, J.R.; Hastie, T. A working guide to boosted regression trees. *J. Anim. Ecol.* **2008**, *77*, 802–813. [CrossRef] [PubMed]

50. Aertsen, W.; Kint, V.; Van Orshoven, J.; Özkan, K.; Muys, B. Comparison and ranking of different modelling techniques for prediction of site index in Mediterranean mountain forests. *Ecol. Model.* **2010**, *221*, 1119–1130. [CrossRef]

51. Cao, D.S.; Xu, Q.S.; Liang, Y.Z.; Zhang, L.X.; Li, H.D. The boosting: A new idea of building models. *Chemom. Intell. Lab.* **2010**, *100*, 1–11. [CrossRef]

52. Naghibi, S.A.; Pourghasemi, H.R.; Abbaspour, K. A comparison between ten advanced and soft computing models for groundwater qanat potential assessment in Iran using R and GIS. *Theor. Appl. Climatol.* **2018**, *131*, 967–984. [CrossRef]

53. França, S.; Cabral, H.N. Predicting fish species richness in estuaries: Which modelling technique to use? *Environ. Model. Softw.* **2015**, *66*, 17–26. [CrossRef]

54. Al-Abadi, A.M. A novel geographical information system-based Ant Miner algorithm model for delineating groundwater flowing artesian well boundary: A case study from Iraqi southern and western deserts. *Environ. Earth Sci.* **2017**, *76*, 534. [CrossRef]

55. Frattini, P.; Crosta, G.; Carrara, A. Techniques for evaluating the performance of landslide susceptibility models. *Eng. Geol.* **2010**, *111*, 62–72. [CrossRef]

56. Pham, B.T.; Bui, D.T.; Dholakia, M.B.; Prakash, I.; Pham, H.V. A comparative study of least square support vector machines and multiclass alternating decision trees for spatial prediction of rainfall-induced landslides in a tropical cyclones area. *Geotech. Geol. Eng.* **2016**, *34*, 1807–1824. [CrossRef]

57. Bennett, N.D.; Croke, B.F.; Guariso, G.; Guillaume, J.H.; Hamilton, S.H.; Jakeman, A.J.; Pierce, S.A. Characterising performance of environmental models. *Environ. Model. Softw.* **2014**, *40*, 1–20. [CrossRef]
58. Zhou, J.; Li, X.; Mitri, H.S. Classification of rockburst in underground projects: Comparison of ten supervised learning methods. *J. Comput. Civ. Eng.* **2016**, *30*, 04016003. [CrossRef]

 *applied sciences*

Article

A Single Point-Based Multilevel Features Fusion and Pyramid Neighborhood Optimization Method for ALS Point Cloud Classification

Yong Li [†], Guofeng Tong *, Xiance Du [†], Xiang Yang, Jianjun Zhang and Lin Yang

College of Information Science and Engineering, Northeastern University, Shenyang 110819, China; liyong@stumail.neu.edu.cn (Y.L.); newneulab@163.com (X.D.); yangxwkm@gmail.com (X.Y.); junneu@126.com (J.Z.); 1700922@stu.neu.edu.cn (L.Y.)
* Correspondence: tongguofeng@ise.neu.edu.cn
† These authors contributed equally to this work.

Received: 18 January 2019; Accepted: 21 February 2019; Published: 6 March 2019

Abstract: 3D point cloud classification has wide applications in the field of scene understanding. Point cloud classification based on points can more accurately segment the boundary region between adjacent objects. In this paper, a point cloud classification algorithm based on a single point multilevel features fusion and pyramid neighborhood optimization are proposed for a Airborne Laser Scanning (ALS) point cloud. First, the proposed algorithm determines the neighborhood region of each point, after which the features of each single point are extracted. For the characteristics of the ALS point cloud, two new feature descriptors are proposed, i.e., a normal angle distribution histogram and latitude sampling histogram. Following this, multilevel features of a single point are constructed by multi-resolution of the point cloud and multi-neighborhood spaces. Next, the features are trained by the Support Vector Machine based on a Gaussian kernel function, and the points are classified by the trained model. Finally, a classification results optimization method based on a multi-scale pyramid neighborhood constructed by a multi-resolution point cloud is used. In the experiment, the algorithm is tested by a public dataset. The experimental results show that the proposed algorithm can effectively classify large-scale ALS point clouds. Compared with the existing algorithms, the proposed algorithm has a better classification performance.

Keywords: ALS point cloud; multi-scale; classification; large scene

---

1. Introduction

Airborne Laser Scanning (ALS) can capture large-scale point clouds of urban scenes. The point cloud classification of outdoor scenes can provide high-precision semantic maps for autonomous driving, improve the accuracy of vehicle positioning, and reconstruct a three-dimensional model of the city, which plays an important role in urban planning and dynamic management. In addition, it can improve the efficiency of resource utilization. Effectively labeling the correct class for all points in the scene is an important basis for the widespread adoption of point clouds [1–4]. However, a laser point cloud has a huge data number, high redundancy, and uneven scene distribution, which may lead to huge challenges in the point cloud classification. Therefore, it is of great significance to classify the three-dimensional point cloud in large outdoor scenes.

Currently, the number of point clouds with manual labeling in outdoor large scenes is not enough. However, machine learning can learn and classify point clouds in the case of less sample training data, and the speed is faster. At present, the point cloud classification methods can be mainly divided into two strategies: the single point-based classification and object-based classification methods.

The point-based point cloud classification is the classification of each individual point in a point cloud; this strategy uses points as the basic unit to extract features, train models, and predict class labels. There are three main steps: the neighborhood selection, feature extraction, and single point classification based on the features and classifiers.

(1) Neighborhood selection. In the neighborhood selection process, the commonly used point cloud neighborhood forms are: K nearest neighbors [5], radius neighborhoods [6], and column neighborhoods [7]. The parameter of neighborhood estimation is highly dependent on prior knowledge, and it is greatly affected by the change of the point cloud density [8,9]. For example, Hackel et al. constructed a multi-level scale pyramid, and a total of 144-dimensional features such as eigenvalues of the covariance matrix were extracted for each point in each pyramid. Subsequently, Random Forest was used for training and finally classified outdoor road scenes [10], and a better classification effect was obtained. Therefore, the selection based on multi-scale neighborhood is an important method for extracting single point effective features.

(2) Features extraction. The local feature of a point cloud is an abstract depiction of the environment around a given point. It is difficult to classify a point cloud by a single local feature. The common practice is to fuse multiple point cloud local features for classification. Normal and curvature are simple local features that clearly show some local information about the point cloud, such as the fact that the normal direction can represent the partial tangent plane of the point cloud, and the curvature can represent the smoothness of the point. For example, Fanxuan et al. [11] optimized the matching accuracy of point pairs in point clouds based on the curvature information, and the registration accuracy of point clouds was further improved. Geometric features are also common local features, also known as shape descriptors. For example, in the spin image [12], the main idea is to set up an image with the normal vector as the center; the image rotates around the axis. The number of 3D points encountered by each pixel in the point cloud is taken as its gray value. Finally, a two-dimensional array representing the local information of the three-dimensional space, that is, the rotated image feature, is obtained. The 3D Shape Context (3DSC) feature [13] is based on the specified point to construct a spherical region. In the support region, the grid is divided into three coordinate directions: the radial direction, direction angle and elevation angle. Following this, a feature histogram is constructed by entering the number of points in the grid. The 3DSC is simple in construction, strong in discrimination and insensitive to noise, but it is time-consuming. The Unique Shape Context (USC) descriptor [14] improves the 3DSC to avoid ambiguities in the classification. Point Feature Histograms (PFH) [15] are local features, which construct a feature histogram with the angles and distances of the normal vectors of any two points in the specified point neighborhood. The descriptor can accurately describe the local features of the points, but the computation is large and the real-time performance is poor. Fast Point Feature Histograms (FPFH) [16] are a simplification of PFH, which greatly reduce the time consumption while retaining most of the description performance of PFH. FPFH have an excellent performance, and are widely used in the field of point cloud classification, segmentation and registration [17]. Although these features can express the local features of the point cloud, they do not take into account the characteristics of the ALS point cloud, which has the characteristics of relative sparse, rich elevation information, as well as a horizontal and vertical distribution.

(3) Single point classification based on features and classifiers. Currently, machine learning is an important method for classification problems. The single-point classification based on machine learning takes the feature vector of the point as the input and the class label of the single point as the output. Common machine learning algorithms can accomplish this classification task, such as AdaBoost [18], Random Forest [19] and Support Vector Machine (SVM) [20]. This kind of method uses a classifier to learn the local features of each point, after which the parameters in the classifier are determined based on the training dataset. Finally, the test set is classified by the classifier. This classification strategy can more accurately segment the boundary regions between different adjacent objects, and this method has a better performance in detail. However, due to the extremely large number of points, the calculation amount is large. Thus, the model training is slow, and there are also

some misclassifications of local regions. However, there are always some errors in the final classification results. Therefore, the initial classification results are required to further optimize according to the characteristics of the point cloud.

In order to solve the above-mentioned problems, an ALS point cloud classification algorithm based on a single point multi-feature fusion is proposed. This kind of algorithm is based on the point as the basic processing unit, and the classification process assigns labels to each point in the point cloud to realize a point cloud classification. The proposed method extracts the local features of each point by constructing a multi-scale neighborhood space, along with two new features: a normal angle distribution histogram (NAD) and latitude sampling histogram (LSH) are proposed. Following this, SVM is used for training and classification. However, since each point is classified, there is a problem regarding some edge points being misclassified. In this regard, the initial classification results are further optimized according to the neighborhood classification optimization of multi-scale pyramids. Experiments prove that the classification algorithm has a higher accuracy.

The main contributions of this paper are as follows:

(1) Two local features are proposed, that is, the NAD histogram and the LSH histogram. The differences of different objects in the normal distribution, and the difference of the neighborhood points around different objects in the horizontal and vertical directions of the three-dimensional space, can be fully utilized to more effectively represent the characteristics of different objects.

(2) A multilevel single-point features fusion method based on a multi-neighborhood space and multi-resolution is proposed. The multi-scale space is constructed by changing the resolution of the point cloud and the number of the neighborhood. The features of the multi-scale are extracted from each single point, and the features are fused. Following this, the SVM classifier is used to classify the features and the better classification results have also been achieved.

(3) A fast optimization method for classification results based on a multi-scale pyramid is proposed. By changing the resolution of the point cloud, a multi-scale pyramid is constructed, and the neighbor points are further re-selected. After this, the misclassifications are eliminated according to the initial classification results of the neighbors for a post-processing optimization.

2. Method

As shown in Figure 1, the algorithm flow is given as follows. In the training part, for the point cloud scene shown in Figure 1a, multiple features of each point are first extracted. Multi-scale and multiple features are fused to a fusion feature. The SVM classifier model is then trained using the fusion features of the training set. In the test part, for the point cloud scene shown in Figure 1b, the fusion feature is first obtained. As shown in Figure 1c, the test points are initially classified using the trained SVM classification model. Following this, the point clouds of different resolutions are obtained by down-sampling the point cloud, in order to construct a multi-scale point cloud pyramid. The corresponding classification labels of the different neighboring points in the different scales are searched. Finally, the label which has the most number in the neighbors is taken as the class label of the current point. The final point cloud classification result is obtained, as shown in Figure 1d.

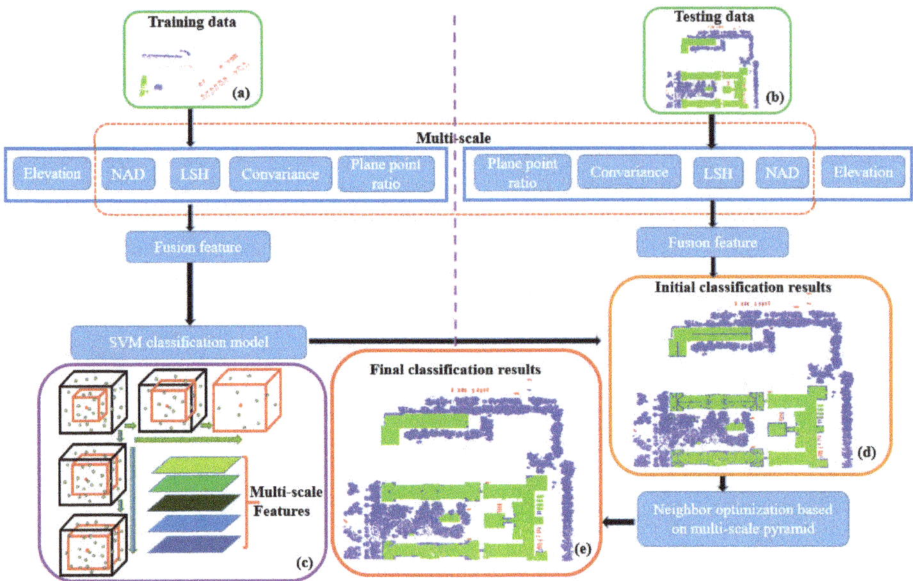

Figure 1. Flowchart of the proposed method. NAD: normal angle distribution histogram;LSH: latitude sampling histogram; SVM: Support Vector Machine.

2.1. Point Feature Extraction

In the classification task of the 3D point cloud data, the feature extraction of the point cloud plays a crucial role. It can seriously affect the final classification result. A well-behaved feature descriptor should reflect obvious differences between different types of points in the point cloud. At the same time, the descriptor should be robust and have a strong anti-interference ability. It is difficult for a single feature descriptor to have the above characteristics, so that a plurality of feature descriptor fusion methods are at present widely used. In the single point-based classification algorithm, this paper uses a variety of feature fusion methods to improve the accuracy of the classification algorithm. The specific features are as follows:

2.1.1. Feature Description

1. Elevation feature

The height is a very intuitive feature in a 3D point cloud. Generally speaking, points with large height are buildings, trees or objects with larger elevation values in the real world. When the elevation value is small, the probability is greater if the point is a vehicle point. Thus, the elevation feature is set to:

$$F_z = [Z_i, \frac{1}{Z_i}]$$

(1)

where Z_i is the distance of the i-th point from the estimated ground to the elevation value.

2. Normal angle distribution histogram

In the large scale scene, the normal direction of different objects has obvious differences. For artificial objects, such as buildings and vehicles, since the surface is relatively regular, almost all points are in the same direction, pointing in the direction of the vertical plane. However, due to the scattered distribution of the whole point cloud, the normal direction of the point cloud has a large scattered nature, and the direction does not point to a fixed direction in a uniform way. Therefore, we calculate the histogram of each point and its own normal angle distribution value in the local neighborhood point set to express the relationship between the normal of the point and the normal of the points in the neighboring region. The angle between the two normal vectors in three-dimensional space should be between $[0, \pi]$. But considering that the normal of the point on the plane can have opposite directions when the angle is larger than $\pi/2$, the corresponding angle is set to $\pi - \pi/2$. Following this, the angle of the normal vectors is defined as $[0, \pi/2]$. Considering efficiency and resolving power, we divide this interval into equal D_n parts, that is, construct a D_n dimensional histogram. After this, the number of points falling within each cell is taken as the value of the interval in the histogram. Finally, the normalization process is performed to form a histogram of the normal angle distribution, called NAD. This feature can distinguish different classes of points based on the normal angular distributions. The specific calculation formulas are as follows:

$$\Delta = \frac{\pi/2 - 0}{D_n} \tag{2}$$

$$\theta_j = a\cos(v \cdot v_j) \tag{3}$$

$$h(x_i) = \frac{n(i * \Delta \leq \theta_j \leq (i+1) * \Delta)}{N} \; (i = 1, \ldots, D_n) \tag{4}$$

$$F_{NAD} = [h(x_1), h(x_2), \ldots, h(x_{D_n})] \tag{5}$$

where v and v_j represent the normal vectors of the current point and the j-th neighbor point, respectively. $a\cos(\cdot)$ represents the inverse cosine function. N represents the number of neighbors for the current point. $n(i * \Delta \leq \theta_j \leq (i+1) * \Delta)$ denotes the number of points for the normal angle at the range $[i * \Delta, (i+1) * \Delta]$. F_{NAD} denotes the final normal angle eigenvalue vector of the normal angle distribution. The histogram of the normal angle distributions for the randomly selected building points and tree points are shown in Figure 2.

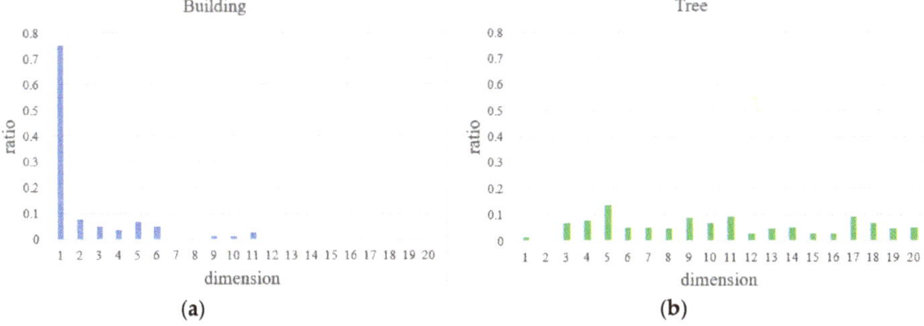

Figure 2. Normal angle distribution histogram. (**a**) Normal angle distribution histogram of a building point. (**b**) Normal angle distribution histogram of a tree point.

3. Latitudinal sampling histogram

In the outdoor large scene environment, as for almost all points belonging to different objects, the surrounding neighborhood points have great differences in the latitudinal distribution in the three-dimensional space. For example, a building surface, which is parallel to the ground, has its neighborhood points mainly distributed near the "equator". For the points belonging to the trees, the distribution of the neighborhood points is more random and extensive, hardly concentrated in a certain latitude interval. Therefore, the selected point is regarded as the center of the sphere, and the distribution histogram of the neighborhood points in the latitudinal direction is counted. Following this, the feature of the point can be expressed. The feature is called LSH. The LSH feature can be used to distinguish different classes of points according to the distribution of neighborhood points in the latitude direction. The LSH has the advantages of anti-occlusion, without interference from the local coordinate system, as well as high efficiency. In this paper, D_l spaces are equally divided along the latitude direction. Following this, the number of points falling into each cell is counted to form a feature vector of the D_l dimension. The specific calculation formulas are:

$$\Delta = \frac{\pi - 0}{D_l} \tag{6}$$

$$\theta_j = a\cos\left(z \cdot (p_j - p)\right) \tag{7}$$

$$f(x_i) = \frac{n\left(i * \Delta \leq \theta_j \leq (i+1) * \Delta\right)}{N} \quad (i = 1, \ldots, D_l) \tag{8}$$

$$F_{LSH} = \left[f(x_1), f(x_2), \ldots, f(x_{D_l})\right] \tag{9}$$

where p and p_j represent the three-dimensional coordinates of the current point and its j-th neighbor point, respectively. $z = (0,0,1)$ represents the unit vector of the positive direction of the z axis. $n(i * \Delta \leq \theta_j \leq (i+1) * \Delta)$ represents the number of points in $[i * \Delta, (i+1) * \Delta]$ of the neighborhood points along the latitudinal direction. F_{LSH} represents the final feature vector of LSH. The LSHs of the randomly selected building points and tree points are compared, as shown in Figure 3.

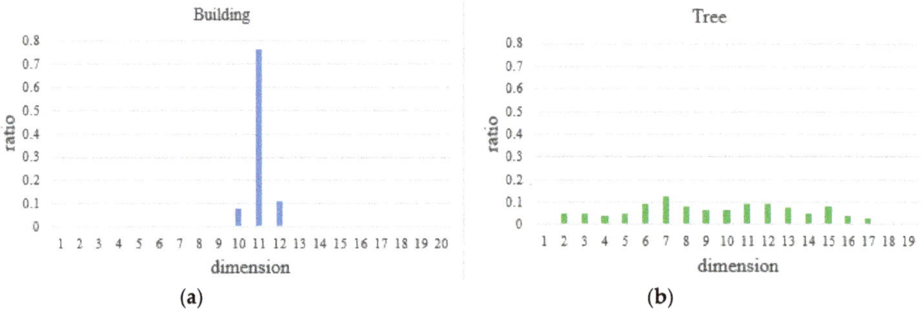

Figure 3. Latitudinal sampling histogram. (**a**) Latitudinal sampling histogram of a building point. (**b**) Latitudinal sampling histogram of a tree point.

4.   Convariance feature

First, a covariance matrix for the selected point neighborhood is constructed. After this, eigenvalues of the covariance matrix are calculated as: $\lambda_2 \geq \lambda_1 \geq \lambda_0 \geq 0$, and the corresponding eigenvectors are calculated as: v_2, v_1, v_0. Here, the covariance feature (CF) is obtained according to the relationship among the eigenvalues, as follows:

Sum of eigenvalues:

$$F_{sum} = \lambda_1 + \lambda_2 + \lambda_3 \tag{10}$$

Full variance:

$$F_{omn} = (\lambda_1 \cdot \lambda_2 \cdot \lambda_3)^{\frac{1}{3}} \tag{11}$$

Anisotropy:

$$F_{ani} = (\lambda_1 - \lambda_3)/\lambda_1 \tag{12}$$

Planarity:

$$F_{pla} = (\lambda_2 - \lambda_3)/\lambda_1 \tag{13}$$

Linearity:

$$F_{lin} = (\lambda_1 - \lambda_2)/\lambda_1 \tag{14}$$

Sphericity:

$$F_{sph} = \lambda_3/\lambda_1 \tag{15}$$

Following this, the final total covariance feature is: $F_{cov} = \left[F_{sum}, F_{omn}, F_{ani}, F_{pla}, F_{lin}, F_{sph}\right]$.

5. Plane point ratio

In outdoor large scale scenes, the classes of objects are complex and the surface shapes are also different. However, a considerable part of the surface of artificial objects exhibits planar characteristics, such as buildings, vehicles, etc. Meanwhile vegetation does not have planar characteristics, so the plane point ratio of the local point cloud can be used as a local feature to classify point clouds. The covariance feature can also reflect the planar characteristics to a certain extent, but it is greatly interfered by noise. For this reason, the Random Sample Consensus (RANSAC) [21] is employed to fit the local neighborhood of the selected point. After this, the ratio of the plane points, called PPR (Plane Point Ratio), is calculated.

RANSAC is a method used to find the subset of data that is the best match for the model from the data set with random samples that are noisy but sufficient. The points matched with the model are called the inner points, and the points unmatched with the model are called the outer points. The plane is fitted using RANSAC as follows.

(1) Select three points randomly from all the neighborhood points and calculate the current model parameters. The model is as follows:

$$ax + by + cz + d = 0 \tag{16}$$

(2) Determine whether each point is an inlier, and then determine the inlier rate ω of the current model:

$$J_i = \begin{cases} 1, & d_i \leq T_d \\ 0, & d_i > T_d \end{cases} \tag{17}$$

$$\omega = \frac{1}{N}\sum_{i=1}^{N} J_i \tag{18}$$

where d_i is the distance from the i-th point to the plane. T_d is a fixed threshold. J_i indicates whether it is an inlier or not. N is the number of neighborhood points.

(3) If the current inlier rate is larger than the previous optimal inlier rate, the optimal inlier rate is updated.

(4) To find the optimal model, repeat steps (1) to (3) k times until the probability reaches P:

$$1 - P \leq \left(1 - \omega^3\right)^k \tag{19}$$

The termination condition is:

$$k \geq \frac{\log(1 - P)}{\log(1 - \omega^3)} \tag{20}$$

When the RANSAC iteration is completed, the optimal inlier rate is the ratio of the plane points: $F_{plane} = [\omega]$.

2.1.2. Single Point Multi-Scale Multi-Feature Fusion

Since the features of the single point are dependent on the selected neighborhood space, different neighborhood spaces have different expression capabilities for different classes of point clouds.

Additionally, the structure descriptions of point clouds with different resolutions also have certain differences. The local feature description of the single scale for a point is relatively single, and there are some noise points in the point cloud, which can make the simple feature of the single scale unable to accurately describe the feature of the single point. Therefore, a multilevel features fusion method based on the multi-neighborhood space and multi-resolution is proposed. As shown in Figure 1c, the proposed method constructs the multi-scale space by changing the resolution and the number of neighborhoods of the point cloud. Following this, multi-scale features for each single point in the point cloud are extracted. Because the elevation feature is not affected by the scale changing, we select NAD, LSH, CF and PPR features to construct the multi-scale features. We extract the features of a single point in each scale by choosing μ neighborhoods with different resolutions and υ different neighborhood sizes under the original resolution. The multi-scale features of each point are expressed respectively as F'_{NAD}, F'_{LSH}, F'_{cov}, and F'_{plane}. Considering the validity of the features and the efficiency of the calculation, this generally results in $2 \leq \mu + \upsilon \leq 5$. In addition, the description of the single point feature only represents one characteristic of the point cloud. Therefore, it is necessary to fuse multiple features. After fusing the features, the multilevel features are expressed as follows:

$$F = \left[F_z, F'_{NAD}, F'_{LSH}, F'_{cov}, F'_{plane} \right] \tag{21}$$

Because we aim at an ALS point cloud, the extracted elevation features are only two-dimensional and play an important role in the point cloud classification. In addition, when the point cloud features are extracted, the values of each feature have been normalized to [0, 1]. In order to reflect the role of the non-zero feature value, the feature should be normalized again according to Formula (22) when the extracted feature F is sparse. While the extracted feature F is not sparse, there is no need to normalize the feature. Therefore, the constructed feature is $X = \left[F_z^*, F_{NAD}^*, F_{LSH}^*, F_{cov}^*, F_{plane}^* \right]$.

$$F_{i,j}^* = \frac{F_{i,j} - \min(F_j)}{\max(F_j) - \min(F_j)} \tag{22}$$

where $F_{i,j}^*$ is the value of the i-th row and the j-th column in the normalized feature matrix F^*. $F_{i,j}$ is the value of the i-th row and the j-th column in the feature matrix F. F_j is the vector of the j-th column (for all the points) in the feature matrix F.

2.2. Point Cloud Classification Based on SVM

SVM [22] is achieved by maximizing the classification interval in the feature space. For non-linear data, SVM maps them into a high-dimensional feature space by a kernel function, which make the data into linear separable data in a high-dimensional feature space. Following this, it realizes a classification by maximizing the interval. In view of the excellent generalization ability of SVM, we use SVM as a classifier for the single point classification in point cloud data. As we know, the correlation between the point cloud single point feature and neighbor points features, and the Gauss kernel function only has one parameter σ and a low model complexity. Thus, in the absence of prior knowledge, the Gauss kernel function is often better than other kernels. Therefore, we choose the Gauss kernel function as the kernel function. Here, the Gauss kernel function of the SVM classifier is defined as follows:

The fused feature space is X. The selected n d-dimensional feature samples $\{x_1, x_2, \ldots, x_n\}$:

$$x_i = (x_{i1}, x_{i2}, \ldots, x_{id})' \epsilon \mathbb{R}^d \tag{23}$$

After the feature transformation, the feature space is Z. We map data in the X space to the Z space $z_i = (z_{i1}, z_{i2}, \ldots, z_{id}) \epsilon \mathbb{R}^d$ via the mapping function $\phi(x)$. The function $K(x, z)$ satisfies the condition

$K(x, z) = \phi(x) \cdot \phi(z)$, and the function $K(x, z)$ is a kernel function, while $\phi(x)$ is a mapping function. The Gauss kernel function is as follows:

$$K(X, z) = \sum_{i}^{n} K(x_i, z) = \sum_{i}^{n} \exp\left(-\frac{\|x_i - z\|^2}{2\sigma}\right) \tag{24}$$

The corresponding decision function is:

$$f(z, \alpha^*, b^*) = sign(\sum_{i}^{n} y_i \alpha^* \exp\left(-\frac{\|x_i - z\|^2}{2\sigma}\right) + b^*) \tag{25}$$

The SVM classifier is trained by the features of the training set, and the test set is classified by the trained classifier. The initial classification results for the point cloud in Figure 1b are shown in Figure 1d.

2.3. Neighborhood Optimization Based on Multi-Scale Pyramid

After the initial classification, the point clouds are basically classified correctly. Due to noise and other reasons, there are still some misclassifications in some details (such as edges). As shown in Figure 1d, most of the points in the scene have been correctly classified, and only a small part of them are misclassified. They mainly concentrate on edges and other places, and most of the points around the misclassified points are correctly classified. Therefore, it is necessary to further optimize the initial classification results to achieve a more accurate classification of the point clouds. Because local information is used as a feature to classify point clouds, the feature extraction relies heavily on a local region selection. In addition, the single point is taken as the basic unit of classification. Each point has its own characteristics, but because the two neighboring points are very close to each other and their neighborhoods are also very close, the extracted features will be very similar, which leads them to be more likely to be classified into the same class. Therefore, the neighbors of the misclassified points are also often misclassified. It is difficult to correct the misclassified points if only the points in the smaller local regions are used for the optimization. Therefore, we propose a classification results optimization method based on the multi-scale pyramid. The specific method is as follows:

First, voxel filters with different radius scales are used to down-sample the point cloud after an initial classification, as shown in Formula (24). Each minimum voxel scale is twice as large as the last down-sampling, and sparse point clouds are gradually obtained. Following this, the q-level pyramid is constructed, and the initial classes of all the points in each level are retained. According to the characteristics of the point cloud down-sampling reflecting the structure information of the shape, the scale pyramid is constructed on three scales of $q = 3$ in this paper.

Following this, the corresponding k-d tree is constructed from the point cloud in each layer of the pyramid. For each point in the original point cloud, a k-d tree is used to search for the radius of the nearest neighbors in the point cloud after the down-sampling. The class labels of the m point clouds searched within the radius of the l-th level are $L^l = \left\{L_1^l, \ldots, L_m^l\right\}, L_i^l \in \{1, \ldots, c\}, i = 1, \ldots, m, l = 1, \ldots, q$; the radius parameters are different when each layer of the point cloud chooses its nearest neighbor. The method of calculation is as follows:

$$r = k \cdot Presolution \tag{26}$$

In the formula, r represents the scale radius parameter. *Presolution* is the resolution used by the current down-sampling point cloud. k is a fixed ratio threshold.

Finally, the initial labels of all the nearest neighbors in the q levels are counted. The discriminant function $1\left\{L_i^l = C\right\}$ represents the fact that when L_i^l belongs to class C, its value is 1; otherwise, its value is 0. This is used to count the number of the initial labels belonging to each class. As shown in Formula (27), the mode label C^* is selected as the new class label for the current point.

$$C^* = \begin{array}{c} \text{argmax} \\ C = L_i^l \in \{1,\dots,c\} \end{array} \sum_{l=1}^{q} \sum_{i}^{m} 1\left\{L_i^l = C\right\} \tag{27}$$

Not only do the optimized point cloud classification results avoid a situation where the nearest neighbor is also misclassified, but they also solve the problem of too many far points in a large scale, thus achieving better results. The optimized point cloud classification results are shown in Figure 1e.

3. Experimental Results and Analysis

In order to verify the performance of the point cloud classification algorithm based on single point multilevel features, we use two urban scenes' ALS data for a qualitative and quantitative comparison and analysis. This section begins by briefly introducing the experimental data, before the classification performance of the proposed method is compared with the other methods on the datasets.

3.1. Experimental Dataset

In this paper, we use two sets of dataset published in Ref [23]. The data was collected in Tianjin, China. The density of the test region point cloud is about $20 \sim 30/m^2$. The data set contains both large objects (buildings and trees) and small objects (cars). It contains different roof shapes, buildings of different heights, and dense and overlapping cars and trees. Table 1 lists the number of each class point for the two scenes. Figure 4a,b shows the training data of scene 1 and scene 2, respectively.

All the programs are run on a computer with an Intel Core i7-7700K CPU, 4.20 GHz with 24-GB RAM. The algorithm is implemented on the C++ platform based on PCL 1.8.0. Each set of data training and testing takes about 6.5 min. However, the feature extraction and optimization process can be implemented in parallel. Therefore, the efficiency of the proposed method can be further improved, and the speed of the point cloud processing will be greatly improved.

Table 1. The experimental dataset.

Scene	Training Points			Test Points		
	Tree	**Building**	**Car**	**Tree**	**Building**	**Car**
Scene 1	68,802	37,128	5380	213,990	200,549	7816
Scene 2	39,743	64,952	4584	73,207	156,186	7409

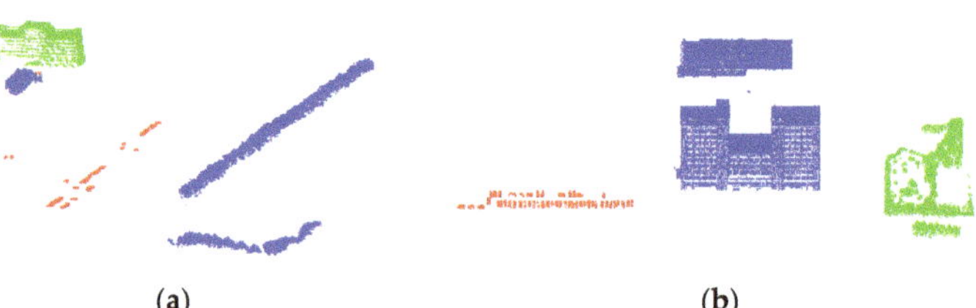

(a)   (b)

Figure 4. The training data of the ALS points. (**a**) scene 1, and (**b**) scene 2. (The figures are captured from the ALS points shown in Cloudcompare (http://www.cloudcompare.org/). The red points are cars, green points are buildings and blue points are trees.)

3.2. Experimental Comparison and Analysis

In order to verify the performance of the proposed algorithm, we compare it with the other seven methods shown in Table 2. Method 1 is the proposed method, which uses the features without NAD and LSH. Method 2 uses the $F_z + F_{NAD} + F_{LSH} + F_{cov} + F_{plane}$ feature fusion in single-scale without a post-processing optimization. Method 3 uses the single-scale feature and multi-scale pyramid optimization. Method 4 is the algorithm proposed in [24]. In this method, each feature uses geometry, strength, and statistics information. Following this, the JointBoost method is used to select features and classify points. Method 5 is the classification method based on the multi-scale Spin Image feature and F_{cov} feature fusion used in Ref [23]. Method 6 [25] constructs a multilevel point set using a linear transformation, and it uses Spin Image and F_{cov} features. Following this, K-means is used to construct an LDA (Latent Dirichlet Allocation) model of a multilevel point set dictionary. Method 7 [23,26,27] constructs a multilevel point set using an exponential transformation. The Spin Image and F_{cov} features are used for dictionary learning, constructing an LDA model of point sets.

Table 2. Main features of the proposed method and other comparison methods. SVM: Support Vector Machine; LDA: Latent Dirichlet Allocation; DD-SCLDA: Discriminative Dictionary based Sparse Coding and LDA.

Method	Scale	Feature Expression	Post-Processing Optimization	Classifier
Our method	Multi-scale	$F_z+F_{NAD}+F_{LSH}+F_{cov}+F_{plane}$	Multi-scale pyramid	SVM
Method 1	Multi-scale	$F_z+F_{cov}+F_{plane}$	Multi-scale pyramid	SVM
Method 2	Single scale	$F_z+F_{NAD}+F_{LSH}+F_{cov}+F_{plane}$	None	SVM
Method 3	Single scale	$F_z+F_{NAD}+F_{LSH}+F_{cov}+F_{plane}$	Multi-scale pyramid	SVM
Method 4	Multi-scale	Geometry, strength, and statistical features	Regional growth	JointBoost
Method 5	Multi-scale	Spin Image feature and F_{cov}	None	AdaBoost
Method 6	Multi-scale	LDA Model of the Spin Image feature and F_{cov} based on Multi-Level Point Sets	None	AdaBoost
Method 7	Multi-scale	DD-SCLDA Model of the Spin Image feature and F_{cov} based on Multi-Level Point Sets	None	AdaBoost

Accuracy, precision, and recall are often used to evaluate the effect of a point cloud classification [27]. The precision rate is the proportion of true positive samples in a positively predicted sample. The recall rate is the proportion of positive samples that are predicted to be successful in the actual positive samples. The accuracy rate is the ratio of all the correctly predicted samples in relation to the overall samples. In order to consider both P_a (precision rate) and R (recall rate), F_1-score values (such as Equation (28)) are generally used to represent the classification quality of the scene. In order to better evaluate the effects of each algorithm, we use the above metrics to evaluate the classification performance.

$$F_1 - score = \frac{2(R \times P_a)}{R + P_a} \tag{28}$$

The classification results of our method and of other comparison methods on scene 1 and scene 2 are shown in Table 3. Table 3 lists the precision, recall, accuracy and F_1-score statistics for the eight methods in the two scenes. It can be seen from Table 3 that the comparison between the proposed method and Method 1 shows that the accuracy of the proposed method has significantly improved. It also shows that the proposed NAD and LSH features have certain effects. By comparing Method 2 and Method 3, the proposed multi-scale pyramid optimization algorithm can effectively improve the classification accuracy. Comparing the proposed method with Method 3, the proposed multi-scale strategy has a significant effect on the improvement of the classification results.

In addition, the proposed method is compared with the methods given in other references. Method 4 and Method 5 classify the point cloud based on the single-point. Method 6 and Method 7 classify the point cloud based on the point set (object). It can be seen from the comparison of Table 3 that the proposed method has a high accuracy rate as a whole, and that it also maintains a high recall rate.

Table 3. Classification results of precision/recall, accuracy and F_1-score.

Scene 1	Tree (%)	Building (%)	Car (%)	Accuracy (%)	F_1-Score (%)
Our method	99.2/90.6	91.1/99.3	92.9/48.2	94.6	94.5/94.9/59.5
Method 1	99.2/84.9	86.8/99.3	99.9/42.7	91.9	91.5/92.7/59.8
Method 2	93.2/78.7	82.1/94.6	63.3/30.4	86.4	85.3/87.9/41.1
Method 3	96.9/81.7	84.1/97.7	98.8/23.2	89.3	88.7/90.4/37.6
Method 4	89.7/98.1	97.9/89.1	65.2/46.6	92.9	93.7/93.3/54.4
Method 5	85.7/92.9	92.0/83.8	56.9/54.7	87.9	89.2/87.7/55.8
Method 6	94.8/93.8	93.5/92.3	41.2/66.7	92.6	94.3/92.9/50.9
Method 7	93.1/96.0	95.2/92.6	73.3/62.2	93.7	94.5/93.9/67.3
Scene 2	**Tree (%)**	**Building (%)**	**Car (%)**	**Accuracy (%)**	**F_1-score (%)**
Our method	92.4/94.3	98.5/97.9	73.0/68.4	95.8	93.4/98.2/70.6
Method 1	83.2/92.9	98.5/92.8	62.6/65.7	92.0	87.8/95.6/64.1
Method 2	77.3/94.3	98.3/88.9	71.7/60.0	89.6	85.0/93.4/65.3
Method 3	91.3/92.6	96.6/96.6	63.2/55.5	93.4	91.9/96.6/59.1
Method 4	86.8/91.2	96.8/95.5	44.1/34.8	92.2	88.9/96.1/38.9
Method 5	73.9/91.2	93.6/88.2	29.5/25.4	87.2	81.6/90.8/27.3
Method 6	90.3/93.9	97.6/96.5	49.4/42.0	94.1	92.1/97.0/45.4
Method 7	94.7/94.5	98.1/97.7	53.9/60.5	95.5	94.6/97.9/57.0

For scene 1, the accuracy of the proposed method is the highest, and the value of the precision/recall is relatively high (the classification result of the proposed method is shown in Figure 1e). From the classification results evaluation of the three kinds of objects by the F_1-score, one can see that the classification effect on cars for the proposed method is not as good as for Method 7. Meanwhile, the tree and building classes can basically be classified correctly.

For scene 2, the precision/recall of trees of the proposed method are lower than for Method 7. Meanwhile, the precision/recall of buildings and cars are the maximum compared with other methods. According to the classification result of the F_1-score value, the proposed method has a better effect than the other methods, except that the tree classification performance is worse than for Method 7. Considering the accuracy, precision/recall rate and the F_1-score evaluation comprehensively, the proposed method has a better classification performance than the other methods, and the proposed method has the highest overall accuracy for both scenes. This proves that the proposed point cloud classification method based on point multilevel features is effective, and that it can accurately classify the ALS point cloud data in large scale scenes.

In order to more intuitively compare the classification performance of each method, Figure 5 shows the performance of the eight classification methods in scene 2. Figure 5 shows that the proposed method can classify most points correctly. Compared with other comparison methods, the classification accuracy of the proposed method is higher. From the comparison between Figures 5c–e and 5b, the classification effect of the proposed method is obviously better than that of the other three algorithms, especially in the buildings and trees. It can be seen from the comparison between Figure 5f,g and Figure 5b that Method 4 and Method 5 have more misclassifications for cars and buildings, and that the performance of the proposed method is significantly better than that of the other two methods. In comparing Figure 5h,i with Figure 5b, one can see that Method 6 and Method 7 have a similar classification performance to that of the proposed method. However, a certain number of architectural edge points are classified incorrectly, and the top part of the trees is also classified incorrectly. One can see from the comparison between Figures 5f–i and 5b that the compared methods have some misclassifications for the edge points and for two objects that are overlapping regions. However, the proposed method has fewer misclassifications in those regions than for the other methods. This proves that the proposed feature descriptors and post-optimization strategies can improve the classification results.

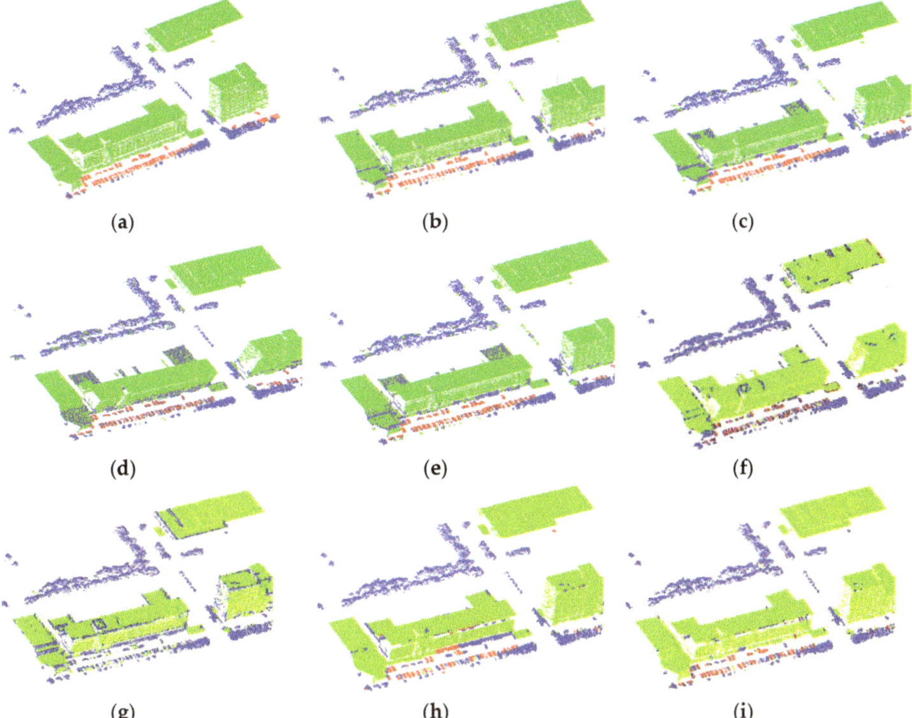

Figure 5. Scene 2 classification results. (**a**) ground truth, (**b**) proposed method, (**c**) method 1, (**d**) method 2, (**e**) method 3, (**f**) method 4, (**g**) method 5, (**h**) method 6, and (**i**) method 7. ((**f–i**) are taken from Ref [23]. The red points are cars, green points are buildings and blue points are trees.)

3.3. Sensitivity of the Parameters

In this part, we focus on the D_m in NAD, D_l in LSH and the number of neighborhood scales ($\mu + \upsilon$). Here, we select the parameters shown in Table 4; the data of scene 1 is used to compare the influence of different parameters on the proposed method. In order to evaluate the classification effect of the three kinds of objects as a whole, we average the F_1–score values of the three object classification results to obtain the mean value mF_1, which is used as the overall classification effect evaluation metric. As shown in Table 4, considering the results of mF_1 and the accuracy in combination, in comparing the parameters of the first three rows one can see that when D_m is 15, the classification effect is better; however, the value of D_m is not sensitive to the classification effect. According to the results of rows 2, 4 and 5, the classification effect of D_l is obviously improved at 15. However, when the value of D_l is too large, the classification effect is reduced. Therefore, the value of D_l is relatively sensitive to the classification result. According to the results of rows 4, 6, 7 and 8–10, the classification effect is improved when the value of the scale $\mu + \upsilon$ is increased. However, when the scale exceeds 4, the classification effect will be reduced to some extent. Therefore, the value of the scale is sensitive to the results of the classification and needs to be within a reasonable range. One can see from Table 4 that the tree and building classes are relatively less affected by the changes of the D_m and D_l, and that the results are susceptible to the size of the scale. The increasing values of D_m and D_l would likely cause a change in the car classification effect. Considering the overall effect of the classification and the factors of accuracy and feature dimension, we select $D_m = 15$, $D_l = 15$ and ($\mu + \upsilon$) = 3 as the optimal parameter values.

Table 4. Parameters comparison based on mF_1 and Accuracy.

	D_m	D_l	μ+υ	Tree (%)	Building (%)	Car (%)	Accuracy (%)	mF_1 (%)
1	10	10	3	94.0	94.4	60.7	93.9	82.9
2	15	10	3	94.0	94.5	60.7	94.0	83.1
3	20	10	3	93.9	94.5	60.3	94.0	82.9
4	15	15	3	94.7	95.0	63.5	94.6	84.4
5	15	20	3	94.3	94.7	60.2	94.3	83.1
6	15	15	4	94.7	95.0	62.6	94.6	84.1
7	15	15	5	94.7	95.0	62.6	94.6	83.8
8	20	20	2	93.5	94.1	54.5	93.5	80.7
9	20	20	3	94.5	94.9	59.5	94.4	83.0
10	20	20	5	94.5	94.9	59.0	94.4	82.8

4. Conclusions

The classification of the ALS point cloud is an important technology for urban planning, digital city and intelligent transportation. We propose a multilevel features fusion and pyramid neighborhood optimization ALS point cloud classification method based on a single point. The proposed method presents two local features, i.e., the NAD and LSH. They are fused with the covariance and elevation features. Following this, the multilevel features are constructed by changing the point cloud resolution and the neighborhood size. The fused features are used to train a classification model based on the Gaussian kernel function SVM for an initial classification. Finally, the point cloud classification is optimized based on the initial classification result using a multi-scale pyramid. The optimized classification results have a higher accuracy. The experimental results prove the effectiveness of the proposed method via the experiments on the two sets of public ALS datasets.

Author Contributions: Conceptualization, Y.L. and X.D.; methodology, Y.L. and X.D.; software, Y.L. and X.D.; validation, G.T., Y.L. and X.D.; data curation, G.T., J.Z. and X.D.; writing—original draft preparation, Y.L. and X.Y.; writing—review and editing, Y.L., L.Y. and X.Y.; visualization, L.Y. and X.Y.; supervision, G.T.; project administration, G.T.; funding acquisition, G.T.

Funding: This research was funded by the National Natural Science Foundation of China (No. 61175031), the National High Technology Research and Development Program of China (863 Program) (No. 2012AA041402), the National Key Technology Research and Development Program of the Ministry of Science and Technology of China (No. 2015BAF13B00-5). The APC was funded by Guofeng Tong.

Acknowledgments: The authors would like to thank Lihao Cao in Northeastern University for helping to check the grammar and spelling of the paper.

Conflicts of Interest: The authors declare no conflict of interest.

References

1. Maturana, D.; Chou, P.-W.; Uenoyama, M.; Scherer, S. Real-time semantic mapping for autonomous off-road navigation. In *Field and Service Robotics*; Hutter, M., Siegwart, R., Eds.; Springer: Cham, Germany, 2018; pp. 335–350. ISBN 978-3-319-67361-5.

2. Li, Y.; Tong, G.F.; Yang, J.C.; Zhang, L.Q.; Peng, H.; Gao, H.S. A Summary of Key Technologies for 3D Point Cloud Scene Data Acquisition and Scene Understanding. *Laser Optoelectron. Prog.* **2019**, *56*, 040002.

3. Yuan, L.W.; Yu, Z.Y.; Luo, W.; Hu, Y.; Feng, L.Y.; Zhu, A.-X. A hierarchical tensor-based approach to compressing, updating and querying geospatial data. *IEEE Trans. Knowl. Data Eng.* **2015**, *27*, 312–325. [CrossRef]

4. Chen, D.; Zhang, L.Q.; Mathiopoulos, P.T.; Huang, X.F. A methodology for automated segmentation and reconstruction of urban 3-D buildings from ALS point clouds. *IEEE J. Sel. Top. Appl. Earth Observ. Remote Sens.* **2014**, *7*, 4199–4217. [CrossRef]

5. Linsen, L.; Prautzsch, H. Local versus global triangulations. In Proceedings of the 22th Annual Conference of the European Association for Computer Graphics, EUROGRAPHICS 2001, Manchester, UK, 5–7 September 2001.

6. Lee, I.; Schenk, T. Perceptual organization of 3D surface points. *Int. Arch. Photogramm. Remote Sens. Spat. Inf. Sci.* **2002**, *34*, 193–198.

7.	Filins, S.; Pfeifer, N. Neighborhood systems for airborne laser data. *Photgramm. Eng. Remote Sens.* **2005**, *71*, 743–755. [CrossRef]

8.	He, E.; Chen, Q.; Wang, H.; Liu, X. A curvature based adaptive neighborhood for individual point cloud classification. *Int. Arch. Photogramm. Remote Sens. Spat. Inf. Sci.* **2017**, *42*, 219–225. [CrossRef]

9.	Weinmann, M.; Jutzi, B.; Hinz, S.; Mallet, C. Semantic point cloud interpretation based on optimal neighborhoods, relevant features and efficient classifiers. *ISPRS J. Photogramm. Remote Sens.* **2015**, *105*, 286–304. [CrossRef]

10.	Hackel, T.; Wegner, J.D.; Schindler, K. Fast Semantic Segmentation of 3D Point Clouds with Strongly Varying Density. *ISPRS Ann. Photogramm. Remote Sens. Spat. Inf.* **2016**, *III*, 177–184. [CrossRef]

11.	Zeng, F.X.; Li, L.; Diao, X.P. Iterative closest point algorithm registration based on curvature features. *Laser Optoelectron. Prog.* **2017**, *54*. [CrossRef]

12.	Johnson, A.E.; Hebert, M. Using spin images for efficient object recognition in cluttered 3D scenes. *IEEE Trans. Pattern Anal. Mach. Intell.* **1999**, *21*, 433–449. [CrossRef]

13.	Frome, A.; Huber, D.; Kolluri, R.; Bülow, T.; Malik, J. Recognizing objects in range data using regional point descriptors. *Eur. Conf. Comput. Vis.* **2004**, *3023*, 224–237. [CrossRef]

14.	Tombari, F.; Salti, S.; Di Stefano, L. Unique shape context for 3D data description. In Proceedings of the International Workshop on 3D Object Retrieval (3DOR 10)—In Conjunction with ACM Multimedia, Firenze, Italy, 25 October 2010; pp. 57–62.

15.	Rusu, R.B.; Blodow, N.; Marton, Z.C.; Beetz, M. Aligning point cloud views using persistent feature histograms. In Proceedings of the 21st IEEE/RSJ International Conference on Intelligent Robots and Systems (IROS 2008), Nice, France, 22–26 September 2008.

16.	Rusu, R.B.; Blodow, N.; Beetz, M. Fast point feature histograms (FPFH) for 3D registration. In Proceedings of the IEEE International Conference on Robotics and Automation (ICRA), Kobe, Japan, 12–17 May 2009; pp. 3212–3217. [CrossRef]

17.	Jeong, J.; Lee, I. Classification of LIDAR Data for Generating a High-Precision Roadway Map. *Int. Arch. Photogramm. Remote Sens. Spat. Inf. Sci.* **2016**, *XLI-B3*, 251–254. [CrossRef]

18.	Lodha, S.K.; Fitzpatrick, D.M.; Helmbold, D.P. Aerial lidar data classification using AdaBoost. In Proceedings of the International Conference on 3-D Digital Imaging and Modeling (3DIM), Montreal, QC, Canada, 21–23 August 2007; pp. 435–442. [CrossRef]

19.	Babahajiani, P.; Fan, L.; KÄMÄRÄINEN, J.K.; Gabbouj, M. Urban 3D segmentation and modelling from street view images and LiDAR point clouds. *Mach. Vis. Appl.* **2017**, *28*, 679–694. [CrossRef]

20.	Zhang, J.X.; Lin, X.G.; Ning, X.G. SVM-based classification of segmented airborne lidar point clouds in urban areas. *Remote Sens.* **2012**, *5*, 3749–3775. [CrossRef]

21.	Fischler, M.A.; Bolles, R.C. Random Sample Consensus: A Paradigm for Model Fitting with Applications to Image Analysis and Automated Cartography. *Commun. ACM.* **1981**, *24*, 381–395. [CrossRef]

22.	Chang, C.-C.; Lin, C.-J. LIBSVM: A library for support vector machines. *ACM Trans. Intell. Syst. Technol.* **2011**, *2*, 27. [CrossRef]

23.	Zhang, Z.X.; Zhang, L.Q.; Tong, X.H.; Wang, Z.; Guo, B.; Huang, X.F.; Wang, Y.B. A Multi-Level Point Cluster-based Discriminative Feature for ALS Point Cloud Classification. *IEEE Trans. Geosci. Remote Sens.* **2016**, *54*, 4714–4726.

24.	Guo, B.; Huang, X.; Zhang, F.; Sohn, G. Classification of airborne laser scanning data using JointBoost. *ISPRS J. Photogramm. Remote Sens.* **2015**, *100*, 71–83. [CrossRef]

25.	Wang, Z.; Zhang, L.; Fang, T.; Mathiopoulos, P.T.; Tong, X.; Qu, H.; Xiao, Z.; Li, F.; Chen, D. A Multiscale and Hierarchical Feature Extraction Method for Terrestrial Laser Scanning Point Cloud Classification. *IEEE Trans. Geosci. Remote Sens.* **2015**, *53*, 2409–2425. [CrossRef]

26.	Zhang, Z.; Zhang, L.; Tong, X.; Guo, B.; Zhang, L.; Xing, X. Discriminative dictionary learning-based multi-level point-cluster features for ALS point cloud classification. *IEEE Trans. Geosci. Remote Sens.* **2016**, *54*, 7309–7322. [CrossRef]

27.	Zhang, Z.X. ALS Point Cloud Classification Based on Multilevel Point Cluster Features. Ph.D. Thesis, Beijing Normal University, Beijing, China, 2017.

Article

Study of the Relationship between Urban Expansion and PM$_{10}$ Concentration Using Multi-Temporal Spatial Datasets and the Machine Learning Technique: Case Study for Daegu, South Korea

Yun-Jae Choung [1,2] and Jin-Man Kim [3,*]

[1]   GIS Research Center, Geo C&I Co., Ltd., Daegu 41165, Korea; choung12osu@gmail.com
[2]   Global Land Satellite Information Center, Kyungpook National University, Daegu 41566, Korea
[3]   Geotechnical Engineering Research Institute, Korea Institute of Civil Engineering and Building Technology, Goyang-si, Gyeonggi-do 10223, Korea
*   Correspondence: jmkim@kict.re.kr; Tel.: +82-31-910-0221

Received: 30 January 2019; Accepted: 11 March 2019; Published: 15 March 2019

Abstract: To protect the population from respiratory diseases and to prevent the damages due to air pollution, the main cause of air pollution should be identified. This research assessed the relationship between the airborne particulate concentrations (PM$_{10}$) and the urban expansion in Daegu City in South Korea from 2007 to 2017 using multi-temporal spatial datasets (Landsat images, measured PM$_{10}$ data) and the machine learning technique in the following steps. First, the expanded urban areas were detected from the multiple Landsat images using support vector machine (SVM), a widely used machine learning technique. Next, the annual PM$_{10}$ concentrations were calculated using the long-term measured PM$_{10}$ data. Finally, the degrees of increase of the expanded urban areas and of the PM$_{10}$ concentrations in Daegu from 2007 to 2017 were calculated by counting the pixels representing the expanded urban areas and computing variation of the annual PM$_{10}$ concentrations, respectively. The experiment results showed that there is a minimal or even no relationship at all between the urban expansion and the PM$_{10}$ concentrations because the urban areas expanded by 55.27 km^2 but the annual PM$_{10}$ concentrations decreased by 17.37 μg/m^3 in Daegu from 2007 to 2017.

Keywords: coarse particle; particulate matter 10 (PM$_{10}$); landsat image; machine learning; support vector machine

---

1. Introduction

Urban expansion, also called "urban sprawl," is defined as "the spreading of urban development (e.g., houses, shopping centers) on undeveloped lands near a city" or "the rapid expansion of the geographic extent of cities and towns, often characterized by low-density residential housing, single-use zoning, and increased reliance on private automobiles for transportation" [1,2]. In general, urban expansion has a close relationship with urban development, infrastructure improvement, population growth, etc. [3].

Coarse particle, defined as particulate matter 10 (PM$_{10}$), consists of particles with a diameter of 10 μm or less [4,5]. PM$_{10}$ is one of the main components of air pollution, and it also results in various environmental impacts (e.g., atmospheric pollution) and human health impacts (e.g., chronic respiratory diseases) [6,7]. In particular, exposure to a high PM$_{10}$ concentration can cause a number of significant health impacts, ranging from coughing to high blood pressure, heart attack, stroke, and lung cancer [8].

Previous studies found that urban expansion generally has a significant impact on air pollution because more human activities that can cause air pollution (e.g., vehicular traffic) are expected in urban areas [9]. Stone (2007) assessed the relationship between urban expansion and air quality [10]. Cho and Choi (2014) investigated the effect of compact urban development on air quality [11]. Liu et al. (2018) assessed the relationship between urban air pollution and urban form, seasonality, and city size [12].

To protect the population from respiratory diseases and to prevent public health disasters due to PM_{10} concentration, the main causes of PM_{10} concentration in the city should be identified. Limited research has been conducted, however, to identify the main causes of PM_{10} concentration in each city in South Korea. In general, urban expansion is regarded as the main cause of urban air pollution owing to the urban development activities accompanying it. This research aims to assess the relationship between urban expansion and PM_{10} concentration by monitoring the 10 years (from 2007 to 2017) of urban expansion and the annual PM_{10} concentrations using multi-temporal spatial datasets acquired in Daegu, South Korea and the machine learning technique. First, the expanded urban areas were detected from the multi-temporal Landsat satellite images using the machine learning technique. Then the annual PM10 concentrations were calculated using the long-term measured PM10 data. Finally, the relationship between urban expansion and PM_{10} concentration was assessed by counting the pixels representing the expanded urban areas and computing variation of the annual PM10 concentrations, respectively.

2. Study Area and Datasets

Daegu Metropolitan City in South Korea was selected as the study area in this research for the following reasons. First, the urban areas of Daegu have been significantly expanded of late [13]. Second, there are long-term measured coarse particle (PM_{10}) data acquired by the 11 air quality monitoring stations (AQMSs) in Daegu, which can be used for the study [14]. Daegu has operated these AQMSs since 1973 for sustainably monitoring the air quality condition of the city. Figure 1 shows the locations of the 11 AQMSs in Daegu, South Korea.

The multi-temporal Landsat satellite images acquired on 13 May 2007 ("first Landsat image") and on April 29, 2017 ("second Landsat image") were used in this study for the following reasons. First, the urban areas in Daegu significantly expanded during such periods due to the city's urban development policy. Second, both Landsat images were less affected by the prevailing atmospheric conditions then. Figure 2 shows one section each of the first and second Landsat images.

The first Landsat image was acquired by the Landsat-5 thematic mapper (TM) sensor, and it consists of seven bands (blue: 450–520 nm; green: 520–600 nm; red: 630–690 nm; near-infrared: 770–900 nm; short-wave infrared 1: 1550–1750 nm; short-wave infrared 2: 2080–2350 nm; and thermal: 10,400–12,500 nm) [15]. The second Landsat image was acquired by the Landsat-8 operational land imager (OLI) and the thermal infrared sensor (TIRS), and it consists of nine bands (coastal aerosol: 435–451 nm; blue: 452–512 nm; green: 533–590 nm; red: 636–673 nm; near-infrared: 851–879 nm; short-wave infrared 1: 1566–1651 nm; short-wave infrared 2: 2107–2294 nm; thermal infrared 1: 10,600–11,190 nm; and thermal infrared 2: 11,500–12,510 nm) [15]. Both Landsat images were georeferenced to the coordinate system universal transverse mercator (UTM), zone 52 N, based on the 1984 datum world geodetic system (WGS).

To measure the annual PM_{10} concentrations during the same period for the monitoring of the urban expansion in Daegu using the Landsat 1 and 2 images, the measured PM_{10} data between 2007 and 2017 acquired through the Daegu atmospheric information system (https://air.daegu.go.kr/) were also utilized.

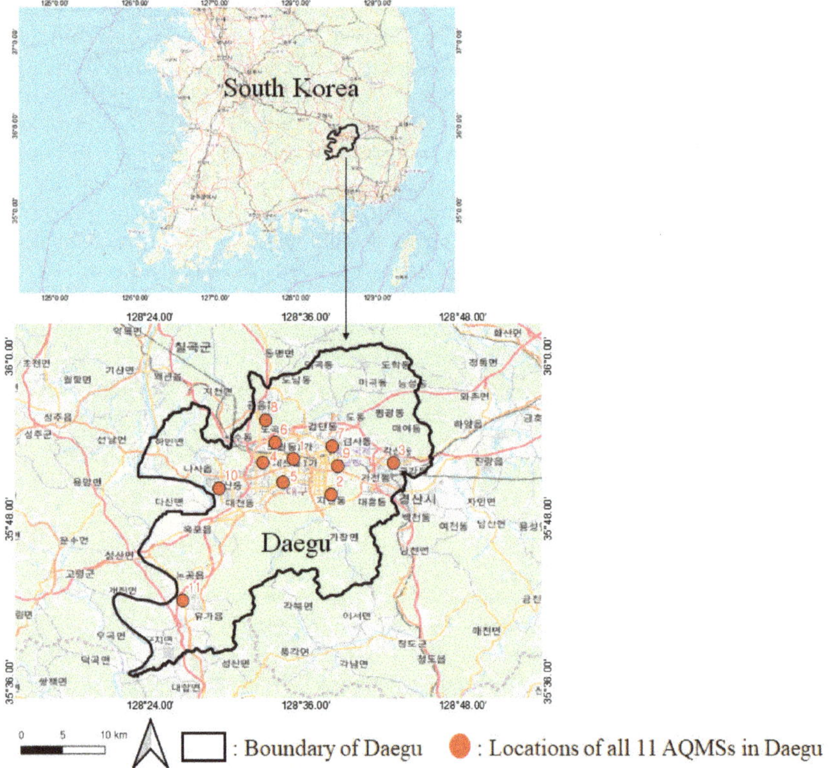

Figure 1. Locations of the 11 air quality monitoring stations (AQMSs) in Daegu, South Korea.

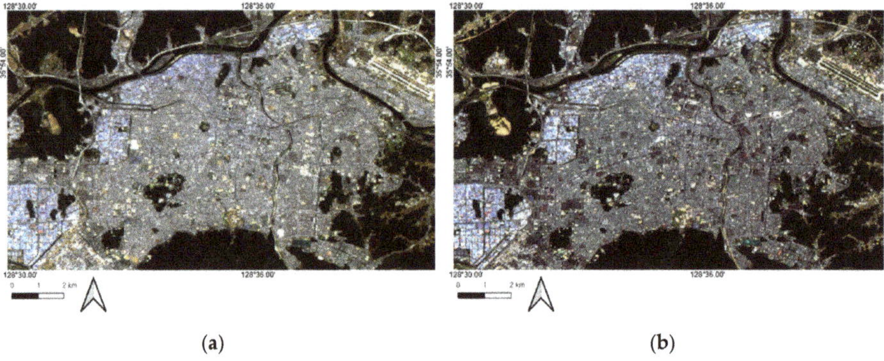

(a)   (b)

Figure 2. One section each of the multi-temporal Landsat images utilized in this research: (**a**) a section of the first Landsat image (acquired on 13 May 2007); and (**b**) a section of the second Landsat image (acquired on 29 April 2017).

3. Methodology

Figure 3 presents a flowchart of the procedure that was employed to assess the relationship between urban expansion and PM_{10} concentration in Daegu from 2007 to 2017 using the given datasets.

As can be seen in Figure 3, in the first step of the proposed methodology, two urban maps were generated, respectively, from the first and second Landsat images, using the support vector machine

(SVM) technique, a widely used machine learning technique. Then the extent of urban expansion was detected using the generated first and second urban maps. In the next step, the annual PM_{10} concentrations were calculated using the measured PM_{10} data acquired by each AQMS of Daegu. Finally, the relationship between the urban expansion and the PM_{10} concentration rate in Daegu was assessed using the two calculated statistics: the increase of the expanded urban areas and the increase of the annual PM_{10} concentrations in Daegu from 2007 to 2017.

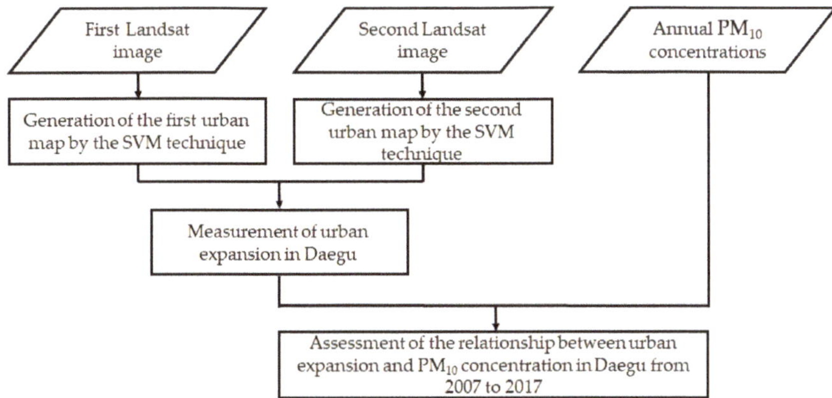

Figure 3. Flowchart showing the procedure that was employed to assess the relationship between urban expansion and PM_{10} concentration in Daegu from 2007 to 2017 using the given datasets.

3.1. Generation of the Urban Maps by the SVM Technique

This section illustrates the procedure for generating urban maps through the SVM technique, a widely used machine learning algorithm. Machine learning is defined as ""the ability of a machine to improve its performance based on previous results" [16]. The machine learning technique has been widely used of late in remote sensing applications for classifying land uses and detecting the significant features from the remote sensing datasets, due to its advantages for high-value classification [17,18]. SVM, a widely used machine learning technique for finding the linear hyperplane that maximizes the margins between the two clusters in n-dimensional spaces, has been widely used in remote sensing applications due to its superior advantages over the other machine learning techniques for classifying land uses, detecting significant features, and avoiding classification errors [19]. Hence, in this research, the SVM technique was used to generate urban maps, which distinguish the urban areas from the non-urban areas (water, soil, vegetation, etc.). Figure 4 shows the first and second urban maps separately generated from the first and second Landsat images, respectively, through the SVM technique.

3.2. Detection of the Expaned Urban Areas in Daegu from 2007 to 2017

In the next step of the proposed methodology, the expanded urban areas in Daegu were detected using the two urban maps that had been generated. The expanded urban areas from 2007 to 2017 were detected by intersecting the pixels representing the non-urban areas in the first urban map and the pixels representing the urban areas in the second urban map. Figure 5 shows the locations of the AQMSs and expanded urban areas in Daegu from 2007 to 2017 detected using the two generated urban maps.

As can be seen in Figure 5, all the AQMSs in Daegu were located near the expanded urban areas detected using the two generated urban maps.

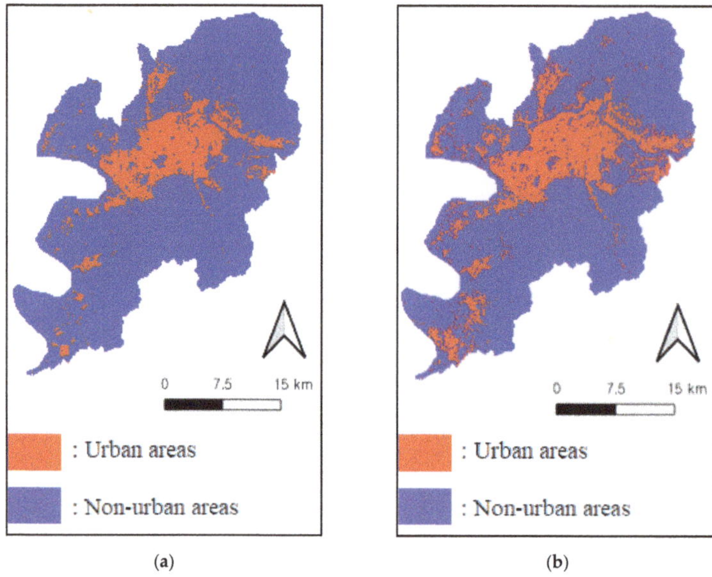

(a)   (b)

Figure 4. First and second urban maps: (**a**) first urban map generated from the first Landsat image through the SVM technique; and (**b**) second urban map generated from the second Landsat image through the SVM technique.

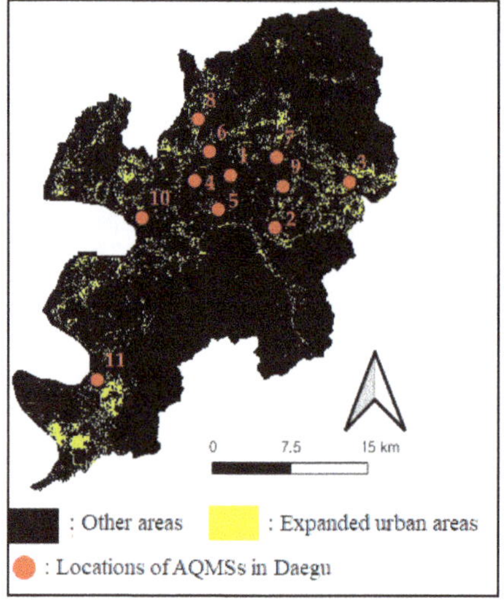

Figure 5. Locations of the AQMSs and expanded urban areas in Daegu from 2007 to 2017 detected using the two generated urban maps.

3.3. Calculation of the Statistics for the Annual PM_{10} Concentrations

In this section, the calculation of the statistics for the annual PM_{10} concentrations in each AQMS in Daegu from 2007 to 2017 is described. Figure 6 presents time series graphs showing the annual PM_{10} concentrations in each AQMS in Daegu from 2007 to 2017.

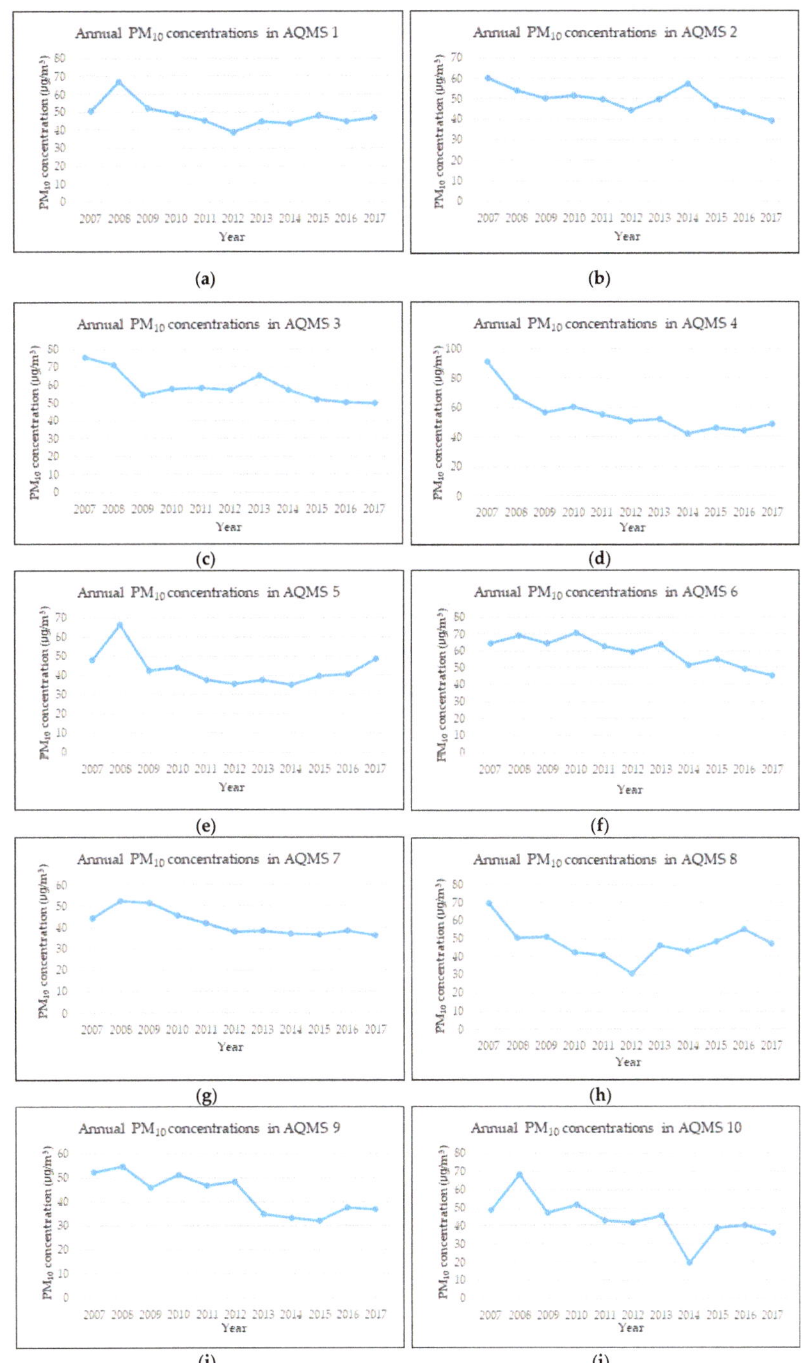

Figure 6. *Cont.*

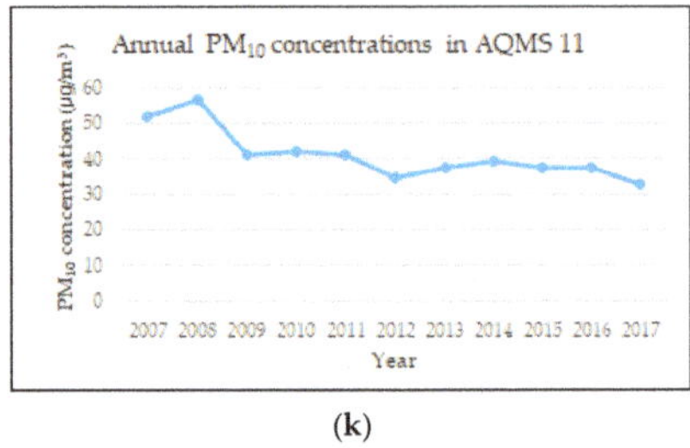

(**k**)

Figure 6. Time series graphs showing the annual PM$_{10}$ concentrations in each AQMS in Daegu from 2007 to 2017. AQMS 1(**a**) to AQMS 11(**k**).

4. Results and Discussions

4.1. Accuracies of the Generated Urban Maps

In this section, the degrees of accuracy of the first and second urban maps separately generated from the first and second Landsat images, respectively, are assessed using the 100 checkpoints generated through manual digitization. Figure 7 shows examples of the checkpoints generated through manual digitization for measuring the degrees of accuracy of the first and second urban maps.

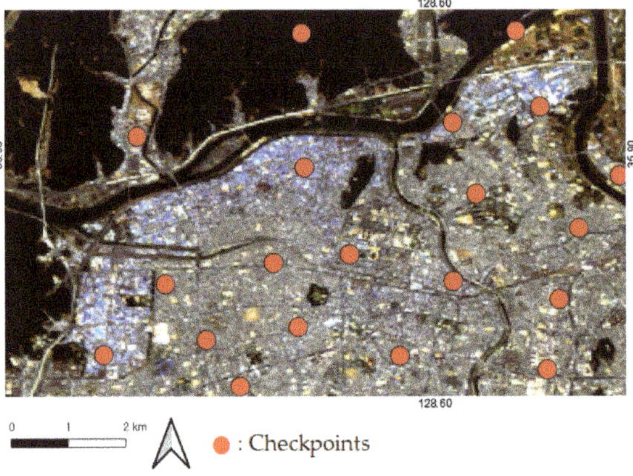

Figure 7. Examples of the checkpoints generated through manual digitization for measuring the degrees of accuracy of the first and second urban maps.

Table 1 shows the statistical results showing the degrees of accuracy of the first and second urban maps separately generated from the first and second Landsat images, respectively.

As can be seen in Table 1, the first and second urban maps separately generated from the first and second Landsat images, respectively, through the SVM technique had high accuracy in identifying the urban areas in the entire Daegu area. There were a few misclassification errors, however, in both urban maps because some urban features (e.g., man-made features) were misclassified as non-urban features (e.g., soil, water, vegetation), or vice versa. These misclassification errors were generally caused by the similar reflectance characteristics of the different materials owing to the shades, etc. [18].

Table 1. Statistical results showing the degrees of accuracy of the first and second urban maps separately generated from the first and second Landsat images, respectively: (**a**) degree of accuracy of the first urban map generated from the first Landsat image; and (**b**) degree of accuracy of the second urban map generated from the second Landsat image.

(a)			
Overall Accuracy		**97%**	
Producer's Accuracy (Error of Omission)		**User's Accuracy (Error of Commission)**	
Urban areas	94%	Urban areas	100%
Non-urban areas	100%	Non-urban areas	94%

(b)			
Overall Accuracy		**99%**	
Producer's Accuracy (Error of Omission)		**User's Accuracy (Error of Commission)**	
Urban areas	100%	Urban areas	98%
Non-urban areas	98%	Non-urban areas	100%

4.2. Relationship between the Urban Expansions and the PM_{10} Concentrations in Daegu from 2007 to 2017

Discussed in this section is the relationship between the urban expansion and the annual PM_{10} concentration rate in Daegu from 2007 to 2017 determined by calculating the following statistics: those showing the increase of the expanded urban areas in Daegu from 2007 to 2017 and those showing the annual PM_{10} concentration changes in each AQMS, each year and each season in Daegu from 2007 to 2017 (see Table 2).

As can be seen in Figure 6, Table 2a,b, the urban areas expanded by 55.27 km^2 in the entire Daegu area from 2007 to 2017, but the annual PM_{10} concentrations in all the AQMSs in Daegu only slightly increased by 0.45 µg/m^3 in only one station or sharply decreased by 4.12~42.28 µg/m^3 in the other 10 stations, within the same period. Table 2c shows that the annual PM_{10} concentrations had decreased in Daegu City from 2007 to 2017 by 17.37 µg/m^3.

Table 2b,c show that the annual PM_{10} concentrations were measured high in AQMSs 3, 4, and 6, while they were measured low in AQMSs 5, 7, 9, 10 and 11. Table 2c also shows that the highest annual PM_{10} concentrations was most frequently measured in AQMS 6 from 2007 to 2017 while the lowest annual PM_{10} concentrations most frequently measured in AQMS 11 during the same period. Based on Table 2b,c results, we assume that there are the number of other facilities (e.g., industrial factories) that emit the PM_{10} particles near AQMSs 3, 4, and 6, while there are few facilities emitting the PM_{10} particles near AQMS 5, 7, 9, 10, and 11.

Table 2c,d show that the annual PM_{10} concentrations generally measured high in spring and winter compared to summer and autumn. In addition, the annual PM_{10} concentrations decreased in all the seasons by 31.11 µg/m^3 for spring, by 8 µg/m^3 for summer, by 8.78 µg/m^3 for autumn, and by 18.77 µg/m^3 for winter. Based on Table 2d, we assume that the climate factors (e.g., air temperature, air pressure, rainfall, humidity, wind speed, and wind direction) can be significant for the annual PM_{10} concentrations.

The above experiment results show that there is a minimal or no relationship at all between the urban expansion and the PM_{10} concentrations rate in Daegu, which means that the urban expansion that occurred in Daegu from 2007 to 2017 was not the main cause of the rise in the PM_{10} concentration rate in Daegu, South Korea during the same period. We assume that, however, the types of facilities and climate factors can influence on the annual PM_{10} concentrations.

Table 2. Statistics showing the increase of the expanded urban areas in Daegu from 2007 to 2017 and the annual PM_{10} concentration changes in each air quality monitoring stations (AQMS) in Daegu within the same period: (a) statistics showing the total number of urbanized areas in Daegu from 2007 to 2017; (b) statistics showing the annual PM_{10} concentration changes in each AQMS in Daegu from 2007 to 2017; (c) statistics showing the range of variability of the annual PM_{10} concentrations in each year from 2007 to 2017; and (d) statistics showing the annual PM_{10} concentrations in each season (spring: March, April, and May; summer: June, July, and August; autumn: September, October, and November; and winter: December, January, and February) from 2007 to 2017.

(a)

Total Areas of the Urban Areas in the First Urban Map (km²)	Total Areas of the Urban Areas in the Second Urban Map (km²)	Increase of the Expanded Urban Areas in Daegu from 2007 to 2017 (km²)
148.08	203.35	+ 55.27

(b)

AQMS ID	Maximum ($\mu g/m^3$)	Minimum ($\mu g/m^3$)	Average ($\mu g/m^3$)	Standard Deviation	Variation of Annual PM_{10} Concentration (2017 vs 2007) ($\mu g/m^3$)
AQMS 1	67.28	38.79	48.07	7.35	−4.12
AQMS 2	60.21	38.80	49.61	6.30	−21.41
AQMS 3	75.32	49.30	58.80	8.40	−26.02
AQMS 4	91.14	41.99	55.85	13.79	−42.28
AQMS 5	66.16	34.64	42.95	8.97	+0.45
AQMS 6	70.65	45.23	59.54	8.34	−19.29
AQMS 7	52.22	36.29	41.85	5.78	−7.88
AQMS 8	56.75	32.78	47.61	9.72	−22.75
AQMS 9	54.86	31.78	43.03	8.43	−15.59
AQMS 10	68.53	19.66	43.78	11.85	-12.81
AQMS 11	56.75	32.78	41.05	7.26	−19.35

(c)

Year	Maximum ($\mu g/m^3$)	AQMS ID for Maximum	Minimum ($\mu g/m^3$)	AQMS ID for Minimum	Average ($\mu g/m^3$)	Standard Deviation
2007	91.14	AQMS 4	44.17	AQMS 7	59.72	14.31
2008	71.04	AQMS 3	50.36	AQMS 8	61.58	7.83
2009	64.37	AQMS 6	41.21	AQMS 11	50.52	6.59
2010	70.65	AQMS 6	42.27	AQMS 11	50.51	8.66
2011	62.63	AQMS 6	37.21	AQMS 5	47.42	8.15
2012	59.51	AQMS 6	30.74	AQMS 8	43.61	9.44
2013	65.03	AQMS 3	34.63	AQMS 9	46.73	10.34
2014	57.22	AQMS 2	19.66	AQMS 10	41.63	11.02
2015	54.55	AQMS 6	31.78	AQMS 9	43.48	7.13
2016	54.98	AQMS 8	31.32	AQMS 11	43.6	5.71
2017	49.30	AQMS 3	32.78	AQMS 11	42.35	6.21

(d)

Year	Spring ($\mu g/m^3$)	Summer ($\mu g/m^3$)	Autumn ($\mu g/m^3$)	Winter ($\mu g/m^3$)
2007	81.58	41.55	48.47	64.32
2008	73.63	52.11	55.88	64.10
2009	51.74	42.37	45.51	62.87
2010	57.39	39.86	50.21	58.59
2011	59.77	37.59	42.20	50.59
2012	50.63	33.98	41.76	48.14
2013	52.92	38.99	39.90	55.02
2014	50.01	33.82	35.30	47.49
2015	50.51	34.59	31.14	57.61
2016	53.33	32.71	40.02	48.49
2017	50.47	33.55	39.69	45.55

5. Conclusions and Future Works

In this research, an experiment was performed to assess the relationship between the urban expansion and the PM_{10} concentration rate in Daegu from 2007 to 2017 by calculating the expanded urban areas and the annual PM_{10} concentration changes in each AQMS, each year and each season. The experiment results showed that there is a minimal or no relationship at all between the urban expansion that occurred in Daegu from 2007 to 2017 and the rise in the PM_{10} concentration rate in the same city during the same period because the urban areas significantly expanded but the annual PM_{10} concentrations sharply decreased.

This research proved that the urban expansion that occurred in Daegu from 2007 to 2017 was not the main cause of the rise in the PM_{10} concentration rate in Daegu during the same period. This research, however, also suggested that the other factors, such as types of facilities or climate factors, can be significant for the annual PM_{10} concentrations. To protect the public health, it is necessary to identify the main causes of PM_{10} concentrations, hence, further research, will be carried out to identify the main causes (e.g., the climate factors and the type of facilities) of PM_{10} concentrations in general. In addition, research will be conducted to identify the main cause of $PM_{2.5}$ concentrations that also causing serious air pollution problems.

Author Contributions: Y.-J.C. performed the experiments, analyzed the data and wrote the manuscript; J.-M.K. proposed the method, designed the experiments and modified the manuscript.

Funding: This research was supported by a grant (18SCIP-B065985-06) from Smart Civil Infrastructure Research Program funded by Ministry of Land, Infrastructure and Transport (MOLIT) of Korea government and Korea Agency for Infrastructure Technology Advancement (KAIA).

Conflicts of Interest: The authors declare no conflict of interest.

References

1. Merriam-Webster. Urban Sprawl. Available online: https://www.merriam-webster.com/dictionary/urban%20sprawl (assessed on 22 November 2018).
2. Encyclopedia Britanica. Urban Sprawl. Available online: https://www.britannica.com/topic/urban-sprawl (assessed on 22 November 2018).
3. Conserve Energy Future. What Is Urban Sprawl. Available online: https://www.conserve-energy-future.com/causes-and-effects-of-urban-sprawl.php (assessed on 22 November 2018).
4. Kang, D.; Kim, J.-E. Fine, Ultrafine and Yellow Dust: Emerging Health Problems in Korea. *J. Korean Med. Sci.* **2014**, *29*, 621–622. [CrossRef] [PubMed]
5. United States Environmental Protection Agency. Particulate Matter (PM) Basics. Available online: https://www.epa.gov/pm-pollution/particulate-matter-pm-basics (assessed on 22 November 2018).
6. Ghorani-Azam, A.; Riahi-Zanjani, B.; Balali-Mood, M. Effects of air pollution on human health and practical measures for prevision in Iran. *J. Res. Med. Sci.* **2016**, *21*, 65. [CrossRef] [PubMed]
7. World Health Organization (WHO) Working Group. *Health Aspects of Air Pollution with Particulate Matter, Ozone and Nitrogen Dioxide*; World Health Organization: Bonn, Germany, 2003.
8. Yin, H.; Xu, L.; Cai, Y. Monetary Valuation of PM_{10}—Related Health Risks in Beijing China: The Necessity for PM_{10} Pollution Indemnity. *Int. J. Environ. Res. Public Health* **2015**, *12*, 9967–9987. [CrossRef] [PubMed]
9. Pourahmad, A.; Baghavand, A.; Shahraki, S.Z.; Givehchi, S. The Impact of Urban Sprawl up on Air Pollution. *Int. J. Environ. Res.* **2007**, *1*, 347–353. [CrossRef]
10. Stone, B. Urban Sprawl and Air Quality in Large US Cities. *J. Environ. Manag.* **2008**, *86*, 688–698. [CrossRef] [PubMed]
11. Cho, H.S.; Choi, M.J. Effects of Compact Urban Development on Air Pollution: Empirical Evidence from Korea. *Sustainability* **2014**, *6*, 5968–5982. [CrossRef]
12. Liu, Y.; Wu, J.; Yu, D.; Ma, Q. The relationship between urban form and air pollution depends on seasonality and city size. *Environ. Sci. Pollut. Res.* **2018**, *25*, 15554–15567. [CrossRef] [PubMed]
13. Daegu Metropolitan City. The Document for the Urban Planning Policy of Daegu. Available online: http://ebook.daegu.go.kr/Viewer/RBL5N2AFGRNW (accessed on 22 November 2018).
14. Daegu Atmospheric Information System. PM10 Statistics. Available online: https://air.daegu.go.kr/open_content/ko/index.do (assessed on 22 November 2018).
15. United States Geological Survey (USGS). Landsat Missions. Available online: https://www.usgs.gov/land-resources/nli/landsat (assessed on 22 November 2018).
16. DicionaryCom. Machine Learning. Available online: https://www.dictionary.com/browse/machine-learning (assessed on 22 November 2018).

Appl. Sci. **2019**, *9*, 1098

17. Choung, Y.-J.; Jo, M.-H. Comparison between a Machine-learning-based Method and a Water-index-based Method for Shoreline Mapping Using a High-Resolution Satellite Image Acquired in Hwado Island, South Korea. *J. Sens.* **2017**, *2017*, 8245204. [CrossRef]
18. Lary, D.J.; Alavi, A.H.; Gandomi, A.H.; Walker, A.L. Machine learning in geosciences and remote sensing. *Geosci. Front.* **2016**, *7*, 3–10. [CrossRef]
19. Mountrakis, G.; Im, J.; Ogole, C. Support vector machines in remote sensing: A review. *ISPRS J. Photogramm. Remote Sens.* **2011**, *66*, 247–259. [CrossRef]

Article

Deep Fusion Feature Based Object Detection Method for High Resolution Optical Remote Sensing Images

Eric Ke Wang [1], Yueping Li [2,*], Zhe Nie [2], Juntao Yu [1], Zuodong Liang [1] and Xun Zhang [1] and Siu Ming Yiu [3]

[1]   Harbin Institute of Technology, Shenzhen 518055, China; wk_hit@hit.edu.cn (E.K.W.); yujuntao@stu.hit.edu.cn (J.Y.); liangzuodong@stu.hit.edu.cn (Z.L.); zhangxun@stu.hit.edu.cn (X.Z.)
[2]   School of Computer Engineering, Shenzhen Polytechnic, Shenzhen 518055, China; niezhe@szpt.edu.cn
[3]   Department of Computer Science, University of Hong Kong, Pokfulam Road, Hong Kong, China; smyiu@cs.hku.hk
*   Correspondence: liyueping@szpt.edu.cn; Tel.: +86-755-2603-3248

Received: 31 January 2019; Accepted: 27 February 2019; Published: 18 March 2019

Abstract: With the rapid growth of high-resolution remote sensing image-based applications, one of the fundamental problems in managing the increasing number of remote sensing images is automatic object detection. In this paper, we present a fusion feature-based deep learning approach to detect objects in high-resolution remote sensing images. It employs fine-tuning from ImageNet as a pre-training model to address the challenge of it lacking a large amount of training datasets in remote sensing. Besides, we improve the binarized normed gradients algorithm by multiple weak feature scoring models for candidate window selection and design a deep fusion feature extraction method with the context feature and object feature. Experiments are performed on different sizes of high-resolution optical remote sensing images. The results show that our model is better than regular models, and the average detection accuracy is 8.86% higher than objNet.

Keywords: high-resolution; optical remote sensing; object detection; deep learning; transfer learning

---

1. Introduction

Object detection for remote sensing images is an important research field. With the development of remote sensing technology, information carried by remote sensing images is more abundant than before. The applications of object detection in remote sensing images are more and more popular, such as city planning and environmental exploration.

However, object detection for remote sensing images is a more difficult job since remote sensing images are quite different from regular images. Objects in regular images have some properties: many objects rarely appear in one image; the main objects are regularly located in the image center, occupying the main parts and significantly different from the background.

However, one high-resolution optical remote sensing image contains more objects with more shapes and texture information than a regular image, and the objects may be scattered in the whole image. Besides, the object to be detected is relatively small and close to the background. If we zoom out from a remote sensing image to a small size for a global view, we would lose many details, and the objects may almost be invisible. Therefore, object detection for remote sensing images is harder work than for regular images to some extent.

At present, object detection methods in remote sensing images are mainly based on traditional image processing technology with machine learning, which requires rich experience and complete prior knowledge. Furthermore, most of them are only effective in a specific environment, so they have poor scalability. With the advent of deep learning technology, we introduce deep learning into the field of object detection for remote sensing images. Nevertheless, deep learning is still in its infancy for remote sensing images. One of its biggest problems is the decency on labeled datasets.

However, with the fast improvement of deep learning-based object detection on regular images, many labeled regular image datasets have appeared in recent years. Therefore, in this paper, we present a novel transfer deep learning approach to detect objects in high-resolution remote sensing images. It employs transfer learning to supply the gap that deep learning on remote sensing images lacks labeled training datasets. Besides, we improved the candidate window selection process and designed a deep feature extraction method with context scene feature fusion and detection. Finally, the proposed approach is validated on different scales of high-resolution optical remote sensing images.

This paper is organized as follows: Section 2 introduces the related works. Section 3 describes the framework of our algorithm. Our size-scalable region proposal algorithm is given in Section 4. Our deep feature extraction method with context scene feature fusion and detection is proposed in Section 5. Section 6 concludes the paper.

2. Related Works

The development of object detection in high-resolution remote sensing images has been in three stages: template matching-based, knowledge testing-based, and machine learning-based.

Weber et al. [1] performed template matching by extending the hit-or-miss transformation in a multivariate fashion to detect storage tanks and the shoreline. Sirmacek et al. [2] demonstrated the detection of urban buildings by using the SIFT feature to represent a two-building template. Knowledge testing-based object detection transforms detection into the hypothesis testing problem by establishing a set of knowledge and rules. Yokoya et al. [3] used buildings and shadows to detect buildings of arbitrary shape automatically. Machine learning-based methods have been the main direction of object detection in the remote sensing field. For example, Tao et al. [4] described the airplane objects by a set of key point SIFT feature and a graph model to detect them. Sun et al. [5] constructed a bag-of-words model by clustering SIFT feature points to represent targets and classified them by SVM. Gan et al. [6] performed ship detection on remote sensing images by extracting the HOG feature of the sliding window and extracted the continuous window feature by rotating the sliding window to achieve certain rotation invariance. Mostafa et al. [7] clustered the urban railroad point clouds into three classes of rail track, contact cable, and catenary cable by a template matching approach. Aytekin et al. [8] automatically selected a representative subset of texture features by the AdaBoost algorithm and used them to identify airport runways and then detect the airport. Felzenszwalb et al. [9] obtained good results in general object detection by combining the pyramid HOG feature with the partial deformation model and training an SVM with hidden variables. Some scholars have applied weak-supervised object detection algorithms to remote sensing images [10].

However, these methods mostly relied on a variety of well-designed prior knowledge or shallow features, etc. [11,12]. That is, they require a wealth of experience and a tedious trial and error process, and they have limitations.

In the field of object detection on regular images, Krizhevsky et al. [13] proposed an image classification algorithm based on deep learning, which automatically extracts the higher level features of the images by a convolutional neural network (CNN), which made the image classification task more concise and the classification accuracy significantly better.

Thanks to the development of deep learning and the development of region proposal algorithms, the object detection field has also made great breakthroughs. One of the main challenges of object detection in remote sensing images is how to reduce the computational complexity. In view of the large scale of remotely-sensed images, there will be a large number of candidates if the conventional multi-scale scanning window exhaustive strategy is used to obtain the region of interest, which makes the subsequent feature extraction and classification cost too much to achieve fast detection. Therefore, how to reduce the search space in this field is a key problem. In recent years, many region proposal algorithms have been proposed [14–17]. These algorithms can be divided into two categories:

(1) segmentation and combination methods: divide the input image into fragments and then combine these fragments by some bottom-up strategy to generate regions of interest; (2) the window scoring method: define the scoring criteria of the probability of the candidate window containing the object, scoring each possible window by the sliding window method, and selecting the candidate with a high score. There are two more regularly-used region proposal algorithms: the selective search algorithm [14] and the binarized normed gradients (BING) algorithm [15].

In 2014, Girshick et al. [18] proposed the R-CNN (region-based convolutional network) framework for object detection, in which a region proposal algorithm was designed to obtain the candidate window instead of the sliding window strategy to improve detection efficiency, and then, the CNN was employed to extract high-level features before an SVM classification was used. The proposed R-CNN framework greatly improved the detection accuracy and brought much inspiration to the object detection field, and many object detection algorithms based on the deep learning have been proposed [19–21]. There are more research works on object detection using deep learning. For example, Kong et al. [22] proposed the HyperNet network structure, which combines the multi-level features of the deep network and merges them to select the region and detect the object, which resulted in a more accurate localization. Ouyang et al. [23] proposed to combine CNN with the deformation model, which made the process of objection detection more sensitive through multiple models, multi-stage cascade, and other integrated approached. Redmon et al. [24] considered the objection detection as a regression prediction problem. They designed the YOLO network structure, of which the network input is the whole map. This original map was divided into 7×7 grids. This structure greatly improves the detection speed and real-time detection. However, their method is not effective with respect to objects that are located close to one another, and the objects have an irregular aspect ratio. Meanwhile, the deep learning method in the field of object detection in remote sensing images is still in a relatively nascent stage.

3. The Overall Idea of Our Method

In object detection in remote sensing images, template matching-based methods are simple and easy to implement, but the template design becomes more and more complicated when directions and shapes of objects vary greatly. Knowledge-based object detection methods can gain better detection performance through abundant a priori knowledge, but how to define the a priori knowledge and rules is a hard problem, which usually requires much experience. While machine learning-based object detection methods are based on shallow feature extraction methods, such as HOG, SIFT, and other classic features presented in object detection on regular images, and have a better detection effect for some specific scenes, when the remote sensing background is complex and the objects are diversified, the scalability of these methods is poor. Deep learning has a great advantage in automatically learning deep-level features, but it is still at a relatively early stage for object detection in remote sensing images. Meanwhile deep learning requires a large amount of labeled data, which are less available in remote sensing images. Aiming at these shortcomings, we propose to employ abundant labeled regular image datasets to assist the object detection in the remote sensing images through the transfer learning method and explore the validity of the transfer learning. As shown in Figure 1, from the detection process, the framework of the proposed approach can be divided into three steps: rapid candidate region proposal, deep feature extraction of the candidate window, context scene feature fusion, classification, and post-processing. The training process is mainly to train the models used in the steps of the detection stage: (a) train the candidate region proposal model; (b) combine transfer learning to train the deep feature extraction network; (c) combine transfer learning to train the context scene feature extraction network; (d) train the classification model.

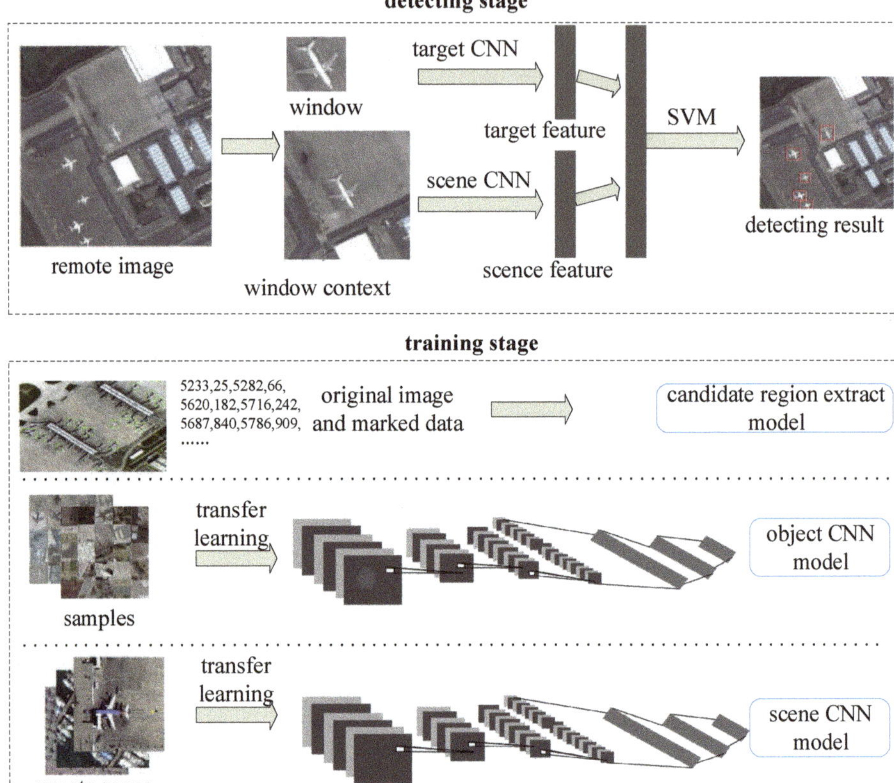

Figure 1. Proposed object detection framework.

In particular, we improve the stages of candidate region proposal and feature extraction:

1. For the stage of candidate region proposal, by analyzing the particularity of the remote sensing image object detection task, we select the BING algorithm to improve it by integrating multiple weak feature scoring to extend to large-scale images. The experiment shows that the improved algorithm achieves a better detection rate and more accurate object coverage when obtaining the same number of candidate windows.

2. For the stage of feature extraction, we employ CNN to extract deep features of the candidate windows and the windows' context scene, respectively, and then fuse the two kinds of features for detection, which improves the detection performance. In addition, we solve the problem of the insufficient annotations on remote sensing images by transfer learning, which reduces the risk of over-fitting and improves the network's ability of feature expression on remote sensing objects and scenes.

In the stage of classification and post-processing, we employ the faster linear SVM with the hard negative mining method to reduce the impact of overfull negative samples. Finally, we filter out duplicates by the non-maximum suppression algorithm to further optimize the test results.

In the following sections, we mainly discuss the two stages: candidate region proposal and feature extraction, and analyze the results of the experiments respectively.

4. Size-Scalable Candidate Region Proposal Algorithm

In the field of object detection, the traditional methods mostly employ the multi-scale sliding window method, which is an exhaustive search strategy. In order to guarantee the detection speed, only the simple feature of the candidate window can be extracted to be classified, which may lead to higher false detections. Therefore, a good region proposal algorithm plays an important role in the whole object detection process. In addition, since the size of remote sensing images is usually large, the region proposal algorithm needs to be scalable to the large image size. However, the segmentation-based selective search algorithm used in the classical deep learning-based object detection framework R-CNN needs to build a graph model: each pixel of the image acts as a node, and it includes a large number of similarity calculations and adjacent regions. This means that the algorithm needs to maintain many intermediate results, which has a heavy cost in time and in memory. Therefore, it becomes one of the bottlenecks of the whole framework. Besides, when the image size is very large, the problem is particularly critical, and memory is more likely to be insufficient in that case. The BING algorithm is based on the sliding window scoring mechanism with an accelerating optimization, and it is considered to be the fastest region proposal algorithm [25]. Furthermore, when increasing the image size, the speed and the memory usage of the algorithm increase linearly at most, and this is acceptable. In addition, because the candidate windows obtained by the BING algorithm contain probabilistic scores, we can select the appropriate number of candidate windows when necessary.

4.1. BING Algorithm

The binarized normed gradients (BING) algorithm is a very simple, but highly efficient region proposal algorithm, which is essentially a two-stage cascade classifier. The first stage uses a multi-scale sliding window to scan the image, and each window is scaled to a uniform size of 8×8, while a linear scoring model is used:

$$s_l = \langle w, g_l \rangle \tag{1}$$

$$l = (i, x, y) \tag{2}$$

where s_l is the filter score, g_l is the window feature, and $l, i, (x, y)$ are the location size and top left corner coordinates of the window, respectively.

The windows with a higher score for each size are selected as the possible candidate windows. During this period, the number of true objects in each size is counted, and the sliding window score is only applied to the size with the objects' number exceeding a certain threshold. The first-stage linear model w is trained by a linear SVM, and the NG features (norm of the gradients) of the real object windows and random sampling background windows are taken as positive and negative samples, respectively.

In the second stage, considering the different possibilities for different window sizes containing a object, such as the square window of 64×64 is more likely to contain an object than the 5×128 one, a score calibrator is trained for each size. We update the score for each candidate window:

$$o_l = v_i \times s_l + t_i \tag{3}$$

where $v_i, t_i \in R$ are the calibration coefficients that are learned for different sizes. This step is necessary only if you need to reorder the candidate windows obtained in one stage. The learning of the parameters v_i and t_i is also performed by the linear SVM.

The biggest contribution of the BING algorithm is the improvement of detection speed. In order to speed up the feature extraction and scoring process, the algorithm uses the idea of model binary approximation [26,27]. The linear model w is approximated by a set of binary basis vectors α_j:

$$w \approx \sum_{j=1}^{N_w} \beta_j \alpha_j \tag{4}$$

where N_w denotes the number of basis vectors, $\alpha_j \in \{-1,1\}^{64}$ denotes a base vector, and β_j denotes the corresponding coefficients.

Further, representing each α_j by using a binary vector and its complement: $\alpha_j = \alpha_j^+ - \overline{\alpha_j^+}$, where $\alpha_j^+ \in \{0,1\}^{64}$. These transforms allow subsequent scoring calculations for a binarized feature just using fast BITWISE andand bit countoperations, as shown in Equation (5):

$$w, b \approx \sum_{j=1}^{N_w} \beta_j \left(2a_j^+, b - |b| \right) \tag{5}$$

However, the NG features are real numbers, and how to binarize the NG features to speed up the calculation is one of the difficulties of the algorithm. This algorithm approximates the NG feature values (each saved as a BYTEvalue) using the top N_g binary bits of the BYTE values. Thus, a 64-dimensional NG feature g_l can be approximated by N_g binarized normed gradient (BING) features:

$$g_l \approx \sum_{k=1}^{N_g} 2^{8-k} b_{k,l} \tag{6}$$

Therefore, the score for whether or not an image window contains an object can be approximated as:

$$s_l = w, g_l \approx \sum_{j=1}^{N_w} \beta_j \sum_{k=1}^{N_g} 2^{8-k} \left(2a_j^+, b - |b_{k,l}| \right) \tag{7}$$

where $2^{8-k} \left(2a_j^+, b - |b_{k,l}| \right)$ can be computed using fast CPU atomic operation: BITWISE and POPCNTSSEoperators, which speeds up the calculation.

4.2. pBING Algorithm with Multiple Weak Feature Scoring

In practice, we found that the ambiguous objects were missed by the original BING algorithm. The possible reason for this phenomenon is that only a simple NG feature is used in the algorithm, and in order to speed up the calculation, the NG feature serves as the BING feature. Thus, some information would be lost in the process, which further reduces the distinguishing degree of the feature. However, in order to preserve the time and space complexity advantages of the BING algorithm, we are not able to use too complex features such as SIFT and HOG features due to the parallel optimization strategy of the algorithm. Thanks to the AdaBoost algorithm [28] that inspired us: integrating multiple weak classifiers is used to obtain a strong classifier. Therefore, we improved the BING algorithm by integrating multiple weak feature scoring models.

We improved the first stage of the candidate window scoring model, and the second stage of score correction was same as the original algorithm. For convenience, we name the improved BING algorithm as the pBING algorithm. Figure 2 shows the flowchart of the pBING algorithm, and the shaded part is our improvement. For each training sample, we extract multiple weak feature channels and train a scoring model for each weak feature channel; in the detection process, we integrate the score of multiple weak feature scoring models as the final score of the candidate window, in which a simple linear weighting method is adopted, and the weights are determined by the accuracy of each model. Each scoring model still uses an efficient linear SVM algorithm. The score of each candidate window in each scoring model is: $s_{kl} = \langle w_k, g_{kl} \rangle$, where s_{kl} denotes the score of model k for window l, w_k denotes the parameter of model k, and g_{kl} denotes the feature k of window l. The final window score for the first stage is as follows:

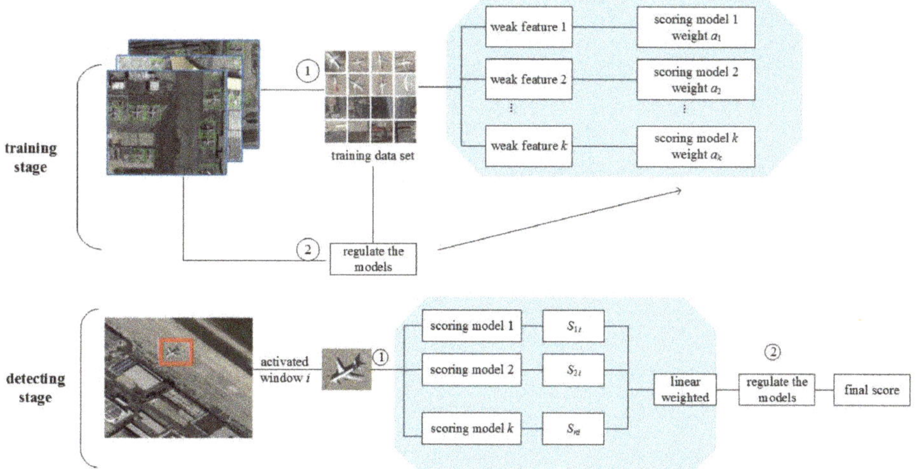

Figure 2. pBING algorithm flowchart.

The NG feature map used in the BING algorithm captures the edge intensity information of the original image by computing the gradient magnitude of each pixel, but this feature is simple and susceptible to noise. In this paper, the NG feature map is replaced by the Sobel feature map with better edge information capture. At the same time, the local binary pattern (LBP) feature map and the difference of Gaussians (DoG) feature map are introduced.

The Sobel feature map uses 3×3 Sobel operators to convolute the original image to obtain the horizontal and vertical direction of the approximate gradient, as shown in Formula (8). Then, it computes the gradient amplitude through Formula (9).

$$
\begin{aligned}
G_x &= \begin{bmatrix} -1 & 0 & +1 \\ -2 & 0 & +2 \\ -1 & 0 & +1 \end{bmatrix} \otimes A \\
G_y &= \begin{bmatrix} +1 & +2 & +1 \\ 0 & 0 & 0 \\ -1 & -2 & -1 \end{bmatrix} \otimes A
\end{aligned}
\tag{8}
$$

$$
G = \sqrt{G_x^2 + G_y^2}
\tag{9}
$$

where A represents the original image matrix, G_x and G_y represent the horizontal and vertical gradient of the image, respectively, and G represents the gradient magnitude matrix, which is the obtained feature map. $\otimes$ represents the convolution operation.

LBP can be used to describe the local texture features, often used for face classification, pedestrian detection [29], and so on. We can get the LBP feature map by computing the LBP code for each pixel point. In order to simplify the calculation, we use the simplest 3×3 LBP operator to calculate in the gray image.

The DoG feature map is obtained by subtracting two different degrees of blurred images from the original image. The blurred image is obtained by the Gaussian kernel convolution of different standard deviation parameters on the gray image. Two Gaussian blurred image subtractions can increase the visibility of edges and other details, and the DoG algorithm does not enhance noise because Gaussian blur suppresses high-frequency noise. The two-dimensional Gaussian kernel function is defined as follows:

$$G_{\sigma_1}(x,y) = \frac{1}{\sqrt{2\pi\sigma_1^2}} exp\left(-\frac{x^2+y^2}{2\sigma_1^2}\right) \tag{10}$$

Then, the Gaussian filtering of the two blurred images is expressed as:

$$\begin{aligned} g_1(x,y) &= G_{\sigma_1}(x,y) * f(x,y) \\ g_2(x,y) &= G_{\sigma_2}(x,y) * f(x,y) \end{aligned} \tag{11}$$

Therefore, the DoG feature map is obtained by subtracting the two images $g_1(x,y)$ and $g_2(x,y)$, where σ_1 and σ_2 are two Gaussian kernel parameters, respectively. When the DoG is used for different purposes, the ratio of the two Gaussian kernel parameters is different. When used for image enhancement, usually $\sigma_2{:}\sigma_1$ is set to 4:1 or 5:1. In this paper, $\sigma_2 = 2.0$ and $\sigma_1 = 0.5$.

4.3. Experiments

4.3.1. Dataset and Evaluation

Two different scales of remote sensing image datasets are used in this paper:

(1) SMALL-FIELD-RSIs: Each image was cropped from the Google Earth software, and all airplane objects were manual marked. As shown in Figure 3, for each airplane object, a minimum enclosing rectangle was used. Each bounding box was represented by its top-left and bottom-right coordinates: (x_1, y_1, x_2, y_2). The dataset contained 980 high-resolution optical remote sensing images, and the image size was about 1300 × 800 pixels, with the spatial resolution of the image being about 0.6 m. A total of 7452 airplane objects were marked.

(2) LARGE-FIELD-RSIs: Each image was downloaded from the commercial professional remote sensing software, and the map level was 19, the scale 1:2257, the data source being from QuickBird satellite, and the spatial resolution 0.6 m. Compared to Google Earth software, the software can download any size of high-resolution optical remote sensing images with no watermark or other labels, and the images are relatively clearer. There were 110 images, and the average size of all images was about 5000 × 5000 pixels. A total of 3380 airplane objects were manually marked. As shown in Figure 4, the left is a high-resolution remote sensing image of an airport (International Airport, Shenzhen, China), and the right is an enlarged view of the red area on the left. As can be seen, for the entire image, the airplane object was very small. When zooming out to a relatively small size, the airplane objects were almost invisible. Table 1 shows the statistical details of the two datasets.

Figure 3. The annotation method of the airplane object.

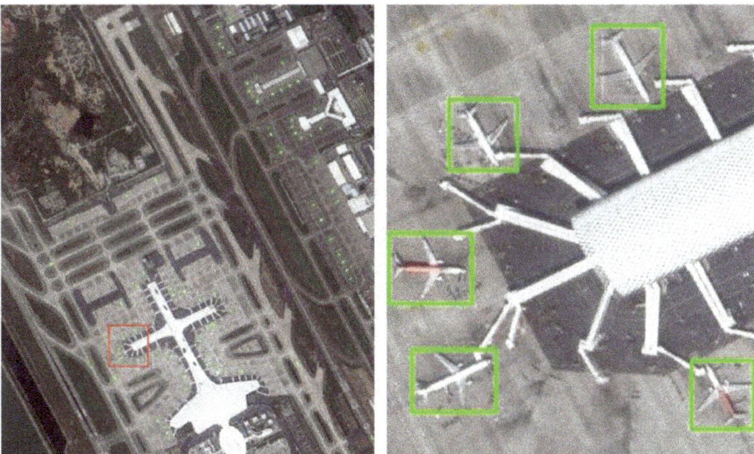

Figure 4. The high-resolution remote sensing image of Baoan Airport.

Table 1. The experimental dataset.

Dataset Name	Spatial Resolution	Image Size (Pixels)	Number of Images	Number of Objects
SMALL-FIELD-RSIs	0.6 m	1300×800	980	7452
LARGE-FIELD-RSIs	0.6 m	5000×5000	109	3380

In the region proposal algorithm, we evaluated the improved algorithm by the DR-#WIN curve and MABO-#WIN curve, where DR refers to the detection rate and MABO refers to mean average best overlap, while #WIN refers to the number of candidate windows proposed.

4.3.2. Results and Analysis

The experiments in this section were divided into two parts: (1) we briefly analyzed the applicability of the selective search algorithm for candidate region proposal in remote sensing images; (2) we evaluated the result of the BING algorithm and pBING algorithm. In the experiment, each dataset was divided into a training set and a test set, where the test set was 20%. The DR and MABO below were the average of all the test image detection results. Table 2 shows the average number of candidate windows and running times of the selective search algorithm on different sizes of remote sensing images. This experiment was performed in fast mode using only two color spaces and two similarity functions. As in the remote sensing image, when the spatial resolution of the image is determined, the size range of the object in the image can be determined. Using this prior knowledge, we filtered out the irrational size of the candidate windows, and the filtered results are shown in the third row of Table 2. It can be found that the average number of candidate windows generated by the algorithm increased significantly, and the running time of the algorithm increased sharply with the increase of the image size. In addition, in the experimental process, we found that when the image size increased to 2000×2000 pixels, the machine was out of memory, and the MATLAB compiler was in a stuck situation. Therefore, this algorithm has a poor scalability for image size.

Table 2. The performance analysis for the selective search algorithm.

remote image scale (pixel)	500×500	800×800	1000×1000	1500×1500
average number of candidate windows without constraints	3161	7123	9060	13,810
average number of candidate windows after filtering	3039	6270	8469	12,782
average computing time (s)	2.7	7.6	14.3	29.8

In summary, the selective search algorithm had higher complexity, and the number of candidate windows was larger, which was not scalable for the image size. The following are the performances of BING algorithm and pBING algorithm when extracting candidate regions in two high-resolution remote sensing images.

As shown in Figure 5, when the number of candidate windows was 1000, the DR of the pBING algorithm was 97.21% on the SMALL-FIELD-RSIs, which was higher than the original algorithm (95.74%). The MABO score increased from 63.43–65.30%, which indicates that the improved algorithm had a better quality of the candidate region proposal on the remote sensing images. In addition, it shows that when the number of candidate windows was 2000, the detection rate of pBING algorithm in this dataset was as high as 98.9%. Therefore, as the input of the second stage, the number of candidate windows in this stage can be 2000, that is, for each image, we output the first 2000 candidate windows of the pBING detection results to further classify.

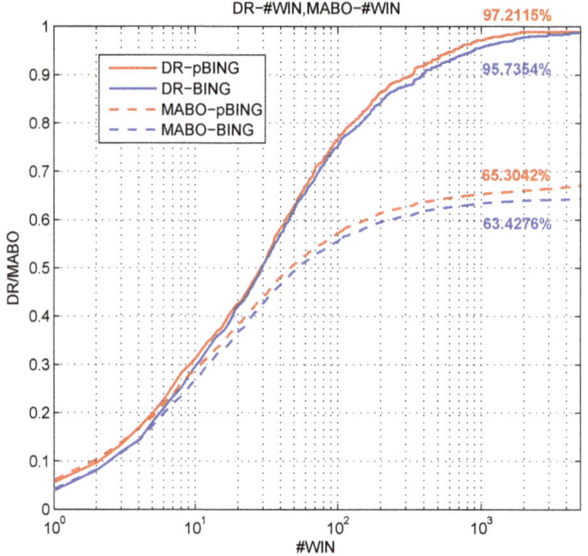

Figure 5. Tradeoff between the number of windows (#WIN) and DR/MABO on the dataset SMALL-FIELD-RSIs.

Figure 6 shows the performance of the two algorithms on dataset LARGE-FIELD-RSIs. Since the image of the dataset is relatively large, we mark DR and MABO of the algorithm when the given number of candidate windows is 8000. Figure 7 shows the details. It appears that DR and MABO are increasing as the number of candidate regions increases. Obviously, that is because that a remote sensing image with large size usually has more objects, and search space of objects' possible locations significantly increases, and then more candidate windows are needed in order to achieve a certain detection rate. As shown in Figure 7, in order to ensure that the follow-up classification to achieve a certain rate of recall, in this stage we select the first 9000 candidates as the input of the second stage. However, since the image size is not fixed in this dataset: the average size of images varies from 2000 pixels to 9000 pixels. In order to further reduce the number of candidate windows, we select the number of output candidate according to the image size.

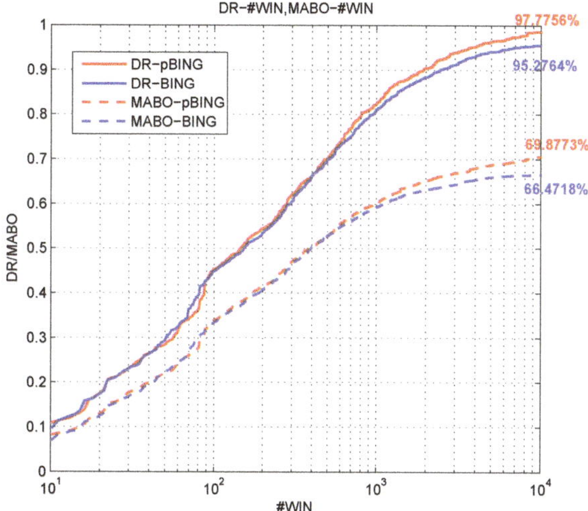

Figure 6. Tradeoff between #WIN and the detection rate (DR)/mean average best overlap (MABO) on the dataset LARGE-FIELD-RSIs.

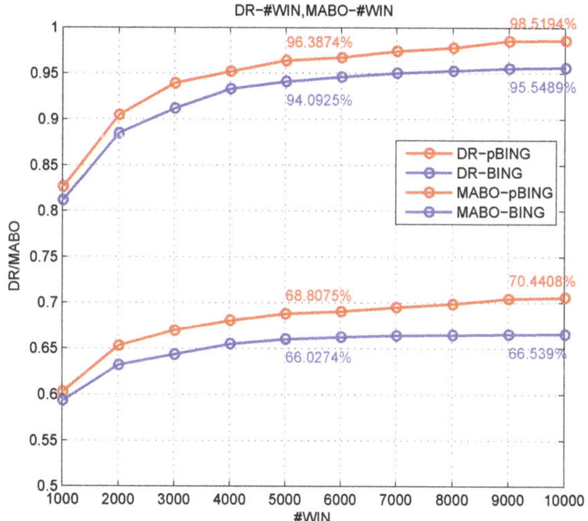

Figure 7. Tradeoff between #WIN and DR/MABO on the dataset LARGE-FIELD-RSIs when #WIN varied from 1000–10,000.

Figure 8 shows the average running time of the BING and pBING algorithms on two datasets to obtain the candidate region proposal. It shows that when the size of the remote sensing image was large, the time required to obtain the candidate region was greatly increased, but the test time was still less than 1 s, indicating that the time complexity of the algorithm did not increase sharply with the image size expanding. In addition, it reveals that the pBING algorithm took more time to acquire candidates, about three-times slower. However, the algorithm itself is very fast; even if the time expansion of the original was three-times, it is still within the acceptable range.

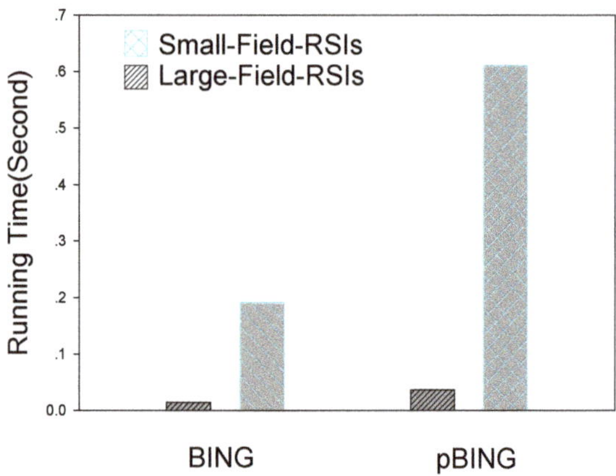

Figure 8. Average running time of the BING algorithm and the pBING algorithm on different datasets.

5. Deep Feature Extraction with Context Scene Feature Fusion and Detection

5.1. Feature Extraction Algorithm CNN

After obtaining the candidate windows that may contain objects, we needed to further determine whether the windows actually contained objects, that is the detection task can be converted into the classification task. To classify the candidate windows, the distinguishing features of each window need to be extracted first. As a base problem of the computer vision field, many classic feature extraction methods have been proposed, such as color histogram, gray level co-occurrence matrix, Haar-like features, HOG features, and SIFT features. These features were designed by scholars according to experience and related prior knowledge and have achieved good results in many fields. However, at the ImageNet LSVRC-2010 competition, Hinton et al. [13] used deep learning to extract features automatically, reducing the top-5 error rate of the image classification task by 15.3%, far beyond the algorithms with traditional feature extraction methods. The reason is that the traditional feature representation can only obtain the shallow features of the image, and it has certain limitations; however, deep learning can automatically learn higher level feature representation from the original image, which has better distinguishing ability and versatility.

Deep learning is developing rapidly, and the most regularly used in the field of object detection is the convolutional neural network (CNN). Compared with traditional neural networks, CNN achieves weight sharing by introducing the convolution layer, which makes the network structure sparser and reduces the complexity of the model. The convolutional algorithm can obtain the feature map of different aspects by difficult convolution kernel parameters. The convolution operation makes the obtained feature map have translation invariance. The convolution layer is usually followed by a pooling layer, which downsamples the obtained feature map layer, preserving useful information while reducing the amount of data to be further processed. The pooling layer usually uses max-pooling to get the maximum response of local features, so that the obtained features have better rotation and light invariance. CNN can learn a higher level feature representation from low-level features through a deep network structure by stacking multiple convolution layers. For different objects, the learned low-level features by CNN differ slightly, which are usually some edge information, and through multi-layer network learning, abstract features of different objects can be obtained finally.

After the multiple convolution layer, a fixed-length feature vector is obtained through the full-connection layer and output to the classifier. Generally, the softmax classifier is employed in the

CNN, and it outputs the probability that the image belongs to each class. CNN combines feature learning with the classification task, which makes the extracted feature more task related.

Model of Neurons

CNN is a type of multi-layer sensor for which each layer is composed of a two-dimensional plane, and each plane is composed of various independent neurons. In the network, some simple and complex cell are marked as cell C and cell S, which is inspired by the vision concept from biology. In the visual cortex, there are two kinds of related cells, simple ones (cell S) and complex ones (cell C). Cell S responds to the stimulation for the modes like margins of images in its maximum receptive field, while cell C has a bigger receptive field, which can locate the modes of stimulation in a spatial way. The merging of C cells forms convolutional layers, denoted as U_C, while the merging of S cells forms the downsampling layers, which can be denoted as U_S. Any intermediate layer in the network is composed of the S-layer and C-layer with series connection. Regularly, U_S is the layer to extract the feature, while U_C is the layer of feature mapping.

In CNN, only the input of cell S is variable, while other inputs are fixed. The first layer can be denoted by $U_{sl}(k_l, n)$, which means a cell s output on the k_l S-plane, and a cell c output on the k_l C-plane can be denoted by $U_{cl}(k_l, n)$. n represent two-dimensional coordinates.

$$U_{sl}(k, n) = r_l(k) \times$$
$$\varphi \left[\frac{1 + \sum\limits_{k_{l-1}}^{K_{l-1}} \sum\limits_{v \in A_l} a_l(v, k_{l-1}, k) U_{cl-1}(k_{l-1}, n + v)}{1 + \frac{r_l(k)}{r_l(k)+1} b_1(k) U_{vl}(n)} - 1 \right] \tag{12}$$

In the above neuron model formula, $a_l(v, k_{l-1}, k)$ and $b_l(k)$ represent the connection coefficients of positive input and negative input, respectively; $r_l(k)$ is a constant that controls the option of feature extraction; the bigger it is, the worse is at tolerating noise and feature distortion.

Process of convolution: Employ a trainable filter f_X to process the convolution on input images (the c_1 layer is the input, and the inputs of subsequent layers are the outputs of the forward layer), based on an activation function (usually sigmoid) with a offset b_X, to get convolutional layer C_X. M_j is the value of the input feature map:

$$X_j^l = f\left(\sum_{i \in M_j} X_i^{l-1} * k_{ij}^l + b_j^l \right) \tag{13}$$

Down-sampling process: Each m adjacent pixels sum up to be one pixel (mcan be set), and use j as the weight, add offset b_j, then use the activation function sigmoid to generate feature mapping. The mapping from one plane to another plane can be a convolutional operation, and the layer can be a fuzzy filter functioning as double feature extraction. Spatial resolution decreases with the hidden layer going forward, while the plane number increases for better feature extraction. For the sampling layer, if there are N input features, then there would be N output features, but the size of each feature changes. The details formula are as follows. $down()$ denotes the down-sampling function.

$$X_j^l = f(\beta_j^l \times down(X_i^{l-1}) + b_j^l) \tag{14}$$

5.2. Context Information

Considering that a specific kind of object can only appear in certain scenes, this a priori knowledge is particularly obvious in remote sensing images; for example, ships can only appear in the port or the sea, and an airplane can only appear on the parking apron or runway. In regular images, the context of the same kind of object varies greatly as a result of different locations, angles, or distances when taking photos. As shown in Figure 9, the car's context scene may be an avenue, the ground, a house, or even water. However, in remote sensing images, the spatial relationship between the object and the background is relatively fixed due to the fixed angle and height of satellites. In addition, as shown in Figure 10, compared to regular images, the objects in high-resolution remote sensing images are usually very small, the details having useful information are few, and objects may not be clear. Therefore, in the remote sensing image, the context scene information of the candidate window may be helpful to determine the objects.

Figure 9. Example of the context scene of a car in regular images.

Figure 10. Airplane objects in high-resolution remote sensing images.

5.3. Deep Feature Extraction and Context Scene Feature Fusion

As objects in remote sensing images have a strong background context, for example, the airplane objects may appear on the runway, parking apron, but not in the forest, the port, etc., and the airplanes may be side by side with another one, how to describe this a priori knowledge and apply them to the detection algorithm comprise a difficult problem. In order to utilize contextual information, conventional algorithms usually define structure and matching constraints, but this manual approach is too subjective and not extensible to different problems.

In this paper, we extract the scene context feature of the candidate window and fuse this feature with the candidate window feature to classify, while making the classifier automatically learn the constraint between the object and the context scene.

In Section 2, it is indicated that the feature extraction based on CNN can avoid the unmanageability and subjectivity of the manual design feature and can obtain a deeper feature representation of the object. Therefore, the feature of the object window and the context scene are both extracted from CNN. For the convenience of description, we named the two networks as objNetand sceNet. In the training process, we trained objNet and sceNet respectively. In the testing process, as shown in Figure 11, we extracted the feature of the candidate window, this being the context scene by the corresponding network, and then fused the two kinds of features to classify. For the feature fusion, taking into account the detection rate, we simply merged the two features and used the faster linear SVM classification.

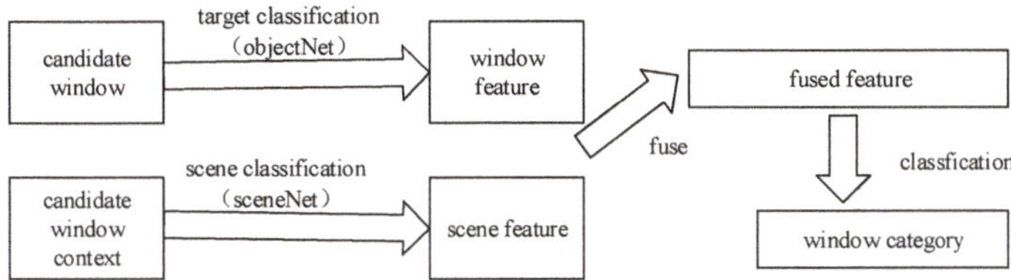

Figure 11. Window feature and context scene feature fusion.

For the scene context of the candidate region, as shown in Figure 12, we extended the candidate region from the center point and obtained a 256 × 256 pixel region as its context scene. When the object was at the edge of the image, we made the scene bounding box a minimum translation, so that it did not exceed the image area.

Figure 12. The definition of the context scene for the candidate region.

For the sceNet network, we used the classic AlexNet network structure, and modified the final output layer number. As shown in Figure 13, the network consisted of eight layers, of which the first five layers were the convolution layer, which can be regarded as multi-stage feature extraction. The latter three layers were fully-connected layers. The parameters of each layer are shown in Table 3. In the detection process, the scene bounding box of the candidate region was input to the network for feed forward calculation, and the output of the fifth layer was used as the scene feature of the candidate region.

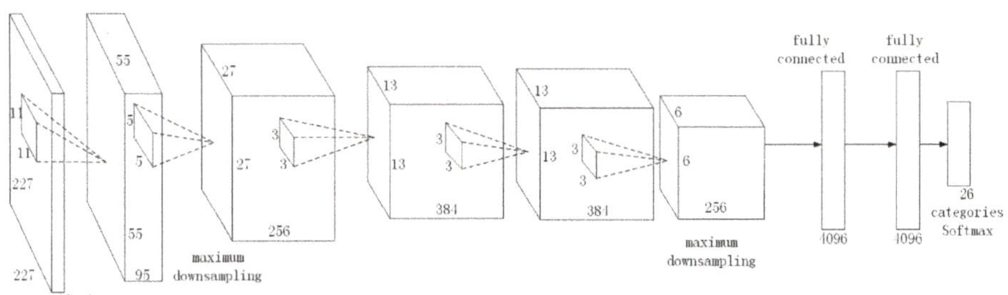

Figure 13. sceNet network structure.

Table 3. The parameters of each layer in the sceNet network.

		1st Layer	2nd Layer	3rd Layer	4th Layer	5th Layer
convolution layer	window size	11×11	5×5	3×3	3×3	3×3
	number of convolution kernels	96	256	384	384	256
	stride	4	1	1	1	1
	pad	0	2	1	1	1
pooling layer	window size	3×3	3×3	–	–	3×3
	stride	2	2	–	–	2

For the feature extraction of the candidate window, the structure of the objNet network is shown in Figure 14. Taking into account that the airplane object is very small and the sizes are more concentrated in 64×64 pixels, so the network input layer using 64×64, other sizes of candidate windows need to be scaled. Since the object to be detected in this paper is only an airplane, the output of the network is two classes: airplane or background. Compared with the classification of the remote sensing scene using AlexNet, the input size of the objNet network was smaller, and the outputs were fewer, so the network can be considered to need relatively simple feature representation when performing object discrimination. Therefore, when designing the objNet network, we modified the size of the convolution and pool layer windows and reduced the number of convolution cores and neurons in the fully-connected layer. The simplified network had fewer parameters and could reduce the risk of over-fitting properly. However, this does not mean that the network was as simple as possible. In practice, it is found that when the network is too simple, the network classification accuracy is high in the training phase, but it is not good when using the network for detection. The possible reason is that the oversimplified network is not strong enough to abstract the features, leading to poor generalization ability. As shown in Figure 14, the final objNet network consisted of eight layers, and the first five layers were the convolution layers, which can be seen as multi-stage feature extraction. The latter three layers were the fully-connected layers, which can be seen as a classifier. The parameters of each layer are shown in Table 4. In the detection process, the candidate region was input to the network for feed forward calculation, taking the output of the fifth layer of the pool as the feature of the candidate region.

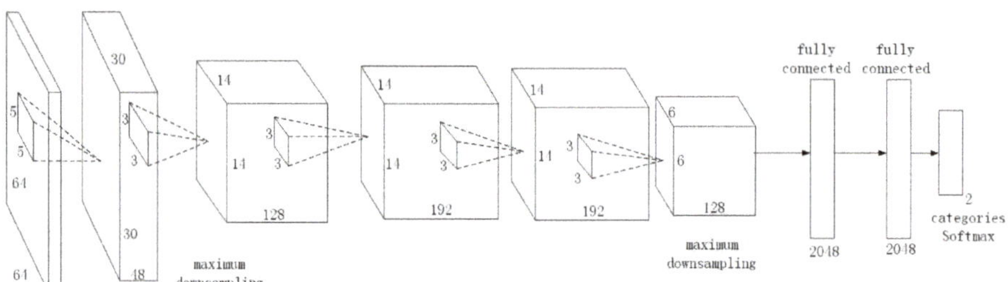

Figure 14. objNet network structure.

Table 4. The parameters of each layer in objNet network.

		1st Layer	2nd Layer	3rd Layer	4th Layer	5th Layer
convolution layer	window size	5×5	3×3	3×3	3×3	3×3
	number of convolution kernels	48	128	192	192	28
	stride	1	1	1	1	1
	pad	0	0	1	1	0
pooling layer	window size	2×2	2×2	–	–	2×2
	stride	2	2	–	–	2

5.4. Training of CNN and Transfer Learning

5.4.1. The Training Process of CNN

The training process of the CNN is shown in Figure 15. It was mainly carried out by iterative updating of the forward propagation and back propagation stages. Specifically, the first stage randomly selected a sample (X_b, Y_b) from the training set and input X_b to the network for feedforward calculation, then obtained the predicted output O_b. At this stage, the data were transformed step-by-step from the input layer through a series of hidden layer levels and finally transmitted to the output layer, which is essentially the process of input multiplication with the weight matrix for each layer, as in Formula (15):

$$O_b = F_n \left(\ldots \left(F_2 \left(F_1 \left(X_b W^1 \right) W^2 \right) \ldots \right) W^n \right) \tag{15}$$

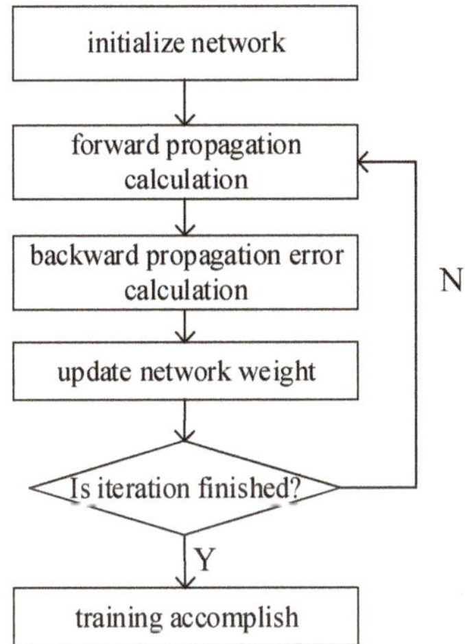

Figure 15. The training process of CNN.

In the second stage, namely backward propagation stages: first, we calculated the error between predicted output O_b and real label Y_b, and then, we continued to pass the error back to the front layer, and each layer updated its weight matrix by minimizing the error of small methods according to the current error situation. For the CNN, the weights usually are updated by the mini-batch gradient descent method, which is an optimization algorithm between the batch gradient descent and the stochastic gradient descent method. When the data volume is large, the mini-batch preserves the advantages of speed in the stochastic gradient descent method, while avoiding the problem of severe congestionin the stochastic gradient descent method. The mini-batch gradient descent method updates the parameters iteratively by randomly selecting small batches of data, as shown in Formula (16), where m represents the number of training samples per iteration in parallel, which is limited by the memory size. For an objNet network, the size of the input sample is small and the network structure is relatively simple, so the value was 1024, and for the sceNet network, it was 64 in the experimental environment used in this section.

$$\omega_k \to \omega_k' = \omega_k - \frac{\eta}{m} \sum_j \frac{\partial C_{X_j}}{\partial \omega_k}$$
$$b_l \to b_l' = b_l - \frac{\eta}{m} \sum_j \frac{\partial C_{X_j}}{\partial b_l} \tag{16}$$

where ω_k and b_l are weight parameters and bias parameters in the network, respectively; η is the learning rate; C_{X_j} is the cost loss of the sample X_j.

Before training the network, appropriate training data should be found. For the scene feature extraction network sceNet, we needed to use the remote sensing scene classification data to train. For the remote sensing scene classification, there are two standard datasets: UCMerced-LandUse [29] and WHU-RS [30]. To increase the training set, this paper combined the two datasets, and a detailed description of this dataset can be seen in Section 5.4.1. For the object feature extraction network objNet, we needed to use the sub-images containing the object as the positive sample and the sub-images without the object as the negative sample. In order to make the training process and detection process the same distribution, the network used the pBING algorithm's output on the training set as the training data, in which a candidate window having an ≥ 0.5 intersection-over-union (IoU) overlap with a ground-truth box was labeled as the positive sample and the rest as the negative sample.

5.4.2. Transfer Learning

Although CNN can automatically extract deep features, the network has many parameters to optimize and usually requires large-scale data (big data) in order to form a better network; if not, it easily over-fits. While the reality is the deficiency of labeled training datasets of remote sensing images, in order to fill the gap and make the model have stronger generalization ability, many methods have been proposed, such as data augmentation technology, dropout [31], and so on. However, these methods are still not enough for small remote sensing dataset. Considering the abundant regular image datasets available, we employed transfer learning technology for deep learning, which can break the" deep learning with big data" limitation.

For deep learning, an appropriate applicable transfer learning method is model transfer: firstly, pre-training network parameters through the source field data, then applying these parameters in the object domain, and finally, fine-tuning the network parameters to get better performance. If transfer learning is not employed, it demands initializing the network parameters and then starting to train the entire network using training data. An inappropriate initialization will make the network convergence slow and easily fall into the local minimum. In addition, because of the deepening of the network layer, the problem of gradient disappearance easy occurs: when the hidden layer near the output layer has been trained well, the parameter update near the input layer becomes slow or even stagnates. However, this does not mean that the network is optimal, because the first few layers of the network may not learn anything and may be just a random combination and numerical transformation of the input, but not really a dissociation of features, resulting in the entire network being a linear transformation of higher levels at work. Especially for high-dimensional data such as images, the network does not have good feature dissociation due to the degradation of the lower layer, so that the network is only performing local numerical learning on the input image and the model easily over-fits. Once the input image has changed, such as the direction or color of the airplanes, the network may not identify the object. This problem is mainly due to the fact that in the deep network, the learning rate of different layers is not the same, and the closer to the output layer, the faster the learning rate. This is because the gradient loss of the front layer is based on the product of the gradient loss, and when the number of layers is larger, the gradient loss becomes smaller and smaller.

The problem of gradient disappearance in deep learning networks is an essential problem brought by gradient descent, which is a big obstacle in deep learning. When the training data are large, the parameters can be initialized by a Gaussian distribution or other optimization methods, and the whole network can be fully trained by adjusting the learning rate and some regularization methods, as well as training for a long enough time. However, when the training data are small, the gradient disappearance and over-fitting problem will become more serious. In addition, since the problem in the deep learning network is mainly that the first few layers may not be fully trained, if the first few layers can be initialized by the parameters of other fully-trained networks, which puts the network in a better initial state, this would contribute to the optimization of the network and could accelerate the training process of the network. From another point of view, the first few layers of the deep learning network usually learn the edges of the image, the color, the texture, and other primitive features of the images, which for many visual tasks are typical. Therefore, the network parameters of lower layers can be shared among different image classification tasks. this is equivalent to transferring the feature extraction knowledge carried by these parameters to the object domain, and then, we only need to continue training the network through the training data of the object domain, that is to correct parameter deviation between the object domain and the source domain.

The following details the transfer learning scheme used in the two CNNs used in this paper.

For pre-training of the remote sensing scene classification network sceNet, we can use the scene recognition task in the regular image field as the source field. For the regular scene recognition task, Zhou et al. [32] established a large-scale regular scene dataset, Places, and published the trained network model. By practice and theory, this paper has demonstrated that the deep features learned on the Place dataset are more effective compared with those learned on the ImageNet 2012 dataset. However, for the remote sensing scene classification task, it is necessary to further validate whether the transfer effect of this model is optimal. Figure 16 shows the flowchart of transfer learning on the sceNet network. Firstly, we used the regular scene classification task to pre-train sceNet and then transferred the learned parameters to the remote sensing scene classification. The last layer of parameters was not transferred, just random initialization, and then, we used the remote sensing scene classification data to continue to train the network; then, the network learned the parameters of the last layer through back propagation and corrected the transferred parameters of the first few layers.

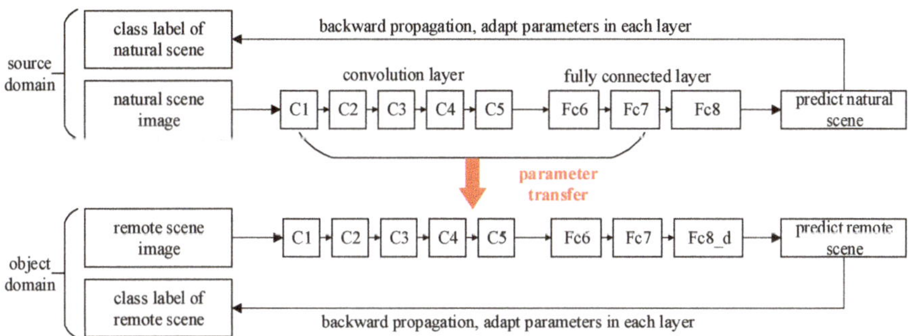

Figure 16. The transfer learning flowchart of the sceNet network.

5.4.3. Training the Classifier and Hard Negative Mining Method

For the classifier, we employed the simple linear SVM. For the training of the SVM classifier, we chose the real object sub-images as the positive sample and the sub-images with $\geq$0.3 IoU overlap with a ground-truth box as negatives. For each training sample, we extracted the object feature by the objNet network and the corresponding context feature by sceNet network, and then, we merged the two features as the input of the SVM classifier.

In general, when training a classifier, the more training samples, the better. However, there is a regular problem in the field of object detection, which is the extreme imbalance of positive and negative samples, that is the positive samples with objects are relatively small, while negative samples with background are very numerous. For remote sensing images, especially large-sized images, the problem is more obvious. Too many negative samples will lead to a very slow training process for classifying algorithms, and will even be detrimental to the performance of the final classifier.

For example, for SVM, many negative samples far away from the separating plane are almost useless for optimization. In addition, too many negative samples will make the algorithm's memory requirement too large. If a negative set with a similar number as the positive set is randomly selected, the algorithm cannot guarantee the best effect on the whole training set. If manually selecting the negative set, the cost is too large, and the subjectivity is too strong. Therefore, it is very important to search for a small representative negative set in the negative sample space. The usual strategy is to initialize a small hard negative set $C_t \in D$ randomly (D denotes the entire negative sample space), train an initial model β_t with all positive samples, and classify the negative sample set C_t. Remove the easy negative sample while searching for hard negative samples from D to add to the C_t, until the memory limit or a threshold L. The iterative updating model β_t and hard negative sample set C_t, until C_t no longer changes or the iteration number reaches a certain limit, stop training. In practice, the hard negative mining method converges very quickly; usually, only a single pass over all images is required.

5.5. Results and Analysis

5.5.1. Dataset

The two high-resolution remote sensing datasets used for object detection are described in Section 4.3.1. This section introduces the remote sensing scene classification datasets used in the training of sceNet networks and the two regular image datasets used for transfer learning. For remote sensing scene classification, there are two public datasets: UCMerced-LandUse [33] and Dataset-WHU-RS [33]. UCMerced-LandUse contains 21 different scene categories, each category containing 100 high-resolution remote sensing images of 256 × 256 pixels. The Dataset-WHU-RS dataset contains 19 scene categories, a total of 950 images of 600 × 600 pixels. In order to expand the training samples, this paper will simply divide one remote sensing image with 600 × 600 pixels into nine 256 × 256 pixel sub-images. In the end, the two remote sensing datasets are merged together, and the data of the same category are merged. In addition, fine-grained similar scenes such as sparse density residential, medium density residential area, and intensive residential area are merged. After merging, there were 26 scene classes, averaging about 320 images per class. In addition, in order to further increase the training data, simple horizontal, vertical flip operations were used.

Next, the regular image scene classification dataset Places used in the transfer learning of sceNet network is introduced briefly. This dataset Places is a large-scale natural scene dataset, containing 205 categories and a total of 2.5 million images.

The data used in the transfer learning of objNet network were mainly extracted from the ImageNet 2012 dataset, in which positive samples were all images in two categories of airplane and military airplane and some airplane images crawled from the Internet, a total of 12,300 images; for negative samples, we picked a category that may appear in the remote sensing images, such as ships, harbors, mountains, etc., and removed the other 980 categories, such as sharks, hens, caps, etc., which might be useless for object identification in remote sensing images. Figure 17 shows examples of the positive and negative samples of the dataset. For the convenience of the following description, this dataset is called NATURE-PLANE.

5.5.2. Environment and Evaluation

The experiment in this section was performed on Caffe. Caffe is widely used in the deep learning domain because of its advantages of being clear, simple, fast, and fully open source. The platform had two NVIDIA GeForce GTX 980 video cards, 16 GB memory, CPU i5-4460. For a detection result with IoU overlap with a real object coincidence degree no less than a threshold (usually set to 0.5), the detection result is considered correct; besides, if there are multiple detections, then only one is considered right, while the rest are false detections. In this paper, we used the precision and recall curve (PR curve) and the average precision (AP) to evaluate the detection performance synthetically. The evaluation method and code used the PASCAL VOC2007 standard. Accuracy and recall are defined as Formulas (17) and (18), respectively:

$$P = \frac{TP}{TP + FP} \tag{17}$$

$$R = \frac{TP}{TP + FN} \tag{18}$$

where TP is true positive, the number of true boxes, that is the number of objects correctly detected; FP is false positive, the number of false positives, that is the number of false detection results; FN is false negative, the number of false negative cases, that is the true number of objects that were missed.

Figure 17. Examples of the NATURE-PLANE dataset used in the transfer learning of the objNet network.

The average detection accuracy of AP can be measured by a single value, which is a representative comprehensive evaluation of the index, called the area under the PR curve, as shown in Formula (19).

$$AP = \int_0^1 P(R)\, dR \tag{19}$$

$$IoU = \frac{DetectionResult \cap GroundTruth}{DetectionResult \cup GroundTruth} \tag{20}$$

5.5.3. Result Analysis

In order to prove the validity of the transfer learning, this section firstly gives the classification accuracy of the object and scene feature extraction network in the training process and then gives the influence of the feature extraction on the detection effect before and after the transfer learning. Furthermore, the effectiveness of scene feature fusion is illustrated by contrasting the detection performance before and after the context scene feature fusion. Finally, we compare the other algorithms to prove the validity of the proposed remote sensing object detection algorithm based on deep learning with scene feature fusion.

Because the sceNet network uses the classic AlexNet network structure, there are many trained parameter models based on the network, which can be used for transfer learning. During the training process, the parameter models of AlexNet, CaffeNet, Places205-AlexNet, and Hybrid-AlexNet were used for transfer learning in this paper. AlexNet and CaffeNet were trained on the ImageNet2012 dataset, and CaffeNet has a very similar architecture to AlexNet, except for two small modifications: training without data augmentation and exchanging the order of pooling and normalization layers. Places205-AlexNet is a parameter model trained on the Places dataset. Hybrid-AlexNet was trained on the dataset combining the Place dataset with the ImageNet2012 dataset for a total of 3.6 million images in 1183 categories. Table 5 shows the classification accuracy of the remote sensing scene using different transfer learning models, where "AlexNet-RSI" indicates the learned network model without transfer learning and "xx-TL" denotes the network transferred from different models.

Table 5. The classification accuracy of the network transferred from different models.

	AlexNet-RSI	AlexNet-TL	CaffeNet-TL	Places205 -AlexNet-TL	Hybrid -AlexNet-TL
accuracy	89.76%	94.04%	93.87%	92.68%	**94.81**%

Table 5 reveals that the accuracy of the network trained with transfer learning was much higher than that of a network trained directly using remote sensing scene data. After transfer learning, the accuracy of the Hybrid-AlexNet network trained on the Place and ImageNet 2012 datasets was the highest, so we used the Hybrid-AlexNet-TL model to extract the feature of the object context scene. In addition, by visualizing the convolution kernel parameters of the first convolution layer, as shown in Figure 18, this shows that the convolution kernel of the network with transfer learning learned more edge features, and without transfer learning, the first layer of the network simply learned some simple fuzzy color information.

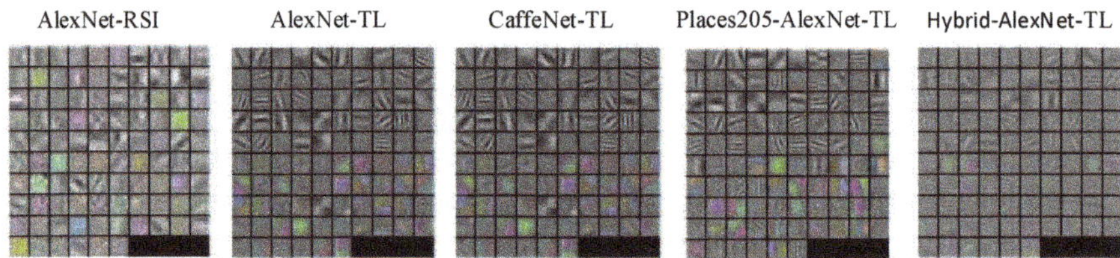

Figure 18. Visualization of the first-level convolutions in different network models.

For the training of object classification network objNet, because the network structure was designed in this paper, there was no trained model for transfer learning, so it was necessary to pre-train the transferable model parameters. Therefore, we used the NATURE-PLANE dataset introduced in Section 5.4.1 to pre-train objNet. However, in practice, it appears that if we pre-train the objNet directly using ImageNet2012's complete data and resume training using the NATURE-PLANE dataset, a better classification result could be obtained. Table 6 lists the classification accuracy of different pre-trained objNet networks, where objNet-RSI denotes the network model obtained without using transfer learning.

Table 6. The accuracy of objNet based on different transfer learning methods.

	Accuracy	
	SMALL-FIELD-RSIs	**LARGE-FIELD-RSIs**
objNet-RSI	93.51%	94.24%
objNet-TL1	96.63%	95.52%
objNet-TL2	97.23%	96.86%

After the training of the objNet network and sceNet network was complete, we used the two networks for feature extraction and classification detection. We first compared the detection performance before and after using transfer learning in objNet when there was no scene feature fusion. Then, to fuse the scene features, in the fusion, we also compared the effect of sceNet before and after transfer. That is, there was in total four groups of experiments, and the four groups of experiments had a progressive relationship, in which the fourth set of experiments was our proposed algorithm. In order to simplify the following description, we named each experiment and the configuration of each set of experiments as listed in Table 7.

Table 7. List pf the configurations of each experiment.

Algorithm	Scene Feature Fusion	sceNet Transfer Learning	objNet Transfer Learning
objNet	×	–	×
objNet-TL	×	–	√
objNet-TL-sceNet	√	×	√
objNet-TL-sceNet-TL (ours)	√	√	√

Figures 19 and 20 show the comparison of the results of the four experiments on two different dataset sizes. From the results of the experiments objNet and objNet-TL, it was revealed that the transfer learning of object the feature extraction network could improve the whole detection performance significantly. Before transfer learning, the accuracy of the curve decreased rapidly with the recall increasing, and the curve decreased slowly after transfer learning, which shows that the extracted feature of network after transfer learning made the classifier discriminate better, which means the transfer learning was effective. It can be concluded from the experiment objNet-TL-sceNet-TL that the detection efficiency on the two remote sensing datasets was better than that for the experiment without context scene feature fusion, indicating the effectiveness of the scene feature fusion. However, compared with objNet-TL-sceNet and objNet-TL, it is shown that when transfer learning was not employed, the improvement was not obvious, the possible reason for which being that the sceNet network has too many parameters for the limited remote sensing scene data, and if we directly trained the network using limited data without transfer learning, it would easily to over-fit, so that the extracted features would not be representative.

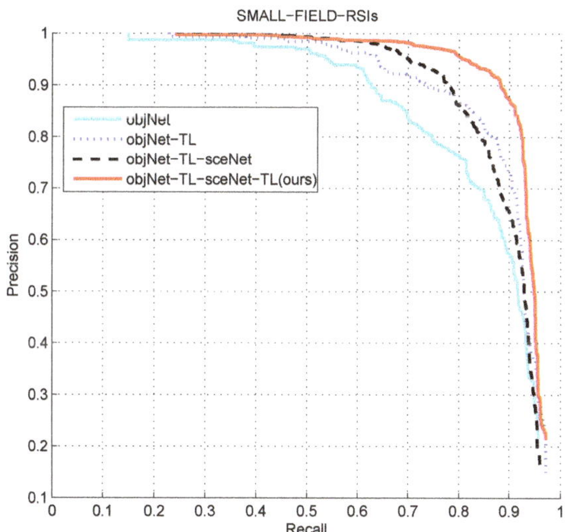

Figure 19. The performance comparison of four experiments on the SMALL-FIELD-RSIs dataset.

Finally, we compared the proposed algorithm objNet-TL-sceNet-TL with the other algorithms. Firstly, in order to prove the effectiveness of the deep features, we used the HOG algorithm [34] to extract the feature of the candidate region and used the SVM algorithm and the hard negative mining method to train the detector; we called it HOG-SVM. In addition, this paper compared the R-CNN algorithm [14]. This algorithm achieved a breakthrough in the PASCAL VOC2007 object detection task. It first uses the selective search algorithm to generate about 2000 candidate regions for each image and then uses the AlexNet network to extract features, and finally classifies each region using linear SVM classification. From the analysis of Section 4.3.2, we can see that the selective search algorithm is not suitable for large-sized remote sensing images. Therefore, this paper replaced the candidate region proposal with the pBING algorithm proposed in this paper. Furthermore, we compared the detection performance of the two methods of R-CNN: directly obtaining classification results by the AlexNet network and extracting features by AlexNet, then classifying by SVM. For convenience, the two algorithms are called pBING-AlexNet and pBING-AlexNet-SVM, respectively. Figures 21 and 22 show the comparison of the detection performance of each algorithm on two different sizes of remote sensing images.

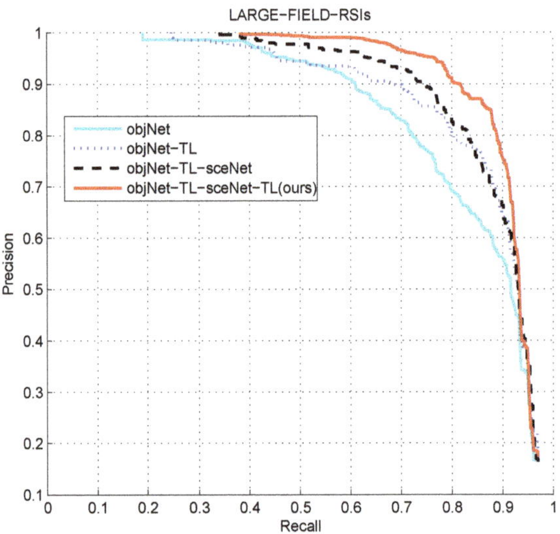

Figure 20. The performance comparison of four experiments on the LARGE-FIELD-RSIs dataset.

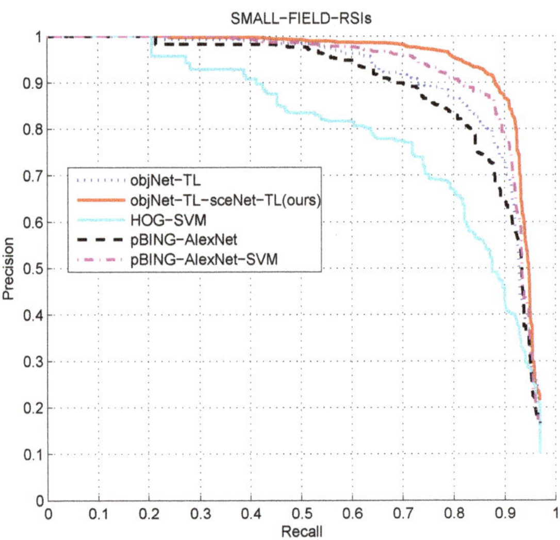

Figure 21. The performance comparison of different algorithms on the SMALL-FIELD-RSIs dataset.

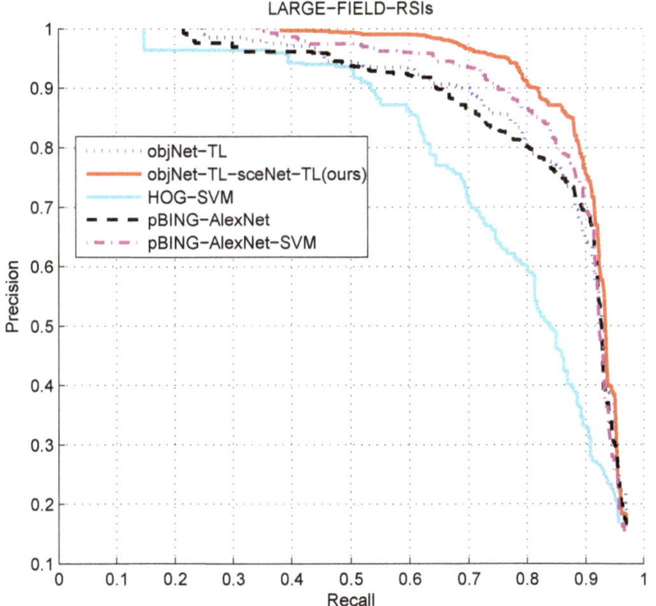

Figure 22. The performance comparison of different algorithms on the LARGE-FIELD-RSIs dataset.

Table 8. The average detection accuracy (average precision (AP)).

Algorithm	SMALL-FIELD-RSIs	LARGE-FIELD-RSIs
objNet	83.38%	82.30%
objNet-TL	87.59%	85.19%
objNet-TL-sceNet	87.67%	86.86%
objNet-TL-sceNet-TL (ours)	**90.77%**	**89.12%**
HOG-SVM	77.52%	76.69%
pBING-AlexNet	85.63%	84.54%
pBING-AlexNet-SVM	88.93%	86.91%

Figures 23 and 24 show the results of the objNet-TL-sceNet-TL algorithm on the SMALL-FIELD-RSIs and LARGE-FIELD-RSIs datasets, respectively. In each image, the red rectangles indicate the real objects marked by the dataset and the blue rectangles the final detection result of our algorithm. It appears that the vast majority of airplane objects can be correctly detected. It is noted that our algorithm can detect one airplane, which was not marked (missed by a human) on the dataset, and it is shown by the blue arrow in Figure 23. The comparison details of average precision (AP) can be checked in Table 8.

However, we found that if the airplane object was ambiguous and small, it may be missed. One missed airplane is pointed out by the red arrow in Figure 24. The reason is that our candidate region proposal algorithm was not strong enough for objects that are too small and ambiguous, and it needs to be further improved in the future.

Figure 23. Examples of detection results on the SMALL-FIELD-RSI dataset.

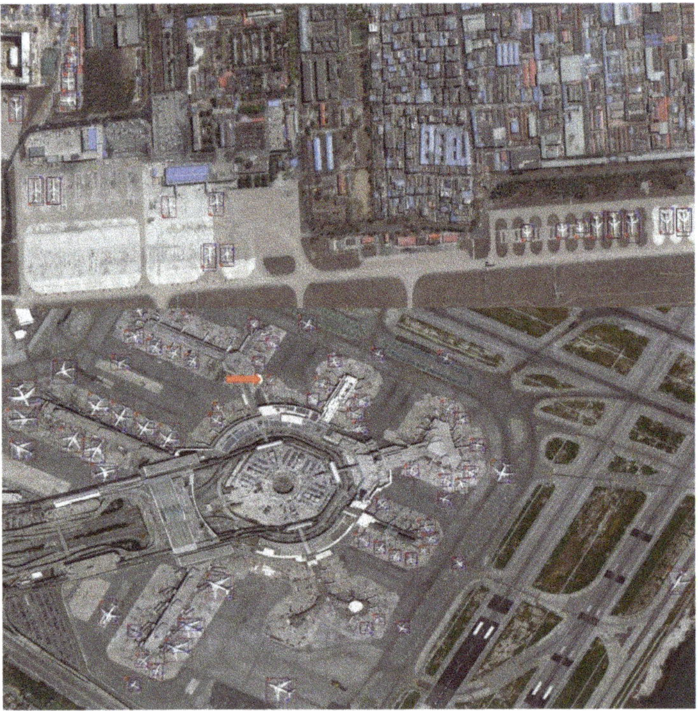

Figure 24. Examples of detection results on the LARGE-FIELD-RSIs dataset.

6. Conclusions

In this paper, we present a deep fusion feature approach to detect objects in high-resolution remote sensing images. Our method is composed of three main steps, which are the candidate region generation, deep feature extraction with fine-tuning, and the SVM classification with deep features. Hence, we re-structured the paper in terms of the three steps. For candidate region generation, we improved the binarized normed gradients algorithm and developed the pBING method. For deep feature extraction, the object feature and scene feature were both extracted for each candidate region, by utilizing the AlexNet model. As the label data in remote sensing are very scarce, we utilized the fine-tuning notion and pre-trained AlexNet on the ImageNet database, then fine-tuned the model with labeled remote sensing data. Finally, the object feature and scene feature were utilized to train an SVM for classification. After introducing the three main steps, we reported the experimental results, which validated the effectiveness of the developed pBING method, the fine-tuning strategy, and the overall detection model.

Author Contributions: E.K.W. is the main writer of the paper. Y.L. responds to the algorithms. Z.N. responds to design the pBing algorithm, J.Y., Z.L. and X.Z. respond to make experiments. S.M.Y. responds to the English Check.

Funding: This research was supported in part by the National Natural Science Foundation of China (No. 61572157), Grant No. 2016A030313660 and 2017A030313365 from the Guangdong Province Natural Science Foundation, and JCYJ20160608161351559, KQJSCX70726103044992, JCYJ20170811155158682, and JCYJ20160428092427867 from the Shenzhen Municipal Science and Technology Innovation Project. The authors thank the reviewers for their comments.

Conflicts of Interest: The authors declare no conflicts of interest.

References

1. Weber, J.; Lefevre, S. A multivariate hit-or-miss transform for conjoint spatial and spectral template matching. In Proceedings of the International Conference on Image and Signal Processing, Cherbourg-Octeville, France, 1–3 July 2008; pp. 226–235.
2. Sirmacek, B.; Unsalan, C. Urban-area and building detection using SIFT keypoints and graph theory. *IEEE Trans. Geosci. Remote Sens.* **2009**, *47*, 1156–1167. [CrossRef]
3. Yokoya, N.; Iwasaki, A. Object localization based on sparse representation for remote sensing imagery. In Proceedings of the International Geoscience and Remote Sensing Symposium, Quebec City, QC, Canada, 13–18 July 2014; pp. 2293–2296.
4. Tao, C.; Tan, Y.; Cai, H.; Tian, J. Airport detection from large ikonos images using clustered sift keypoints and region information. *IEEE Geosci. Remote Sens. Lett.* **2011**, *8*, 128–132. [CrossRef]
5. Sun, H.; Sun, X.; Wang, H.; Li, Y.; Li, H. Automatic target detection in high-resolution remote sensing images using spatial sparse coding bag-of-words mode. *IEEE Geosci. Remote Sens. Lett.* **2012**, *9*, 109–113. [CrossRef]
6. Gan, L.; Liu, P.; Wang, L. Rotation sliding window of the hog feature in remote sensing images for ship detection. In Proceedings of the 8th International Symposium on Computational Intelligence and Design, Hangzhou, China, 12–13 December 2015; pp. 401–404.
7. Arastounia, M.; Oude Elberink, S. Application of template matching for improving classification of urban railroad point clouds. *Sensors* **2016**, *16*, 2112. [CrossRef] [PubMed]
8. Aytekin, O.; Zongur, U.; Halici, U. Texture-based airport runway detection. *IEEE Geosci. Remote Sens. Lett.* **2013**, *10*, 471–475. [CrossRef]
9. Felzenszwalb, P.F.; Girshick, R.B.; McAllester, D.; Ramanan, D. Object detection with discriminatively trained part-based models. *IEEE Trans. Pattern Anal. Mach. Intell.* **2010**, *32*, 1627–1645. [CrossRef]
10. Han, J.; Zhang, D.; Cheng, G.; Guo, L.; Ren, J. Object detection in optical remote sensing images based on weakly supervised learning and high-level feature learning. *IEEE Trans. Geosci. Remote Sens.* **2015**, *53*, 3325–3337. [CrossRef]
11. Mottaghi, R.; Chen, X.; Liu, X.; Cho, N.-G.; Lee, S.-W.; Fidler, S.; Urtasun, R.; Yullie, A. The role of context for object detection and semantic segmentation in the wild. In Proceedings of the IEEE Conference on Computer Vision and Pattern Recognition, Columbus, OH, USA, 23–28 June 2014; pp. 891–898.

12. Ren, X.; Ramanan, D. Histograms of sparse codes for object detection. In Proceedings of the IEEE Conference on Computer Vision and Pattern Recognition, Portland, OR, USA, 23–28 June 2013; pp. 3246–3253.

13. Krizhevsky, A.; Sutskever, I.; Hinton, G.E. Imagenet Classification with Deep CNNs. In Proceedings of the Advances in Neural Information Processing Systems, Lake Tahoe, NV, USA, 3–6 December 2012; pp. 1097–1105.

14. Uijlings, J.; Sande, K.E.; Gevers, T.; Smeulders, A.W.M. Selective search for object recognition. *Int. J. Comput. Vis.* **2013**, *104*, 154–171. [CrossRef]

15. Cheng, M.; Zhang, Z.; Lin, W.-Y.; Torr, P. BING: Binarized Normed Gradients for Objectness Estimation at 300fps. In Proceedings of the Computer Vision and Pattern Recognition, Columbus, OH, USA, 24–27 June 2014; pp. 3286–3293.

16. Zitnick, C.L.; Dollar, P. Edge boxes: Locating object proposals from edges. In Proceedings of the European Conference on Computer Vision, Zurich, Switzerland, 6–12 September 2014; pp. 391–405.

17. Humayun, A.; Li, F.; Rehg, J.M. RIGOR: Reusing inference in graph cuts for generating object regions. In Proceedings of the Computer Vision and Pattern Recognition, Columbus, OH, USA, 24–27 June 2014; pp. 336–343.

18. Girshick, R.; Donahue, J.; Darrell, T.; Malik, J. Rich feature hierarchies for accurate object detection and semantic segmentation. In Proceedings of the Computer Vision and Pattern Recognition, Columbus, OH, USA, 24–27 June 2014; pp. 580–587.

19. Simonyan, K.; Zisserman, A. Very deep convolutional networks for large-scale image recognition. *arXiv* **2014**, arXiv:1409.1556.

20. Oquab, M.; Bottou, L.; Laptev, I.; Sivic, J. Learning and transferring mid-level image representations using convolutional neural networks. In Proceedings of the IEEE Computer Vision and Pattern Recognition, Columbus, OH, USA, 24–27 June 2014; pp. 1717–1724.

21. Girshick, R.; Donahue, J.; Darrell, T.; Malik, J. Region-based convolutional networks for accurate object detection and segmentation. *IEEE Trans. Pattern Anal. Mach. Intell.* **2016**, *38*, 142–158. [CrossRef] [PubMed]

22. Kong, T.; Yao, A.; Chen, Y.; Sun, F. HyperNet: Towards Accurate Region Proposal Generation and Joint Object Detection. Accurate region proposal generation and joint object detection. In Proceedings of the Computer Vision and Pattern Recognition, Las Vegas, NV, USA, 26 June–1 July 2016; pp. 845–853.

23. Ouyang, W.; Wang, X.; Zeng, X.; Qiu, S.; Luo, P.; Tian, Y.; Li, H.; Yang, S.; Wang, Z.; Loy, C.-C.; et al. DeepID-Net: Deformable deep convolutional neural networks for object detection. In Proceedings of the Computer Vision and Pattern Recognition, Boston, MA, USA, 7–12 June 2015; pp. 2403–2412.

24. Redmon, J.; Divvala, S.; Girshick, R.; Farhadi, A. You only look once: Unified, real-time object detection. In Proceedings of the Computer Vision and Pattern Recognition, Boston, MA, USA, 7–12 June 2015; pp. 779–788.

25. Hosang, J.; Benenson, R.; Dollar, P.; Schiele, B. What makes for effective detection proposals? *IEEE Trans. Pattern Anal. Mach. Intell.* **2016**, *38*, 814–830. [CrossRef] [PubMed]

26. Hare, S. Efficient online structured output learning for keypoint-based object tracking. In Proceedings of the IEEE Conference on Computer Vision and Pattern Recognition, Providence, RI, USA, 16–21 June 2012; pp. 1894–1901.

27. Zheng, S.; Sturgess, P.; Torr, P.H.S. Approximate structured output learning for constrained local models with application to real-time facial feature detection and tracking on low-power devices. In Proceedings of the IEEE International Conference and Workshops on Automatic Face and Gesture Recognition, Shanghai, China, 22–26 April 2013; pp. 1–8.

28. Kegl, B. The return of AdaBoost.MH: Multi-class hamming trees. *arXiv* **2013**, arXiv:1312.6086.

29. Wang, X.; Han, T.X.; Yan, S. An HOG-LBP human detector with partial occlusion handling. In Proceedings of the International Conference on Computer Vision, Lisboa, Portugal, 5–8 February 2009; pp. 32–39.

30. Srivastava, N.; Hinton, G.; Krizhevsky, A.; Sutskever, I.; Salakhutdinov, R. Dropout: A simple way to prevent neural networks from overfitting. *J. Mach. Learn. Res.* **2014**, *15*, 1929–1958.

31. Zhou, B.; Garcia, A.L.; Xiao, J.; Torralba, A.; Olivia, A. Learning deep features for scene recognition using places database. In Proceedings of the Advances in Neural Information Processing Systems, Montreal, QC, Canada, 7–12 December 2015; pp. 487–495.

32. Yang, Y.; Newsam, S. Bag-of-visual-words and spatial extensions for land-use classification. In Proceedings of the Advances in Geographic Information Systems, San Jose, CA, USA, 2–5 November 2010; pp. 270–279.

33. Xia, G.S.; Yang, W.; Delon, J.; Gousseau, Y.; Sun, H.; Maitre, H. Structural high-resolution satellite image indexing. In Proceedings of the ISPRS TC VII Symposium-100 Years ISPRS, Vienna, Austria, 5–7 July 2010; pp. 298–303.
34. Dalal, N.; Triggs, B. Histograms of oriented gradients for human detection. In Proceedings of the Computer Vision and Pattern Recognition, San Diego, CA, USA, 20–25 June 2005; pp. 886–893.

 *applied sciences*

Article

Land Subsidence Susceptibility Mapping Using Bayesian, Functional, and Meta-Ensemble Machine Learning Models

Hyun-Joo Oh [1], Mutiara Syifa [2], Chang-Wook Lee [2,*] and Saro Lee [3,4,*]

[1] Geo-Environmental Hazard Research Center, Korea Institute of Geoscience and Mineral Resources (KIGAM), 124, Gwahak-ro Yuseong-gu, Deajeon 34132, Korea; ohj@kigam.re.kr

[2] Division of Science Education, Kangwon National University, Chuncheon Campus, 1 Gangwondaehakgil, Chuncheon-si, Gangwon-do 24341, Korea; mutiarasyifa@kangwon.ac.kr

[3] Geoscience Platform Research Division, Korea Institute of Geoscience and Mineral Resources (KIGAM), 124, Gwahak-ro Yuseong-gu, Daejeon 34132, Korea

[4] Department of Geophysical Exploration, Korea University of Science and Technology, 217 Gajeong-ro Yuseong-gu, Daejeon 34113, Korea

* Correspondence: cwlee@kangwon.ac.kr (C.-W.L.); leesaro@kigam.re.kr (S.L.); Tel.: +82-33-250-6731 (C.-W.L.); +82-42-868-3057 (S.L.)

Received: 25 January 2019; Accepted: 22 March 2019; Published: 25 March 2019

Abstract: To effectively prevent land subsidence over abandoned coal mines, it is necessary to quantitatively identify vulnerable areas. In this study, we evaluated the performance of predictive Bayesian, functional, and meta-ensemble machine learning models in generating land subsidence susceptibility (LSS) maps. All models were trained using half of a land subsidence inventory, and validated using the other half of the dataset. The model performance was evaluated by comparing the area under the receiver operating characteristic (ROC) curve of the resulting LSS map for each model. Among all models tested, the logit boost, which is a meta-ensemble machine leaning model, generated LSS maps with the highest accuracy (91.44%), i.e., higher than that of the other Bayesian and functional machine learning models, including the Bayes net (86.42%), naïve Bayes (85.39%), logistic (88.92%), and multilayer perceptron models (86.76%). The LSS maps produced in this study can be used to mitigate subsidence risk for people and important facilities within the study area, and as a foundation for further studies in other regions.

Keywords: land subsidence; Bayes net; naïve Bayes; logistic; multilayer perceptron; logit boost

---

1. Introduction

Coal mining was once the driving force of the national industry and economic development in Korea, but this situation changed as demand for coal decreased. Gangwon Province was once Korea's largest coal mining area but most of its mines were closed in the early 1990s. Among the environmental problems that follow mine closures, land subsidence events can threaten human life and damage property and infrastructure, including buildings, houses, railroads, and roads [1–4]. Recovery of surface structures following land subsidence is difficult and costly; therefore, it is necessary to predict land subsidence susceptibility (LSS) zones before subsidence occurs, and to implement management strategies in these zones [3].

Generally, prediction of subsidence susceptibility zones requires the input of several environmental factors and the application of perdition models [5]. Several previous studies have developed quantitative and qualitative models that have been successfully applied in various hazard susceptibility zones worldwide [3–11]. These include logistic regression (LR) [3], frequency ratio (FR) [3,6], weight of evidence (WOE) [3], evidential belief function (EBF) [4], artificial neural network

(ANN) [3,5,7,8], support vector machine (SVM) [9], random forest (RF) [10], and fuzzy logic (FL) [8,11] models. Single LSS mapping models can be combined to form ensemble models, which provide more precise and meaningful results [9]. Ensemble models based on machine learning have recently improved the prediction accuracy and performance of single classifiers [12]. The main advantages of this approach are the ability to represent complex relationships between influential factors, and to incorporate spatial data of various scales [13].

Based on existing studies, probability and statistical models using geographical information systems (GIS) have been applied extensively to predict the susceptibly of geohazards, such as landslides, floods, subsidence, and rockfalls [3,14–16]. Recently, data mining and machine learning models for addressing nonlinear problems have been developed, which have been applied frequently and had their performances compared in landslide susceptibility mapping [17–20]. In ground subsidence hazard mapping, ground subsidence hazard maps around abandoned underground coal mines (AUCMs) have been constructed by integrating the adaptive neuro-fuzzy inference system and GIS [21]. In addition, a fuzzy operator, decision tree with the CHAID and QUEST algorithms, and the frequency ratio have been applied to construct subsidence susceptibility maps at AUCMs in Korea [2,11]. In this study, we investigated the performance of some models that have never been applied to land subsidence prediction. Therefore, in this study, we generated LSS maps for a South Korean district containing abandoned subsurface coal mines using machine learning methods, including a logit boost meta-ensemble model, two Bayesian models (Bayes net and NB models) and two functional models (logistic and multilayer perceptron models). The reliability and accuracy of all models were assessed by comparing their area under the receiver operating characteristic (ROC) curves. Data processing was performed using WEKA 3.9.2 and ArcGIS 10.5 software to produce five machine learning algorithms.

2. Land Subsidence in the Study Area

The study area, Hwajeon, is located in the city of Taebaek, South Korea (Figure 1), at 37°11′07″–37°11′07″ N, 128°56′40″–128°57′43″ E. Underground coal mining activities were carried out in Taebaek for nearly 20 years. The coal seams in this area were irregularly disturbed and inclined with various widths by reverse and thrust faults [22]. Therefore, the slant-chute block caving method was mainly used. About 10 million tons of coal were mined from the study area between 1953–1991 [22], and coal was transported to other areas by railroad beginning in 1973 (Figure 1). Since 1990, most of the coal mines have been closed due to reduced coal demand. However, the abandoned underground coal mines are currently causing land subsidence in the study area [11,21–23]. Additionally, infrastructure has been damaged by the land subsidence, as shown in photographs in a previous report [11].

Subsidence is caused by a variety of contributing factors, including geological discontinuities, presence of water, mining depth, and weak overburden [24,25]. The two forms of subsidence caused by underground coal mining are trough and sinkhole subsidence [25]. In the study area, a very irregular sinkhole occurred due to many complex underground coal mine pits excavated via slant–chute block caving in combination with the aforementioned factors [22]. After a mine cavity is excavated, roof stability becomes unstable over time due to changes in the strength and stress of the roof strata. Under such conditions, additional contributing factors can lead to the occurrence of sinkholes [25]. The Coal Industry Promotion Board [11,26] has reported 24 land subsidence events within the study area. Figure 1 shows a representative land subsidence from location S1 to location S6 of a subsidence event reported in 1999. Table 1 provides a description of the land subsidence. Locations S1 to S5 of this land subsidence mainly occurred along railways and at elevations above 800 m. Location S6 occurred in residential areas and at a lower elevation than S1–S5. Also, the depth of subsidence of S6 is the deepest (508 mm). Some photographs providing evidence of the land subsidence have been published [11,23].

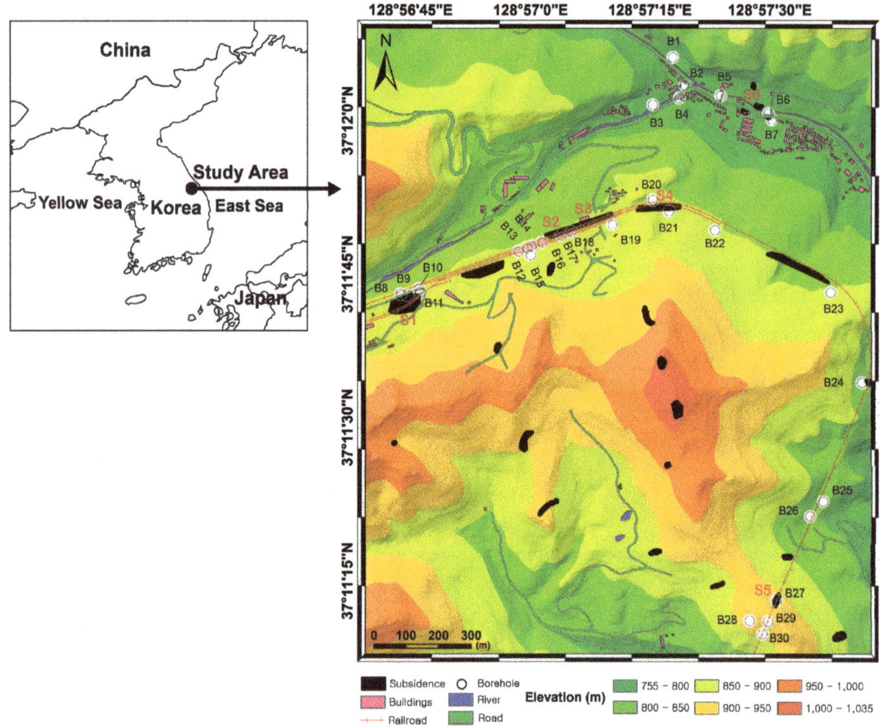

Figure 1. The study area in Taebaek, South Korea.

Table 1. Description of representative land subsidence in the study area [22].

Location	Structure	Elevation (m)	Mining Depth (m)	Thickness (m) and Slope of Coal Seam	Subsidence Depth (mm)	Other
S1	Railway	885	20–30	1–2 40–50°	90	-The coal seam is oblique to the railroad. -Shallow depth of mine -Sinkhole-type subsidence
S2	Railway	885	0	–	72	-Progression of cavity by mining -Subsidence by limestone cavity
S3	Railway	885	30–50	1–2 20°	329	-Subsidence along railway
S4	Railway	885	40–65	2 20°	223	-Shallow depth of mine -Coal bonanza
S5	Tunnel Railway	810	30–260	105 50–70°	65	-The tunnel is located above the mine cavity. -Vertical cracks and leakage in tunnel
S6	Road	765	60–98	3 20°	508	-Residential area and elementary school -Differential subsidence

3. Construction of Spatial Database

It is necessary to determine the factors affecting the land subsidence of a coal mine area. The lithology of the overburden rocks, geological discontinuities, ground slope, scope of the mined cavity, extent and depth of mining, mechanical characteristics of the rock mass rating (RMR), and flow of groundwater are considered the main factor [11,25,27,28]. Spatial data for all of these factors may be difficult to collect and may not be available. The available spatial databases used in this study were constructed using ArcGIS 10.5.

The surface geology with cross section lines was constructed using a digital geological map with 1:50,000 scale [29] published by the Korea Institute of Geoscience and Mineral Resources (KIGAM). The geological formations include the Manhan, Jangseong, Hambaegsan, Dosagog, and Alluvium horizons (Figure 2a, Figure 3). Most of the coal was mined from the Jangseong Formation with a thickness of 80–15 m [22,30]. This formation includes four to five cyclothems consisting of dark-gray sandstone, black shale, and coal seam (Table 2). The land use was constructed from a digital land characteristics map with 1:5,000 scale [31] supplied by the National Geographic Information Institute (NGII). Land use for the study area was classified into 10 categories: wood land, railroad, river, field, plot, road, school, hybrid land, brook and unclassified area (Figure 2b). The rate of land subsidence compared to the area of each category was higher in the railroad and school classes [21]. The surface slope was calculated from a digital elevation model (DEM) constructed from a digital elevation contour line with 1:5000 scale [32] published by NGII (Figure 2c). Surface slope was considered an affecting factor because land subsidence can change surface slope, differential horizontal strain, and vertical displacement [33]. Distance from drift was calculated from a digital drift map provided by the Mine Reclamation Corporation (MIRECO) [26] (Figure 2d). The map is important because it identifies the areas of mining activity in this region. Geological discontinuities are considered to be factors affecting land subsidence, but no geological lineaments appear in the study area on the available 1:50,000 geological map. Therefore, geomorphological lineament was visually extracted from an IKONOS satellite image by a field geologist (Figure 2e). If the location is near a lineament, the value of distance from lineament is low.

The borehole data in the study area, provided by the Mine Reclamation Organization (MIRECO) in 1996 [26], were collected from 29 boreholes (Figure 1 and Table 3). The depths of the boreholes ranged from a minimum of 19.5 m to a maximum of 200 m. The data included hydrologic properties and rock mass information [34]. The depth of groundwater, rock mass rating (RMR), and permeability were obtained from 16, 19, and 6 boreholes, respectively (Table 3 and Figure 2f,g,h). The maximum depth of groundwater was 42.5 m. On the railroad, the upper part of the railroad had a deeper groundwater depth and lower elevation than the lower part of the railway. The RMR was classified as classes 1–5, representing very good, good, fair, poor, and very poor, respectively. In this study, the RMR ranged from 2–4.5. The lowest RMRs appeared in the northwest and southeast portions of the railroad. Permeability was classified as classes 1–6, representing very highly (>1 cm/s), highly (1–10^{-2} cm/s), moderately (10^{-2}–10^{-3} cm/s), slightly (10^{-3}–10^{-5} cm/s), and very slightly (10^{-5}–10^{-7} cm/s) permeable and practically impermeable ($<10^{-7}$ cm/s), respectively. In this study, the permeability grade ranged from 4–4.5 (slightly permeable). The groundwater data were collected from a report published in May 1996 by the Coal Industry Promotion Board. Borehole point data should be converted into raster data for spatial analysis, and the accuracy of a raster map depends on the number of data points. However, the available borehole data were limited in this study. Therefore, raster maps from the limited borehole data were constructed using an inverse distance weighting (IDW) interpolation method, which is useful for predicting values at unmeasured locations where data are insufficient [11].

Eight control factors influencing land subsidence were constructed with 2 m × 2 m grid data, resulting in 775 columns and 860 rows, for a total of 666,500 cells within the study area. In total, 24 land subsidence areas as 24 vector-type polygons were converted to 2 m × 2 m grid data for a total of 3863 cells with a value of 1. The 3863 cells of land subsidence were randomly classified into training and validation sets, with a 50% (1931 cells) and 50% (1932 cells) distribution, respectively, to evaluate model performance.

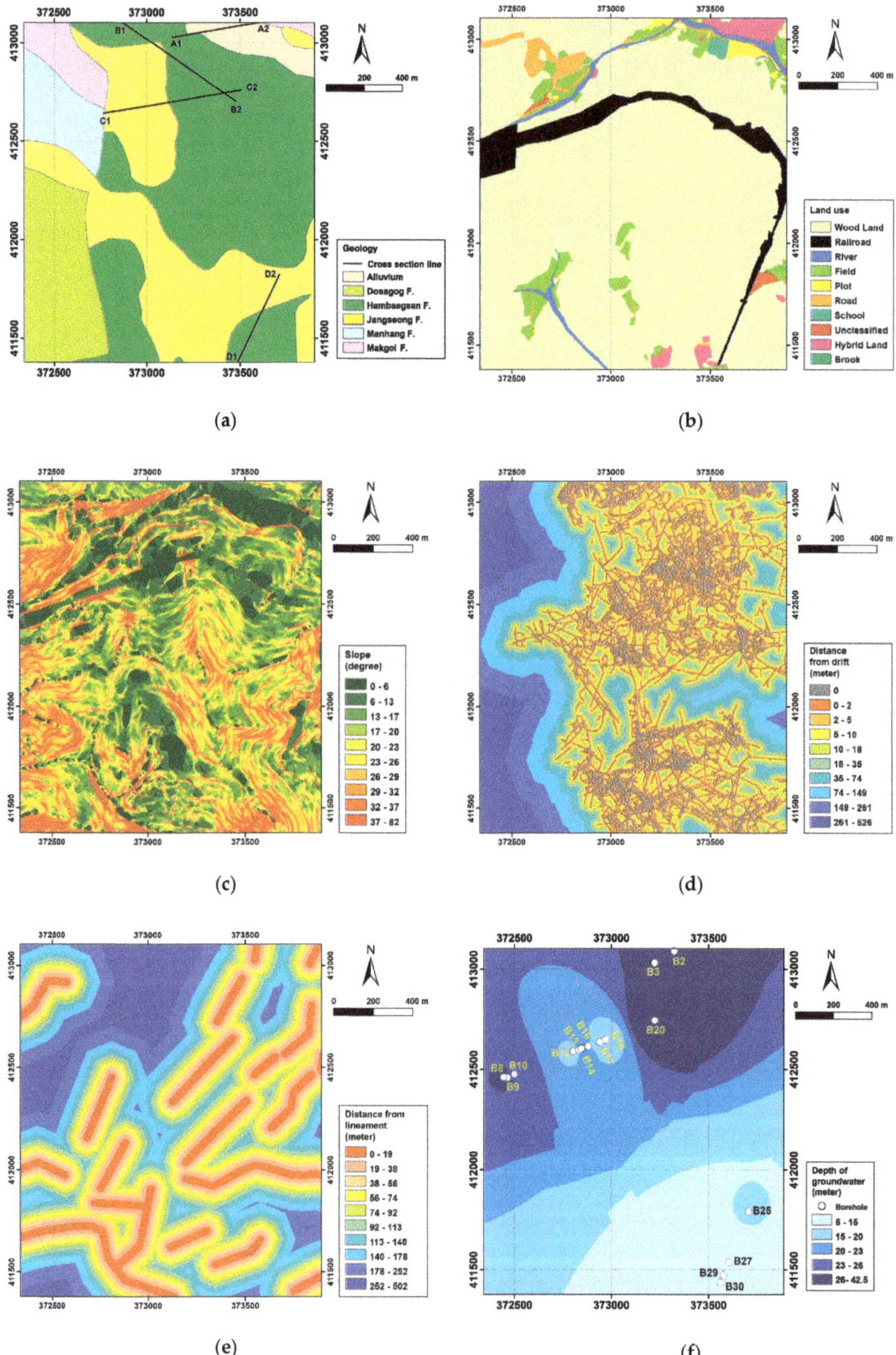

Figure 2. *Cont.*

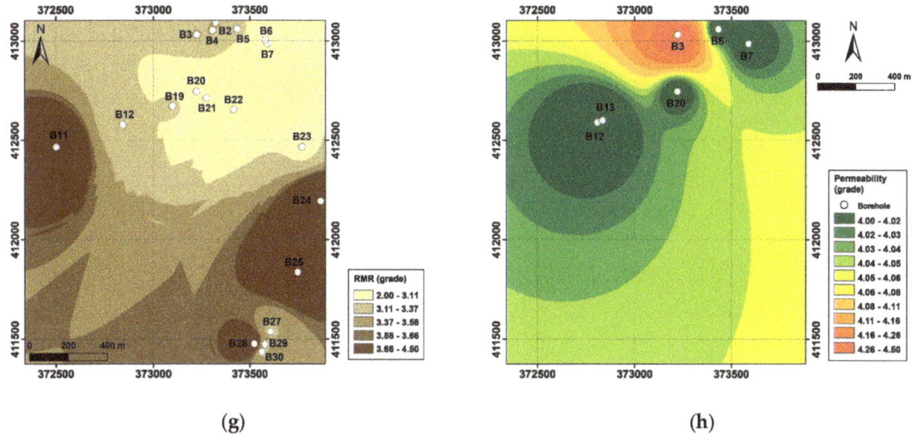

(g)   (h)

Figure 2. Eight factors influencing coal mine subsidence were used as input data in this study: (**a**) Geology, (**b**) land use, (**c**) slope, (**d**) distance from drift, (**e**) distance from lineament, (**f**) groundwater depth, (**g**) rock mass rating (RMR), and (**h**) permeability.

Table 2. Description of geological stratigraphy in Taebaek [30].

Geological Age	Formation	Thickness (m)	Description
Quaternary	Alluvium (Qa)	~20	- Gravel, sand, and clay
Permian	Dosagog (Pd)	250–350	- Mainly milky white–light green coarse–very coarse sandstone with greenish-gray–gray shale interbeds. Intercalations of pinkish sandstone, purple shale, and grayish-green sandy shale in the upper part. The sandstone is less compact than that of the Hambaegsan Formation.
Permian	Hambaegsan (Ph)	70–250	- Mainly milky white–light gray coarse sandstone with some interbeds of black shale with thickness of 2–3 m. Some pebbly sandstones occur at the base.
Permian	Jangseong (Pj)	80–150	- Four–five cyclothems consisting of dark-gray sandstone, black shale, and coal seam. Abundant plant fossils occur in the shale above the coal seam, the most valuable anthracite bed, of the 3rd–4th cyclothem from the bottom.
Carboniferous	Geumcheon (Cg)	50–100	- Mainly dark-gray–black shale and dark-gray fine sandstone intercalated with dark-gray limestone lenses and two to three thin coal seams
Carboniferous	Manhang (Cm)	250–300	- Mainly purple, greenish-gray, or light-green shale and light-green–green or light-gray medium–very coarse sandstone intercalated with three–four limestone lenses. Conglomerates with a thickness of a few meters occur at the base in some places.
Ordovician	Makgol (Om)		- In the upper part, gray–dark gray limestone intercalated with dolomite

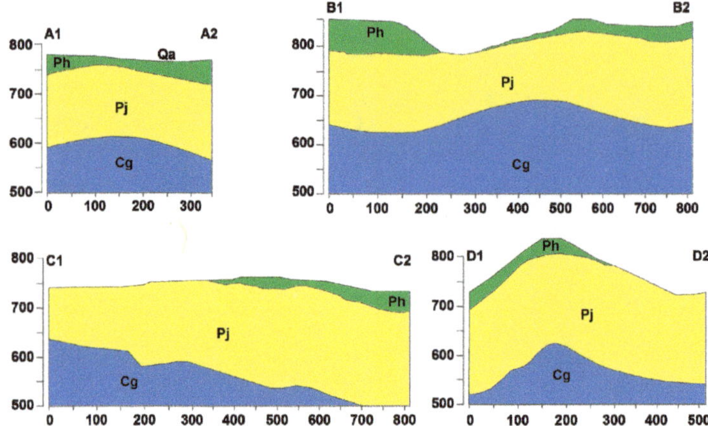

Figure 3. Geological cross sections in the study area.

Table 3. Borehole data in the study area.

ID	Depth of Borehole	Depth of Groundwater (m)	RMR (grade)	Permeability (grade)	Geology
B1	50.0	32.0	3.4	-	Alluvium-Hambaegsan
B2	50.0	27.2	3.4	4.5	Alluvium-Hambaegsan
B3	30.0	-	3.4	-	Alluvium-Hambaegsan
B4	60.2	-	3.4	4	Alluvium-Hambaegsan
B5	86.3	-	2.0	-	Alluvium-Hambaegsan
B6	80.0	-	2.0	4	Alluvium-Hambaegsan
B7	33.0	27.5	-	-	Jangseong
B8	20.5	27.7	-	-	Jangseong
B9	40.0	26.1	-	-	Jangseong
B10	35.5	-	4.4	-	Jangseong
B11	30.0	15.7	-	4	Jangseong
B12	40.5	21.6	-	4	Jangseong
B13	41.1	29.4	-	-	Jangseong
B14	22.0	-	3.2	-	Jangseong
B15	35.7	20.0	-	-	Jangseong
B16	40.8	20.0	-	-	Jangseong
B17	50.5	14.7	-	-	Jangseong
B18	58.0	-	3.2	-	Jangseong
B19	54.0	42.5	2.5	4	Hambaegsan-Jangseong
B20	60.0	-	3.0	-	Hambaegsan-Jangseong
B21	115.0	-	3.0	-	Hambaegsan-Jangseong
B22	80.0	-	3.0	-	Hambaegsan-Jangseong
B23	80.0	-	4.5	-	Hambaegsan-Jangseong
B24	84.0	-	4.3	-	Jangseong
B25	80.4	18.0	-	-	Jangseong
B26	19.5	5.0	3.3	-	Hambaegsan
B27	200.0	-	4.3	-	Hambaegsan-Jangseong
B28	40.0	5.0	3.3	-	Hambaegsan-Jangseong
B29	35.0	5.5	3.3	-	Hambaegsan-Jangseong

4. Methods

As shown in Figure 4, the mapping process consisted of five steps: (a) Spatial database construction, (b) random categorization of land subsidence locations into training and validation datasets at a ratio of 1:1, (c) selection of land subsidence conditioning factors, (d) application of machine learning methods to map LSS, and (d) validation and comparison of the five models.

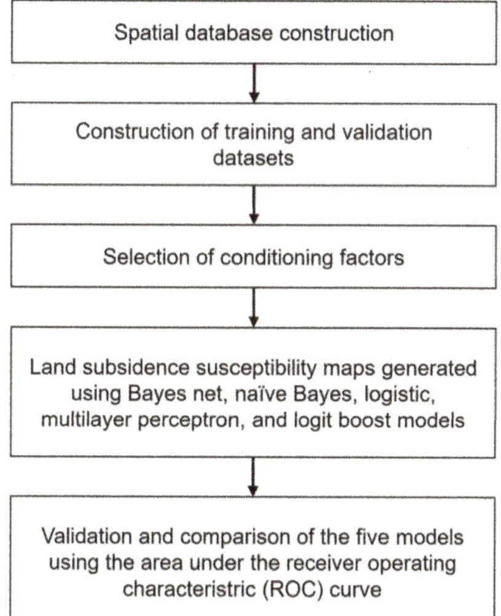

Figure 4. Flowchart for the generation of land subsidence susceptibility (LSS) maps using various machine learning models including Bayes net, naïve Bayes (NB), logistic, multilayer perceptron, and logit boost models.

4.1. Models

4.1.1. Bayes Net (BN)

The BN algorithm applies Bayes' theorem to produce graphical representations of the probability distribution [35]. BN is commonly used to model complex systems [36]. BN has not yet been used to model land subsidence; however, Pham et al. (2016) [37] applied this algorithm to evaluate landslide risk. The distinct universal probability of a subsidence event for a set of input factors can be estimated as follows:

$$P_B(X_1, X_2, \ldots, X_n) = \Pi_{i=1}^{n} = P_B\left(X_1 \,|\, \Pi_{x_i}\right) = \Pi_{i=1}^{n} \theta_{x_i | \Pi_{x_i}} \tag{1}$$

where $X = (X_1, X_2, \ldots, X_n)$ represents the subsidence input factors, $P_B\left(X_1 \,|\, \Pi_{x_i}\right) = \theta_{x_i | \Pi_{x_i}}$ is a common probability distribution for input factors X_i, and n is the number of subsidence input factors [37].

4.1.2. Naïve Bayes (NB)

The NB algorithm is a classification system that applies Bayes' theorem under the assumption of conditional independence for all attributes [10,38]. The NB classifier is easy to build, without any need for complicated iterative parameter-estimation schemes [38]. The NB algorithm estimates the probability $P(y_j/x_i)$ for all possible output classes as shown in Equation (2). The class with the largest posterior probability is predicted as follows:

$$y = \text{argmax}\, P\left(y_j\right) \prod_{i=1}^{n} P(x_i/y_j) \tag{2}$$

{subsidence, no subsidence}

where x_i is the input factor, y_j is the output class, $P(y_j)$ is the prior probability, and $P(y_j/x_i)$ is the conditional probability.

The conditional probability is calculated as

$$P\left(\frac{x_i}{y_j}\right) = \frac{1}{\sqrt{2\pi}\sigma}e^{-(x_i-\mu)^2/2\sigma^2} \tag{3}$$

where μ is the mean and σ is the standard deviation of x_i.

4.1.3. Logistic Regression (LR)

LR is a statistical technique that allows the predictor to analyze several types of variables [39–41]. LR does not require the normality assumption, which is an advantage over linear and log-linear regression. The inclusion of multiple parameters offers the user the ability to select the best predictors for use in the model [39]. The LR model is formulated as follows [42]:

$$f(x) = \text{logit}\,(P) = \ln\left[\frac{P}{1-P}\right] == c_0 + c_1 x_1 + \cdots + c_n x_n \tag{4}$$

$$P = \frac{1}{1+e^{-f(x)}} = \frac{1}{1+e^{-(c_0 + c_1 x_1 + \cdots + c_n x_n)}} \tag{5}$$

where $x_1, x_2, \ldots, x_n$ are the input factors, c_0 is the model intercept, and $c_1, \ldots, c_n$ are the regression coefficients to be approximated. In this study, P is the probability of subsidence occurrence and $1 - P$ is probability that subsidence will not occur. The function $f(x)$ is represented as logit (P).

4.1.4. Multilayer Perceptron (MLP)

MLP is an artificial neural network classifier that is widely used in various fields [12,43]. MLP neural nets consist of three structures: Input, hidden, and output layers. In this study, the input layers represent factors that affect land subsidence, and the inputs are processed to become outputs within the hidden layers. The classification results, dividing land subsidence and non-subsidence, are shown in the output layers [12,44]. Two processes are required to train data from MLP neural nets: 1) Forward propagation of the inputs through the hidden layers to obtain output and compare output values to initial values, and 2) adjustment of the connection weights using differences between subsequent values to generate the best results [44,45]. In this study, $t = t_i$, $i = 1, 2, \ldots, 8$ is a vector containing eight land-subsidence conditioning factors, and $\phi = \phi_j$, $j = 1, 2$ represents the land subsidence and non-subsidence classes. The MLP neural net function is then determined as follows:

$$\phi = f(t) \tag{6}$$

where $f(t)$ is an unknown function that is improved during the training process by adjustable network weights for a given network architecture.

An advantage of MLP is that the user is not required to decide the relative importance of the various input measurements; most inputs can be selected during the training process, based on weight adjustment [46]. Additionally, MLP does not require assumptions about the distribution of the training dataset.

4.1.5. Logit Boost (LB)

LB is a famous machine-learning algorithm introduced by Friedman et al., 2000 [47] that effectively reduces bias and variance; it is a slight modification of the most popular boosting method (AdaBoost) for handling noisy data [48], which reduces training errors and improves classification accuracy [49]. LB has been widely applied in binary classification problems [50], medical science [51], and computer science [52]; however, it has not yet been applied to land-subsidence problems [53].

In the current study, we create a vector $x_i = x_1, x_2, \ldots, x_n$, where n is the number of input factors; $y = [1, 0]$ represents two output classes (subsidence or non-subsidence). The LB algorithm is trained in the following steps [47]:

1. Assign weights $\omega_i = \frac{1}{n}$, $i = 1, 2, \ldots, n$, $f(x) = 0$ and probability estimates $p_e(x_i) = 1/2$.
2. For m = 1, 2, ..., m, repeat the following steps:

   a. Compute the working response and weights:

   $$r_i = \frac{[y_i^* - p_e(x_i)]}{[p_e(x_i)(1 - p(x_i))]}$$

   $$\omega_i = p_e(x_i)(1 - p(x_i))$$

   b. Fit the function by weighted least-squares regression of r_i to x_i using weights ω_i.
   c. Update the function as:

   $$f(x) \leftarrow f(x) + \frac{1}{2}f_m(x)$$

   $$p(x) \leftarrow \frac{e^{f(x)}}{e^{f(x)} + e^{-f(x)}}$$

3. Output the classifier.

$$\text{sign}[f(x)] = \text{sign}\left[\sum_{m=1}^{M} f_m(x)\right]$$

$$= \begin{cases} 1 \ (\text{subsidence}) \ \text{if} \ f(x) < 0 \\ -1 \ (\text{non subsidence}) \text{if} \ f(x) \geq 0 \end{cases}$$

4.2. Model Evaluation and Comparison

During the modeling and validation phases, model efficiency should be evaluated and compared [44]. We quantitatively evaluated and compared the efficiency of the models according to the area under the ROC curve (%). This technique has been applied to assess risk models of various hazards including subsidence [9], landslides [54], and sinkholes [55]; it is a standard method to quantitatively evaluate the quality of probabilistic and statistical models [56]. The x and y axes of the curve are sensitivity and specificity, respectively [56], and the area under the curve ranges from 0.5–1, with higher values indicating higher model accuracy and prediction capability.

5. Results

5.1. LSS Mapping

Figure 5 shows the LSS maps produced by the five algorithms: Bayes net (Figure 5a), NB (Figure 5b), logistic (Figure 5c), multilayer perceptron (Figure 5d), and logit boost (Figure 5e). To generate the LSS maps, we used the LSS index (LSSI) to classify susceptibility events into four classes: Very high (5% of total area), high (5%), moderate (5%), and low (85%). The probability of land subsidence was predicted for each class, and subsidence hazard was predicted for residential areas. The susceptibility indexes from the five algorithms were similar. The region with very high susceptibility appeared from the western part of the region to the eastern part as railroad area, which is marked by the red color. In the Bayes net result, the very high susceptibility area did not appear as often as in the other models. In the middle of the region, the Bayes net result has a low index, whereas the rest of the models have a very high or high index. Some very high indexes also appear in the northeastern part of the region, as elementary school area, but most of the region has a low susceptibility index rank for subsidence.

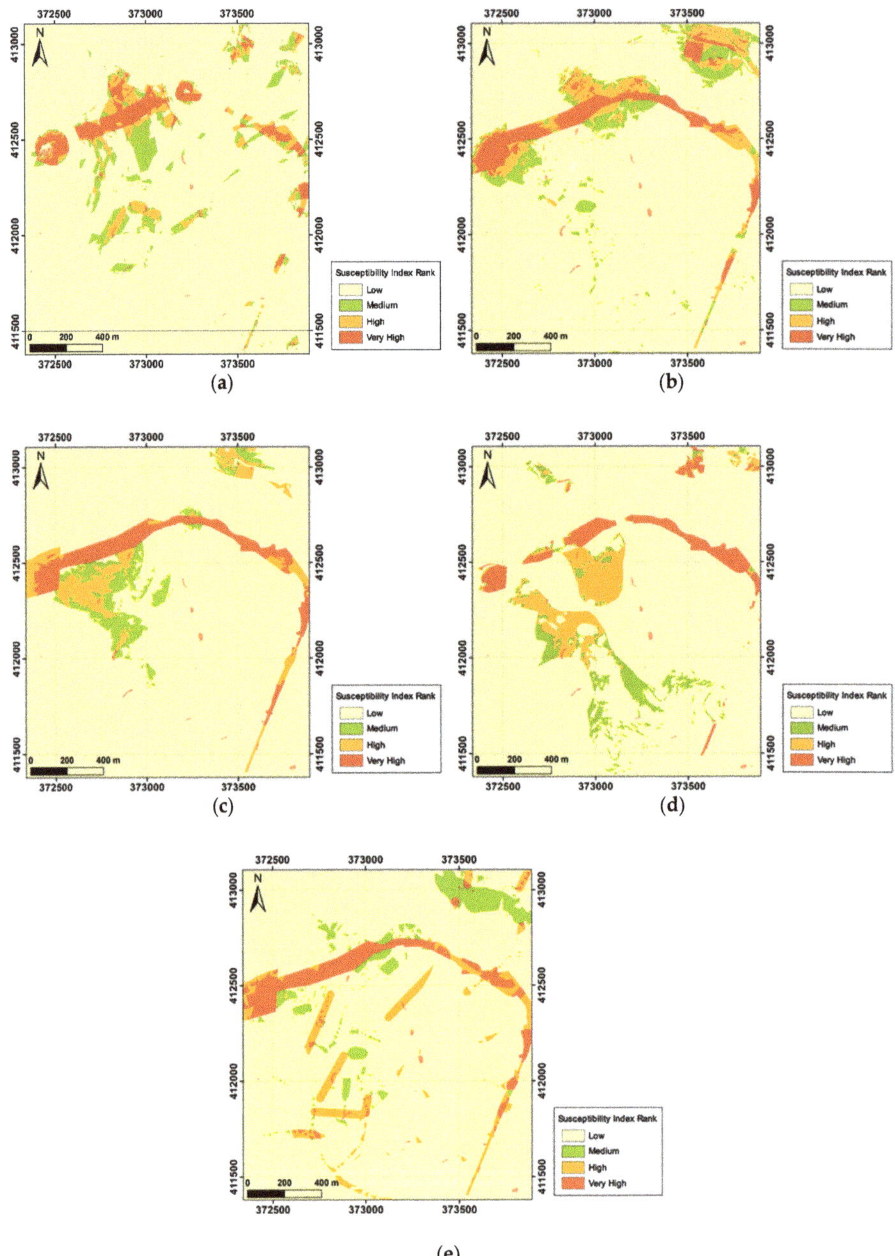

Figure 5. LSS maps generated using the five algorithms: (**a**) Bayes net, (**b**) NB, (**c**) logistic, (**d**) multilayer perceptron, and (**e**) logit boost.

However, there are some differences for the medium-susceptibility index rank, marked by the green color. The area with medium susceptibility of land subsidence is spreading and has a different pattern in each model result. For example, the NB and logit boost results show the northern part of the region is mostly covered by the medium susceptibility index. In contrast, the multilayer perceptron shows the medium index in the southern part of the region. Meanwhile, in the Bayes net and logistic models, the medium index is diffusely distributed from the northern to the middle part of the study area.

5.2. Validation

The land subsidence susceptibility (LSS) analysis results were validated by comparison with 1932 land subsidence cells (i.e., 50% of the total subsidence data) that had not been used in the analysis. A quantitative comparison among all models of the receiver operating characteristic (ROC) curves for model performance is shown in Figure 6. The land subsidence susceptibility index (LSSI) values of all cells were sorted in descending order, divided into 100 classes [57], and associated with the cumulative number of subsidence events for each class (Figure 6). The model with the highest area under the ROC curve was considered to be the model with the best predictive performance. The area under the curve values for the Bayes net, naïve Bayes (NB), logistic, multilayer perceptron, and logit boost models were 0.8640, 0.8539, 0.8892, 0.8676, and 0.9144, respectively; thus, the respective LSS mapping accuracy rates were 86.42, 85.39, 88.92, 86.76, and 91.44%. Although all models had sufficient performance, the different applied models had different prediction performances using same training data. In particular, the logit boost model had a higher predictive accuracy (by about 2.52, 4.68, 5.02, and 6.05%, respectively) than the logistic, multilayer perceptron, Bayes net, and NB. Therefore, model reliability followed the order logit boost > logistic > multilayer perceptron > Bayes net > NB. The percentage differences of the validation result are discussed in Section 6.

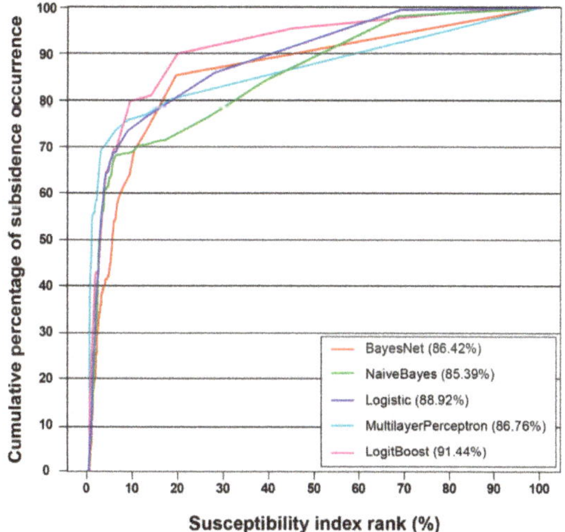

Figure 6. Susceptibility index rank (x-axis) and subsidence occurrence (y-axis) of the five algorithms.

6. Discussion

Recently, there has been great interest within the hazard prediction community toward improving the performance of hazard susceptibility models. In various fields, machine learning techniques have been shown to be effective in terms of performance [58–62]. In particular, ensemble learning has improved machine learning results by combining several models [17,63,64]. The results of different applied models under the same conditions (i.e., study area, input data, ratio of training, and validation datasets) can be compared to the quantitative accuracy values of the area under the ROC to present the predictive power of the model. Models with similar (different) accuracy values can be said to have similar (differing) performances. Therefore, the reliabilities of the models can be ordered according to the accuracies of the models.

In this study, the logit boost model, based on ensemble machine learning, had a 91.44% accuracy and a predictive accuracy that was higher (by 2.52–6.05%) than those of the logistic, multilayer perceptron, Bayes net, and NB based on machine learning. Similarly, a previous study [2] found that a decision tree model (the CHAID algorithm) produced LSS maps with higher accuracy (94.01%) than the QUEST decision tree (90.37%) and frequency ratio (86.70%). The other algorithms examined in the current study also exhibited high accuracy. Thus, the Bayes net, NB, logistic, and multilayer perceptron models can also be used as alternative models for mapping land subsidence hazard risk. Even though the logit boost model, as an ensemble model, had not been used to predict land subsidence in previous research, the results of the current study indicate that it can achieve high accuracy.

However, some limitations of the models might be a consideration for future studies. For example, the Bayes net model assumes no missing values, and this model also needs to be updated, especially for estimating the conditional probabilities [65]. The benefits and drawbacks of the machine learning models are influenced by several factors, such as the availability of datasets, characteristics of the study area, and condition of the region [18]. The use of Bayesian algorithms, such as the Bayes net and Naïve Bayes, has not been fully verified in natural hazard assessments [18]. According to Mezaal [66], the multilayer perceptron algorithm also has limitations, such as overlearning and high computational complexity.

It has been reported that the sinkhole subsidence attributable to underground mining is caused by shallow depth, weak overburden, geological discontinuities, solution of rocks, rainfall, groundwater, and earthquakes [25]. However, this study used a spatial database obtained from previous studies due to the limitation of available data. No further surveys or new surveys on land subsidence have been conducted in the study area for 14 years. If real-time monitoring data and additional data are obtained in the study area, a 4D underground subsidence model [67] with 3D geological modeling could be constructed to predict land subsidence hazard areas accurately. Thus, continuous monitoring and detailed new surveying for causative factors are essential in the study area. The maps produced in this study can be used as basic data for policymakers and further research. Future studies should develop alternative models and methods to determine the relative influence of factors affecting LSS, so that these methods can be applied in other regions.

7. Conclusions

Land subsidence is a hazardous effect of coal mine abandonment, including that in Korea. To prevent damage and loss of life in the Taebaek region, it is necessary to predict areas with high subsidence risk effectively. In this study, we used Bayesian (i.e., Bayes net and NB), functional (i.e., logistic, multilayer perceptron), and meta-ensemble (i.e., logit boost) machine learning models to perform LSS assessments. Although all models had sufficient performance, the logit boost meta-ensemble machine learning model had the highest accuracy (91.44%) among the five models. The logit boost model also had higher predictive accuracy (by 2.52%, 4.68%, 5.02%, and 6.05%, respectively) than the logistic, multilayer perceptron, Bayes net, and NB models. According to previous studies [11,57] in the same study area, the fuzzy operator with 84.40–88.98% accuracy, frequency ratio with 86.70% accuracy, CHAID decision tree with 94.01% accuracy, and QUEST decision tree with 90.37% accuracy have been applied to the subsidence hazard assessment, but the five models used in this study had been rarely applied. Based on these case studies, the land subsidence hazard rating can be applied to future policy decisions using additional data.

Author Contributions: Conceptualization, S.L. and C.-W.L.; methodology, S.L. and H.-J.O.; software, H.-J.O. and M.S.; validation, H.-J.O. and M.S.; formal analysis, H.-J.O. and M.S.; investigation, S.L. and H.-J.O.; resources, S.L. and H.-J.O.; writing—original draft preparation, H.-J.O. and M.S.; writing—review and editing, S.L. and C.-W.L.; visualization, H.-J.O.; supervision, S.L. and C.-W.L.; project administration, S.L. and C.-W.L.; funding acquisition, S.L. and C.-W.L.

Funding: This research was supported by the Basic Research Project of the Korea Institute of Geoscience and Mineral Resources (KIGAM), funded by the Ministry of Science, ICT, and Future Planning of Korea. This work was supported by a National Research Foundation of Korea (NRF) grant from the Korea government (MSIP) (number NRF-2017R1A2B4003258).

Conflicts of Interest: The authors declare no conflict of interest.

References

1. Ghorbanzadeh, O.; Rostamzadeh, H.; Blaschke, T.; Gholaminia, K.; Aryal, J. A New Gis-Based Data Mining Technique Using an Adaptive Neuro-Fuzzy Inference System (Anfis) and K-Fold Cross-Validation Approach for Land Subsidence Susceptibility Mapping. *Nat. Hazards* **2018**, *94*, 497–517. [CrossRef]
2. Lee, S.; Park, I. Application of decision tree model for the ground subsidence hazard mapping near abandoned underground coal mines. *J. Environ. Manag.* **2013**, *127*, 166–176. [CrossRef] [PubMed]
3. Oh, H.-J.; Lee, S. Integration of ground subsidence hazard maps of abandoned coal mines in Samcheok, Korea. *Int. J. Coal Geol.* **2011**, *86*, 58–72. [CrossRef]
4. Pradan, B.; Abokharima, M.H.; Jebur, M.N.; Tehrany, M.S. Land Subsidence Susceptibility Mapping at Kinta Valley (Malaysia) Using the Evidential Belief Function Model in Gis. *Nat. Hazards* **2014**, *73*, 1019–1042. [CrossRef]
5. Lee, S.; Park, I.; Choi, J.K. Spatial prediction of ground subsidence susceptibility using an artificial neural network. *Environ. Manag.* **2012**, *49*, 347–358. [CrossRef] [PubMed]
6. Oh, H.-J.; Ahn, S.-C.; Choi, J.-K.; Lee, S. Sensitivity analysis for the GIS-based mapping of the ground subsidence hazard near abandoned underground coal mines. *Environ. Earth Sci.* **2011**, *64*, 347–358. [CrossRef]
7. Pishro, M.; Khosravi, S.; Tehrani, S.M.; Mousavi, S.R. Modeling and zoning of land subsidence in the southwest of Tehran using artificial neural networks. *Int. J. Hum. Cap. Urban Manag.* **2016**, *1*, 159–168.
8. Rafie, M.; Samimi Namin, F. Prediction of subsidence risk by FMEA using artificial neural network and fuzzy inference system. *Int. J. Min. Sci. Technol.* **2015**, *25*, 655–663. [CrossRef]
9. Tien Bui, D.; Shahabi, H.; Shirzadi, A.; Chapi, K.; Pradhan, B.; Chen, W.; Khosravi, K.; Panahi, M.; Bin Ahmad, B.; Saro, L. Land Subsidence Susceptibility Mapping in South Korea Using Machine Learning Algorithms. *Sensors* **2018**, *18*, 2464. [CrossRef]
10. Ilia, I.; Loupasakis, C.; Tsangaratos, P. Land subsidence phenomena investigated by spatiotemporal analysis of groundwater resources, remote sensing techniques, and random forest method: The case of Western Thessaly, Greece. *Environ. Monit. Assess.* **2018**, *190*, 623. [CrossRef]
11. Choi, J.-K.; Kim, K.-D.; Lee, S.; Won, J.-S. Application of a fuzzy operator to susceptibility estimations of coal mine subsidence in Taebaek City, Korea. *Environ. Earth Sci.* **2010**, *59*, 1009–1022. [CrossRef]
12. Pham, B.T.; Tien Bui, D.; Prakash, I.; Dholakia, M.B. Hybrid integration of Multilayer Perceptron Neural Networks and machine learning ensembles for landslide susceptibility assessment at Himalayan area (India) using GIS. *CATENA* **2017**, *149*, 52–63. [CrossRef]
13. Kanevski, M.; Parkin, R.; Pozdnukhov, A.; Timonin, V.; Maignan, M.; Demyanov, V.; Canu, S. Environmental data mining and modeling based on machine learning algorithms and geostatistics. *Environ. Model. Softw.* **2004**, *19*, 845–855. [CrossRef]
14. Baillifard, F.; Jaboyedoff, M.; Sartori, M. Rockfall hazard mapping along a mountainous road in Switzerland using a GIS-based parameter rating approach. *Nat. Hazards Earth Syst. Sci.* **2003**, *3*, 435–442. [CrossRef]
15. Samanta, S.; Pal, D.K.; Palsamanta, B. Flood susceptibility analysis through remote sensing, GIS and frequency ratio model. *Appl. Water Sci.* **2018**, *8*, 66. [CrossRef]
16. Lee, S.; Pradhan, B. Landslide hazard mapping at Selangor, Malaysia using frequency ratio and logistic regression models. *Landslides* **2007**, *4*, 33–41. [CrossRef]
17. Kadavi, P.R.; Lee, C.-W.; Lee, S. Application of Ensemble-Based Machine Learning Models to Landslide Susceptibility Mapping. *Remote Sens.* **2018**, *10*, 1252. [CrossRef]
18. Pham, B.T.; Prakash, I.; Khosravi, K.; Chapi, K.; Trinh, P.T.; Ngo, T.Q.; Hosseini, S.V.; Bui, D.T. A comparison of Support Vector Machines and Bayesian algorithms for landslide susceptibility modelling. *Geocarto Int.* **2018**, 1–23. [CrossRef]

19. Darabi, H.; Choubin, B.; Rahmati, O.; Torabi Haghighi, A.; Pradhan, B.; Kløve, B. Urban flood risk mapping using the GARP and QUEST models: A comparative study of machine learning techniques. *J. Hydrol.* **2019**, *569*, 142–154. [CrossRef]
20. Truong, X.L.; Mitamura, M.; Kono, Y.; Raghavan, V.; Yonezawa, G.; Truong, X.Q.; Do, T.H.; Tien Bui, D.; Lee, S. Enhancing Prediction Performance of Landslide Susceptibility Model Using Hybrid Machine Learning Approach of Bagging Ensemble and Logistic Model Tree. *Appl. Sci.* **2018**, *8*, 1046. [CrossRef]
21. Park, I.; Choi, J.; Jin Lee, M.; Lee, S. Application of an adaptive neuro-fuzzy inference system to ground subsidence hazard mapping. *Comput. Geosci.* **2012**, *48*, 228–238. [CrossRef]
22. Bang, G.M.; Choi, S.S.; Oh, S.H.; Sin, J.S.; Jeon, M.G.; Woo, S.U.; Heo, J.E.; Cheon, M.N.; Jo, N.S.; Kim, B.C.; et al. *A Detailed Survey of Monitoring in Whajeon Region*; Coal Industry Promotion Board: Seoul, Korea, 1999; p. 681.
23. Kim, J.N. *A Case Study on Stability Analysis of Ground Subsidence in Abandoned Mine Area—Focused on the Vicinity of the Chunjeon Station*; Seoul National University of Science and Technology: Seoul, Korea, 2011.
24. Canbulat, I.; Zhang, C.; Black, K.; Johnston, J.; McDonald, S. Assessment of Sinkhole Risk in Shallow Coal Mining. In Proceedings of the 10th Triennial Conference of Mine Subsidence: Adaptive Innovation for Managing Challenges, Pokolbin, Australia, 5–7 November 2017.
25. Singh, K.B.; Dhar, B.B. Sinkhole subsidence due to mining. *Geotech. Geol. Eng.* **1997**, *15*, 327–341. [CrossRef]
26. Board, C.I.P. *Ground Stability Investigation for Hwajeon (Korean Edn)*; Coal Industry Promotion Board: Seoul, Korea, 1996; pp. 9–84.
27. Sahu, P.; Lokhande, R.D. An Investigation of Sinkhole Subsidence and its Preventive Measures in Underground Coal Mining. *Procedia Earth Planet. Sci.* **2015**, *11*, 63–75. [CrossRef]
28. Lee, F.T.; Abel, J.F. *Subsidence from Underground Mining; Environmental Analysis and Planning Considerations*; USGS: Reston, VA, USA, 1983. [CrossRef]
29. The Geological Society of Korea. *Geologic Altas of Taebaegsan Region: Homyeong Geological Sheet*; The Geological Society of Korea: Seoul, Korea, 1962.
30. Seo, H.K.; Kim, D.S.; Park, S.H.; Park, J.S.; Bae, D.J.; Yu, Y.S.; Lee, D.Y.; Im, S.B.; Jang, Y.H.; Jo, M.J. *Geologic Atlas of the Samcheog Coalfield*; Korea Institute of Geoscience and Mineral Resources: Daejeon, Korea, 1979; p. 22.
31. National Geographic Information Institute. *Digital Land Characteristics Map with 1:5000 Scale: No. of index: 37816018, 37816019, 37816028, 37816029*; National Geographic Information Institute: Suwon, Korea, 1996.
32. National Geographic Information Institute. *Digital Topographical Map with 1:5000 Scale: No. of index: 37816018, 37816019, 37816028, 37816029*; National Geographic Information Institute: Suwon, Korea, 1996.
33. Tripathi, N.; Singh, R.S. Underground Coal Mine Subsidence Impacts Forest Ecosystem. 2010. Available online: https://www.researchgate.net/publication/275207695 (accessed on 25 January 2019).
34. Öge, İ.F.; Çırak, M. Relating rock mass properties with Lugeon value using multiple regression and nonlinear tools in an underground mine site. *Bull. Eng. Geol. Environ.* **2017**. [CrossRef]
35. Marcot, B.; Steventon, J.; Sutherland, G.; McCann, R. Guidelines for Developing and Updating Bayesian Belief Networks Applied to Ecological Modeling and Conservation. *Can. J. Forest Res.* **2006**, *36*, 3063–3074. [CrossRef]
36. Song, Y.; Gong, J.; Gao, S.; Wang, D.; Cui, T.; Li, Y.; Wei, B. Susceptibility assessment of earthquake-induced landslides using Bayesian network: A case study in Beichuan, China. *Comput. Geosci.* **2012**, *42*, 189–199. [CrossRef]
37. Pham, B.T.; Pradhan, B.; Tien Bui, D.; Prakash, I.; Dholakia, M.B. A comparative study of different machine learning methods for landslide susceptibility assessment: A case study of Uttarakhand area (India). *Environ. Model. Softw.* **2016**, *84*, 240–250. [CrossRef]
38. Tien Bui, D.; Pradhan, B.; Lofman, O.; Revhaug, I. Landslide susceptibility assessment in vietnam using support vector machines, decision tree, and nave bayes models. *Math. Probl. Eng.* **2012**, *2012*. [CrossRef]
39. Erener, A.; Mutlu, A.; Sebnem Düzgün, H. A comparative study for landslide susceptibility mapping using GIS-based multi-criteria decision analysis (MCDA), logistic regression (LR) and association rule mining (ARM). *Eng. Geol.* **2016**, *203*, 45–55. [CrossRef]
40. Ozdemir, A.; Altural, T. A comparative study of frequency ratio, weights of evidence and logistic regression methods for landslide susceptibility mapping: Sultan Mountains, SW Turkey. *J. Asian Earth Sci.* **2013**, *64*, 180–197. [CrossRef]

41. Mertler, C.A.; Reinhart, R.V. *Advanced and Multivariate Statistical Methods: Practical Application and Interpretation: Sixth Edition*; Routledge: New York, NY, USA, 2016; pp. 1–374. [CrossRef]

42. Hosmer, D.W., Jr.; Lemeshow, S.; Sturdivant, R.X. *Applied Logistic Regression*; John Wiley & Sons: Hoboken, NJ, USA, 2013; Volume 398.

43. Haykin, S.S. *Neural Networks and Learning Machines/Simon Haykin*; Prentice Hall: New York, NY, USA, 2009.

44. Pham, B.T.; Tien Bui, D.; Pourghasemi, H.R.; Indra, P.; Dholakia, M.B. Landslide susceptibility assesssment in the Uttarakhand area (India) using GIS: A comparison study of prediction capability of naïve bayes, multilayer perceptron neural networks, and functional trees methods. *Theor. Appl. Climatol.* **2017**, *128*, 255–273. [CrossRef]

45. Bui, D.T.; Pradhan, B.; Revhaug, I.; Nguyen, D.B.; Pham, H.V.; Bui, Q.N. A novel hybrid evidential belief function-based fuzzy logic model in spatial prediction of rainfall-induced shallow landslides in the Lang Son city area (Vietnam). *Geomat. Nat. Hazards Risk* **2015**, *6*, 243–271. [CrossRef]

46. Gardner, M.W.; Dorling, S.R. Artificial neural networks (the multilayer perceptron)—A review of applications in the atmospheric sciences. *Atmos. Environ.* **1998**, *32*, 2627–2636. [CrossRef]

47. Friedman, J.; Tibshirani, R.; Hastie, T. Additive Logistic Regression: A Statistical View of Boosting (With Discussion and a Rejoinder by the Authors). *Ann. Stat.* **2000**, *28*, 337–407. [CrossRef]

48. Zhang, G.; Fang, B. LogitBoost classifier for discriminating thermophilic and mesophilic proteins. *J. Biotechnol.* **2007**, *127*, 417–424. [CrossRef] [PubMed]

49. Song, J.; Lu, X.; Liu, M.; Wu, X. Stratified Normalization LogitBoost for Two-Class Unbalanced Data Classification. *Commun. Stat. Simul. Comput.* **2011**, *40*, 1587–1593. [CrossRef]

50. Fraz, M.M.; Remagnino, P.; Hoppe, A.; Uyyanonvara, B.; Rudnicka, A.R.; Owen, C.G.; Barman, S.A. An ensemble classification-based approach applied to retinal blood vessel segmentation. *IEEE Trans. Biomed. Eng.* **2012**, *59*, 2538–2548. [CrossRef]

51. Cai, Y.-D.; Feng, K.-Y.; Lu, W.-C.; Chou, K.-C. Using LogitBoost classifier to predict protein structural classes. *J. Theor. Biol.* **2006**, *238*, 172–176. [CrossRef]

52. Lutz, R.W. Logitboost with trees applied to the wcci 2006 performance prediction challenge datasets. In Proceedings of the 2006 IEEE International Joint Conference on Neural Network, Vancouver, BC, Canada, 16–21 July 2006; pp. 1657–1660.

53. Pham, B.T.; Tien Bui, D.; Dholakia, M.B.; Prakash, I.; Pham, H.V. A Comparative Study of Least Square Support Vector Machines and Multiclass Alternating Decision Trees for Spatial Prediction of Rainfall-Induced Landslides in a Tropical Cyclones Area. *Geotech. Geol. Eng.* **2016**, *34*, 1807–1824. [CrossRef]

54. Conforti, M.; Pascale, S.; Robustelli, G.; Sdao, F. Evaluation of prediction capability of the artificial neural networks for mapping landslide susceptibility in the Turbolo River catchment (Northern Calabria, Italy). *CATENA* **2014**, *113*, 236–250. [CrossRef]

55. Ozdemir, A. Sinkhole susceptibility mapping using logistic regression in Karapınar (Konya, Turkey). *Bull. Eng. Geol. Environ.* **2016**, *75*, 681–707. [CrossRef]

56. Zweig, M.H.; Campbell, G. Receiver-operating characteristic (ROC) plots: A fundamental evaluation tool in clinical medicine. *Clin. Chem.* **1993**, *39*, 561–577.

57. Park, I.; Lee, S. Spatial prediction of landslide susceptibility using a decision tree approach: A case study of the Pyeongchang area, Korea. *Int. J. Remote Sens.* **2014**, *35*, 6089–6112. [CrossRef]

58. Korup, O.; Stolle, A. Landslide Prediction from Machine Learning. *Geol. Today* **2014**, *30*, 26–33. [CrossRef]

59. McGaughey, W.J.; Laflèche, V.; Howlett, C.; Sydor, J.L.; Campos, D.; Purchase, J.; Huynh, S. Automated, Real-Time Geohazard Assessment in Deep Underground Mines. In *Proceedings of the Eighth International Conference on Deep and High Stress Mining*; Wesseloo, J., Ed.; Australian Centre for Geomechanics: Perth, Australia, 2017; pp. 521–528.

60. Tayfur, G.; Singh, V.P.; Moramarco, T.; Barbetta, S. Flood Hydrograph Prediction Using Machine Learning Methods. *Water* **2018**, *10*, 968. [CrossRef]

61. Karpatne, A.; Ebert-Uphoff, I.; Ravela, S.; Babaie, H.A.; Kumar, V. Machine Learning for the Geosciences: Challenges and Opportunities. *IEEE Trans. Knowl. Data Eng.* **2018**. [CrossRef]

62. Canli, E.; Mergili, M.; Thiebes, B.; Glade, T. Probabilistic Landslide Ensemble Prediction Systems: Lessons to Be Learned from Hydrology. *Nat. Hazards Earth Syst. Sci.* **2018**, *18*, 2183–2202. [CrossRef]

63. Mojaddadi, H.; Pradhan, B.; Nampak, H.; Ahmad, N.; Ghazali, A.H.B. Ensemble Machine-Learning-Based Geospatial Approach for Flood Risk Assessment Using Multi-Sensor Remote-Sensing Data and Gis. *Geomat. Nat. Hazards Risk* **2017**, *8*, 1080–1102. [CrossRef]
64. Chen, W.; Sun, Z.; Han, J. Landslide Susceptibility Modeling Using Integrated Ensemble Weights of Evidence with Logistic Regression and Random Forest Models. *Appl. Sci.* **2019**, *9*, 171. [CrossRef]
65. Bouckaert, R.R. *Bayesian Network Classifiers in Weka for Version 3-5-7*; The University of Waikato: Waikato, New Zealand, 2008; Volume 11, pp. 369–387.
66. Mezaal, M.; Pradhan, B.; Shafri, H.; Md Yusoff, Z.; Al-Zuhairi, M. Optimized Neural Architecture for Automatic Landslide Detection from High-Resolution Airborne Laser Scanning Data. *Appl. Sci.* **2017**, *7*, 730. [CrossRef]
67. Mira Geoscience. *Geohazmap Workflow Earth Modelling Solutions for Mining. Montreal: Mira Geoscience*; Mira Geoscience: Montreal, QC, Canada, 2007.

Article

Fusion Network for Change Detection of High-Resolution Panchromatic Imagery

Wahyu Wiratama and Donggyu Sim *

Department of Computer Engineering, Kwangwoon University, Seoul 139701, Korea; wiratama@kw.ac.kr
* Correspondence: dgsim@kw.ac.kr; Tel.: +82-2941-6470

Received: 28 January 2019; Accepted: 2 April 2019; Published: 5 April 2019

Abstract: This paper proposes a fusion network for detecting changes between two high-resolution panchromatic images. The proposed fusion network consists of front- and back-end neural network architectures to generate dual outputs for change detection. Two networks for change detection were applied to handle image- and high-level changes of information, respectively. The fusion network employs single-path and dual-path networks to accomplish low-level and high-level differential detection, respectively. Based on two dual outputs, a two-stage decision algorithm was proposed to efficiently yield the final change detection results. The dual outputs were incorporated into the two-stage decision by operating logical operations. The proposed algorithm was designed to incorporate not only dual network outputs but also neighboring information. In this paper, a new fused loss function was presented to estimate the errors and optimize the proposed network during the learning stage. Based on our experimental evaluation, the proposed method yields a better detection performance than conventional neural network algorithms, with an average area under the curve of 0.9709, percentage correct classification of 99%, and Kappa of 75 for many test datasets.

Keywords: change detection; convolutional network; deep learning; panchromatic; remote sensing

---

1. Introduction

Change detection is a challenging task in remote sensing, used to identify areas of change between two images acquired at different times for the same geographical area. Such detection has been widely used in both civilian and military fields, including agricultural monitoring, urban planning, environment monitoring, and reconnaissance. In general, change detection involves a preprocessing step, feature extraction, and classification or clustering algorithm to distinguish changed and unchanged pixels. To obtain a good performance, the selected classification or clustering algorithm plays an important role in the field of change detection.

In prior studies, statistical approaches have been proposed to identify a change [1–3]. A corresponding maximal invariant statistic is obtained by analyzing a suitable group of transformations leaving problem invariant [2]. Then, a general problem of testing equality among M covariance metrices in the complex-valued Gaussian case is analyzed for synthetic aperture radar (SAR) change detection. A sample coherence magnitude as a change metric has been proposed by [3]. A new maximum-likelihood temporal change estimation and complex reflectance change detection is used for SAR coherent temporal change detection. Currently, a classification or clustering is becoming one approach to be used for change detection in remote sensed images by employing supervised or unsupervised learning, respectively. Feature selection and feature extraction are important aspects in this approach. Several detection algorithms using two images have been proposed with different features for different types of applications [3–19]. The methods used for change detection have mostly been designed to extract changed features such as in a difference image (DI) [3–9], a local change vector [10], or a texture vector [11–13]. A DI is a common feature used to represent a change in information through the subtraction of temporal images. Local change vectors have also been used by applying neighbor pixels to avoid a direct subtraction based on the log ratio. This method computes a mean value of the log ratio of temporal neighbor pixels. A texture vector [11–13] is employed to extract statistical characteristics. These changed features are then fed into a classification or clustering algorithm to determine changed/unchanged pixels. Some unsupervised change detection methods have been proposed based on the fuzzy c-mean (FCM) algorithm [14,16]. Such approaches are useful when labels in the training stages are unavailable. The learning algorithms in the aforementioned studies are based on observed data without any additional information, therefore, their application leads to overfitting for invariant changes. Furthermore, they do not yield a reasonably good performance in the change detection rates because they do not incorporate accurate information without supervision. Therefore, supervised change detection methods, such as a support vector machine (SVM) [11,16–18], have been proposed. The basic SVM can apply a binary classification to changed or unchanged pixels with texture information or using a change vector analysis. These algorithms are not perfect in terms of incorporating accurate and full statistical characteristics for large multi-dimensional data. Furthermore, they do not yield the best detection performance for new datasets.

Recently, a deep convolution neural network (DCNN) was developed to produce a hierarchy of feature-maps such as learned filters. The aforementioned DCNN can automatically learn a complex feature space from a huge amount of image data. A DCNN can achieve a superior performance compared to conventional classification algorithms. A restricted Boltzmann machine (RBM) [19], a convolutional neural network (CNN) [20–22], and deep belief networks (DBNs) [23] have been proposed for use in change detection. Such change detection algorithms based on deep learning yield a relatively good performance in terms of the detection accuracy. However, most can be categorized into front-end differential change detection using low-level features such as a difference image as a feature input of their networks, resulting in sensitivity to several deteriorated conditions caused by geometric/radiometric distortions, different viewing angles, and so on. This front-end differential change detection conducts an early feature extraction of two image inputs into a single-path network. In contrast, back-end differential detection methods by employing dual-path networks have been proposed for fusing higher-level features with a long short-term memory (LSTM) model [24] to avoid the use of low-level difference features such as a difference image. In addition, a Siamese convolutional network (SCN) [25–27] and dual-DCN (dDCN) [28] were also proposed to detect changed areas by measuring the similarity with high-level network features. These algorithms achieve a relatively good performance, although false negatives are still observed.

To reduce false positives and false negatives in change detection, a fusion network incorporating low- and high-level feature spaces in neural networks was proposed in this paper. For low-level differential features, the difference image is fed into the front-end differential DCN (FD-DCN). For a high-level differential feature, a back-end differential dDCN (BD-dDCN) is employed. In addition, a two-stage decision algorithm is incorporated for post-processing to enhance the detection rate during the inference stage. The intersection and union operations are employed to validate the change map. First, an intersection operation is used to avoid false positives. The second-stage decision operates a union by considering the local information of the first decision. This stage is developed to validate and repair the change map from the first decision. In addition, this study introduces a new loss function that combines a contrastive loss and weighted binary cross entropy loss function to optimize high- and low-level differential features, respectively. In our experiment, we found that the proposed algorithm can yield a better performance than existing algorithms by achieving an average area under the curve (AUC) of 0.9709, a percentage correct classification (PCC) of 99%, and a Kappa of 75 for several test datasets.

This work contributes three main key features as follows. (1) Unlike the mentioned existing works above, we propose a fusion network by combining a front- and back-end networks to perform the low- and high-level differential detection in one structure. (2) A combining loss function between contrastive loss and binary cross entropy loss is proposed to accomplish fusion of the proposed networks in training stage. (3) The two-stage decision as a post-processing is presented to validate and ensure the changes prediction at the inference stage to obtain better the final change map.

This paper is organized into five sections. In Section 2, related studies are briefly described. Section 3 presents the proposed algorithm in detail. Section 4 describes and analyzes the experiment results. Finally, we provide some concluding remarks regarding this research.

2. Deep Convolutional Network and Related Studies on Change Detection

Deep neural architectures with hundreds of hidden layers have been developed to learn high-level feature spaces. The recently developed convolutional neural network (CNN) is a deep learning architecture that has been shown to be effective in image recognition and classification [29]. The CNN architecture employs multiple convolutional layers, followed by an activation function, resulting in the development of feature maps. The rectified linear unit (ReLU) is widely used as the activation function in many CNN architectures. To progressively gather global spatial information, the feature maps are sub-sampled by the pooling layer. The final feature maps are connected to a fully connected layer to produce the class probability outputs (P_{class}), as shown in Figure 1. During the training stage, an objective loss such as cross-entropy is computed. All of the weighting parameters of the network are updated to reduce the cost function using the back-propagation algorithm.

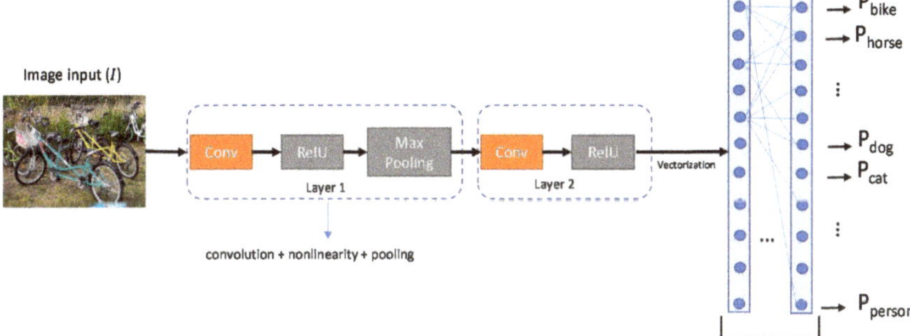

Figure 1. Convolutional neural network (CNN) architecture.

The related studies on change detection based on deep learning can be categorized into two categories based on the type of network that is used: A front-end differential network (FDN) and a back-end differential network (BDN). The front-end network uses low-level differential features such as a DI or joint feature (JF) as the feature input of the network, as shown in Figure 2a. In this case, a network with a single-path architecture receives the extracted DI as low-level differential features of the temporal images to identify changed pixels. Several studies based on an FDN have been proposed to improve the performance of the change detection rate. In addition, a deep neural network (DNN) is applied to detect objects from synthetic aperture radar (SAR) data [30]. The differential feature of temporal data is employed instead of a DI. This feature is used to solve the initial weight problem through pre-training using the restricted Boltzmann machine (RBM) algorithm. These pre-trained weights are then fed into the initial weights of the DNN during the training stage. In contrast, unsupervised change detection has been proposed by combining DBNs with a feature change analysis [23]. The feature maps of temporal input images are obtained using the DBN. The magnitude and direction of these feature maps are analyzed to distinguish the types of feature changes using an unsupervised fuzzy C-means algorithm. Other unsupervised systems have been proposed by combining a sparse autoencoder (SAE), unsupervised clustering, and a CNN to overcome the change detection problem without supervision [20]. First, a DI is computed using a log-ratio operator. The feature maps of the DI are extracted through the SAE and clustered into change classes as the labels for the training CNN. Next, some feature maps extracted by the SAE are taken as the training data for the CNN. In addition, an autoencoder and multi-layer perceptron (MLP) are combined to identify changed pixels [31]. Change detection using faster R-CNN has been proposed for high-resolution images [32]. This work detects changed areas with bounding boxes. The DI is extracted and then fed into faster R-CNN to detect changed locations. Each of these deep learning algorithms tackles the change detection problem using a front-end differential network. This network identifies changes by observing low-level feature such as the DI, which is sensitive to various distortions, including geometric and radiometric distortions, and different viewing angles. Another approach of FDN to detect the changes has been proposed by joining feature inputs (JF) [23]. Two temporal images are concatenated and they are fed into DBN to avoid a DI for change detection. However, by joining the features in the early network causes both low-level differential inputs to be dependently learned in the single network. It is for global change detection, resulting in more false positives.

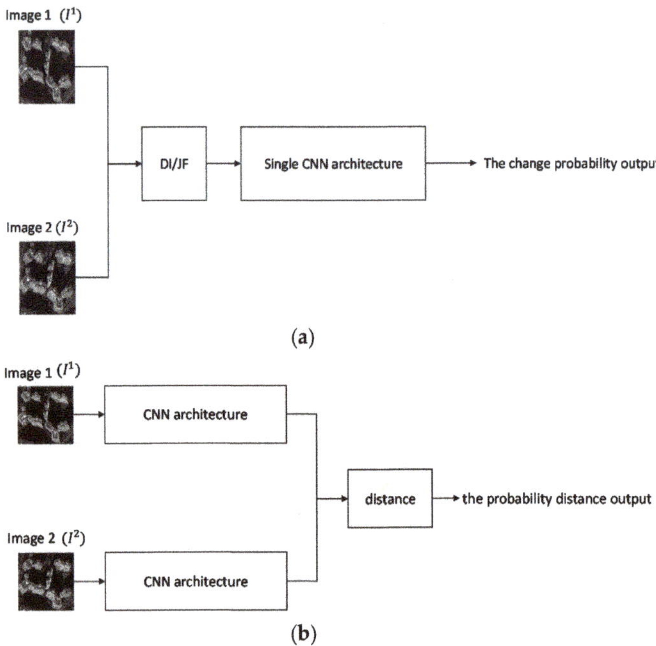

Figure 2. Front-end differential network (FDN) and back-end differential network (BDN) architectures: (**a**) Difference image (DI)/Joint features (JF) + single CNN, and (**b**) Dual-CNN.

Alternative algorithms for change detection were introduced by employing a high-level differential feature with a dual-path network, as shown in Figure 2b. Siamese CNN (SCNN) was proposed to detect changed areas for multimodal remote sensing data [27]. This architecture was employed to learn the different characteristics between multimodal remote sensing data. This approach learns the feature map of temporal images in each path network. The Euclidean distance was employed to measure the similarity at the back-end of the network. A similar method was developed based on an SCNN for optical aerial images [25]. A deep CNN was proposed by producing a change detection map directly from two images [33]. A change map was evaluated using the pixel-wise Euclidean distance from high-dimensional feature maps. Another method was proposed that incorporates a deep stacked denoising autoencoder (SDAE) and feature change analysis (FCA) for multi-spatial resolution change detection [34]. In the aforementioned study, denoising autoencoders were stacked to learn local and high-level features for unsupervised learning. Then, the inner relationship between the multi-resolution image pair was exploited by building a mapping neural network to identify any change representations. A dual-dense convolutional network was presented by incorporating information from neighbor pixels [28]. In the aforementioned study, a dense connection was used to enhance the features of the changed map information. All of the above-mentioned BDN architectures yield good performances by inspecting high-level differential features, which can reduce false positives. However, a BDN can achieve higher sensitivity and specificity through high-level differential features.

Although a high-level differential network can improve the sensitivity and specificity, the false negative rate is still too high for practical applications. The FDN architecture can achieve a relatively higher true-positive rate regardless of the number of false positives. In addition, the BDN architecture can reduce the false-positive rate by producing some false negatives caused by strict decision criteria in high-level differential features. In this work, an FDN and a BDN were fused to employ the advantages of both. A post-processing step was then employed during the inference stage to obtain the final decision for change detection.

3. Proposed Fusion Network for Change Detection with Panchromatic Imagery

In general, a change detection system involves a pre-processing step to reduce geometric and radiometric distortions for better results. A radiometric correction is applied to remove atmospheric effects for a time-series image analysis. In this study, a radiometric correction was applied by converting digital numbers (DNs) into a radiance value. Then, the top-of-atmosphere (TOA) reflectance values were computed using the gain and offset values provided by a satellite provider. In addition, to ensure that the pixels in the image were in their proper geometric position on the Earth's surface, a geometric correction was applied. The parameters (polynomial coefficients) of the polynomial functions were estimated using least square fitting with ground control points (GCPs) identified in an unrectified image and corresponding to their real coordinates. A digital elevation model (DEM), namely, shuttle radar topography mission DEM (SRTM DEM), was then used to correct the optical distortion and terrain effect. The corrected images were then incorporated into the proposed network to detect changes.

To achieve a change detection, the proposed network employs a fusion network by fusing the FDN and BDN architectures. Dual outputs were generated to solve low-level differential and high-level differential problems. For the training stage, a contrastive loss function and a weighted binary loss function were combined to optimize the proposed fusion network parameters. In addition, a pre-processing step was applied to validate and ensure false changes during the inference stage. Intersection and union operations were then applied from the dual outputs of the proposed network. According to the proposed change detection, the false-positive and false-negative rates could be reduced, resulting in high sensitivity and specificity for a proper change detection. Symbols used in the proposed method are tabulated in Table 1.

Table 1. Symbols used in the proposed fusion network for change detection.

Symbol	Description
I^1 and I^2	Cropped temporal input image in time 1 and 2, respectively
N^1 and N^2	Patch network 1 and 2 correspond to the back-end network
N^3	Patch network 3 correspond to the front-end network
F^i_{l,d_r}	Feature maps of the l-th layer at the r-th dense block and the i-th network
P^1 and P^2	Outputs of N^1 and N^2, respectively
D	Dissimilarity distance
O	Change detection probability output of N^3
H^i_{l-1,d_r}	Incorporation process of a batch normalization (BN), a 3×3 convolution, and ReLU of the $(l-1)$-th layer at the r-th dense block and the i-th network
$[F^i_{0,d_r}, F^i_{1,d_r}, \ldots, F^i_{l-1,d_r}]$	A concatenation of the feature-maps of all of previous layers, layer 0, $\ldots$, and layer $(l-1)$
L	Proposed loss function
E_c	Contrastive loss function
E_B	Weighted binary cross entropy loss function
Y	Ground truth
Ls and L_D	Partial loss function for a pair of similar and dissimilar pixels, respectively
m	Margin value
α	Weighted loss
W	Proposed weighted function
C and U	Changed and unchanged numbers of pixels, respectively
N	The number of full dataset
β_c and β_u	Penalization weights for false-negative and false-positive errors, respectively
M_1	Change map for first prediction
M_2	Change map for second prediction
Nb	Local information of M_1
T	Tested temporal images
m and n	Size of T
s	Size of I

3.1. Fusion Network for Change Detection

For a change detection, an FDN architecture is commonly used for identifying changed pixels. Such an architecture uses low-level differential features that are relatively sensitive to noises. It is caused by direct low-level features comparison, which misalignments of geometric error and a different angle view are very influential. This FDN assigns a DI or JF to a single path network. They conduct dependent learning of both low-level features together which lead to hard learning for invariant changes and above-mentioned noisy conditions. Thus, this approach would produce a global change detection, resulting in true positives and more false positives. In addition, BDN architectures are designed to avoid low-level differential features, thereby reducing the false-positive detection rate. These architectures apply strict identification for a high-level differential, which may cause some false negatives. Therefore, an FDN is suitable in terms of the true-positive detection rate, and a BDN is extremely reliable for overcoming false positives. To obtain a proper change detection, a fusion network architecture is proposed by fusing an FD-DCN and a BD-dDCN with a dense-connectivity of the convolution layers, as shown in Figure 3. There are three branch networks, N^1, N^2, and N^3, receiving two temporal images (I^1 and I^2) in which N^1 and N^2 correspond to the back-end network, and N^3 refers to the front-end network by concatenating these two inputs (I^1 and I^2). A dense convolutional connection was employed in the proposed fusion network to enhance the feature representation [35]. This dense architecture is very effective at covering invariant change representations by reusing all preceding feature maps of the network. The proposed network was designed using dual outputs, namely, the dissimilarity distance (D) and change probability (O) at the last layer, corresponding to the back-end and front-end networks, respectively. Let us assume that the feature maps of the l-th layer at the r-th dense block and the i-th network are computed as:

$$F^i_{l,d_r} = H^i_{l-1,d_r}([F^i_{0,d_r}, F^i_{1,d_r}, \ldots, F^i_{l-1,d_r}]), \quad r = 0,1; \ i = 1,2,3, \tag{1}$$

where $[F^i_{0,d_r}, F^i_{1,d_r}, \ldots, F^i_{l-1,d_r}]$ indicates a concatenation of the feature-maps of all of previous layers, layer 0, ..., and layer $(l-1)$. In addition, $H(\cdot)$ incorporates a batch normalization (BN), a 3×3 convolution, and ReLU. A pair of temporal images were cropped into two patches (40×40) (I^1 and I^2) by sliding the window and fed into N^1 and N^2, respectively. The dissimilarity distance (D) was then computed based on the Euclidean distance, which is defined as follows:

$$D = \|P^1 - P^2\|_2 \tag{2}$$

where P^1 and P^2 are the outputs of N^1 and N^2 activated by sigmoid function, respectively. The proposed method applies a pixel-wise change detection by inspecting the neighboring pixels. The 40×40 patch images identify a change corresponding to the center pixel of the patch. Thus, when the value of D is close to 1, the center of I is assigned to a changed pixel. In addition, I^1 and I^2 were concatenated to be fed into N^3. The same dense convolution architecture was employed in this branch network to generate the change detection probability (O). The dual outputs (D and O) are a result of this fusion network. In addition, a post-processing step during the inference stage was proposed based on these outputs (D and O) to achieve a proper prediction.

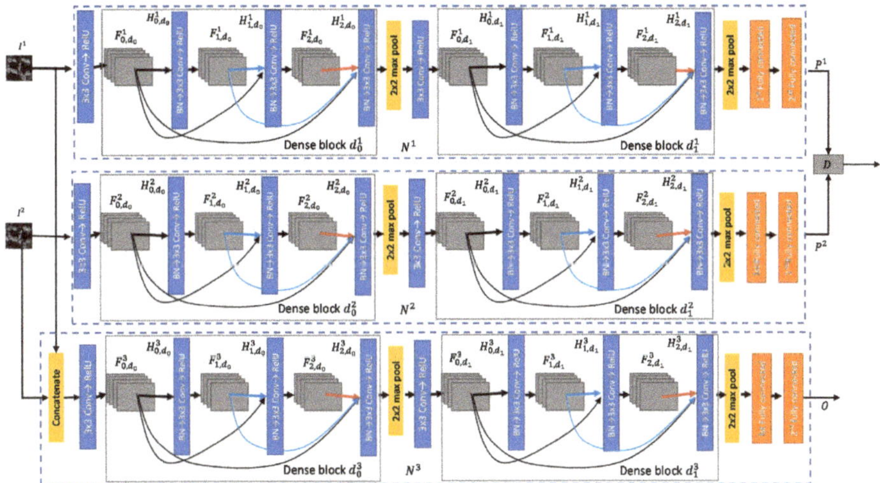

Figure 3. The proposed fusion network architecture for change detection.

3.2. Training of the Proposed Fusion Network for Change Detection

During the training stage, this paper introduced a loss function (L) by combining the contrastive loss (E_c) [36] and weighted binary cross entropy loss (E_B) as defined by:

$$L = \alpha E_c + (1 - \alpha) E_B \tag{3}$$

where α is a weight loss. Given a training set consisting of 40×40 image pairs and a binary label of the ground truth (Y), the proposed network was trained using backpropagation. Here, E_c was applied to optimize the parameters of N^1 and N^2, and is as computed as follows [36]:

$$E_c = \sum_i (1 - y_i) L_S(D_i) + (y_i) L_D(D_i) \tag{4}$$

where $y = 1$ is a changed pixel and $y = 0$ is an unchanged pixel. In addition, Ls is a partial loss function for a pair of similar pixels, and L_D is a partial loss function for a pair of dissimilar pixels, as defined by [36]:

$$Ls = \frac{1}{2}(D_i)^2 \tag{5}$$

$$L_D = \frac{1}{2}(max\{0, m - D_i\})^2 \tag{6}$$

The value of m is set to 1 as the margin value. In addition, E_B was used to optimize the parameters of N^3, as defined by:

$$E_B = -\sum_i W_i(y_i \log(O_i) + (1 - y_i)\log(1 - O_i)) \tag{7}$$

where W is the proposed weighted function used to penalize the false-positive and false-negative errors. Thus, W is computed by:

$$W_i = y_i\left(\beta_c\left(1 - \frac{C}{N}\right)\right) + (1 - y_i)\left(\beta_u\left(1 - \frac{U}{N}\right)\right) \tag{8}$$

where β_c and β_u are penalization weights for false-negative and false-positive errors, respectively. Moreover, C and U are the changed and unchanged numbers of pixels in the full dataset (N), respectively.

The proposed network was trained using a stochastic gradient descent (SGD) with the training parameters, including 0.001, 1×10^{-6}, and 0.9 as the learning rate, decay, and momentum, respectively. In addition, the epoch number was set to 30. The value of α was set to 0.7 to further penalize E_c. It was to prevent false positives, which are possible in a back-end network. The goal of prediction through the front-end was to obtain better true-positive rates regardless of the number of false positives. Thus, the false negatives were penalized ten times more than false positives, namely, $\beta_c = 10$ and $\beta_u = 1$.

3.3. Dual-Prediction Post-Processing for Change Detection

During the inference stage, post-processing was introduced using dual-prediction for change detection. In the counting rule, binary hypotheses can be passed to a fusion center, which then decides which one of the two hypotheses is true [37]. The proposed algorithm employed a hard-logical rule using an AND and OR operation with the same probability output thresholds to predict a changed pixel. This aimed to validate and ensure the change detection based on the proposed fusion network outputs (D and O). There were two steps to applying this post-processing. First, an intersection operation was employed to obtain a strict prediction and avoid false positives. Assume that ($m \times n$) images (T) will be tested using the proposed fusion network, resulting in an ($m \times n$) change map (M_1). This prediction was conducted by sliding in the raster scan order, as shown in Figure 4. The inputs (I^1 and I^2) with the central pixel position, x and y, were assigned to the proposed fusion network to generate the values of D and O. If D and O identified a changed pixel, then $M_1(x, y)$ was set to a value of 1; otherwise, it was set 0. This was performed for the entire image T.

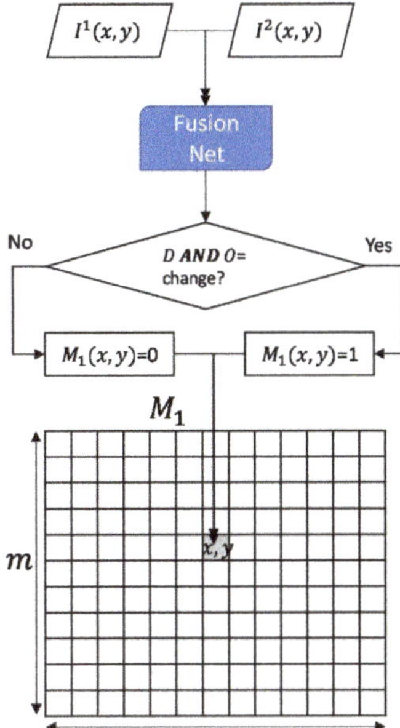

Figure 4. First prediction flowchart.

Then, the second prediction was performed to ensure the first prediction, as shown in Figure 5. Let us assume that ($m \times n$) M_2 was a change map for the second prediction. Initially, a prediction noise was investigated by analyzing the local information from M_1 by computing Nh, as defined by:

$$Nb(x,y) = \sum_{i=x-\frac{q}{2}}^{x+\frac{q}{2}} \sum_{j=y-\frac{q}{2}}^{y+\frac{q}{2}} M_1(i,j). \tag{9}$$

where $Nb(x,y)$ computes the local information $M_1(x,y)$ using a $q \times q$ window. If the value of $Nb(x,y)$ is greater than the input size s (40) divided by 4, then the second prediction is applied, otherwise, $M_2(x,y)$ is assigned to 0. A union operation was operated from D and O for the second prediction. When it returned the changed pixel, $M_2(x,y)$ was assigned a value of 1, otherwise, it was assigned a value of 0. The final change map was obtained based on the result of M_2.

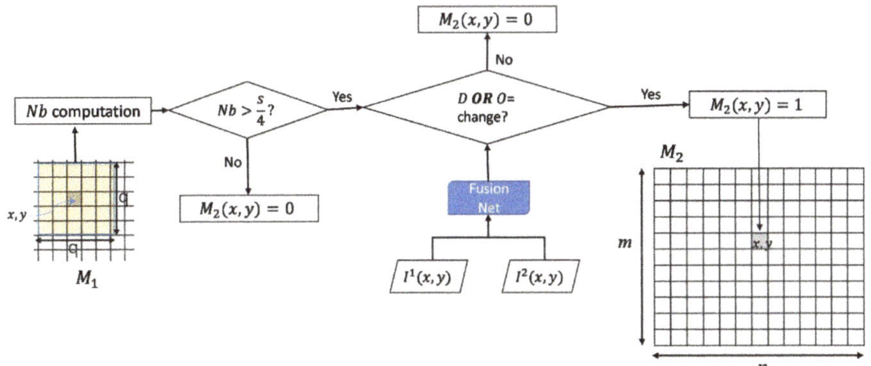

Figure 5. Second prediction flowchart.

4. Experimental Evaluation and Discussion

This study used a dataset of panchromatic imageries, which provided 0.7 GSD captured by the KOMPSAT-3 sensor. For the training dataset, this study used a scene of overlapped images (1214 × 886) over Seoul, South Korea, as shown in Figure 6. These images were cropped into a 40 × 40 sliding patch, and the center pixels of the cropped patch pair were labeled based on the ground truth.

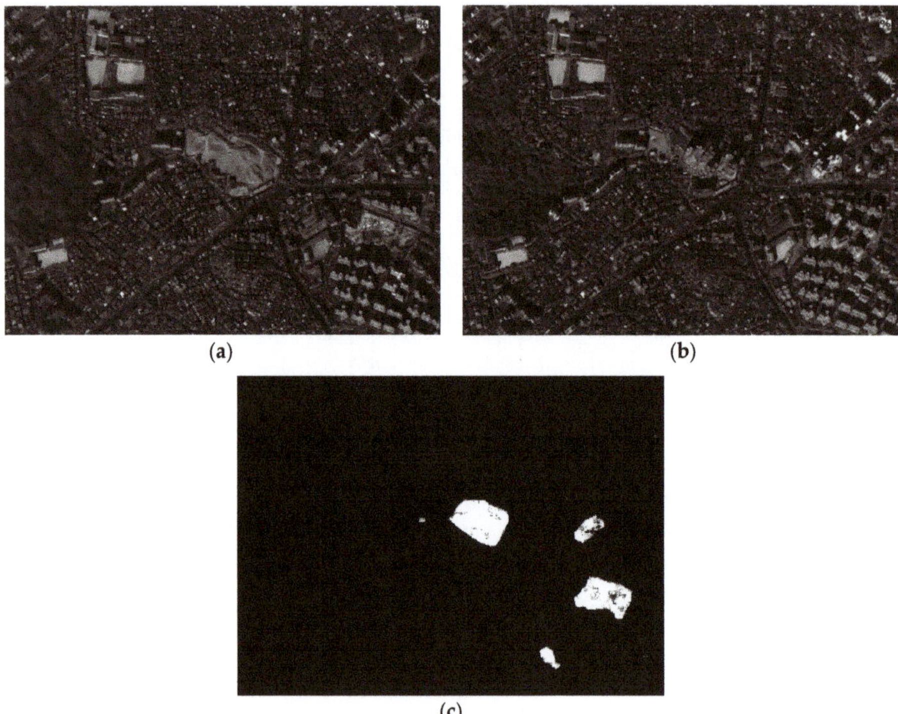

(a)　　　　　　　(b)

(c)

Figure 6. Training dataset: (**a**) Image acquired in March 2014, (**b**) image acquired in December 2015, and (**c**) the ground truth.

Figure 6 shows an area containing completed changes and changes under contraction. In addition, these images have many tall buildings, roads, houses, and forests to be trained for solving the misalignment and viewing angle problems. In our experiments, to assess the effectiveness of the proposed change detection system, three areas of the panchromatic datasets were used, namely, Areas 1, 2, and 3, as shown in Figure 7.

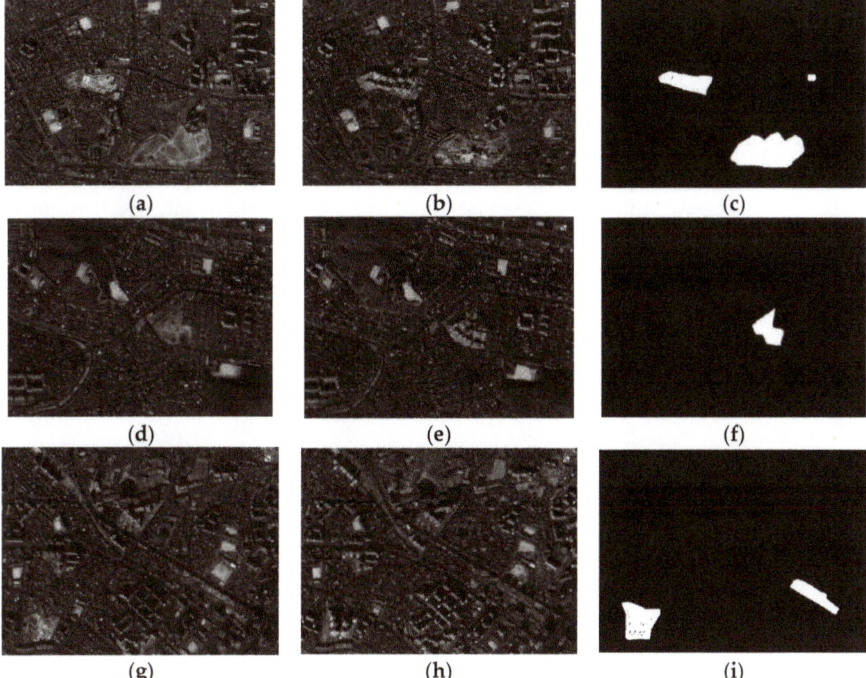

Figure 7. Experiment dataset: (**a**) Input image for Area 1 (March 2014), (**b**), input image for Area 1 (October 2015), (**c**) ground truth for Area 1, (**d**) input image for Area 2 (March 2014), (**e**) input image for Area 2 (October 2015), (**f**) ground truth for Area 2, (**g**) input image for Area 3, (March 2014), (**h**) image input for Area 3 (October 2015), and (**i**) ground truth for Area 3.

The images in Figure 7 were acquired in March 2014 and October 2015 over different areas of Seoul, South Korea. Each image pair had been radiometrically corrected and had a geometric misalignment of approximately ±6 pixels. In addition, it also had a different angle view, which cannot be resolved without precise 3D building models. Area 1 was located in a downtown part of Seoul, and contained areas changed through building construction. Moreover, the urban area had tall buildings and roads. These datasets included several factors of geometric distortion, misalignments, and different viewing angle effects, which could lead to many false changes. In addition, Area 2 represented a downtown area near a forest. These two images were acquired in different seasons. It was difficult to achieve robustness to seasonal changes for practical applications. Area 3 had many tall buildings, making it difficult to achieve an accurate detection rate owing to the different viewing angles.

In this study, the receiver operating characteristic (ROC) curve, AUC, PCC, and Kappa coefficient were used to quantitatively evaluate the performance of the proposed method. Moreover, to evaluate the effectiveness of the proposed method, it was compared with conventional algorithms having FDN and BD-dDCN architectures [28]. A DI and JF were incorporated into a single-path CNN architecture (DI + CNN and JF + CNN). These architectures included eight depth convolutional, two pooling, and two fully connected layers, which were the same as the proposed depth layers. In addition, Dual-DCN [28] was also compared to the proposed method.

Figure 8 shows an ROC curve, which indicates that the proposed method could achieve a better AUC compared to the existing algorithms. For Area 1, the proposed method yielded an AUC of 0.9904, which means that it could identify changes approximating the ground truth. It had a slightly higher dual-DCN of 0.9878. The FDN architectures provided an AUC lower than the proposed algorithm which JF + CNN and DI + CNN gave an AUC of approximately 0.9509 and 0.7060, respectively. Furthermore, the proposed method significantly outperformed the other algorithms with regard to the AUC for Areas 2 and 3 because it could properly detect the change events with the incorporation of low- and high-level differential features. Table 2 summarizes the PCC and Kappa values of the different methods applied for the three areas. The proposed method showed a higher PCC in Areas 1 and 3. The dual-DCN achieved a slightly higher PCC than the proposed method in Area 2. However, in terms of the Kappa value, the proposed fusion network outperformed all other existing algorithms. The proposed method achieved a Kappa value of 75.16 on average, which means that it yielded a good agreement in terms of the results.

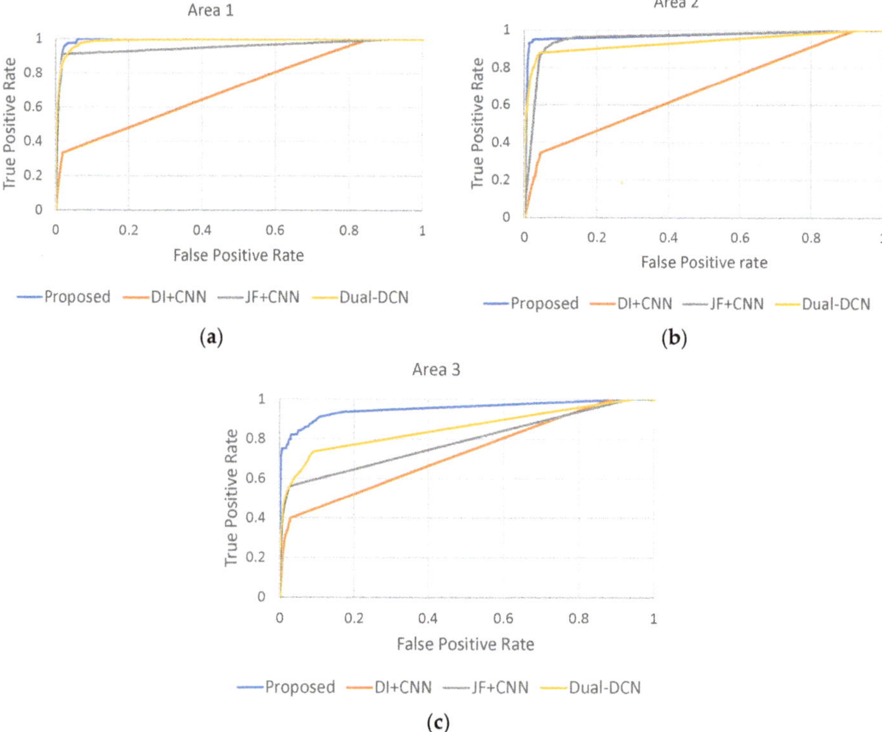

Figure 8. Receiver operating characteristic (ROC) for (**a**) Area 1, (**b**) Area 2, and (**c**) Area 3.

Table 2. Quantitative assessment of the existing and proposed algorithms.

Algorithm	Area 1			Area 2			Area 3		
	AUC	PCC	Kappa	AUC	PCC	Kappa	AUC	PCC	Kappa
DI + CNN	0.7060	0.9458	36.8938	0.6764	0.9571	11.8939	0.7213	0.9855	33.2651
JF + CNN	0.9509	0.9775	79.7190	0.9536	0.9570	29.7251	0.7847	0.9732	47.6066
Dual-DCN	0.9878	0.9774	78.4277	0.9546	0.9922	60.0070	0.8515	0.9751	50.7542
Proposed	0.9904	0.9782	80.7942	0.9707	0.9902	65.9929	0.9517	0.9892	78.6898

Figure 9 shows the change map results when applying the existing and proposed algorithms. Visually, the proposed method achieved a much better result than the existing algorithms. In Area 1, the proposed fusion network nearly approximated the ground truth. It could reduce the number of false positives while preserving the true positives. The proposed network produced a cleaner change map than the existing algorithms regarding false positives. Moreover, the proposed algorithm yielded reasonably good results for Areas 2 and 3. The proposed method significantly reduced the number of false positives and enhanced the true positives. This is caused by the proposed fusion network, which was designed and trained for low- and high-level differential problems. In addition, a post-processing step was employed to validate and repair the change map.

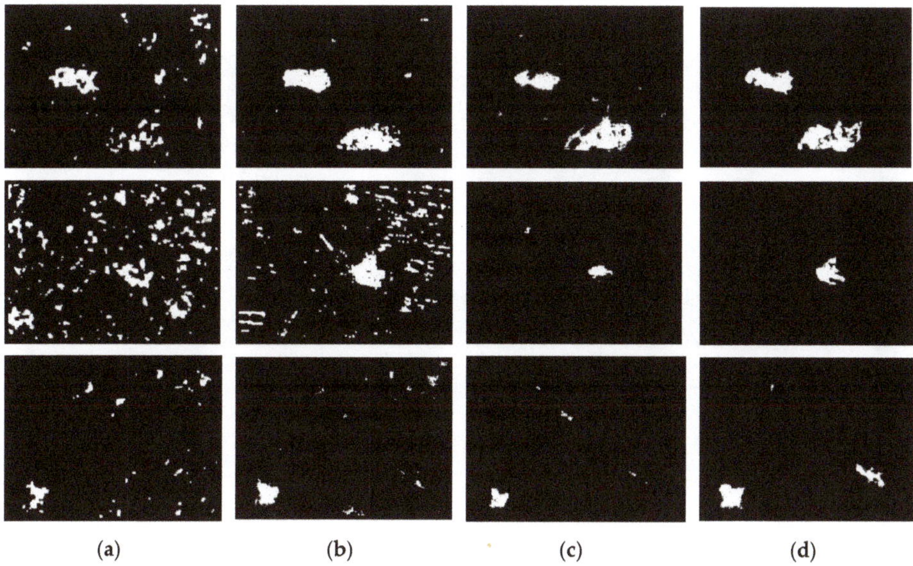

(a)   (b)   (c)   (d)

Figure 9. Detection results for three areas when using the existing and proposed algorithms: (a) DI + CNN, (b) JF + CNN, (c) dual-DCN, and (d) the proposed fusion network.

To evaluate the effectiveness of the proposed two-stage decision, the proposed algorithm also was compared to each individual network output (D and O) and the other decision method between two outputs of the proposed fusion network based on the mean operation. In addition, another single output fusion network (SOFN) architecture was designed same as the proposed fusion network architecture by fusing D and O outputs for more comparisons. This network was trained with the binary cross entropy loss function by the same training parameters. The objective and subjective evaluation are presented in Table 3 and Figure 10, respectively.

Table 3. Quantitative assessment of single output decision and proposed algorithms.

Network Outputs	Area 1			Area 2			Area 3		
	AUC	PCC	Kappa	AUC	PCC	Kappa	AUC	PCC	Kappa
D	0.9206	0.9655	65.3497	0.8154	0.9854	33.8273	0.8410	0.9794	55.1115
O	0.9357	0.9607	61.9808	0.8948	0.9879	50.0476	0.8667	0.9436	31.3879
Mean	0.9886	0.9781	78.4481	0.9588	0.9875	52.7712	0.9165	0.9803	59.5712
SOFN	0.9685	0.9661	65.7595	0.8903	0.9897	52.6660	0.8658	0.9820	61.2094
Proposed	0.9904	0.9782	80.7942	0.9707	0.9902	65.9929	0.9517	0.9892	78.6898

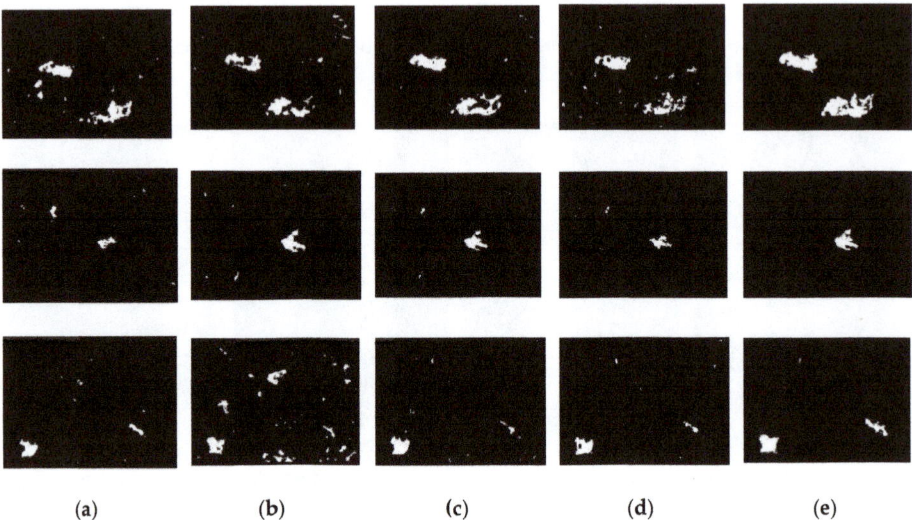

(a)   (b)   (c)   (d)   (e)

Figure 10. Detection results for three areas when using an individual network output and the proposed algorithms: (**a**) D, (**b**) O, (**c**) Mean, (**d**) SOFN, and (**e**) the proposed fusion network with a two-stage decision.

According to Table 3, the proposed two-stage decision shows better performance compared to individual outputs and mean operation. In term of AUC, PCC, and Kappa, the proposed gave significantly better results than that by individual outputs (D and O). Figure 10 shows that the output O produced more true positives regardless of the number of false positives. However, the output D can reduce the false-positive rate. This condition makes the proposed two-stage decision working as the goal that detection rates can be accelerated by the combining of two network outputs with a two-stage decision. In addition, the proposed algorithm still outperformed the mean operation between two network outputs for all areas. SOFN with the single output also gave worse results than the proposed one caused by no validation decision of post-processing for change detection. The proposed fusion network was employed with a two-stage decision to obtain a better prediction rate.

Regarding time complexity, the proposed fusion network consumed more computational complexity than the existing algorithm by a factor of approximately two over the dual-path network and three with the single-path network. It was due to the proposed architecture designed with more network paths. In addition, the proposed two-stage decision required an additional prediction process in the inference stage. Let us see that the general total time complexity of dense convolutional network [35] was $O(K^2)$ run-time complexity for a depth K network [38]. Dual-DCN [28] employed dual-path dense convolutional network with the depth of 6 that produced a run-time complexity of $O(2 \cdot 6^2)$. The proposed fusion network included three-path dense convolutional networks with the same depth by fusing back- and front-end differential network architectures, resulting in a run-time complexity of $O(3 \cdot 6^2)$. In the inference stage, a two-steps decision for the proposed made the run-time be $O(2 \cdot (3 \cdot 6^2))$ that gave it an expensive computational complexity while producing a better result.

5. Conclusions

This paper presented a robust fusion network for detecting changed/unchanged areas in high-resolution panchromatic images. The proposed method learns and identifies the changed/unchanged areas by combining front- and back-end neural network architectures. The dual outputs are efficiently incorporated for low- and high-level differential features with a modified loss function that combines the contrastive and weighted binary cross entropy losses.

In addition, a post-processing step was applied to enhance the sensitivity and specificity from false changes/unchanged detections based on the neighboring information. We found through qualitative and quantitative evaluations that the proposed algorithm can yield a higher sensitivity and specificity compared to the existing algorithms, even under noisy conditions such as geometric distortions and different viewing angles

For further work, the proposed algorithm can be extended for other modalities such as multi-spectrum images, Pan-sharpening, and SAR data. In addition, the proposed algorithm requires expensive time complexity caused by pixel-wise detection with a two-stage decision. To accelerate run-time complexity, block-wise prediction design would also be a focus of future work.

Author Contributions: All authors contributed to the writing of the manuscript. W.W. and D.S. conceived, designed the algorithm, and analyzed the data. W.W. developed it and conducted the experiments and analyzed the data. D.S. supervised the study.

Funding: This research was supported by the Ministry of Science and ICT (MSIT), Korea, under the Information Technology Research Center (ITRC) support program (IITP-2018-2016-0-00288) supervised by the Institute for Information & communications Technology Promotion (IITP) and Basic Science Research Program through the National Research Foundation of Korea (NRF) funded by the Ministry of Science, ICT & Future Planning (NRF-2018R1A2B2008238).

Conflicts of Interest: The authors declare no conflicts of interest.

References

1. Ciuonzo, D.; Salvo Rossi, P. DECHADE: Detecting slight changes with hard decisions in wireless sensor networks. *Int. J. Gen. Syst.* **2018**, *47*, 535–548. [CrossRef]
2. Ciuonzo, D.; Carotenuto, V.; de Maio, A. On multiple covariance equality testing with application to SAR change detection. *IEEE Trans. Signal Proc.* **2017**, *65*, 5078–5091. [CrossRef]
3. Wahl, D.E.; Yocky, D.A.; Jakowatz, C.V.; Simonson, K.M. A new maximum-likelihood change estimator for two-pass SAR coherent change detection. *IEEE Transon. Geosand. Remote Sens.* **2016**, *54*, 2460–2469. [CrossRef]
4. Coppin, P.R.; Bauer, M.E. Digital change detection in forest ecosystems with remote sensing imagery. *Remote Sens. Rev.* **1996**, *13*, 207–234. [CrossRef]
5. Bazi, Y.; Bruzzone, L.; Melgani, F. Automatic identification of the number and values of decision thresholds in the log-ratio image for change detection in SAR images. *IEEE Geosci. Remote Sens. Lett.* **2006**, *3*, 349–353. [CrossRef]
6. Singh, K.K.; Mehrotra, A.; Nigam, M.J.; Pal, K. Unsupervised change detection from remote sensing using hybrid genetic FCM. In Proceedings of the IEEE 2013 Students Conference on Engineering and Systems (SCES), Allahabad, India, 12–14 April 2013; pp. 1–5.
7. Bi, C.; Wang, H.; Bao, R. SAR image change detection using regularized dictionary learning and fuzzy clustering. In Proceedings of the 2014 IEEE 3rd International Conference on Cloud Computing and Intelligence Systems (CCIS), Shenzhen, China, 27–29 November 2014; pp. 327–330.
8. Gong, M.; Zhou, Z.; Ma, J. Change detection in synthetic aperture radar images based on image fusion and fuzzy clustering. *IEEE Trans. Image Process.* **2012**, *21*, 2141–2151. [CrossRef] [PubMed]
9. Gong, M.; Su, L.; Jia, M.; Chen, W. Fuzzy clustering with a modified MRF energy function for change detection in synthetic aperture radar images. *IEEE Trans. Fuzzy Syst.* **2014**, *22*, 98–109. [CrossRef]
10. Johnson, R.D.; Kasischke, E.S. Change vector analysis: A technique for the multispectral monitoring of land cover and condition. *Int. J. Remote Sens.* **1998**, *19*, 411–426. [CrossRef]
11. Gao, F.; Zhang, L.; Wang, J.; Mei, J. Change detection in remote sensing images of damage areas with complex terrain using texture information and SVM. In Proceedings of the International Conference on Circuits and Systems (CAS 2015), Paris, France, 9–10 August 2015.
12. Guo, Z.; Du, S. Mining parameter information for building extraction and change detection with very high-resolution imagery and GIS data. *GISci. Remote Sens.* **2017**, *54*, 38–63. [CrossRef]
13. Huang, S.; Ramirez, C.; Kennedy, K.; Mallory, J.; Wang, J.; Chu, C. Updating land cover automatically based on change detection using satellite images: Case study of national forests in Southern California. *GISci. Remote Sens.* **2017**, *54*, 495–514. [CrossRef]

14. Hao, M.; Zhang, H.; Shi, W.; Deng, K. Unsupervised change detection using fuzzy c-means and MRF from remotely sensed images. *Remote Sens. Lett.* **2013**, *4*, 1185–1194. [CrossRef]

15. Hao, M.; Hua, Z.; Li, Z.; Chen, B. Unsupervised change detection using a novel fuzzy c-means clustering simultaneously incorporating local and global information. *Multimed. Tools Appl.* **2017**, *76*, 20081–20098. [CrossRef]

16. Habib, T.; Inglada, J.; Mercier, G.; Chanussot, J. Support vector reduction in SVM algorithm for abrupt change detection in remote sensing. *IEEE Geosci. Remote Sens. Lett.* **2009**, *6*, 606–610. [CrossRef]

17. Volpi, M.; Tuia, D.; Bovolo, F.; Kanevski, M.; Bruzzone, L. Supervised change detection in VHR images using contextual information and support vector machines. *Int. J. Appl. Earth Obs. Geoinf.* **2013**, *20*, 77–85. [CrossRef]

18. Bovolo, F.; Bruzzone, L.; Marconcini, M. A novel approach to unsupervised change detection based on a semisupervised SVM and a similarity measure. *IEEE Trans. Geosci. Remote Sens.* **2008**, *46*, 2070–2082. [CrossRef]

19. Zhao, J.; Gong, M.; Liu, J.; Jiao, L. Deep learning to classify difference image for image change detection. In Proceedings of the IEEE 2014 International Joint Conference on Neural Networks (IJCNN), Beijing, China, 6–11 July 2014; pp. 411–417.

20. Gong, M.; Yang, H.; Zhang, P. Feature learning and change feature classification based on deep learning for ternary change detection in SAR images. *ISPRS J. Photogram. Remote Sens.* **2017**, *129*, 212–225. [CrossRef]

21. El Amin, A.M.; Liu, Q.; Wang, Y. Convolutional neural network features-based change detection in satellite images. In Proceedings of the First International Workshop on Pattern Recognition, Tokyo, Japan, 11–13 May 2016.

22. Liu, J.; Gong, M.; Zhao, J.; Li, H.; Jiao, L. Difference representation learning using stacked restricted Boltzmann machines for change detection in SAR images. *Soft Comput.* **2016**, *20*, 4645–4657. [CrossRef]

23. Zhang, H.; Gong, M.; Zhang, P.; Su, L.; Shi, J. Feature-level change detection using deep representation and feature change analysis for multispectral imagery. *IEEE Geosci. Remote Sens. Lett.* **2016**, *13*, 1666–1670. [CrossRef]

24. Lyu, H.; Lu, H.; Mou, L. Learning a transferable change rule from a recurrent neural network for land cover change detection. *Remote Sens.* **2016**, *8*, 506. [CrossRef]

25. Zhan, Y.; Fu, K.; Yan, M.; Sun, X.; Wang, H.; Qiu, X. Change detection based on deep siamese convolutional network for optical aerial images. *IEEE Geosci. Remote Sens. Lett.* **2017**, *14*, 1845–1849. [CrossRef]

26. Zhang, W.; Lu, X. The spectral-spatial joint learning for change detection in multispectral imagery. *Remote Sens.* **2019**, *11*, 240. [CrossRef]

27. Zhang, Z.; Vosselman, G.; Gerke, M.; Tuia, D.; Yang, M.Y. Change detection between multimodal remote sensing data using Siamese CNN. *arXiv*, 2018; arXiv:1807.09562.

28. Wiratama, W.; Lee, J.; Park, S.E.; Sim, D. Dual-dense convolution network for change detection of high-resolution panchromatic imagery. *Appl. Sci.* **2018**, *8*, 1785. [CrossRef]

29. Yoo, H.-J. Deep convolution neural networks in computer vision. *IEIE Trans. Smart Process. Comput.* **2015**, *4*, 35–43. [CrossRef]

30. Gong, M.; Jiaojiao, Z.; Jia, L.; Qiguang, M.; Jiao, L. Change detection in synthetic aperture radar images based on deep neural networks. *IEEE Trans. Neural Net. Learning Sys.* **2016**, *27*, 125–138. [CrossRef]

31. De, S.; Pirrone, D.; Bovolo, F.; Bruzzone, L.; Bhattacharya, A. A novel change detection framework based on deep learning for the analysis of multi-temporal polarimetric SAR images. In Proceedings of the 2017 IEEE International Geoscience and Remote Sensing Symposium (IGARSS), Fort Worth, TX, USA, 23–28 July 2017; pp. 5193–5196.

32. Wang, Q.; Zhang, X.; Chen, G.; Dai, F.; Gong, Y.; Zhu, K. Change detection based on Faster R-CNN for high-resolution remote sensing images. *Remote Sens. Lett* **2018**, *10*, 923–932. [CrossRef]

33. El Amin, A.M.; Liu, Q.; Wang, Y. Convolutional neural network features based change detection in satellite images. *Intern. Soc. Opt. Photonics.* **2016**, *10011*, 100110.

34. Zhang, P.; Gong, M.; Su, L.; Liu, J.; Li, Z. Change detection based on deep feature representation and mapping transformation for multi-spatial-resolution remote sensing images. *ISPRS J. Photo Remote Sens.* **2016**, *116*, 24–41. [CrossRef]

35. Huang, G.; Liu, Z.; van Der Maaten, L.; Weinberger, K.Q. Densely connected convolutional networks. In Proceedings of the IEEE Conference on Computer Vision and Pattern Recognition, Honolulu, HI, USA, 21–26 July 2017; pp. 4700–4708.

36. Hadsell, R.; Chopra, S.; Le Cun, Y. Dimensionality reduction by learning an invariant mapping. In Proceedings of the 2006 IEEE Computer Society Conference on Computer Vision and Pattern Recognition, New York, NY, USA, 17–22 June 2006.
37. Viswanathan, R.; Aalo, V. On counting rules in distributed detection. *IEEE Trans. Acous. Speech Signal Process.* **1989**, *37*, 772–775. [CrossRef]
38. Hu, H.; Dey, D.; del Giorno, A.; Hebert, M.; Bagnell, J.A. Log-denseNet: How to sparsify a denseNet. *arXiv*, 2018; arXiv:1711.00002.

applied sciences

Article

Comparison of Machine Learning Regression Algorithms for Cotton Leaf Area Index Retrieval Using Sentinel-2 Spectral Bands

Huihui Mao [1,2], Jihua Meng [1,*], Fujiang Ji [1,2], Qiankun Zhang [1,3] and Huiting Fang [1,2]

[1] Key Laboratory of Digital Earth Sciences, Institute of Remote Sensing and Digital Earth, Chinese Academy of Sciences, Beijing 100101, China; maohh@radi.ac.cn (H.M.); jifj@radi.ac.cn (F.J.); zhang_q_k@163.com (Q.Z.); fanght@yeah.net (H.F.)

[2] University of Chinese Academy of Sciences, Beijing 100049, China

[3] College of Geomatics and Geoinformation, Guilin University of Technology, Guilin 541004, China

* Correspondence: mengjh@radi.ac.cn; Tel.: +86-010-6486-9473

Received: 2 March 2019; Accepted: 2 April 2019; Published: 7 April 2019

Abstract: Leaf area index (LAI) is a crucial crop biophysical parameter that has been widely used in a variety of fields. Five state-of-the-art machine learning regression algorithms (MLRAs), namely, artificial neural network (ANN), support vector regression (SVR), Gaussian process regression (GPR), random forest (RF) and gradient boosting regression tree (GBRT), have been used in the retrieval of cotton LAI with Sentinel-2 spectral bands. The performances of the five machine learning models are compared for better applications of MLRAs in remote sensing, since challenging problems remain in the selection of MLRAs for crop LAI retrieval, as well as the decision as to the optimal number for the training sample size and spectral bands to different MLRAs. A comprehensive evaluation was employed with respect to model accuracy, computational efficiency, sensitivity to training sample size and sensitivity to spectral bands. We conducted the comparison of five MLRAs in an agricultural area of Northwest China over three cotton seasons with the corresponding field campaigns for modeling and validation. Results show that the GBRT model outperforms the other models with respect to model accuracy in average ($\overline{R^2}$ = 0.854, $\overline{RMSE}$ = 0.674 and $\overline{MAE}$ = 0.456). SVR achieves the best performance in computational efficiency, which means it is fast to train, and to validate that it has great potentials to deliver near-real-time operational products for crop management. As for sensitivity to training sample size, GBRT behaves as the most robust model, and provides the best model accuracy on the average among the variations of training sample size, compared with other models ($\overline{R^2}$ = 0.884, $\overline{RMSE}$ = 0.615 and $\overline{MAE}$ = 0.452). Spectral bands sensitivity analysis with dCor (distance correlation), combined with the backward elimination approach, indicates that SVR, GPR and RF provide relatively robust performance to the spectral bands, while ANN outperforms the other models in terms of model accuracy on the average among the reduction of spectral bands ($\overline{R^2}$ = 0.881, $\overline{RMSE}$ = 0.625 and $\overline{MAE}$ = 0.480). A comprehensive evaluation indicates that GBRT is an appealing alternative for cotton LAI retrieval, except for its computational efficiency. Despite the different performance of the ML models, all models exhibited considerable potential for cotton LAI retrieval, which could offer accurate crop parameters information timely and accurately for crop fields management and agricultural production decisions.

Keywords: leaf area index (LAI); machine learning; Sentinel-2; sensitivity analysis; training sample size; spectral bands

---

1. Introduction

Leaf area index (LAI), which characterizes the structure and functioning of vegetation, is usually defined as half of the total green leaf area per unit horizontal ground surface area [1,2]. LAI is one of the most important vegetation biophysical parameters, and a key variable for climate modeling, evapotranspiration modeling and crop modeling, and it is recognized as an Essential Climate Variable (ECV) by the Global Climate Observing System [3–8]. LAI has a wide range of applications regarding agricultural fields, and it has been demonstrated to be an essential indicator for crop growth monitoring and key variables for crop yield forecasting [9–11]. Therefore, it is of special relevance to retrieve LAI in a timely and accurate manner.

Remote sensing techniques provide promising alternatives to obtaining crop biophysical parameters by high temporally and spatially continuous means over large areas. To date, there are mainly three categories of methods developed to retrieve LAI based on optical remote sensing data, which are statistical methods, physically based methods, and hybrid methods [12–14]. Statistical methods can be further divided into parametric and non-parametric regression methods. Parametric regression methods usually consist of an explicit relationship between biophysical parameters and vegetation indices, while non-parametric regression methods define regression models learnt from the training dataset [15]. Non-parametric algorithms can be split into linear and non-linear regression methods; the latter is also commonly referred to as machine learning regression algorithms (MLRAs). While physically-based methods are applications of physical laws establishing cause-effect relationships, a hybrid method combines elements of non-parametric statistics and physically-based methods [13], whereas these two methods are both sophisticated models that demand a large number of parameters, which are usually difficult to obtain in practice. Empirical parametric models typically make use of a limited number of spectral bands [13,16]. However, nonparametric models can make full use of spectral information, and directly learn the input-output relationships from a given training dataset, which makes these models attractive alternatives for crop LAI retrieval.

With the development of remote sensing, more and more optical remote sensing satellites have been launched (e.g., Landsat 8, Sentinel-2, and Chinese GF-1, GF-2 and newly launched GF-6), which ensures the availability of high spatial, high temporal resolution satellite remote sensing data, and correspondingly, high dimensional (spatial, temporal and spectral) of remote sensing data amounts to large data volume, which also poses great challenges for more efficient, robust and accurate algorithms in a wide variety of applications with remote sensing.

Recently, machine learning (ML), a broad subfield of artificial intelligence, has attracted considerable attention in remote sensing applications for classification and regression problems, and encouraging results have been obtained [17–23]. With advances in computer technology and associated techniques, ML has drawn tremendous interest in a variety of fields to address complex problems. ML can be broadly defined as computational methods using experience to improve performance or to make accurate predictions [24]. ML has been extensively applied to biophysical parameter retrievals due to the ability to accurately approximate robust relationships between input-output data, which provides tremendous opportunities for remote sensing-based applications. Considering ML regression algorithms, a more efficient, robust, and accurate model for crop LAI retrieval should be established. Despite the considerable advances in ML for remote sensing applications, challenging problems remain in the selection of MLRAs for crop LAI retrieval among the variety of ML algorithms available, as well as the optimal number of training sample size and spectral bands to different MLRAs.

As for the versatile ML algorithms, artificial neural network (ANN), support vector regression (SVR), Gaussian process regression (GPR) and random forest (RF) are reportedly effective for crop LAI retrieval [25–28]. However, gradient boosting regression tree (GBRT), a highly robust ML algorithm for a wide range of applications, is capable of achieving high levels of accuracy for regression problems [29–31] and to our knowledge, has not been investigated for LAI retrieval. Further studies should be conducted to assess the performance of the GBRT model in crop LAI retrievals.

Many studies have been dedicated to crop LAI retrieval using MLRAs. However, there are a limited number of academic studies involving comparisons of different MLRAs for crop LAI retrievals using remote sensing. Apparently, none of these studies have focused on multispectral remote sensing data, and none of these studies have evaluated the different impact factor together, to conduct a comprehensive comparison.

In addition, the validation of global LAI products are important procedures to ensure the application of the products in a wide range of fields [32]. Regional high-resolution LAI maps can be used as a reference LAI map to validate the global LAI products, which calls for efficient, robust and accurate algorithms for LAI retrieval.

The objective of this study is to compare the performance of five advanced MLRAs (ANN, SVR, GPR, RF and GBRT) for cotton LAI retrieval in a relatively comprehensive manner. We conducted the study over the entire growth period of cotton using Sentinel-2 spectral bands and corresponding ground data. Specifically, the following research questions are addressed:

(1) Which of the five MLRAs perform best with regard to model accuracy?

(2) Which is the fastest model during the training and validation processes?

(3) How does the number of training sample size influence the performance of the five MLRAs?

(4) How does the number of spectral bands influence the performance of the five MLRAs?

(5) Which is the best model in consideration of model accuracy, computational efficiency, sensitivity to training sample size, and sensitivity to spectral bands together?

(6) How accurate are the global LAI products in Northwest China?

2. Materials

2.1. Study Area and Field Campaigns

China is one of the largest cotton producers and importers in the world, and Xinjiang is the primary cotton-growing region in China [33,34]. The chosen study area was on a large agriculture region (6118.08 km^2 or $1,511,810$ acres) in Shihezi ($44°37'$ N, $85°42'$ E), Xinjiang Province, China. The region is located in a temperate continental climate zone. The average field size in the study area is 73139.7 m^2 (18.07 acres), and the majority of fields have a flat topography, which is preferable for decametric remote sensing applications (e.g., Sentinel-2). The annual mean temperature of the study area is 7.39 °C, the annual total precipitation is 206 mm, and the average altitude is 450.8 m.

The field campaigns were conducted in early June 2018, mid-July 2018, and mid-August 2018 (Table 1), with 117 total quadrats obtained. Each quadrat was assigned one leaf area index (LAI) value, obtained as the average leaf area index (LAI) of the three sample points that matches the corresponding Sentinel-2 pixel (Table 2). Each sample point datum was collected using an LAI-2200C Plant Canopy Analyzer (Li-Cor, Inc., Lincoln, NE, USA). The main planted crops in the study area are cotton, grape, spring maize and winter wheat, and cotton holds the largest planting proportion, which is 71.7%. The cotton season is from mid-April to early October. The spring maize season is from mid-April to late September. The winter wheat season is from late September of the previous year to late June. Notably, the study area has completely achieved mechanization for agricultural production and management. Figure 1 shows the location of the study area and field observation sites with Sentinel-2A imagery (20 August 2018). A descriptive statistic of the measured cotton LAI at three observation dates is shown in Figure 2.

Table 1. Sentinel-2 imagery and the corresponding field campaign data.

Satellite	Date (dd/mm/yy)	Field Campaign Date	Quadrats
Sentinel-2B	6 June 2018	1–3 June 2018	23
Sentinel-2A	11 July 2018	8–11 July 2018	58
Sentinel-2A	20 August 2018	18–21 August 2018	36

Table 2. Sentinel-2 satellite imagery spectral band characteristics [35].

Band	Central Wavelength (nm)	Bandwidth (nm)	Spatial Resolution (m)	Purpose
B1	443	20	60	Atmospheric correction (aerosol scattering).
B2 [1]	490	65	10	Sensitive to vegetation senescing, carotenoid, browning and soil background; atmospheric correction (aerosol scattering).
B3	560	35	10	Green peak, sensitive to total chlorophyll in vegetation.
B4	665	30	10	Maximum chlorophyll absorption.
B5	705	15	20	Position of red edge; consolidation of atmospheric corrections/fluorescence baseline.
B6	740	15	20	Position of red edge; atmospheric correction, retrieval of aerosol load.
B7	783	20	20	Leaf Area Index (LAI), edge of the Near Infrared plateau.
B8	842	115	10	LAI.
B8A	865	20	20	NIR plateau, sensitive to total chlorophyll, biomass, LAI and protein; water vapor absorption reference; retrieval of aerosol load and type.
B9	945	20	60	Water vapor absorption, atmospheric correction.
B10	1380	30	60	Detection of thin cirrus for atmospheric correction.
B11	1610	90	20	Sensitive to lignin, starch and forest above ground biomass; snow/ice/cloud separation.
B12	2190	180	20	Assessment of Mediterranean vegetation conditions; distinction of clay soils for the monitoring of soil erosion; distinction between live biomass, dead biomass and soil, e.g., for burn scars mapping.

[1] The spectral bands in bold are ones used in this study.

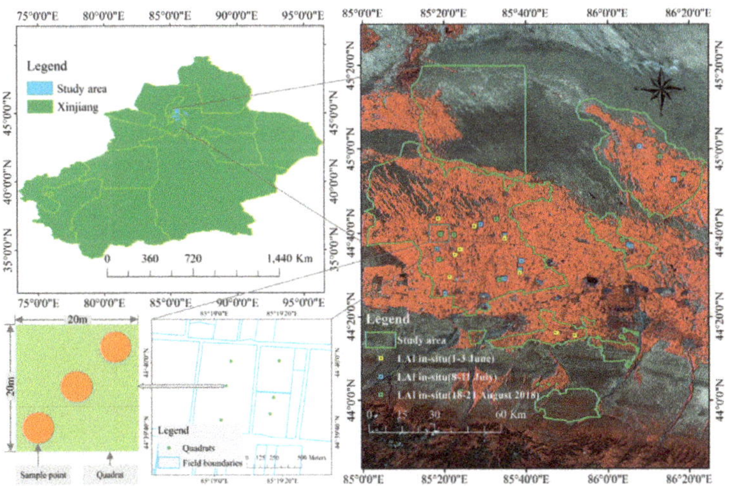

Figure 1. Location of study area and in-situ leaf area index (LAI) quadrats from three field campaigns. The background was the Sentinel-2A image acquired on August 20, 2018 and was shown in a false color band composition of R (8) G (4) B (3), with standard deviation stretch (**right**). The red color in the Sentinel image represents the vegetation (mainly crops and a few trees), the grey color represents the desert and bare land (**right**). The green dots represent the quadrats (including three sample points), and the blue polygons represent agricultural field boundaries (**lower left**).

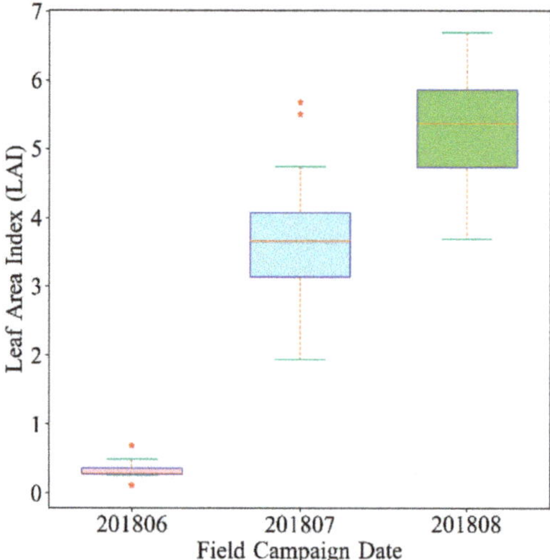

Figure 2. Field campaign cotton LAI descriptive statistics on three observation dates, shown with boxplots. The box extends from the lower to upper quantile values of the field observation data, with a line at the median, the whiskers extend from the box to show the range of the data. Flier points are those past the end of the whiskers, named as outliers (**red dots**).

2.2. Sentinel-2 Data and Preprocessing

The remote sensing data used in this study were Sentinel-2 imagery. Sentinel-2 is a constellation of satellites, Sentinel-2A and Sentinel-2B, which were launched by the European Space Agency (ESA) on June 2015 and March 2017, respectively. Each satellite carries a MultiSpectral Instrument (MSI) that provides a variety of spectral bands covering the visible, near infrared and shortwave infrared bands. The MSI contains four bands at 10 m, six bands at 20 m and three bands at 60 m [36]. It is of great importance that the MSI incorporates three bands in the red-edge region, centered at 705, 740 and 783 nm, and two Shortwave Infrared (SWIR) bands centered at 1610 and 2190 nm at 20 m (S2-20 m). Many studies have revealed that the red-edge bands and SWIR bands have the potential to improve the accuracy of LAI retrievals [37–39], which open great opportunities for crop LAI retrievals considering Sentinel-2's high revisit frequency. The Sentinel-2 spectral band characteristics are shown by Table 2.

The acquired Sentinel-2 imagery products are Level-1C products, which are top of atmosphere (TOA) reflectances [40]. Atmospheric correction was conducted using the Sen2Cor (2.5.5, ESA, Frascati, Italy, 2018) and Sentinel Application Platform (SNAP) toolbox (6.0.1, ESA, Frascati, Italy, 2018) provided by the ESA to produce Level-2A bottom of atmosphere (BOA) products [41,42]. A flowchart of Sentinel-2 preprocessing is presented in Figure 3.

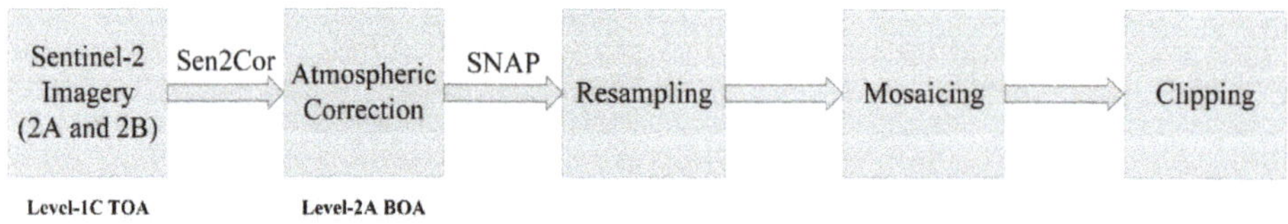

Figure 3. Flowchart of the preprocessing of Sentinel-2 imagery.

In this study, aiming at making full use of the spectral bands of Sentinel-2 (especially the red-edge and SWIR bands), we focus on S2-20 m (the same size as our quadrats) with 10 spectral bands considering the red-edge and SWIR spectral bands of Sentinel-2 imagery. A resampling process was performed using the Nearest Neighbor method with 4 spectral bands (B2, B3, B4 and B8) from 10 m to 20 m in the SNAP toolbox. Three of the atmospheric spectral bands at 60 m were not used in this study because these bands contributed to atmospheric applications, such as aerosol correction (B1), water vapor correction (B9) and cirrus detection (B10) [42]. The reflectance data collected on the Sentinel-2 images were using Extract by Points on the ArcToolbox of ArcGIS Desktop software (10.5, Environmental Systems Research Institute, Redlands, CA, USA, 2016), the points used to extract data on the Sentinel-2 images are the center GPS coordinate of the three sample points that represent one quadrat.

2.3. Global LAI Products

Many global LAI products with different spatial resolution and temporal characteristics has been produced, in which MODIS LAI products retrieved from Terra and Aqua platforms are one of the most famous global LAI products [2,43]. All MODIS products are available at [44]. In addition, GEOV1 LAI products have also been widely used for variety of applications. The GEOV LAI product was downloaded from the Copernicus Global Land Service [45]. More recently, new versions of these two products have been delivered with great improvement in spatial and temporal resolution [46–48]. Table 3 presents the main characteristics of the latest version of these two global LAI products.

Table 3. Main characteristics of two global LAI products under study.

Products	Version	Spatial Resolution	Temporal Resolution	Algorithms	Temporal Coverage	References
MODIS	MCD15 C6	500 m	4-day	RTM 3D [1] (LUT)	2002-present	Myneni, et al. 2015 [46]
GEOV3	V1.0.1	1/3 km	10-day	Neural network (red, NIR)	2014-present	Baret, et al. 2013, 2016 [47,48]

[1] RTM and LUT stands for "Radiative Transfer Model" and "Look Up Table", respectively.

3. Methods

ML algorithms can automatically learn the relationships in any given data between input (reflectances) and output (LAI). To identify the performance of five popular ML algorithms for cotton LAI retrieval, regression models were established based on artificial neural network (ANN), SVR, Gaussian process regression (GPR), random forest (RF) and gradient boosting regression tree (GBRT), at S2-20 m for the whole growth period of cotton (using all 117 quadrats data over the three observation dates).

All the ML models were implemented using the Scikit-learn package [49], which is an open-source Python [50] module project that integrates a wide range of prevalent ML algorithms [51]. All hyperparameter tuning of the models is based on GridSearchCV in the Scikit-learn package, which can evaluate all possible combinations of hyperparameter values using five-fold cross-validation to determine the best combination of hyperparameter values (the hyperparameter combination that has the best accuracy of the model in terms of root-mean-square error (RMSE)). Cross-validation are model validation techniques to obtain reliable and stable models. This study implemented a five-fold cross-validation, basically, the training datasets are split into five smaller sets, and a model is trained using four of the folds as training data, then the resulting model is validated on the remaining part of the data, the processes continues to circulate five times, and finally, the performance measure estimated by five-fold cross-validation is the average of the values computed in the loop. Training and testing sampling distribution has a great impact on machine learning regression algorithms (MLRAs), and some previous studies demonstrated that 70%/30% split option is appropriate for model training and validation [52–54], nonetheless, other studies argue that the 80%/20% split option is preferable [27,55]. In our study, to validate the performance of the ML models, all the datasets were randomly split into

75% (n = 87) for model training, and 25% (n = 30) for model validation. Regression models that achieve satisfactory performance in training datasets may fail to predict unseen datasets, and therefore, a model that performs well at both training and unseen testing datasets is referred to as having excellent generalization ability. This type of model could be used for crop LAI retrieval applications.

3.1. Artificial Neural Network (ANN)

The ANN has been one of the most widely used ML algorithms for wide range of remote sensing applications. ANN is a computational model that is inspired by the human brain. ANN is formed by a collection of interconnected units (neurons) that learn from experience by modifying connections (weights) [56,57]. ANN usually consists of an input layer, hidden layer, and output layer. The backpropagation (BP) ANN used in this study implements a multilayer perceptron (MLP) algorithm that trains with a BP algorithm, while an MLP refers to a feedforward network that generalizes the standard perceptron by having a hidden layer that resides between the input and output layers. It has been demonstrated that MLP can approximate any continuous function to an arbitrary degree of accuracy, given a sufficiently large but finite number of hidden neurons [56,58]. In this study, the input layer and output layer are referred to as the Sentinel-2 spectral bands (reflectances) and the cotton LAI, respectively.

The major tuning hyperparameters for ANN are the number of hidden layers and the number of neurons in the hidden layer. Here, we used the one-hidden-layer network because it was demonstrated to be powerful enough to approximate any measurable function to any desired degree of accuracy [59]. Many studies have been dedicated to the investigation of the optimal number of neurons in the hidden layer, and several empirical equations have been proposed [60–62]. In this study, the number of neurons in the hidden layer is determined by the following equation [62]:

$$N_h = \sqrt{(m+2)N} + 2\sqrt{\frac{N}{m+2}} \tag{1}$$

where N_h is the number of neurons in the hidden layer, m specifies the number of layers, and N denotes the number of input neurons. The number of neurons in the hidden layer was set to ten in this study according to Equation (1). In the input layer, the input variables include 10 spectral bands. Other important hyperparameters were optimized using GridSearch and 5-fold cross-validation. The remainder of hyperparameters for the ANN model remain defaults. The structure of the neural network used in this study is presented in Figure 4. The hyperparameter values adopted in this study are listed in Table 4.

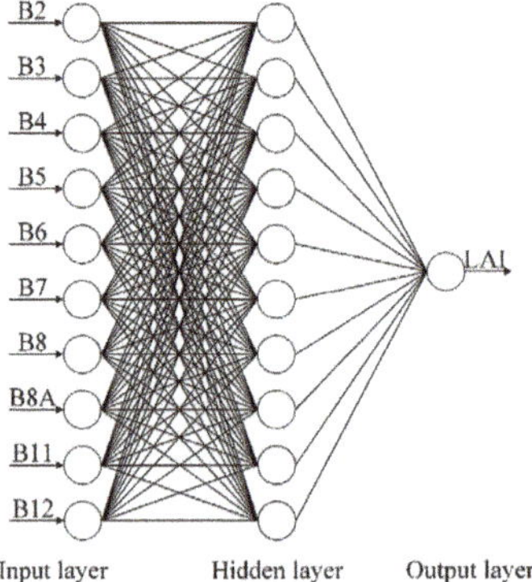

Figure 4. The architecture of backpropagation (BP) neural networks used in this study for cotton LAI retrieval.

Table 4. Parameter settings to determine the optimal hyperparameters for the artificial neural network (ANN) model.

Parameters	Description	GridSearch Values	Searching Results
activation	Activation function for the hidden layer.	'identity', 'logistic', 'tanh', 'relu'	'logistic'
solver	The solver for weight optimization.	'lbfgs', 'sgd', 'adam'	'lbfgs'
alpha	L2 penalty (regularization term) parameter.	1e-4, 1e-2, 0.1, 1	0.1
learning_rate	Learning rate schedule for weight updaters.	'constant', 'invscaling', 'adaptive'	'constant'

3.2. Support Vector Regression (SVR)

SVR is a significant application form of support vector machine (SVM), which was first introduced by Corinna Cortes (b. 1961 in Denmark) and Vapnik [63,64]. SVM is based on the idea of mapping the input space into a new feature space with higher dimensions using the kernel function, after which a hyperplane, known as the decision boundary, is constructed with the maximum margin [65,66]. SVM extension to SVR is realized by introducing an ε-insensitive region around the function, which is referred to as the ε-tube that best approximates the regression function [67]. Given training vectors $x_i \in \mathbb{R}^p, I = 1, \cdots, n$, and a vector $y \in \mathbb{R}^n$, ε-SVR solves the following original problem:

$$\min_{w, b, \xi, \xi^*} \frac{1}{2}\|w\|^2 + C\sum_{i=1}^{n}(\xi_i + \xi_i^*) \tag{2}$$

$$subject\ to \begin{cases} y_i - \langle w, x_i \rangle - b \leq \varepsilon + \xi_i \\ \langle w, x_i \rangle + b - y_i \leq \varepsilon + \xi_i^* \\ \xi_i, \xi_i^* \geq 0,\ i = 1, \cdots, n \end{cases} \tag{3}$$

where $C > 0$ is a regularization parameter that gives more weight to minimizing the flatness or error.

The principal hyperparameter of SVR is the kernel, as it defines the kernel functions of the model. The radial basis function (RBF) was selected as the kernel function because it has been found to be efficient and accurate for regression problems [68,69]. The RBF kernel is described as follows:

$$k(x_i, x_j) = exp\left(-\frac{\|x_i - x_j\|^2}{2\sigma^2}\right) \tag{4}$$

The SVR model is easy to establish because only two hyperparameters must be tuned: Penalty parameter C and the kernel coefficient gamma. The optimal values of the two hyperparameters that were optimized using GridSearch and 5-fold cross-validation are presented in Table 5. The remainder of hyperparameters for the SVR model remain defaults.

Table 5. Parameter settings to determine the optimal hyperparameters for the support vector regression (SVR) model.

Parameters	Description	GridSearch Values	Searching Results
C	Penalty parameter C of the term.	0.1, 0.5, 1, 5, 10, 15, 20, 50, 100, 500	50
gamma	Kernel coefficient for 'rbf', 'poly' and 'sigmoid'.	0.01, 0.1, 0.2, 0.4, 0.8, 1, 1.5, 2, 5, 10	0.2

3.3. Gaussian Processes Regression (GPR)

A Gaussian process is a stochastic process that is formed by a collection of random variables and has a Gaussian probability distribution [70]. The major factor in the Gaussian process is the covariance function known as the kernel function. The learning problem in the Gaussian process amounts to adjusting the covariance hyperparameters. The mean function $m(x)$ and covariance function $k(x, x')$ characterize the Gaussian process $f(x)$, which can be described as follows:

$$m(x) = \mathbb{E}[f(x)] \tag{5}$$

$$k(x, x') = \mathbb{E}\big[(f(x) - m(x))(f(x') - m(x'))\big] \tag{6}$$

$$f(x) \sim GP\big(m(x), k(x, x')\big) \tag{7}$$

For simplicity, we usually consider the mean function to be zero.

Similar to SVR, GPR has been advanced to solve complex nonlinear problems through projection of inputs into high dimensional space by applying highly flexible kernels, where such a technique is referred to as kernel trick. Additionally, GPR models have a great ability to provide the most informative feature (spectral band) from the input dataset. In this study, we used the sum of an RBF and a noise component function kernel:

$$k(x_i, x_j) = \theta_0{}^2 exp\left(-\frac{(x_i - x_j)^2}{2\sigma^2}\right) + \theta_1{}^2\delta_{ij} \tag{8}$$

where θ_0 is the scaling factor, σ is the length-scale, and θ_1 corresponds to the independent noise component.

Beyond the kernel functions, alpha and n_restarts_optimizer, are significant for the GPR model. The results of hyperparameter tuning using GridSearch and 5-fold cross-validation are shown in Table 6 as follows. The remainder of hyperparameters for the GPR model are set as defaults.

Table 6. Parameter settings to determine the optimal hyperparameters for the Gaussian process regression (GPR) model.

Parameters	Description	GridSearch Values	Searching Results
alpha	Value added to the diagonal of the kernel matrix during fitting; larger values correspond to increased noise level in the observations; this can also prevent a potential numerical issue during fitting, by ensuring that the calculated values form a positive definite matrix.	1e-2, 1e-1, 1	1
n_restarts_optimizer	The number of restarts of the optimizer for finding the kernel's parameters which maximize the log-marginal likelihood; the first run of the optimizer is performed from the kernel's initial parameters, the remaining ones (if any) from thetas sampled log-uniform randomly from the space of allowed thetas-values.	0, 1, 2, 4, 8, 10, 12, 16, 20, 32, 64	16

3.4. Random Forest (RF)

RF has been a prevalent ML algorithm for a wide range of fields for classification, regression and other complicated problems. As one of the ensemble learning methods, RF grows a multitude of decision trees as base learners, and combines these trees together to obtain a better performance by averaging the predictions [71,72]. Each tree grows independently with training samples obtained using bootstrap sampling from the original data. Then, m variables out of M input variables are chosen, after which the best of m is used for splitting the node (note that m $\ll$ M). The final prediction comes from the averaging predictions of each independent tree. This kind of technique is referred to as bagging. RF models can provide feature importance estimates, which enables insight into feature selection processes.

Key hyperparameters include n_estimators, max_depth, min_samples_split and min_samples_leaf. Hyperparameter tuning results using GridSearch and 5-fold cross-validation are displayed in Table 7. The remainder of the RF model hyperparameters are set as defaults.

Table 7. Parameter settings to determine the optimal hyperparameters for the random forest (RF) model.

Parameters	Description	GridSearch Values	Searching Results
n_estimators	The number of trees in the forest.	1, 5, 10, 20, 40, 60, 80, 120	40
max_depth	The maximum depth of the tree.	2, 5, 8, 15, 25, 30, 50, None	15
min_samples_split	The minimum number of samples required to split an internal node.	2, 5, 10, 15, 100	2
min_samples_leaf	The minimum number of samples required to be at a leaf node; a split point at any depth will only be considered if it leaves at least min_samples_leaf training samples in each of the left and right branches; this may have the effect of smoothing the model, especially in regression.	1, 2, 5, 10, 50	1

3.5. Gradient Boosting Regression Tree (GBRT)

GBRT, also known as gradient boosting decision tree (GBDT) or multiple additive regression tree (MART), is one of the most widely used ML algorithms for the model's great generalization ability and highly robust performance in practical applications, and this GBRT was introduced by Friedman [73,74]. As one of the ensemble learning methods that combines different weak learners to generate strong learners, GBRT uses boosting techniques that aim to reduce bias, rather than bagging (e.g., RF algorithm), which aims to reduce variance. To evaluate the accuracy of the model, a variety of loss functions can be used during boosting, including least squares, least absolute deviation, Huber and quantile for regression, binomial deviance, multinomial deviance, and exponential loss for classification. GBRT considers additive models of the following form:

$$F(x) = \sum_{m=1}^{M} \gamma_m h_m(x) \tag{9}$$

where $h_m(x)$ represents the basis learners in boosting. Then, GBRT builds the additive model in a forward stagewise fashion:

$$F_m(x) = F_{m-1}(x) + \gamma_m h_m(x) \tag{10}$$

At each stage, the decision tree $h_m(x)$ is chosen to minimize the loss function L, given the current model F_{m-1} and its fitting of $F_{m-1}(x_i)$:

$$F_m(x) = F_{m-1}(x) + \arg \min_h \sum_{i=1}^{n} L(y_i, F_{m-1}(x_i) + h(x)) \tag{11}$$

The initial model F_0 is problem specific. Gradient boosting attempts to solve this minimization problem numerically via the steepest descent. The steepest descent direction is the negative gradient of the loss function evaluated at the current model F_{m-1}, which can be calculated for any differentiable loss function as follows:

$$F_m(x) = F_{m-1}(x) - \gamma_m \sum_{i=1}^{n} \nabla_F L(y_i, F_{m-1}(x_i)) \tag{12}$$

where the step length γ_m is chosen using a line search:

$$\gamma_m = \arg\min_{\gamma} \sum_{i=1}^{n} L\left(y_i, F_{m-1}(x_i) - \gamma \frac{\partial L(y_i, F_{m-1}(x_i))}{\partial F_{m-1}(x_i)}\right) \tag{13}$$

The GBRT algorithm can provide feature relevance information and partial dependence, where this partial dependence shows the dependence among the target response and the most important features (not shown in this study). However, despite the GBRT's demonstrated satisfactory accuracy for versatile domains of regression problems [29,75,76], to our knowledge, this algorithm has not been previously applied to crop LAI retrievals with remote sensing. GBRT should be studied as it might be a promising alternative in crop LAI retrieval with remote sensing.

With regard to GBRT model hyperparameters, loss, n_estimators, max_depth, min_samples_split and min_samples_leaf are selected for hyperparameter tuning using GridSearch and 5-fold cross-validation, the results are exhibited in Table 8. The remainder of the GBRT model hyperparameters are set as defaults.

Table 8. Parameter settings to determine the optimal hyperparameters for the gradient boosting regression tree (GBRT) model.

Parameters	Description	GridSearch Values	Searching Results
loss	Loss function to be optimized.	'ls', 'lad', 'huber', 'quantile'	'lad'
n_estimators	The number of boosting stages to perform; gradient boosting is fairly robust to over-fitting so a large number usually results in a better performance.	1, 5, 10, 20, 40, 60, 80, 120, 300	300
max_depth	Maximum depth of the individual regression estimators; the maximum depth limits the number of nodes in the tree; tune this parameter for best performance; the best value depends on the interaction of the input variables.	2, 3, 5, 8, 15, 25, 30, None	25
min_samples_split	The minimum number of samples required to split an internal node.	2, 5, 10, 15	10
min_samples_leaf	The minimum number of samples required to be at a leaf node; a split point at any depth will only be considered if it leaves at least min_samples_leaf training samples in each of the left and right branches; this may have the effect of smoothing the model, especially in regression.	1, 2, 5, 10, 50	1

3.6. Performance Evaluation

To evaluate the performance of the ML regression models, the root mean square error (RMSE), mean absolute error (MAE) and coefficient of determination (R^2) between the measured and predicted values were used to assess the performance of the models. The RMSE, MAE and R^2 are calculated as follows:

$$RMSE(y, \hat{y}) = \sqrt{\frac{1}{n_{samples}} \sum_{i=0}^{n_{samples}-1} (y_i - \hat{y}_i)^2} \tag{14}$$

$$MAE(y, \hat{y}) = \frac{1}{n_{samples}} \sum_{i=0}^{n_{samples}-1} |y_i - \hat{y}_i| \tag{15}$$

$$R^2(y, \hat{y}) = 1 - \frac{\sum_{i=0}^{n_{samples}-1} (y_i - \hat{y}_i)^2}{\sum_{i=0}^{n_{samples}-1} (y_i - \overline{y})^2} \tag{16}$$

where $\bar{y} = \frac{1}{n_{samples}} \sum_{i=0}^{n_{samples}-1} y_i$, $\hat{y}_i$ is the estimated cotton LAI value, y_i is the measured cotton LAI value and $n_{samples}$ is the number of validation datasets. The higher the R^2, the smaller the RMSE and MAE, and thus the higher the model precision and accuracy.

4. Results

4.1. Performance Evaluation

To avoid skew results caused by the random sampling of training and testing datasets, we performed 20 random repetitions of the five ML regression models (87 samples for training and 30 for testing). The performance of five ML models with respect to R^2, RMSE and MAE is displayed in Figure 5. GBRT surpasses the other models on the average ($\overline{R^2}$ = 0.854, $\overline{RMSE}$ = 0.674 and $\overline{MAE}$ = 0.456), however, GBRT acts less robust to the training/testing random split according to the distribution of R^2, RMSE and MAE. RF delivers the worst accuracy on the average ($\overline{R^2}$ = 0.807, $\overline{RMSE}$ = 0.781 and $\overline{MAE}$ = 0.545), and it also acts less robust to the training/testing random split. Nonetheless, SVR achieves a desirable result while it also acts reasonably robust to the training/testing random split ($\overline{R^2}$ = 0.835, $\overline{RMSE}$ = 0.730 and $\overline{MAE}$ = 0.550), which indicates that SVR is highly stable to the random sampling processes. Overall, all models achieved satisfactory performances, which indicates that ML algorithms are appealing methods for cotton LAI estimation.

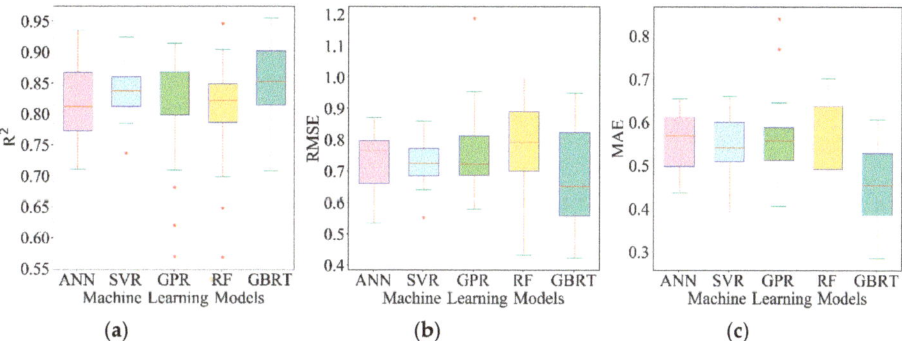

Figure 5. The R^2, RMSE and MAE distribution of 20 repetitions between the predicted LAI values and corresponding measured LAI. (a) R^2; (b) RMSE; (c) MAE.

4.2. Computational Efficiency

Beyond the model accuracy of the five models for the training and testing random split, it is of particular relevance to compare the computational efficiency (time required to the model during training and validation processes) of the models, as it is an important criterion for operational algorithms. All models were implemented in a Python environment on an Intel(R) Xeon(R) CPU E5-2620 v2 @ 2.10 GHz processor and installed memory (RAM) of 32.0 GB. The computational efficiency is recorded from the 20 repetitions in Section 4.1. The averaged results of 20 repetitions are illustrated in Figure 6. Large differences among the five ML models are clearly found. SVR performs incredibly fast (less than 1 s). However, GBRT is frustrating, owing to the large amounts of its hyperparameter tuning processes. In general, SVR is able to deliver near real time operational products with a highly efficient processing speed, whereas the GBRT model is not recommended for this kind of application, because the GBRT model is computationally more demanding. RF, GPR and ANN showed moderate performances in terms of computational efficiency.

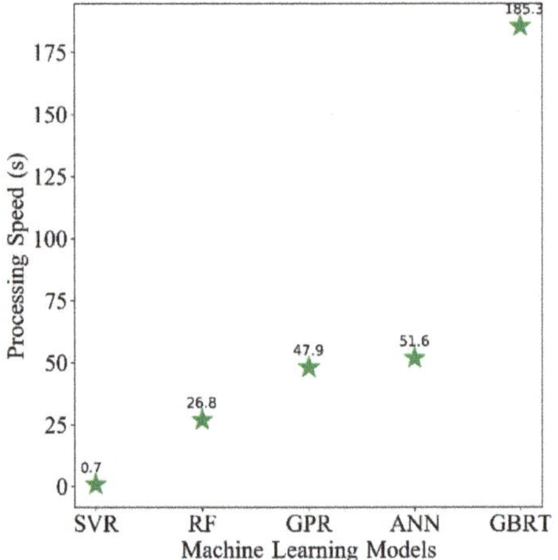

Figure 6. The average processing speed for five ML models.

4.3. Sensitivity to Training Sample Size

We use R^2, RMSE and MAE between the predicted LAI values and measured LAI values to assess the sensitivity of five ML methods to the training sample size. Eight datasets with different sample sizes (17-87) were generated by randomly sampling from the total training datasets (87 samples) at intervals of 10. Table 9 shows the number of training and testing samples of the eight datasets, while the testing datasets keep as the same datasets with 30 samples.

Table 9. Number of training and testing samples of the eight datasets.

Sample Name	Set 1	Set 2	Set 3	Set 4	Set 5	Set 6	Set 7	Set 8
Training Datasets	17	27	37	47	57	67	77	87
Testing Datasets				30				

To evaluate the robustness of five ML models for the training samples, we conducted 10 random repetitions of each model and provided the corresponding standard error. The standard error is given as follows:

$$\sigma_{\overline{x}} = \frac{\sigma}{\sqrt{n}} \tag{17}$$

where σ denotes the standard deviation, n represents the number of repetitions.

Figure 7 shows the changes in R^2, RMSE and MAE as the training sample sizes varied, the standard error is given at each training sample size (filling area of each model), lower values of the filling area corresponds to smaller standard error, and consequently, more robust model performance. According to Figure 7, different sensitivity performances of the sample size variation are found among the ML models. GPR and GBRT show robust performances for the training samples according to the standard error of different models. GBRT produces a remarkable performance for the training samples that is more robust than the other models, GPR behaves suboptimally. ANN and SVR are very sensitive to the training samples. Moreover, all models behave more robust with the growth of training sample size according to the variation of standard error. In summary, the model accuracy improves overall with an increasing training sample size, while GBRT provides the most robust performance for the training samples, and the best model accuracy among the variations in training sample size on average ($\overline{R^2}$ = 0.884, $\overline{RMSE}$ = 0.615 and $\overline{MAE}$ = 0.452, calculated by all the 8 groups of training sample size).

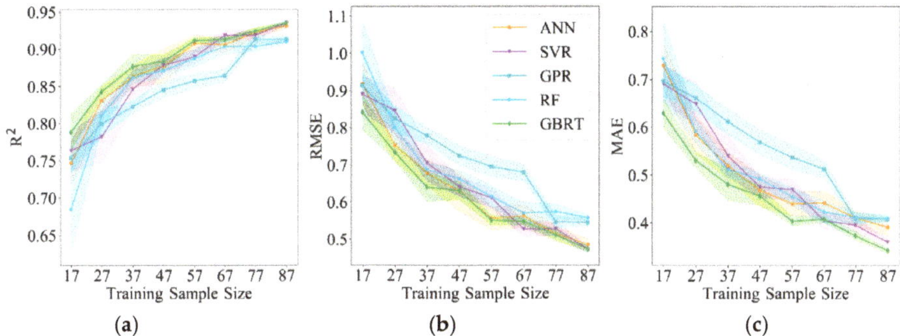

Figure 7. Sensitivity of the ML models to the training sample size (with intervals of 10) in terms of R^2, RMSE and MAE with standard error. (**a**) R^2; (**b**) RMSE; (**c**) MAE.

4.4. Sensitivity to Spectral Bands

Variable and feature selection have been one of the most significant processes in ML, with the aim of improving model accuracy and accelerating model processing. With 10 spectral bands of Sentinel-2 available (B2, B3, B4, B5, B6, B7, B8, B8A, B11, B12), it may not be practical to evaluate the performance of all the possible band combinations. This process could be challenging, as the characteristics of different ML models are distinct, and thus it is challenging to conduct a comparison among the different models under a unified standard. For example, GPR, RF and GBRT could provide feature relevance (spectral bands relevance) information that enables insight into feature selection processes, however, the feature relevance results are subject to the model itself, in other words, we could obtain different feature selection schemes, and such information is practical with single ML model based applications. In addition, ANN and SVR are not able to deliver feature relevance information.

Székely, Rizzo, and Bakirov [77] proposed the distance correlation (dCor) to measure the dependence of random vectors. dCor ranges between 0 and 1, and dCor (**X**, **Y**) = 0 only if **X** and **Y** are independent, which is effective in the feature selection processes [78–80]. To identify the optimal number of Sentinel-2 spectral bands required for cotton LAI retrieval using MLRAs, we used the distance correlation combined with the backward elimination method [81,82]. We also performed 10 random repetitions of each model to assess the model's robustness with the spectral bands. Notably, the distance correlation is subject to the training dataset adopted in this study. Figure 8 displays the distance correlation among the 10 spectral bands and LAI. The ranking of the dCor values between the 10 spectral bands and LAI are identified as B12 > B3 > B4 > B2 > B5 > B11 > B8A > B7 > B8 > B6 (B12: 0.88631, B3: 0.88630, B4: 0.88016, B2: 0.87721, B5: 0.87400, B11: 0.87338, B8A: 0.80219, B7: 0.76812, B8: 0.76585, B6: 0.68998). The visible and SWIR bands occupied the top rankings, while differences among the red-edge bands were distinct. The sensitivity results of different ML models to spectral bands using dCor and the backward elimination method (iteratively removing the spectral band with the lowest dCor value until only one spectral band is left) are shown in Figure 9, similarly, the standard error is given at each spectral bands combination (filling area of each model), lower values of the filling area corresponds to smaller standard error, and consequently, more robust model performance. Differences exist among the sensitivity of the models with the reduction in spectral bands. A turning point occurred at 6 for ANN, the accuracy begins to increase with the reduction in spectral bands before 6, though with some fluctuations, and then, the accuracy begins to decrease dramatically until reaching approximately 3. Similar trends are found among all the models except for GPR, as there is a relatively strong fluctuation at 7. GPR has a stable performance after 4. SVR, GPR and RF are robust for the spectral bands according to the standard error. Overall, model accuracy decreases as the spectral bands decrease, although there are some fluctuations in the models, whereas ANN provides the best model accuracy among the spectral band reductions on average ($\overline{R^2}$ = 0.881, $\overline{RMSE}$ = 0.625 and $\overline{MAE}$ = 0.480, calculated by all the 10 groups of spectral bands).

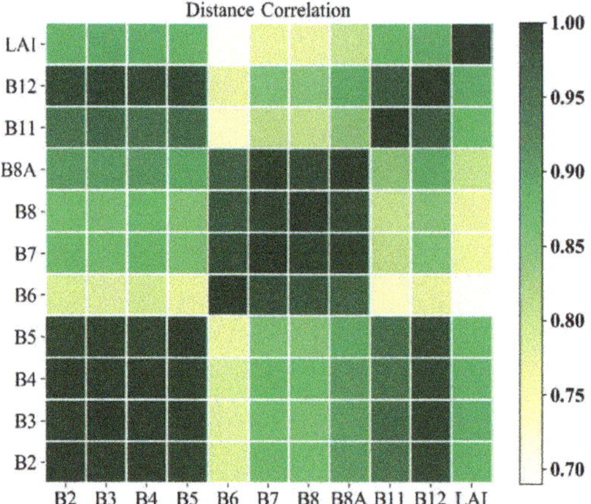

Figure 8. Distance correlation among the Sentinel-2 spectral bands and corresponding cotton LAI.

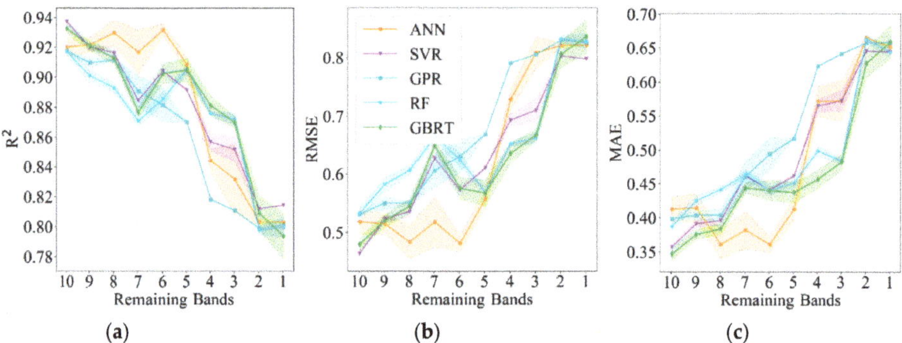

Figure 9. Sensitivity of the ML models to the spectral bands with respect to R^2, RMSE and MAE using the dCor and backward elimination method (the standard error is displayed). The horizontal axis (Remaining Bands) represents the number for the rest of spectral bands after each iterative removing process. (**a**) R^2; (**b**) RMSE; (**c**) MAE.

The minimum number of bands required for cotton LAI retrieval are recognized as 6 for ANN and SVR, 5 for RF and GBRT, and 8 for GPR, in other words, increasing of the spectral bands of Sentinel-2 does not significantly improves the model accuracy. Given the recognized number of spectral bands of each of the ML models, this is vital for associated applications with great amounts of samples, as it may reduce model processing time and desirable model accuracy could be acquired in the meantime.

4.5. Comprehensive Evaluation

In this study, we conducted a comparison of five universal MLRAs for cotton LAI retrieval with regard to model accuracy, computational efficiency, sensitivity to training sample size and sensitivity to spectral bands. In this section, we provide a comprehensive evaluation of five ML algorithms according to the obtained results, which were discussed earlier in this paper. To combine all the metrics, we performed a standardization process on the results. The mean values of the results in the sensitivity analysis (training sample size and spectral bands) were used, and we make RMSE, MAE and processing speed negative for the sake of comprehensive comparison. An alternative standardization is interval scaling, which is as follows:

$$x' = \frac{x - Min}{Max - Min} \tag{18}$$

where *Max* denotes the maximum value of the data, and *Min* represents the minimum value of the data. The results are displayed in Figure 10, and the metrics used in the radar chart and its corresponding implications are displayed in Table 10. Clearly, GBRT shows an outstanding performance from a comprehensive viewpoint, however, GPR shows the worst performance. ANN and SVR show moderate performances.

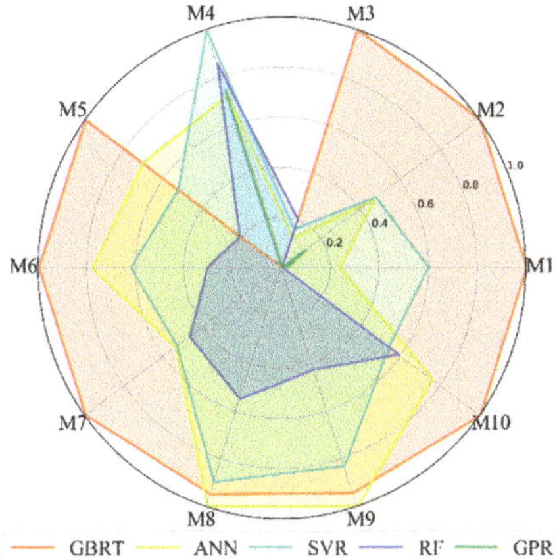

Figure 10. Comprehensive comparison of five machine learning regression algorithms (MLRAs), with different metrics using a radar chart.

Table 10. Metrics used in a radar chart and the corresponding descriptions.

Metrics in the Radar Chart	Corresponding Description
M1	Fitness (R^2)
M2	Accuracy (RMSE)
M3	Accuracy (MAE)
M4	Computational efficiency
M5	Sensitivity to training sample size (R^2)
M6	Sensitivity to training sample size (RMSE)
M7	Sensitivity to training sample size (MAE)
M8	Sensitivity to spectral bands (R^2)
M9	Sensitivity to spectral bands (RMSE)
M10	Sensitivity to spectral bands (MAE)

4.6. Final LAI Maps

The best performing ML regression model has been applied to map cotton LAI with Sentinel-2 imagery, as analyzed in previous sections, we use the GBRT model to map cotton LAI in the study area. The fields in the study area are regular, we conducted a manual vectorization based on Google Earth image data to obtain accurate agricultural fields boundaries, and Landsat 8 image data were collected to update the fields, as Google Earth image data has poor temporal information. Totally, 40,211 fields were collected by this way in the study area. Figure 11 displayed the agricultural fields boundaries. A per-field crop classification was performed to extract cotton fields. The final cotton LAI map was obtained by masking the results using the extracted cotton fields. Figure 12 presents the final cotton

LAI map. From Figure 12, it is clear to find that there is an increasing trend of cotton LAI growth, as revealed by Figure 2.

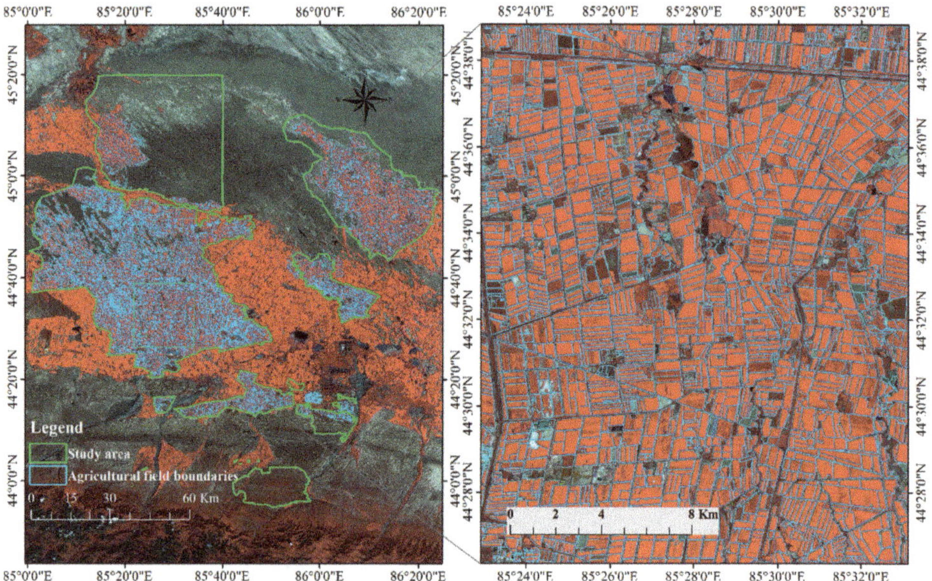

Figure 11. Agricultural fields boundaries collected by manual vectorization.

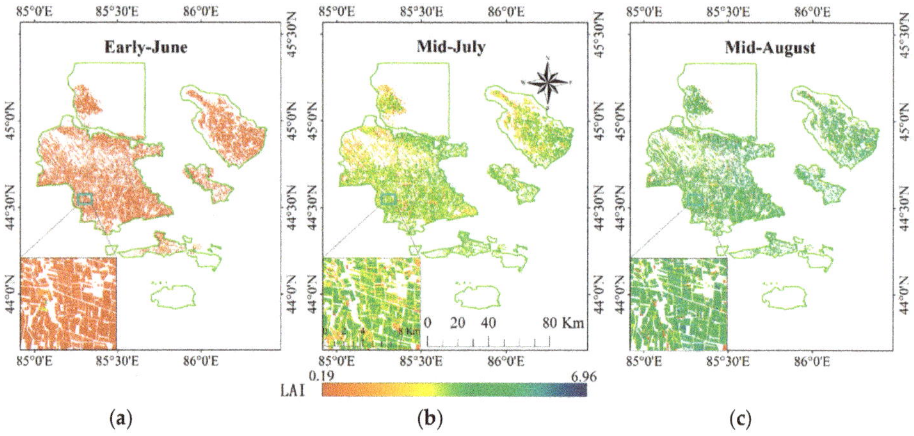

Figure 12. Final cotton LAI map obtained using GBRT model and masked by the extracted cotton fields. (**a**) Early June 2018; (**b**) Mid July 2018;(**c**) Mid August 2018.

In the early June cotton LAI map, some areas have LAI values much larger than the whole map (green color), as well as LAI values much smaller found in mid-July and mid-August (orange color), these fields are some other crops rather than cotton, and it can be ascribed to the misclassification.

4.7. Comparison with Global LAI Products

The validation of moderate-resolution products are key steps to assess the quality of the global LAI products. In this study, we performed a comparison among GEOV3, MODIS LAI products and upscaled Sentinel-2 LAI to assess the quality of these two products in Northwest China. To conduct the comparison of the cotton LAI results obtained using the GBRT model with global LAI products, we choose the closest dates of the products at a spatial resolution of the products to minimize the spatial and temporal differences [2].

All Sentinel-2 LAI maps were resampled to the same spatial resolution of MODIS and GEOV3 LAI products, and then GEOV3 LAI product were also resampled to the same spatial resolution of MODIS LAI product to make a straight comparison of these two products. Figure 13 shows the scatter plots among the comparison of GEOV3, MODIS and Sentinel-2 cotton LAI over three observation dates. Regarding the comparison of the MODIS LAI product and Sentinel-2 LAI maps, it was found that both the MODIS LAI product and the Sentinel-2 LAI suffered from the influence of the background, and are kind of overestimated in some areas in early June 2018. MODIS LAI product underestimated a little bit in mid-August 2018. In terms of GEOV3 LAI product and Sentinel-2 LAI maps, the GEOV3 LAI product overestimated at some areas both in early June and mid-July 2018, and underestimated in mid-August 2018. Finally, with respect to the comparison of the two global LAI products, relatively, large differences are found in early June and mid-July 2018, whereas a good relationship appeared in mid-August 2018.

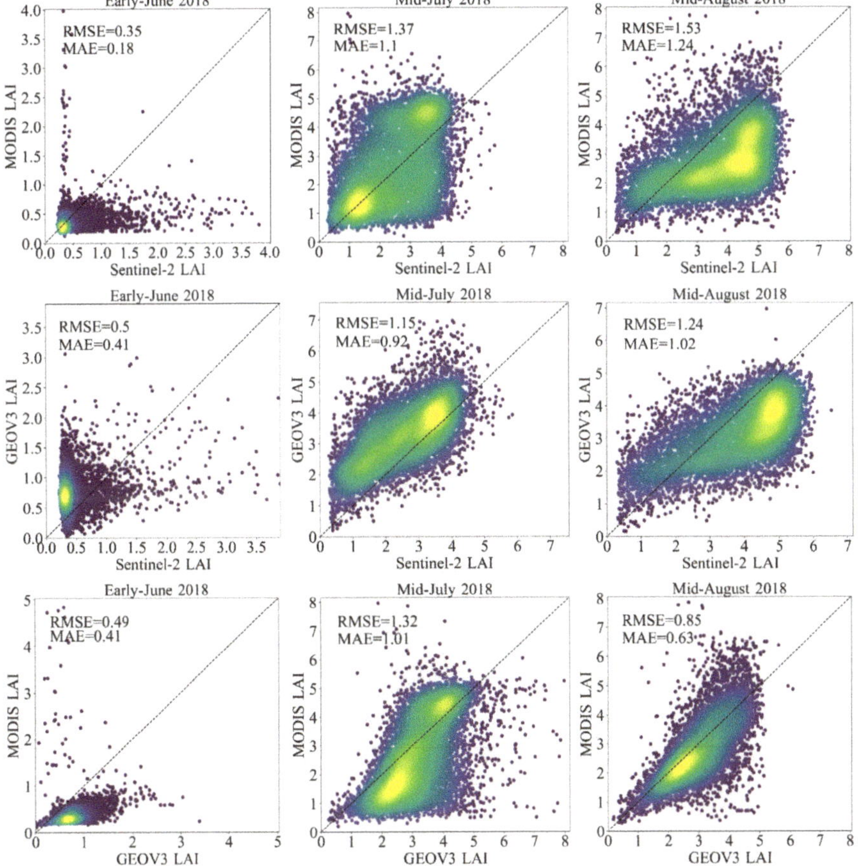

Figure 13. Scatter plots among GEOV3, MODIS LAI products and upscaled Sentinel-2 LAI in the 2018 cotton season.

5. Discussion

Over the last decade, there has been a considerable increase in the introduction of ML algorithms to remote sensing for a wide range of fields. Crop biophysical parameters (e.g., LAI) are key variables for a wide range of applications. With the progress of remote sensing techniques, we are able to acquire high dimensional (spatial, temporal, and spectral) remote sensing data, which demands more efficient, accurate and robust algorithms in a wide variety of applications with remote sensing. In this context, there is great potential for ML algorithms to be used in a wide range of remote sensing applications.

The diversity of available ML algorithms poses great challenges for the selection of MLRAs, as well as the decision of optimal number of training sample size and spectral bands to different MLRAs. Besides, another significant problem that may arise involves hyperparameter tuning in the application of ML algorithms. In general, experience is required to obtain satisfactory results for ML algorithms, and experiments with a large number of datasets may be needed, which are generally not available. Accordingly, it is necessary to perform a comparison of different fashionable ML models to support better remote sensing applications. In this study, we focus on the comparison of five well-known ML algorithms for cotton LAI retrievals with Sentinel-2 imagery, because these algorithms have a wide range of applications in remote sensing. Additionally, the hyperparameter tuning results of five prominent ML models are provided. Our study could provide support for associated remote sensing studies based on ML algorithms.

Furthermore, a comparatively great fluctuation is observed at 7 among all the models except for GPR in Figure 9, the previous removing spectral band is the red-edge band (B7), which indicates that having low dCor values do not necessarily correspond to being less important for LAI retrieval. Notably, ensemble methods (RF and GBRT) and GPR models have a great benefit of delivering feature importance information (not shown in our study), which provides insight into the greatest contributing features of the model. Such information could be used for better model interpretation, and this information is also useful for feature selection processes when applied to models that contain a great number of features.

In related studies, Verrelst, et al. [55] compared four ML algorithms (NN, SVR, KRR and GPR) using simulated Sentinel-2 and 3 data to assess three biophysical parameters (Chlorophyll content, LAI and FVC). GPR outperformed the other regression methods for the majority of Sentinel configurations, whereas in our study, GPR performed worse than the other regression methods. Results that differ from our study may be for the following reasons. First, we used real Sentinel-2 data rather than simulation data, and it may be difficult to represent the performance of real Sentinel-2 data with simulation data. Second, we focus on cotton crops over the entire growth period rather than on various crop types. In addition, there may be different hyperparameter settings between models. Finally, there are differences between the Sentinel-2 bands, and we used 10 spectral bands (SWIR included). Yuan, et al. [83] compared the RF, ANN and SVM regression models for soybean LAI retrieval using unmanned aerial vehicle (UAV) hyperspectral data with different sampling methods. The results showed that RF is suitable for the whole growth period of soybean LAI estimation, while ANN is appropriate for a single growth period. Siegmann and Jarmer [84] compared SVR and RFR for wheat LAI estimation using hyperspectral data, and the results showed that SVR provided the best performance of the entire dataset. Different from previous studies, we considered the GBRT model due to its highly robust performance over a wide range of applications. We further explored the sensitivity of different ML algorithms to training sample size and Sentinel-2 spectral bands. The results show that GBRT achieves the best performance, and GBRT was more robust for the training samples than the other models.

Despite the promising results revealed by five ML regression methods, there are some limitations of our study, and further studies are needed to assess the generalization ability of the five ML regression methods. Firstly, we focus on only one crop type, and the performance of ML regression methods for other crop types has not been validated. It is necessary to evaluate ML algorithms with various crop types, provided there are enough field measurement LAI data of different crop types available.

Secondly, some other crop biophysical parameters (e.g., Chlorophyll content and FVC) demand evaluation as well as ML regression methods, as different results may appear with different ML algorithms and different crop biophysical parameters. Thirdly, there are limitations with Sentinel-2 spectral bands sensitivity analysis, since some spectral bands are highly correlated, and redundancies remain among the spectral bands, so this process would be worth further investigations to explore a better spectral bands sensitivity analysis. Additionally, there are limitations to our quadrats in the field campaigns, as we chose only three sample points to represent a quadrat (20 m × 20 m). Finally, the performance of five MLRAs has not yet been evaluated for large amounts of data (e.g., tens of thousands of data) due to the limited number of samples in our study, however, it remains to be seen whether a favorable performance can be obtained. In consideration of the ability of ML techniques, it is expected to obtain a better accuracy of the regression models provided there are enough samples, and a turning point should be obtained, as has been demonstrated in previous studies [19]. While before the turning point, the model accuracy increased rapidly with the increase in the training sample size, whereas, after the turning ponits, the model behaves more stable, in other words, the model accuracy remmained unchanged with the increasing of training sample szie. Future studies are required to investigate this problem.

With the advance in ML technology, deep learning, a new subfield of ML, has been applied successfully to many remote sensing domains, especially with a large quantity of data [85]. It is not necessary to construct complex ML models (deep learning-based models) in our study, as traditional ML algorithms have the capability of establishing efficient, accurate and robust estimation models for cotton LAI retrieval. However, as there are greater amounts of data available, deep learning techniques may be promising alternatives for handling such large volume datasets and complex relationships among the input datasets, which deserve further study.

6. Conclusions

With the challenges of selecting ML algorithms for crop LAI estimation, the results of this study have great implications for the selection of appropriate ML models from the diversity of available ML algorithms, and at the same time, these same results provide the optimal number of training sample size and spectral bands of Sentinel-2 for each model required for cotton LAI retrieval. Regarding the comparison of different ML models for cotton LAI retrieval employed in our study, the GBRT model outperforms the other ML models according to our results. Our findings increase the potential for cotton LAI retrieval with Sentinel-2 imagery, and may be transferrable to other associated problems related to agricultural remote sensing applications. On the other hand, the GBRT model is computationally demanding, which may be a significant problem with a large scale of data. However, GBRT can challenge the model accuracy and computational efficiency selection problems. Considering the computational efficiency, SVR exhibits considerable computational superiority over the other ML models.

Regarding the sample size sensitivity of ML models, model accuracy increases with the growth of the training sample size, and the GBRT produces the most robust performance for the training samples with respect to the standard error and the best model accuracy on average.

In terms of spectral band sensitivity analysis, the distance correlation results showed that the SWIR and visible bands have great potential to improve the accuracy of cotton LAI retrievals. By using dCor combined with the backward elimination method, the model accuracy decreases with the reduction in spectral bands. SVR, GPR and RF perform robustly with the spectral bands, and ANN provides the best accuracy on average. The minimum number of bands required for cotton LAI retrieval are recognized as 6 (ANN), 6 (SVR), 5 (RF), 5 (GBRT) and 8 for GPR.

A comprehensive evaluation has been employed to identify the performance of five ML models, considering a combination of model accuracy, computational efficiency, sensitivity to training sample size and sensitivity to spectral bands. The comprehensive performance of the models is identified as GBRT > ANN > SVR > RF > GPR.

Despite the different performances of the five ML regression models, MLRAs are promising ways to retrieve cotton LAI with Sentinel-2 imagery because the MLRAs all achieved encouraging accuracies. With profound applications for a diversity of ML algorithms in remote sensing, MLRAs may provide positive effects for remote sensing applications in terms of classification, regression, and other associated problems.

Author Contributions: Conceptualization, J.M. and H.M.; Methodology, H.M.; Formal Analysis, H.M.; Investigation, H.M., F.J., Q.Z. and H.F.; Data curation, H.M.; Writing—Original Draft Preparation, H.M.; Writing—Review and Editing, J.M.; Visualization, H.M.

Funding: This research was funded by GF6 Project under grant No. 30-Y20A03-9003-17/18 and 09-Y20A05-9001-17/18; the National Natural Science Foundation of China (41871261); and the open fund of the Key Laboratory of Oasis Eco-agriculture, Xinjiang Production and Construction Group (201701).

Acknowledgments: We are grateful to Agricultural College of Shihezi University for providing help in field campaigns. We would like to acknowledge the authors who contributed to the development of Scikit-learn project. We acknowledge the Copernicus Global Land Service, and the Land Processes Distributed Active Archive Center (LP DAAC) within the NASA Earth Observing System Data and Information System (EOSDIS) for providing the global LAI products. We are thankful to the three anonymous reviewers for their insightful comments and suggestions helped to clarify and improve the manuscript.

Conflicts of Interest: The authors declare no conflict of interest.

References

1. Chen, J.M.; Black, T.A. Defining leaf-area index for non-flat leaves. *Plant Cell Environ.* **1992**, *15*, 421–429. [CrossRef]
2. Garrigues, S.; Lacaze, R.; Baret, F.; Morisette, J.T.; Weiss, M.; Nickeson, J.E.; Fernandes, R.; Plummer, S.; Shabanov, N.V.; Myneni, R.B.; et al. Validation and intercomparison of global Leaf Area Index products derived from remote sensing data. *J. Geophys. Res.-Biogeosci.* **2008**, *113*. [CrossRef]
3. Asner, G.P.; Scurlock, J.M.O.; Hicke, J.A. Global synthesis of leaf area index observations: Implications for ecological and remote sensing studies. *Glob. Ecol. Biogeogr.* **2003**, *12*, 191–205. [CrossRef]
4. Buermann, W.; Dong, J.R.; Zeng, X.B.; Myneni, R.B.; Dickinson, R.E. Evaluation of the utility of satellite-based vegetation leaf area index data for climate simulations. *J. Clim.* **2001**, *14*, 3536–3550. [CrossRef]
5. Yuan, H.; Dai, Y.J.; Xiao, Z.Q.; Ji, D.Y.; Shangguan, W. Reprocessing the MODIS Leaf Area Index products for land surface and climate modelling. *Remote Sens. Environ.* **2011**, *115*, 1171–1187. [CrossRef]
6. Van den Hurk, B.; Viterbo, P.; Los, S.O. Impact of leaf area index seasonality on the annual land surface evaporation in a global circulation model. *J. Geophys. Res.-Atmos.* **2003**, *108*, 4191. [CrossRef]
7. Cheng, Z.Q.; Meng, J.H.; Wang, Y.M. Improving spring maize yield estimation at field scale by assimilating time-series HJ-1 CCD data into the WOFOST model using a new method with fast algorithms. *Remote Sens.* **2016**, *8*, 303. [CrossRef]
8. Systematic Observation Requirements for Satellite-Based Products for Climate 2011 Update: Supplemental Details to the Satellite-Based Component of the "Implementation Plan for the Global Observing System for Climate in Support of the UNFCCC (2010 Update)". Available online: https://library.wmo.int/index.php?lvl=notice_display&id=12907 (accessed on 13 December 2018).
9. Dong, Y.Y.; Zhao, C.J.; Yang, G.J.; Chen, L.P.; Wang, J.H.; Feng, H.K. Integrating a very fast simulated annealing optimization algorithm for crop leaf area index variational assimilation. *Math. Comput. Model.* **2013**, *58*, 871–879. [CrossRef]
10. Jego, G.; Pattey, E.; Liu, J.G. Using Leaf Area Index, retrieved from optical imagery, in the STICS crop model for predicting yield and biomass of field crops. *Field Crop. Res.* **2012**, *131*, 63–74. [CrossRef]
11. Baez-Gonzalez, A.D.; Kiniry, J.R.; Maas, S.J.; Tiscareno, M.; Macias, J.; Mendoza, J.L.; Richardson, C.W.; Salinas, J.; Manjarrez, J.R. Large-area maize yield forecasting using leaf area index based yield model. *Agron. J.* **2005**, *97*, 418–425. [CrossRef]
12. Liang, S.L. Recent developments in estimating land surface biogeophysical variables from optical remote sensing. *Prog. Phys. Geogr.* **2007**, *31*, 501–516. [CrossRef]

13. Verrelst, J.; Camps-Valls, G.; Munoz-Mari, J.; Rivera, J.P.; Veroustraete, F.; Clevers, J.; Moreno, J. Optical remote sensing and the retrieval of terrestrial vegetation bio-geophysical properties—A review. *ISPRS-J. Photogramm. Remote Sens.* **2015**, *108*, 273–290. [CrossRef]

14. Verrelst, J.; Malenovský, Z.; Van der Tol, C.; Camps-Valls, G.; Gastellu-Etchegorry, J.-P.; Lewis, P.; North, P.; Moreno, J. Quantifying vegetation biophysical variables from imaging spectroscopy data: A review on retrieval methods. *Surv. Geophys.* **2018**, 1–41. [CrossRef]

15. Campos-Taberner, M.; Garcia-Haro, F.J.; Busetto, L.; Ranghetti, L.; Martinez, B.; Gilabert, M.A.; Camps-Valls, G.; Camacho, F.; Boschetti, M. A critical comparison of remote sensing Leaf Area Index estimates over rice-cultivated areas: From Sentinel-2 and Landsat-7/8 to MODIS, GEOV1 and EUMETSAT polar system. *Remote Sens.* **2018**, *10*, 763. [CrossRef]

16. Baret, F.; Buis, S. Estimating canopy characteristics from remote sensing observations: Review of methods and associated problems. In *Advances in Land Remote Sensing: System, Modeling, Inversion and Application*; Liang, S., Ed.; Springer: Dordrecht, The Netherlands, 2008; pp. 173–201. ISBN 978-1-4020-6449-4.

17. Maxwell, A.E.; Warner, T.A.; Fang, F. Implementation of machine-learning classification in remote sensing: An applied review. *Int. J. Remote Sens.* **2018**, *39*, 2784–2817. [CrossRef]

18. Lary, D.J.; Alavi, A.H.; Gandomi, A.H.; Walker, A.L. Machine learning in geosciences and remote sensing. *Geosci. Front.* **2016**, *7*, 3–10. [CrossRef]

19. Wang, T.T.; Xiao, Z.Q.; Liu, Z.G. Performance evaluation of machine learning methods for Leaf Area Index retrieval from time-series MODIS reflectance data. *Sensors* **2017**, *17*, 81. [CrossRef]

20. Noi, P.T.; Kappas, M. Comparison of random forest, k-nearest neighbor, and support vector machine classifiers for land cover classification using Sentinel-2 imagery. *Sensors* **2018**, *18*, 18. [CrossRef]

21. Chen, G.B.; Li, S.S.; Knibbs, L.D.; Hamm, N.A.S.; Cao, W.; Li, T.T.; Guo, J.P.; Ren, H.Y.; Abramson, M.J.; Guo, Y.M. A machine learning method to estimate PM2.5 concentrations across China with remote sensing, meteorological and land use information. *Sci. Total Environ.* **2018**, *636*, 52–60. [CrossRef]

22. Lary, D.J.; Zewdie, G.K.; Liu, X.; Wu, D.; Levetin, E.; Allee, R.J.; Malakar, N.; Walker, A.; Mussa, H.; Mannino, A. Machine learning applications for earth observation. In *Earth Observation Open Science and Innovation*; Mathieu, P.-P., Aubrecht, C., Eds.; Springer: Cham, Switzerland, 2018; pp. 165–218. ISBN 978-3-319-65633-5.

23. Kwon, S.K.; Jung, H.S.; Baek, W.K.; Kim, D. Classification of forest vertical structure in South Korea from aerial orthophoto and lidar data using an artificial neural network. *Appl. Sci.* **2017**, *7*, 1046. [CrossRef]

24. Mohri, M.; Talwalkar, A.; Rostamizadeh, A. *Foundations of Machine Learning*; Dietterich, T., Bishop, C., Heckerman, D., Jordan, M., Kearns, M., Eds.; MIT Press: Cambridge, MA, USA, 2012; ISBN 978-0-262-01825-8.

25. Durbha, S.S.; King, R.L.; Younan, N.H. Support vector machines regression for retrieval of leaf area index from multiangle imaging spectroradiometer. *Remote Sens. Environ.* **2007**, *107*, 348–361. [CrossRef]

26. Karimi, S.; Sadraddini, A.A.; Nazemi, A.H.; Xu, T.R.; Fard, A.F. Generalizability of gene expression programming and random forest methodologies in estimating cropland and grassland leaf area index. *Comput. Electron. Agric.* **2018**, *144*, 232–240. [CrossRef]

27. Verrelst, J.; Alonso, L.; Camps-Valls, G.; Delegido, J.; Moreno, J. Retrieval of vegetation biophysical parameters using Gaussian process techniques. *IEEE Trans. Geosci. Remote Sens.* **2012**, *50*, 1832–1843. [CrossRef]

28. Bacour, C.; Baret, F.; Beal, D.; Weiss, M.; Pavageau, K. Neural network estimation of LAI, fAPAR, fCover and LAIxC(ab), from top of canopy MERIS reflectance data: Principles and validation. *Remote Sens. Environ.* **2006**, *105*, 313–325. [CrossRef]

29. Li, X.; Bai, R.B. Freight Vehicle travel time prediction using gradient boosting regression tree. In Proceedings of the 2016 15th IEEE International Conference on Machine Learning and Applications, Anaheim, CA, USA, 18–20 December 2016; IEEE: Piscataway, NJ, USA, 2016; pp. 1010–1015.

30. Guneralp, I.; Filippi, A.M.; Randall, J. Estimation of floodplain aboveground biomass using multispectral remote sensing and nonparametric modeling. *Int. J. Appl. Earth Obs. Geoinf.* **2014**, *33*, 119–126. [CrossRef]

31. Xiao, Z.B.; Wang, Y.; Fu, K.; Wu, F. Identifying different transportation modes from trajectory data using tree-based ensemble classifiers. *ISPRS Int. Geo-Inf.* **2017**, *6*, 57. [CrossRef]

32. Martinez, B.; Garcia-Haro, F.J.; Camacho-de Coca, F. Derivation of high-resolution leaf area index maps in support of validation activities: Application to the cropland Barrax site. *Agric. For. Meteorol.* **2009**, *149*, 130–145. [CrossRef]

33. Traoré, F. Assessing the impact of China net imports on the world cotton price. *Appl. Econ. Lett.* **2014**, *21*, 1031–1035. [CrossRef]
34. Wang, H.D.; Wu, L.F.; Cheng, M.H.; Fan, J.L.; Zhang, F.C.; Zou, Y.F.; Chau, H.W.; Gao, Z.J.; Wang, X.K. Coupling effects of water and fertilizer on yield, water and fertilizer use efficiency of drip-fertigated cotton in northern Xinjiang, China. *Field Crop. Res.* **2018**, *219*, 169–179. [CrossRef]
35. ESA. GMES Sentinel-2 Mission Requirements Document, Technical Report issue 2 revision 1. Available online: http://esamultimedia.esa.int/docs/GMES/Sentinel-2_MRD.pdf (accessed on 30 March 2019).
36. Jaramaz, D.; Perović, V.; Belanović, S.; Saljnikov, E.; Čakmak, D.; Mrvić, V.; Životić, L. The ESA Sentinel-2 Mission Vegetation Variables for Remote Sensing of Plant Monitoring. In Proceedings of the 2nd International Conference on Regional Development, Spatial Planning and Strategic Governance (RESPAG 2013), Belgrade, Serbia, 22–25 May 2013; Vujošević, M., Milijić, S., Eds.; Institute of Architecture and Urban & Spatial Planning of Serbia (IAUS): Belgrade, Serbia, 2013; pp. 950–961.
37. Delegido, J.; Verrelst, J.; Meza, C.M.; Rivera, J.P.; Alonso, L.; Moreno, J. A red-edge spectral index for remote sensing estimation of green LAI over agroecosystems. *Eur. J. Agron.* **2013**, *46*, 42–52. [CrossRef]
38. Gong, P.; Pu, R.L.; Biging, G.S.; Larrieu, M.R. Estimation of forest leaf area index using vegetation indices derived from Hyperion hyperspectral data. *IEEE Trans. Geosci. Remote Sens.* **2003**, *41*, 1355–1362. [CrossRef]
39. Twele, A.; Erasmi, S.; Kappas, M. Spatially explicit estimation of leaf area index using EO-1 hyperion and landsat ETM+ data: Implications of spectral bandwidth and shortwave infrared data on prediction accuracy in a tropical montane environment. *GISci. Remote Sens.* **2008**, *45*, 229–248. [CrossRef]
40. ESA. Copernicus Open Access Hub. Available online: https://scihub.copernicus.eu/dhus/#/home (accessed on 6 April 2019).
41. Louis, J.; Debaecker, V.; Pflug, B.; Main-Knorn, M.; Bieniarz, J.; Müller-Wilm, U.; Cadau, E.; Gascon, F. SENTINEL-2 SEN2COR: L2A processor for users. In Proceedings of the Living Planet Symposium 2016, Prague, Czech Republic, 9–13 May 2016.
42. Müller-Wilm, U.; Louis, J.; Richter, R.; Gascon, F.; Niezette, M. Sentinel-2 Level-2A prototype processor: Architecture, algorithms and first results. In Proceedings of the ESA Living Planet Symposium 2013, Edinburgh, UK, 9–13 September 2013.
43. Fang, H.L.; Jiang, C.Y.; Li, W.J.; Wei, S.S.; Baret, F.; Chen, J.M.; Garcia-Haro, J.; Liang, S.L.; Liu, R.G.; Myneni, R.B.; et al. Characterization and intercomparison of global moderate resolution leaf area index (LAI) products: Analysis of climatologies and theoretical uncertainties. *J. Geophys. Res.-Biogeosci.* **2013**, *118*, 529–548. [CrossRef]
44. NASA. LAADS DAAC. Available online: https://ladsweb.modaps.eosdis.nasa.gov/ (accessed on 6 April 2019).
45. ESA. Copernicus Global Land Service. Available online: https://land.copernicus.eu/global/ (accessed on 6 April 2019).
46. Myneni, R.; Knyazikhin, Y.; Park, T. MCD15A3H MODIS/Terra+Aqua Leaf Area Index/FPAR 4-day L4 Global 500m SIN Grid V006. NASA EOSDIS Land Processes DAAC. 2015. Available online: http://doi.org/10.5067/MODIS/MCD15A3H.006 (accessed on 6 April 2019).
47. Baret, F.; Weiss, M.; Lacaze, R.; Camacho, F.; Makhmara, H.; Pacholcyzk, P.; Smets, B. GEOV1: LAI and FAPAR essential climate variables and FCOVER global time series capitalizing over existing products. Part1: Principles of development and production. *Remote Sens. Environ.* **2013**, *137*, 299–309. [CrossRef]
48. Baret, F.; Weiss, M.; Verger, A.; Smets, B. ATBD FOR LAI, FAPAR AND FCOVER FROM PROBA-V PRODUCTS AT 300M RESOLUTION (GEOV3). Available online: https://land.copernicus.eu/global/sites/cgls.vito.be/files/products/ImagineS_RP2.1_ATBD-LAI300m_I1.73.pdf (accessed on 30 March 2019).
49. Scikit-Learn Developers. Scikit-learn. Available online: https://scikit-learn.org/stable/index.html (accessed on 6 April 2019).
50. Python Software Foundation. Python. Available online: https://www.python.org/ (accessed on 6 April 2019).
51. Pedregosa, F.; Varoquaux, G.; Gramfort, A.; Michel, V.; Thirion, B.; Grisel, O.; Blondel, M.; Prettenhofer, P.; Weiss, R.; Dubourg, V.; et al. Scikit-learn: Machine learning in Python. *J. Mach. Learn. Res.* **2011**, *12*, 2825–2830.
52. Adelabu, S.; Mutanga, O.; Adam, E. Testing the reliability and stability of the internal accuracy assessment of random forest for classifying tree defoliation levels using different validation methods. *Geocarto Int.* **2015**, *30*, 810–821. [CrossRef]

53. Omer, G.; Mutanga, O.; Abdel-Rahman, E.M.; Adam, E. Performance of support vector machines and artificial neural network for mapping endangered tree species using WorldView-2 data in Dukuduku Forest, South Africa. *IEEE J. Sel. Top. Appl. Earth Observ. Remote Sens.* **2015**, *8*, 4825–4840. [CrossRef]

54. Dube, T.; Mutanga, O.; Adam, E.; Ismail, R. Intra-and-inter species biomass prediction in a plantation forest: Testing the utility of high spatial resolution spaceborne multispectral rapideye sensor and advanced machine learning algorithms. *Sensors* **2014**, *14*, 15348–15370. [CrossRef]

55. Verrelst, J.; Munoz, J.; Alonso, L.; Delegido, J.; Rivera, J.P.; Camps-Valls, G.; Moreno, J. Machine learning regression algorithms for biophysical parameter retrieval: Opportunities for Sentinel-2 and -3. *Remote Sens. Environ.* **2012**, *118*, 127–139. [CrossRef]

56. van Gerven, M. Computational foundations of natural intelligence. *Front. Comput. Neurosci.* **2017**, *11*, 7–30. [CrossRef]

57. Waske, B.; Fauvel, M.; Benediktsson, J.A.; Chanussot, J. Machine learning techniques in remote sensing data analysis. In *Kernel Methods for Remote Sensing Data Analysis*; Camps-Valls, G., Bruzzone, L., Eds.; John Wiley & Sons: Hoboken, NJ, USA, 2009; pp. 3–24. ISBN 978-470-72211-4.

58. Rumelhart, D.E.; Hinton, G.E.; Williams, R.J. Learning representations by back-propagating errors. *Nature* **1986**, *323*, 533–536. [CrossRef]

59. Hornik, K.; Stinchcombe, M.; White, H. Multilayer feedforward networks are universal approximators. *Neural Netw.* **1989**, *2*, 359–366. [CrossRef]

60. Sonmez, H.; Gokceoglu, C.; Nefeslioglu, H.A.; Kayabasi, A. Estimation of rock modulus: For intact rocks with an artificial neural network and for rock masses with a new empirical equation. *Int. J. Rock Mech. Min. Sci.* **2006**, *43*, 224–235. [CrossRef]

61. Madhiarasan, M.; Deepa, S.N. Comparative analysis on hidden neurons estimation in multi layer perceptron neural networks for wind speed forecasting. *Artif. Intell. Rev.* **2017**, *48*, 449–471. [CrossRef]

62. Huang, G.B. Learning capability and storage capacity of two-hidden-layer feedforward networks. *IEEE Trans. Neural Netw.* **2003**, *14*, 274–281. [CrossRef]

63. Cortes, C.; Vapnik, V. Support-vector networks. *Mach. Learn.* **1995**, *20*, 273–297. [CrossRef]

64. Vapnik, V.; Golowich, S.E.; Smola, A. Support vector method for function approximation, regression estimation, and signal processing. In Proceedings of the 9th International Conference on Neural Information Processing Systems, Denver, CO, USA, 3–5 December 1996; Mozer, M.C., Jordan, M.I., Petsche, T., Eds.; MIT Press: Cambridge, MA, USA, 1996; pp. 281–287.

65. Mountrakis, G.; Im, J.; Ogole, C. Support vector machines in remote sensing: A review. *ISPRS-J. Photogramm. Remote Sens.* **2011**, *66*, 247–259. [CrossRef]

66. Basak, D.; Pal, S.; Patranabis, D.C. Support vector regression. *Neural Inf. Process. Lett. Rev.* **2007**, *11*, 203–224.

67. Awad, M.; Khanna, R. *Efficient Learning Machines: Theories, Concepts, and Applications for Engineers and System Designers*; Pepper, J., Weiss, S., Hauke, P., Eds.; Apress: Berkeley, CA, USA, 2015; ISBN 978-1-4302-5990-9.

68. Ramedani, Z.; Omid, M.; Keyhani, A.; Shamshirband, S.; Khoshnevisan, B. Potential of radial basis function based support vector regression for global solar radiation prediction. *Renew. Sust. Energ. Rev.* **2014**, *39*, 1005–1011. [CrossRef]

69. Li, M.; Liu, Y.H. Learning interaction force model for endodontic shaping with support vector regression. In Proceedings of the 2006 IEEE International Conference on Robotics and Automation, Orlando, FL, USA, 15–19 May 2006; IEEE: Piscataway, NJ, USA, 2006; pp. 3642–3647.

70. Rasmussen, C.E.; Williams, C.K. *Gaussian Process for Machine Learning*; Dietterich, T., Bishop, C., Heckerman, D., Jordan, M., Kearns, M., Eds.; MIT Press: Cambridge, MA, USA, 2006; ISBN 0-262-18253-X.

71. Scornet, E. Random forests and Kernel methods. *IEEE Trans. Inf. Theory* **2016**, *62*, 1485–1500. [CrossRef]

72. Breiman, L. Random forests. *Mach. Learn.* **2001**, *45*, 5–32. [CrossRef]

73. Friedman, J.H. Greedy function approximation: A gradient boosting machine. *Ann. Stat.* **2001**, *29*, 1189–1232. [CrossRef]

74. Natekin, A.; Knoll, A. Gradient boosting machines, a tutorial. *Front. Neurorobot.* **2013**, *7*, 21. [CrossRef] [PubMed]

75. Fischer, P.; Etienne, C.; Tian, J.J.; Krauss, T. Prediction of wind speeds based on digital elevation MODELS using boosted regression trees. In Proceedings of the International Conference on Sensors & Models in Remote Sensing & Photogrammetry, Kish Island, Iran, 23–25 November 2015; Arefi, H., Motagh, M., Eds.; Copernicus Gesellschaft Mbh: Göttingen, Germany, 2015; pp. 197–202.

76. Kanungo, T.; Orr, D. Predicting the readability of short web summaries. In Proceedings of the Second ACM International Conference on Web Search and Data Mining, Barcelona, Spain, 9–12 February 2009; Baeza-Yates, R., Boldi, P., Ribeiro-Neto, B., Cambazoglu, B.B., Eds.; ACM: New York, NY, USA, 2009; pp. 202–211.

77. Szekely, G.J.; Rizzo, M.L.; Bakirov, N.K. Measuring and testing dependence by correlation of distances. *Ann. Stat.* **2007**, *35*, 2769–2794. [CrossRef]

78. Li, R.Z.; Zhong, W.; Zhu, L.P. Feature screening via distance correlation learning. *J. Am. Stat. Assoc.* **2012**, *107*, 1129–1139. [CrossRef]

79. Zhong, W.; Zhu, L.P. An iterative approach to distance correlation-based sure independence screening. *J. Stat. Comput. Simul.* **2015**, *85*, 2331–2345. [CrossRef]

80. Kundu, P.P.; Mitra, S. Feature selection through message passing. *IEEE Trans. Cybern.* **2017**, *47*, 4356–4366. [CrossRef] [PubMed]

81. Guyon, I.; Elisseeff, A. An introduction to variable and feature selection. *J. Mach. Learn. Res.* **2003**, *3*, 1157–1182. [CrossRef]

82. Hong, X.; Mitchell, R.J. Backward elimination model construction for regression and classification using leave-one-out criteria. *Int. J. Syst. Sci.* **2007**, *38*, 101–113. [CrossRef]

83. Yuan, H.H.; Yang, G.J.; Li, C.C.; Wang, Y.J.; Liu, J.G.; Yu, H.Y.; Feng, H.K.; Xu, B.; Zhao, X.Q.; Yang, X.D. Retrieving soybean Leaf Area Index from unmanned aerial vehicle hyperspectral remote sensing: Analysis of RF, ANN, and SVM regression models. *Remote Sens.* **2017**, *9*, 309. [CrossRef]

84. Siegmann, B.; Jarmer, T. Comparison of different regression models and validation techniques for the assessment of wheat leaf area index from hyperspectral data. *Int. J. Remote Sens.* **2015**, *36*, 4519–4534. [CrossRef]

85. LeCun, Y.; Bengio, Y.; Hinton, G. Deep learning. *Nature* **2015**, *521*, 436–444. [CrossRef] [PubMed]

Article

Spatial Data Reconstruction via ADMM and Spatial Spline Regression [†]

Bang Liu [1,*, Borislav Mavrin [2], Linglong Kong [2] and Di Niu [1]**

[1] Electrical and Computer Engineering, University of Alberta, 9211-116 Street NW, Edmonton, AB T6G 1H9, Canada; dniu@ualberta.ca

[2] Mathematical and Statistical Sciences, University of Alberta, 632 Central Academic Building, Edmonton, AB T6G 2G1, Canada; mavrin@ualberta.ca (B.M.); lkong@ualberta.ca (L.K.)

* Correspondence: bang3@ualberta.ca

† This paper is an extended version of our paper published in the 2017 IEEE International Conference on Data Mining (ICDM), New Orleans, LA, USA, 18–21 November 2017.

Received: 30 March 2019; Accepted: 22 April 2019; Published: 26 April 2019

Abstract: Reconstructing fine-grained spatial densities from coarse-grained measurements, namely the aggregate observations recorded for each subregion in the spatial field of interest, is a critical problem in many real world applications. In this paper, we propose a novel Constrained Spatial Smoothing (CSS) approach for the problem of spatial data reconstruction. We observe that local continuity exists in many types of spatial data. Based on this observation, our approach performs sparse recovery via a finite element method, while in the meantime enforcing the aggregated observation constraints through an innovative use of the Alternating Direction Method of Multipliers (ADMM) algorithm framework. Furthermore, our approach is able to incorporate external information as a regression add-on to further enhance recovery performance. To evaluate our approach, we study the problem of reconstructing the spatial distribution of cellphone traffic volumes based on aggregate volumes recorded at sparsely scattered base stations. We perform extensive experiments based on a large dataset of Call Detail Records and a geographical and demographical attribute dataset from the city of Milan, and compare our approach with other methods such as Spatial Spline Regression. The evaluation results show that our approach significantly outperforms various baseline approaches. This proves that jointly modeling the underlying spatial continuity and the local features that characterize the heterogeneity of different locations can help improve the performance of spatial recovery.

Keywords: spatial sparse recovery; constrained spatial smoothing; spatial spline regression; alternating direction method of multipliers

---

1. Introduction

The problem of reconstructing fine-grained spatial data from its coarse-grained aggregate observations of each subregions lies in the core of many real world applications. For example, the reconstruction of fine-grained spatial distribution of cell phone activities is of particular interest to telecommunication and information technology companies, where the recovered data can be used for device installation, capacity planning, the study of urban ecology [1–3], population density estimation [4–6], and human mobility prediction [7–11]. However, the companies may only have access to the aggregate mobile traffic volumes on each base station, as either privacy issues or additional technical overhead is involved to get fine-grained spatial data of users. Similarly, it is also highly valuable if we can infer the spatial distribution of population (e.g., the population vote for a certain party) densities based on the total population recorded at polling stations that sparsely scattered at different subregions. Internet media providers or retailers, such as Google, Tencent, Amazon,

Facebook, etc., may want to recover a fine-grained geographical distribution of their users based on the aggregated user counts observed at different points of presence (PoPs) or data centers. Note that, in all the above-mentioned cases, it is impossible or not allowed to track the position of each individual due to either privacy concerns or technical overhead. Therefore, reconstructing the spatial data from coarse aggregation will be highly useful in such cases.

In this paper, we study such spatial sparse recovery problem, that is, to infer the fine-grained distribution of certain spatial data in a region given the aggregate observations recorded for each of its subregions. However, it is an extremely challenging problem and has seldom been studied. A straightforward idea is assuming the density is uniformly distributed within each subregion. Based on the based on the obtained aggregate observation, we can calculate a patched piece-wise constant estimation for each subregion. However, the densities estimated by this method will jump between neighboring subregions and disregard the local continuity or similarity of the studied spatial distribution across subregion boundaries. In addition, the piece-wise constant spatial field given by this approach provides little value for applications such as hot spot discovery. Many spatial data presents local continuity, e.g., Internet activity or cell phone activity. This is because the data often highly depend on underlying factors which are usually smoothly changing, like area functionality, urban geographical features, population density and so on. To exploit the smoothness, we may utilize spatial smoothing techniques such as Thin Plate Splines [12], Soap film smoothing [13], Spline smoothing [14], Bivariate Spline Regression [15], or Spatial Spline Regression [16] developed in statistics to smoothen the patched estimation. However, nearly all existing spatial smoothing techniques [12–16] are designed to recover a spatial field of densities according to sampled observations, e.g., reconstruct a spatial field of temperatures based on the temperature records at some sample points. In contrast, our problem needs to recovery a spatial field based on coarse-grained aggregate observations. Therefore, existing spatial smoothing techniques are not directly applicable to our new problem. Without modification, these smoothing techniques will violate the necessary constraint that the estimated spatial data in each subregion must sum up to its corresponding aggregate observation in the first place, leading to systematic errors.

To overcome the difficulties mentioned above, in this paper, we propose a new technique named Constrained Spatial Smoothing (CSS) for the problem of spatial data reconstruction. Specifically, given a region, we aim to reconstruct a spatial field of densities over that region based on observed aggregate values in patched subregions. Our approach penalizes the "roughness" of the reconstructed spatial field subject to the constraint that the aggregation of discretized values of the spatial field in each patched subregion equals the aggregate observation made in that subregion. It is distinct from previous spatial smoothing techniques due to the additional constraint in our problem. We propose an Alternating Direction Method of Multipliers (ADMM) [17,18] algorithm to decouple the problem into the alternated minimizations of a quadratic program (QP) [19] subproblem and a spatial smoothing subproblem, where we use the QP to iteratively enforce the observation constraints, while solving the spatial smoothing subproblem with a recently proposed finite element technique called Spatial Spline Regression (SSR) [16]. In addition, our approach not only leverages the intrinsic smoothness from local continuity to reconstruct a spatial field, but is also able to incorporate additional external information, such as the number of schools, number of bus stops, population, etc., in the underlying geographical region as a regression add-on component to further enhance recovery performance. Last but not least, our algorithm can be applied to a variety of sparse recovery problem where intrinsic smoothness exists.

Another important contribution of the paper is that we conduct extensive evaluation to compare our proposed algorithms with a variety of baseline methods. In our evaluation, we are trying to reconstruct the mobile phone activity distributions in Milan, Italy from base station observations. The Telecom Italia Big Data Challenge dataset is a multi-source dataset that contains a variety of informations, including aggregation of telecommunication activities, news, social networks, weather, and electricity data from the city of Milan. With the important information about human activities contained in the dataset, especially the cellphone activity records, researchers utilized the data to study different problems, such as modeling human mobility patterns [20–22], population density estimation [4,5], models the spread of diseases [23,24], modeling city structure [3] and city ecology [2], etc. Specifically, our evaluation is based on the Milan Call Detail Records (CDR) dataset, a part of the Telecom Italia Big Data Challenge dataset [25] which contains the phone call and Short Message Service (SMS) activity records of two months in each grid square of 235 m × 235 m in the city of Milan, Italy.

Given the Milan Call Detail Records (CDR) dataset, we consider a region that consists of 2726 grid squares in an irregularly bounded region in the city of Milan. To stress-test the algorithm performance, we assume we only know the aggregate phone activities observed on 100 or 200 base stations and aim to recover the entire spatial field of phone activities. We also use another geographical attribute dataset available from the Municipality of Milan's Open Data website [1] as the additional external attribute data to improve performance. Extensive evaluation shows that our proposed approach achieves significant improvement, compared to various state-of-the-art baseline methods, including the spatial spline regression (SSR) [16] approach. Our technique can recover the fine-grained cell phone activity distribution of 2726 data points only from 200 data points of base stations, with a mean absolute percentage error of 0.309, representing a 26.3% improvement from the SSR baseline scheme.

The remainder of this paper is organized as follows. In Section 2, we formulate the problem of spatial field reconstruction from coarse aggregate observations. In Section 3, we describe existing solutions, including a state-of-the-art Spatial Spline Regression (SSR) technique for spatial smoothing. In Section 4, we propose our Constrained Spatial Smoothing method which respects both the local continuity in the spatial field and the aggregation constraints at the same time. In Section 5, we conduct extensive evaluation in comparison with various other methods through a solid and extensive case study of cell phone activity density estimation in the city of Milan. We discuss related literature in Section 6 and conclude the paper in Section 7.

2. Problem Formulation

In this section, we formally introduce the problem of spatial field reconstruction from coarse aggregations observed at sparse scattered points in that field. Our problem can be formulated as a new type of sparse recovery problems. To ease the presentation, we may use cell phone activity recovery as an example.

Let $\Omega \subset \mathbb{R}^2$ denote an irregularly bounded domain, which is the entire region of interest in our problem. Usually, it excludes the uninhabited areas such as hills, ocean coasts, rivers, and so on. Suppose $f(\mathbf{p})$ is a real-valued function that represents certain spatial densities field (e.g., cell phone activities), where $\mathbf{p} = (x,y) \in \Omega$ denotes different geographical positions in Ω. Let $B = \{B_1, \ldots, B_m\}$ denote m observation points (e.g., base stations) that scattered in Ω. Each point B_i is located in a position $\mathbf{p}_{B_i} \in \Omega$ and in charge of a subregion Ω_{B_i}. In our problem, we are given the *aggregated volume* z_i in Ω_{B_i} that B_i is in charge of. Our goal is to reconstruct the spatial field $f(\mathbf{p})$ based on the observed aggregated volumes z_i.

To give an instance, consider the problem of recover cell phone activity distribution. In this case, each user will connect to a base station (cell tower) that is closest to his/her cell phone. Therefore, we can observe the aggregated volume for each base station

$$z_i = \int_{\Omega_{B_i}} f(\mathbf{p}) \, d\mathbf{p}, \quad i = 1, \ldots, m,$$

where Ω_{B_i} denotes the subregion that B_i is in charge of, and is given by

$$\Omega_{B_i} = \{\mathbf{p} \in \Omega : \|\mathbf{p} - \mathbf{p}_{B_i}\| < \|\mathbf{p} - \mathbf{p}_{B_{i'}}\|, \forall B_{i'} \in B, \, i' \neq i\}.$$

Given the aggregated activity volumes $z_1, \ldots, z_m$ recorded on m base stations, our goal is to reconstruct the entire cell phone activity densities distribution f, which is a spatial field in the domain Ω. We may call $z_1, \ldots, z_m$ base station volumes in this case. However, reconstructing a continuous spatial field is almost computationally infeasible as a personal computer can not handle the continuous nature of Ω_{B_i}.

In reality, we only need to recover f to a certain granularity required by the operator (e.g., 235 m $\times$ 235 m squares in the dataset provided by Telecom Italia Mobile). To fix notations, suppose Ω is discretized into n small grid squares $\mathbf{p}_1, \ldots, \mathbf{p}_n$, where $\mathbf{p}_j = (x_j, y_j) \in \Omega, j = 1, \ldots, n$ are the center positions of each square j in Ω. We can assume the area of each square is $\Delta = 1$ without loss of generality. In addition, the number of aggregate observations is much smaller than the total number of squares to be reconstructed, therefore we have $m \ll n$.

After domain discretization, we can get the aggregate volume on each base station B_i by

$$z_i = \sum_{\mathbf{p}_j \in \Omega_{B_i}} f(\mathbf{p}_j) \cdot \Delta, \quad i = 1, \ldots, m, \tag{1}$$

where the subregion that B_i represents is given by

$$\Omega_{B_i} = \{\mathbf{p}_j : 1 \leq j \leq n, \|\mathbf{p}_j - \mathbf{p}_{B_i}\| < \|\mathbf{p}_j - \mathbf{p}_{B_{i'}}\|, \forall i' \neq i\}. \tag{2}$$

Therefore, our goal is to reconstruct the underlying spatial field f, and especially the activity densities

$$\mathbf{f} := (f(\mathbf{p}_1), \ldots, f(\mathbf{p}_n))^\mathsf{T}$$

in all n grid squares if the desired granularity is on a per-square level, with only access to the aggregated observations z_i in Label (1).

The problem defined above is broadly applicable to characterize a variety of applications other than the recovery of cell phone activity density distribution, e.g., inferring a fine-grained geographical user distribution for a certain app or website based on aggregated user counts collected at sparsely distributed Presence of Points (PoPs) or data centers, and recovering the voter distribution for a certain party based on aggregate voting statistics at different polling stations. The nonessential difference is that the definition of subregion Ω_{B_i}, from which volume z_i is aggregated, is different for each specific application.

Constrained Spatial Smoothing Problem

Denote $\mathbf{z} = (z_1, \ldots, z_m)^\mathsf{T}$. Since all Ω_{B_i} are predetermined, e.g., from Label (2) for the problem of cell phone activity distribution recovery, and z_i are known, reconstructing spatial field $\mathbf{f}$ from (1) is essentially solving a linear system of equations for $\mathbf{f}$, i.e.,

$$\mathbf{z} = \mathbf{Af},$$

where the matrix $\mathbf{A} \in \mathbb{R}^{m \times n}$ is given by

$$A_{ij} = \begin{cases} 1, & \text{if } \mathbf{p}_j \in \Omega_{B_i}, \\ 0, & \text{otherwise.} \end{cases} \tag{3}$$

Since $m \ll n$, i.e., the number of equations is far smaller than the number of the unknowns, reconstructing $f(\mathbf{p}_1), \ldots, f(\mathbf{p}_n)$ from $z_1, \ldots, z_m$ is essentially a sparse recovery problem.

Directly solving the linear system of Equation (1) is infeasible, as it is an underdetermined system which has an infinite number of solutions. However, the spatial property of f can be utilized as constraints to make the sparse recovery problem feasible and has a unique solution. We observe that spatial data usually exhibit local continuity or correlation within domain Ω. For example, in the problem of cell phone activity density recovery, the activity density of a certain location highly depends on the population and activity at that place, e.g., the downtown has more population and cell phone activity than suburban residential areas. In addition, the underlying area functionality and the spatial distributions of human activity density are often slowly changing over the domain Ω rather than suddenly jumping between different subregions.

Therefore, we can formulate our constrained spatial sparse recovery problem as the following:

$$
\begin{aligned}
\underset{f}{\text{minimize}} \quad & \int_{\Omega} (\nabla^2 f)^2 \, d\mathbf{p}, \\
\text{subject to} \quad & \mathbf{z} = \mathbf{Af}, \\
& \mathbf{f} \geq 0,
\end{aligned}
\tag{4}
$$

by taking into account the non-negative property and the local spatial continuity (smoothness) of f. $\nabla^2 f = \frac{\partial^2 f}{\partial x^2} + \frac{\partial^2 f}{\partial y^2}$ is the Laplacian of f, and is utilized to encourage local similarity and penalize the roughness of the spatial field f. It is worth noting that once f is reconstructed, we have not only recovered the densities $\mathbf{f}$ at the square centers $\mathbf{p}_1, \ldots, \mathbf{p}_n$, but can also recover the density $f(\mathbf{p})$ of any point $\mathbf{p} \in \Omega$, e.g., between the centers of two neighboring grid squares, although such a fine-grained recovery may not be needed in every application.

To further improve the recovery performance, we can utilize additional external demographic or social features at each location. In the problem of cell phone activity density reconstruction, cell phone activities are often correlated with the underlying population density and social functionalities (e.g., the percentage of green area, the number of schools, the number of businesses/restaurants, the number of sport facilities, and the number of bus stops, etc.) of the considered regions.

Specifically, suppose $\mathbf{w}_j = (w_{j1}, \ldots, w_{jq})^{\mathsf{T}}$ represents the feature vector consisting of q external feature values of square j. When $\mathbf{w}_j$ is available as additional input, we can estimate the spatial density data in square j by

$$
f(\mathbf{p}_j) = f'(\mathbf{p}_j) + \mathbf{w}_j^{\mathsf{T}} \beta,
\tag{5}
$$

where $f'(\mathbf{p})$ is an underlying spatial field functional that preserves local spatial continuity, while $\mathbf{w}_j^{\mathsf{T}} \beta$ is a linear regression part based on the attributes of square $\mathbf{p}_j$ that allows position-specific variation or jumps.

In the presence of attributes, we can formulate the constrained spatial sparse recovery problem as

$$
\begin{aligned}
\underset{f',\beta}{\text{minimize}} \quad & \int_{\Omega} (\nabla^2 f')^2 \, d\mathbf{p}, \\
\text{subject to} \quad & f(\mathbf{p}_j) = f'(\mathbf{p}_j) + \mathbf{w}_j^{\mathsf{T}} \beta, \quad j = 1, \ldots, n, \\
& \mathbf{z} = \mathbf{Af}, \\
& \mathbf{f} \geq 0.
\end{aligned}
\tag{6}
$$

Once we get the spatial field f' and β, we can reconstruct $f(\mathbf{p}_j)$ for all the squares using (5). For example, we can calculate the cell phone activity at a specific place by the summation of an underlying smooth spatial field $f'(\mathbf{p}_j)$ and a linear regression of location attributes, where the add-on regression helps to model the jump between two subregions if the two regions are quite different and have distinct functionalities or attributes.

3. Patched Estimation and Spatial Spline Regression

In this section, we present some tentative solutions and then show their limitations in solving our constrained spatial sparse recovery problem.

3.1. Patched Piece-Wise Constant Estimation

In our problem, we only have access to the aggregated volumes z_i at locations $\mathbf{p}_{B_i}$. To infer the fine-grained spatial distribution of z_i over subregion Ω_{B_i} that covers the point B_i, a first intuitive heuristic is estimating $f(\mathbf{p}_j)$ as the volume z_i divided by its area by assuming the density is distributed uniformly:

$$\bar{f}(\mathbf{p}_j) = \frac{z_i}{|\Omega_{B_i}|}, \text{ for each } \mathbf{p}_j \in \Omega_{B_i}, \tag{7}$$

where $|\Omega_{B_i}|$ is the area of Ω_{B_i}. This method gives us a patched piece-wise constant estimation. Note that we use *patch* to refer to Ω_{B_i} in this paper, which is the subregion covered B_i.

However, the patched estimation gives an oversimplified solution. The reconstructed spatial field $\bar{f}(\mathbf{p}_j)$ may have jumps on the borders of neighboring patches, which is far from smooth. In reality, the spatial field $f(\mathbf{p}_j)$ should change smoothly over the domain, as the underlying characteristics also change smoothly across different regions. Hence, $f(\mathbf{p}_j)$ should not be constant within each patch Ω_{B_i}.

3.2. Spatial Spline Regression

Given the above observation, we can naturally come up with a second idea, which is learning a smooth estimation of $\bar{f}(\mathbf{p}_j)$ by spatial smoothing techniques. In the following, we introduce the powerful smoothing technique named Spatial Spline Regression (SSR) proposed in Sangalli et al. [16]. We will show how it can be applied to our particular spatial data reconstruction problem, as well as point out its limitations in solving the problem.

Given l data points in Ω, which contains the following information: (1) their positions $\{\mathbf{p}_j\}_{j=1}^l$; (2) the values of these l points: $\{h_j\}_{j=1}^l$; and (3) their feature vectors $\{\mathbf{w}_j\}_{j=1}^l$, SSR is able to fit a smooth spatial field f by minimizing the following equation [14,16], i.e.,

$$\underset{\beta,f}{\text{minimize}} \sum_{j=1}^l \left(h_j - \mathbf{w}_j^\mathsf{T}\beta - f(\mathbf{p}_j)\right)^2 + \lambda \int_\Omega (\nabla^2 f)^2 \, \mathrm{d}\mathbf{p}, \tag{8}$$

where f is assumed to be twice-differentiable over Ω, and $\nabla^2 f = \frac{\partial^2 f}{\partial x^2} + \frac{\partial^2 f}{\partial y^2}$ denotes the *Laplacian* of f to penalize the roughness of f. The hyper parameter λ is used to trade the smoothness of f off for a better approximation to data value h_j.

However, the challenge to solving problem (8) is that it involves searching for a functional f over a possibly non-convex domain Ω that may have strong concavities, complicated boundaries, and even interior holes. Although kernel-based methods [26] are also a commonly used smoothing technique, their major drawback is that, by using uniformly damping weights in distance-based kernels, they tend to link data points across unrelated or weakly related subregions in an irregularly shaped non-convex domain.

We now briefly describe how spatial spline regression [16] can solve problem (8) via finite element analysis for any irregularly shaped domain Ω. SSR splits a domain Ω by transforming it into a triangular mesh with triangulation methods (e.g., Delaunay triangulation [27]). After triangulation, it defines a polynomial function on each triangle, such that the summation of these polynomial functions defined on different pieces closely approximates the desired spatial field f.

Specifically, let $\zeta_1, \ldots, \zeta_K$ denote the vertices of all the small triangles, which are called control points and can be adaptively selected by available data points. Define a piecewise linear or quadratic basis function $\psi_k(x, y)$ called *Lagrangian finite element* with $(x, y) \in \Omega$, associated with each control point ζ_k such that ψ_k evaluates to 1 at ζ_k and is equal to 0 at all other control points. Therefore, according to the *Lagrangian property of the basis*, we can approximate $f(x, y)$ for any $(x, y) \in \Omega$ only using the values of f on the K control points, i.e., $\mathbf{f}_K := (f(\zeta_1), \ldots, f(\zeta_K))^\mathsf{T}$. That is, if we let $\psi(x, y) := (\psi_1(x, y), \ldots, \psi_K(x, y))^\mathsf{T}$ denote the K predefined basis functions, each corresponding to a control point, then we have

$$f(x, y) = \sum_{k=1}^K f(\zeta_k)\psi_k(x, y) = \mathbf{f}_K^\mathsf{T}\psi(x, y). \tag{9}$$

Since $\psi_1(x,y),\ldots,\psi_K(x,y)$ are predefined and known a priori, the variational estimation of f in problem (8) boils down to the estimation of only K scalar values, i.e., $\mathbf{f}_K = (f(\zeta_1),\ldots,f(\zeta_K))^\mathsf{T}$.

In fact, it is shown in Sangalli et al. [16] that with the piece-wise approximation given by (9), solving (8) is simply solving a set of linear equations for $\hat{f}(\zeta_1),\ldots,\hat{f}(\zeta_K)$. The estimator $\hat{f}(x,y)$ for f can then be derived from (9) as

$$\hat{f}(x,y) = \hat{\mathbf{f}}_K^\mathsf{T}\psi(x,y).$$

It is worth noting that commodity triangulation software for finite element analysis is readily available in many free and commercial finite element packages. For example, Delaunay triangulations of a set of data location points (e.g., [27]) V are such that no point in V is inside the circumcircle of any triangle; they maximize the minimum angle of all the triangle angles, avoiding stretched triangles.

Now, we can see that if $l = n$ and we plug $h_j = \bar{f}(\mathbf{p}_j)$, $j = 1,\ldots,n$ into problem (8), we will get a new density surface $\hat{f}$ as a solution to the SSR problem (8) that is a smoothened approximation of the patched estimates $\bar{f}(\mathbf{p}_j)$.

However, SSR given by (8) can not accommodate any constraints, which is the major limitation in solving our problem. Specially, in our case, SSR does not enforce the aggregated volume constraint (1) (or $\mathbf{z} = \mathbf{A}f$ in (4)). Therefore, SSR gives no guarantee that the estimated densities in each patch Ω_{B_i} will sum up to the observed volume z_i on the point B_i. In this way, SSR would likely cause large estimation errors as it violates the constraint.

4. An ADMM Algorithm for Constrained Spatial Smoothing

Our spatial sparse recovery problem (4) is different from (8) from two aspects: the additional constraints and the loss function. As a consequence, we can not directly apply the previous SSR method to solve it. A new approach is needed to handle our new loss function with constraints.

In this section, we propose to utilize the Alternating Direction Method of Multipliers (ADMM) [28], to decompose our constrained optimization problem into two sub-problems that can be solved effectively by SSR and Quadratic Programming (QP), respectively. Algorithm 1 presents the proposed ADMM algorithm to learn our model parameters.

Algorithm 1: Constrained Spatial Smoothing by ADMM

Input: The m observed volume of base stations $\mathbf{z} = (z_1,\ldots,z_m)^\mathsf{T}$, smoothing parameter λ,
penalty parameter β, initialize $\alpha = \alpha^0$, $f = f^0$.

Output: Spatial field and parameters $\hat{f}$, $\hat{\beta}$. Estimation values on n locations
$\hat{f} = (\hat{f}(\mathbf{p}_1),\ldots,\hat{f}(\mathbf{p}_n))$.

1: **for** $iter = 1,\ldots,\mathrm{maxIter}$ **do**

2:      Update f by solving (18) using Quadratic Programming.
3:      Update g by solving (19) using Spatial Spline Regression.
4:      Update α according to (17).
5: **end for**

First, we introduce the following indicator function $\mathbb{1}_\mathbf{f}$,

$$\mathbb{1}_\mathbf{f} = \begin{cases} 0, & \text{if } \mathbf{f} \geq 0 \text{ and } \mathbf{z} = \mathbf{A}\mathbf{f}, \\ \infty, & \text{otherwise.} \end{cases} \tag{10}$$

With the indicator function, the original problem (4) is equivalent to

$$\underset{f}{\text{minimize}} \quad \lambda \int_\Omega (\nabla^2 f)^2 \, d\mathbf{p} + \mathbb{1}_\mathbf{f}, \tag{11}$$

where λ is a hyper parameter that controls the smoothness of f.

Second, we introduce an auxiliary variable **g** that is defined as

$$\mathbf{g} := (g(\mathbf{p}_1), \ldots, g(\mathbf{p}_n))^{\mathsf{T}}. \tag{12}$$

This variable is utilized to split the convex optimization problem into two sub-convex problems. With **g**, we can formulate the problem as the standard ADMM format,

$$\begin{aligned} \underset{f}{\text{minimize}} \quad & \lambda \int_{\Omega} (\nabla^2 f)^2 \, d\mathbf{p} + \mathbb{1}_{\mathbf{g}}, \\ \text{subject to} \quad & \mathbf{f} = \mathbf{g}. \end{aligned} \tag{13}$$

The *augmented Lagrangian* for (13) is

$$\begin{aligned} \text{minimize} \quad \mathcal{L}_\rho(\mathbf{f}, \mathbf{g}, \boldsymbol{\alpha}) = & \lambda \int_{\Omega} (\nabla^2 f)^2 \, d\mathbf{p} + \mathbb{1}_{\mathbf{g}} \\ & + \boldsymbol{\alpha}^{\mathsf{T}} (\mathbf{g} - \mathbf{f}) + \frac{\rho}{2} \|\mathbf{g} - \mathbf{f}\|_2^2, \end{aligned} \tag{14}$$

where $\boldsymbol{\alpha} = (\alpha_1, \ldots, \alpha_n)^{\mathsf{T}}$ is the dual variable, and $\rho > 0$ is the penalty parameter in ADMM. Then, the ADMM consists of the following iterations:

$$\mathbf{g}^{k+1} := \underset{\mathbf{g}}{\operatorname{argmin}} \, \mathcal{L}_\rho(\mathbf{f}^k, \mathbf{g}, \boldsymbol{\alpha}^k), \tag{15}$$

$$\mathbf{f}^{k+1} := \underset{\mathbf{f}}{\operatorname{argmin}} \, \mathcal{L}_\rho(\mathbf{f}, \mathbf{g}^{k+1}, \boldsymbol{\alpha}^k), \tag{16}$$

$$\boldsymbol{\alpha}^{k+1} := \boldsymbol{\alpha}^k + \rho(\mathbf{f} - \mathbf{g}). \tag{17}$$

For the **g**-update step in each iteration, Label (16) is equivalent to

$$\begin{aligned} \underset{\mathbf{g}}{\text{minimize}} \quad & \frac{\rho}{2} \|\mathbf{g}\|_2^2 + (\boldsymbol{\alpha}^{\mathsf{T}} - \rho \mathbf{f}^{\mathsf{T}}) \mathbf{g}, \\ \text{subject to} \quad & \mathbf{g} \geq 0, \\ & \mathbf{z} = \mathbf{A}\mathbf{g}. \end{aligned} \tag{18}$$

We can solve this convex problem efficiently by Quadratic Programming (QP).
For the **f**-update step in each iteration, Equation (15) is equivalent to

$$\underset{f}{\text{minimize}} \quad \left\| \left(\boldsymbol{\alpha}^{\mathsf{T}} + \rho \mathbf{g}^{\mathsf{T}} \right) / 2 - \mathbf{f} \right\|_2^2 + \lambda \int_{\Omega} (\nabla^2 f)^2 \, d\mathbf{p}, \tag{19}$$

which is exactly the form of (8) with $h_j = (\alpha_j + \rho g(\mathbf{p}_j))/2$ and $\mathbf{w}_j = 0$, thus can be solved efficiently by SSR. It should be noted that λ is the penalty parameter which controls the smoothness of f. If it is small, we put little emphasis on the smoothness, and the estimated surface f will be over fitted. If it is too big, the surface will be too smooth, which can cause underfitting.

For the case with attributes, the algorithm does not require major changes. We just need to replace **f** by $\mathbf{f} + \mathbf{W}\boldsymbol{\beta}$ in (19), where $\mathbf{W} := (\mathbf{w}_1, \ldots, \mathbf{w}_n)^{\mathsf{T}}$ represents the attributes and β is the corresponding contributions.

Our proposed ADMM training algorithm is able to efficiently reconstruct the spatial field and fit the covariates for our constrained spatial sparse recovery problem. In **g**-update step, it enforces the constraints by solving a constrained QP with no need to worry about smoothing; in a **f**-update step, it approximates the obtained **g** with a smooth f using the SSR-based smoothing technique. In this way, we decouple the handling of smoothing and constraints which was not possible in pure SSR previously.

5. Performance Evaluation

In this section, we perform an extensive case study of the approach we described above in order to demonstrate its applicability. We picked the cell phone data as an example of how the model can solve empirical problem and compare the model's performance to other approaches.

5.1. Dataset Description

The model in (13) is general and not attached to any particular empirical problem, and it does not contain many implicit assumptions. However, in order to measure its performance, we evaluate the model using real-world data. Due to generality of the proposed learning algorithm, the range of possible data sets is potentially big. For our empirical case study, we chose cell phone data, where there exists a problem of recovering a spatial field from coarse aggregations observed at sparse cell phone towers. We do not overestimate the problem, but rather see this particular data set suitable for an extensive case study.

The Milan Call Description Records (CDR) dataset is a part of the Telecom Italia Big Data Challenge dataset provided by Telecom Italia Mobile. It contains the telecommunications activity records from 1 November 2013 to 31 December 2013 in the city of Milan [25]. The dataset divides Milan into a 100×100 square grid, where each square is size of about 235 m $\times$ 235 m. In the dataset, each record consists of six entries: square ID, incoming call activity, outgoing call activity, incoming SMS activity, outgoing SMS activity, and time-stamp of 10-minute time slot. The values of the four types of activities are normalized to the same scale.

Another dataset we utilized is the Milan geographical attribute dataset available from the Municipality of Milan's Open Data website [1]. This dataset consist of features of central 2726 squares among the whole 10,000 squares. The features of each square include: population, green area percentage, number of sport centers, number of universities, number of businesses, and number of bus stops. Figure 1 shows the area covered by these grid squares. The 2726 squares covers the central part of the Milan city and contains the majority of telecommunication activities in the dataset. We refer to [2] for more detailed description about this dataset. In our experiments, we compare the performance of different approaches on these squares.

The general problem of recovering a spatial field from coarse aggregations observed at sparse points in the field in this particular case study is reformulated into the problem of recovering the distribution of cell phone activities over the whole 2726 square regions given that only aggregated activity observations in base stations are known. We need to further process the Milan CDR dataset to study this problem.

First, we sum up the four types of activities during 1 November 2013 to 28 November 2013 and 1 December 2013 to 28 December 2013, respectively, to come up with the activity volume of each squares during November or December. These two datasets are served as the ground-truth datasets of Milan cell phone activity distributions. Figure 1a,b show the heat maps of activity volumes in each square during November and December.

Second, after we aggregated the two months' activities for each square, we need to set the locations of base stations (BSs). According to [29], there are roughly 200 base stations in Milan. However, the exact locations are not available. Thus, we assume the 200 BSs are randomly distributed according to the following probability distribution

$$\Pr(\text{Set square } i \text{ as BS}) = f(\mathbf{p}_i) / \sum_{j=1}^{N} f(\mathbf{p}_j), \tag{20}$$

where $f(\mathbf{p}_i)$ is the cell phone activity volume in square $i, i = \{1, \dots, N\}, N = 2726$ is the number of squares we are focusing on. Notice that, when we have 200 base station's aggregated observations, they only cover 7.34% of the whole 2726 squares region. This is extremely sparse and makes our problem highly challenging. In addition, we also assume $n_{\text{BS}} = 100$ and choose 100 squares as BSs according to the same probability distribution to stress-test our algorithm's capability under even sparser observations. Figure 2a,b show the base station distributions for $n_{BS} = 200$ and $n_{BS} = 100$, respectively.

After sampling the location of base stations, for each square, we assign the activity of it to its closest base station. When multiple base stations are equidistant from a square, the activity of the square will be evenly distributed among these base stations. We then assume we only know the aggregated activities in base station squares, which is usually the true case in reality. Figure 2c shows the regions split by 100 base stations, where each colour patch is a region charged by one base station. To save space, we don't present the figure for 200 base stations.

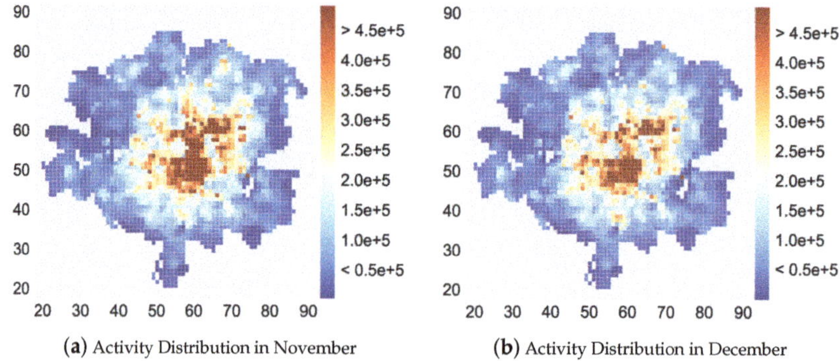

(a) Activity Distribution in November (b) Activity Distribution in December

Figure 1. The cell phone activity distributions of Milan. It shows the metropolitan area of Milan, Italy, and the area covered by the 2726 grid squares. (**a**,**b**) show the heat map of cell phone activities (Call + SMS) during November and December respectively.

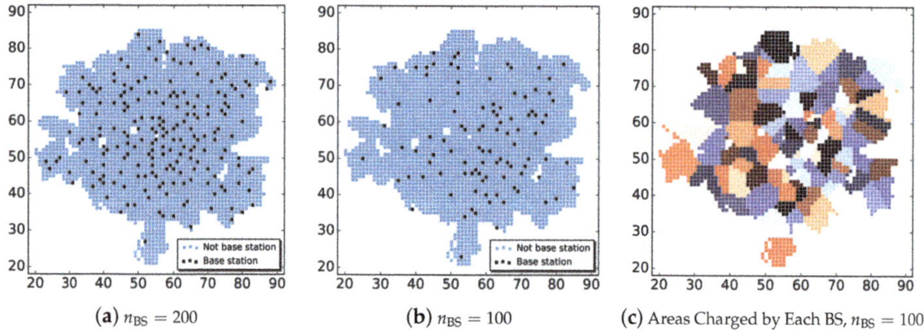

(a) $n_{BS} = 200$ (b) $n_{BS} = 100$ (c) Areas Charged by Each BS, $n_{BS} = 100$

Figure 2. The location distributions of sampled base stations and the areas in charged by them. (**a**,**b**) shows the sampled base station distributions for $n_{BS} = 200$ and $n_{BS} = 100$; (c) shows the areas in charged by different base stations for $n_{BS} = 100$.

5.2. Experimental Setup

Algorithms Evaluated

We test our proposed approach and compare it with three baseline methods. In particular, we evaluate and compare the following models using the aggregated November and December datasets, with number of base stations $n_{BS} = 200$ or $n_{BS} = 100$ for stress testing.

- **Patched Estimation**: estimate the cell phone activity distribution by patched piece-wise constant estimation, that is, assume cell phone activity density is distributed uniformly within each sub-region Ω_{B_i}, i.e., the area covered by base station B_i, and estimate each square's activity volume by (7).

- **Patched Estimation + SSR 1**: first estimate *only base station* activity volumes by (7). Use these sparse points to fit a smooth surface by running Spatial Spline Regression to obtain the estimated cell phone activity in all squares.

- **Patched Estimation + SSR 2**: as opposed to the previous model, get the initial estimation of the activity volume of *all squares* by Patched Estimation. Then, use all these points to fit a smooth surface by running Spatial Spline Regression to obtain the final estimated cell phone activity in all squares.

- **Constrained Spatial Smoothing**: first get the initial estimation of the activity volume of all squares by Patched Estimation, then run Constrained Spatial Smoothing algorithm to get the final activity volumes estimation of all squares.

- **Constrained Spatial Smoothing + Features**: in this case, we incorporate the geographical features into Constrained Spatial Smoothing algorithm.

We set the penalty parameter $\lambda = 1$ when $n_{BS} = 200$ and $\lambda = 10$ when $n_{BS} = 100$, for all methods that utilizes SSR. The geographical features of Milan are only incorporated in the last algorithm described above. In addition, for the implementation of Spatial Spline Regression, we use the *fdaPDE* R Package [30].

To compare different approaches, we evaluate the performance by the Mean Relative Error (MRE) of the produced activity estimates for the true activity values. The relative error of an estimation $\hat{f}(\mathbf{p}_j)$ compared to the true value $f(\mathbf{p}_j)$ is defined as $|\hat{f}(\mathbf{p}_j) - f(\mathbf{p}_j)|/f(\mathbf{p}_j)$.

5.3. Performance Evaluation

5.3.1. Comparison of Different Algorithms

We show the cumulative distribution function (CDF) of Relative Errors given by different approaches in Figures 3 and 4. In addition, we compare the estimation's Mean Relative Error of different approaches in Figure 5. It is quite clear that our proposed algorithms outperform other three baseline approaches significantly in all cases ($n_{BS} = 200$ and $n_{BS} = 100$, data aggregated in November and in December).

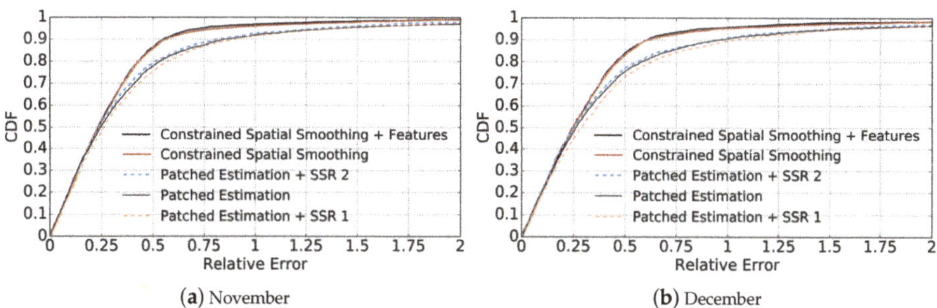

(**a**) November　　　　　　(**b**) December

Figure 3. Comparison of the CDFs of estimation relative errors given by different methods when $n_{BS} = 200$. The legends follow the same order as the curves at relative error $= 0.5$. (**a**) compares the CDFs based on the data aggregated in November; (**b**) compares the CDFs based on the data aggregated in December.

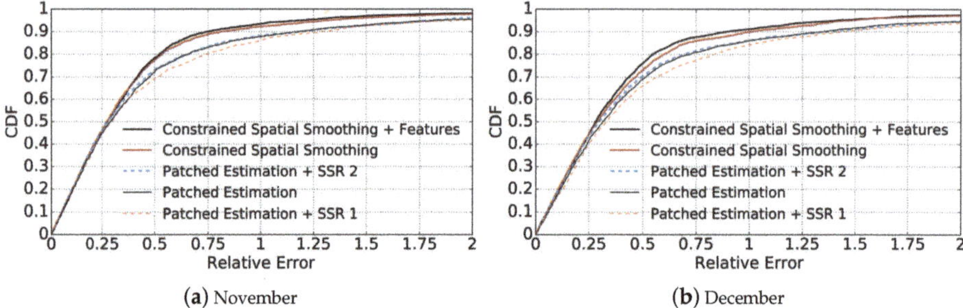

(a) November   **(b)** December

Figure 4. Comparison of the CDFs of estimation relative errors given by different methods when $n_{BS} = 100$ for stress-testing. The legends follow the same order as the curves at relative error $= 0.5$. **(a)** compares the CDFs based on the data aggregated in November; **(b)** compares the CDFs based on the data aggregated in December.

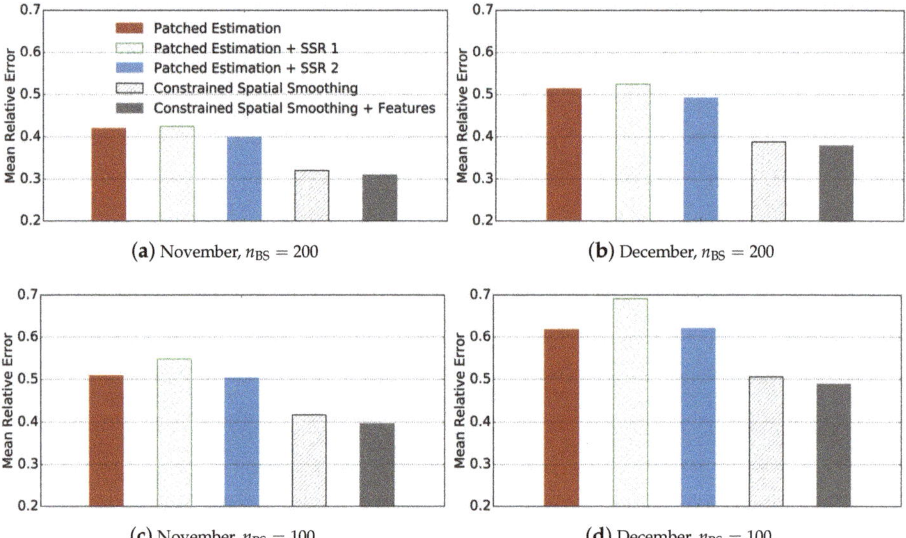

(a) November, $n_{BS} = 200$ **(b)** December, $n_{BS} = 200$

(c) November, $n_{BS} = 100$ **(d)** December, $n_{BS} = 100$

Figure 5. Comparison of the estimation's Mean Relative Error of different methods when $n_{BS} = 200$ or $n_{BS} = 100$ for stress-testing. In each figure, the bars from left to right stands for Patched Estimation, Patched Estimation + SSR 1, Patched Estimation + SSR 2, Constrained Spatial Smoothing, and Constrained Spatial Smoothing + Features respectively. **(a)** we use the data aggregated in November, and set number of base stations to be 200, similarly for **(b–d)**.

By comparing Patched Estimation + SSR 1 with Patched Estimation approach, we can see that using spatial smoothing based on only base station squares' observations leads to worse performance than patched estimation. This can be explained by the smoothing property of SSR and how we set the values of base station squares. As we described, we set the activity value of base stations by averaging the total activity amount of each base station on all squares it covers. Thus, given the activity $\frac{z_i}{|\Omega_{B_i}|}$, ($|\Omega_{B_i}|$ denotes the number of squares within region Ω_{B_i}) of a base station B_i, the true activities of itself and its surrounding squares within region B_i are distributed with a mean of $\frac{z_i}{|\Omega_{B_i}|}$. Given two base stations B_1 and B_2 that are close to each other, with aggregated activities of z_1 and z_2 respectively, the Spatial Smoothing approach will fit a smooth surface between the two base stations. Suppose

$z_1 > z_2$, in this case, overall, the activities of B_1's neighbour squares will be underestimated, and that of B_2 will be overestimated. Therefore, Patched Estimation + SSR 1's performance is not as good as Patched Estimation.

By comparing Patched Estimation + SSR 2 with Patched Estimation and Patched Estimation + SSR 1, we can observe that applying spatial smoothing on the results of patched estimation improves the performance. This proves the rationality and effectiveness of introducing smoothness into the estimated cell phone activity distribution surface.

Our proposed approach achieves much better performance compared with the three baseline methods. By using Constrained Spatial Smoothing instead of applying Spatial Spline Regression directly, we are able to fit a smooth activity distribution while forcing it to match the observations of base station squares (the aggregated activity volumes) at the same time. By comparing Constrained Spatial Smoothing that incorporates additional features of each square with the version without features, we can see that the performance is further improved. The reason is that the heterogeneity of different locations will influence the telecommunication activity distribution, therefore making the distribution not smooth everywhere. Incorporating additional features into our model can help to explain the residuals between estimated smooth distribution and the true activity distribution, therefore further increasing estimation accuracy. By comparing Figure 3 and Figure 4, we also can see that incorporating additional features into Constrained Spatial Smoothing becomes more important when the base stations are more sparse.

The performance of different methods on the December dataset is worse than on the November dataset. This is because there are multiple holidays in December. The cell phone activities will become more irregular than usual during holidays, as discussed in Cici et al. [2] and Ratti et al. [29].

Figure 6a–c show the distribution surfaces of true cell phone activity volumes, estimated volumes by Patched Estimation, and estimated volumes by Constrained Spatial Smoothing with features when $n_{BS} = 200$ using the November dataset. We can see that the Patched Estimation approach fits a stepped surface, while our approach gives a much smoother surface.

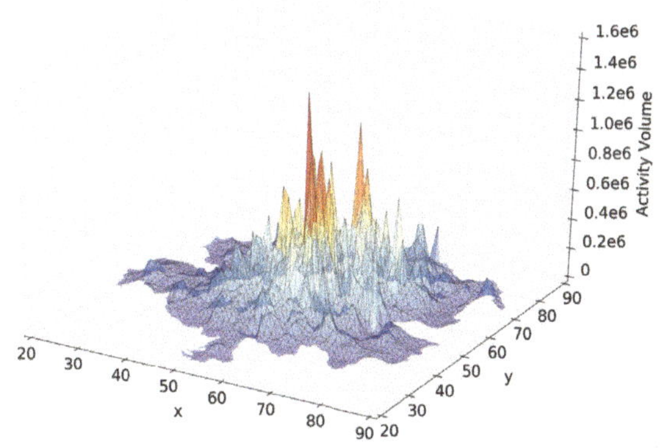

(**a**) Real cell phone activity distribution.

Figure 6. *Cont.*

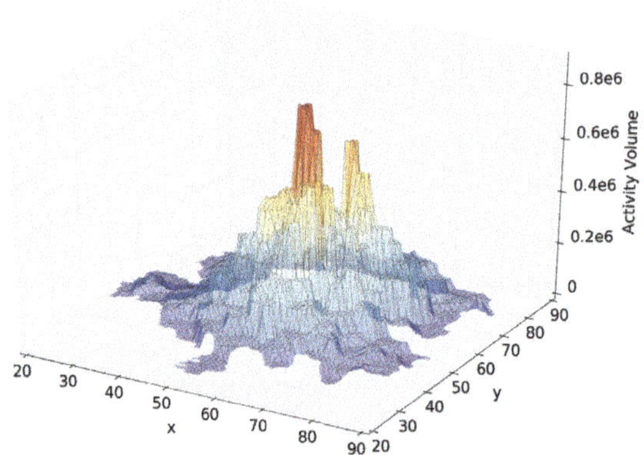

(**b**) Estimated cell phone activity distribution by Patched Estimation.

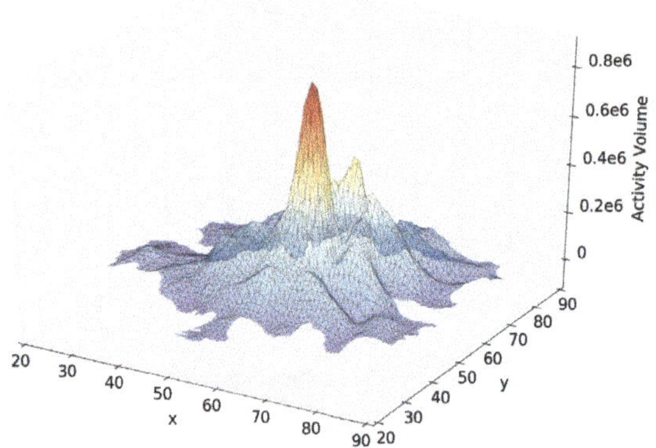

(**c**) Estimated cell phone activity distribution by Constrained Spatial Smoothing + Features.

Figure 6. Comparison of the activity distributions. (**a**) real cell phone activity distribution; (**b**) estimated distribution by Patched Estimation method; (**c**) estimated distribution by our method.

For time efficiency, experiments based on the Milan Call Description Records (CDR) dataset show that the average time for our approach to converge is less than five minutes on a MacBook Pro with a 2 GHz Intel Core i7 processor, and 8 GB memory. This proves that our system is highly efficient and practical.

5.3.2. Impact of Smooth Penalty Parameter λ

Figure 7 shows how the the estimation's Mean Relative Error varies when λ increases from 10^{-4} to 10^3. We make two interesting observations. First, λ around 1~10 usually gives the best performance. Too big or too small λ will decrease the estimation accuracy. This is reasonable, as when λ is too small, we put little emphasis on the smoothness of estimated surface, thus the performance will suffer. If λ is too big, it enforces a smooth surface, which also doesn't match the reality. Second, when we have less base stations, λ that gives the best performance will increase (from 1 to 10). In addition, we can see that the performance of the model with λ between 1~100 does not significantly change when $n_{BS} = 100$. That indicates the following: when the base station distribution is more sparse, the estimation performance is less sensitive to λ when it is around the best value (1 for $n_{BS} = 200$ and 10 for $n_{BS} = 100$).

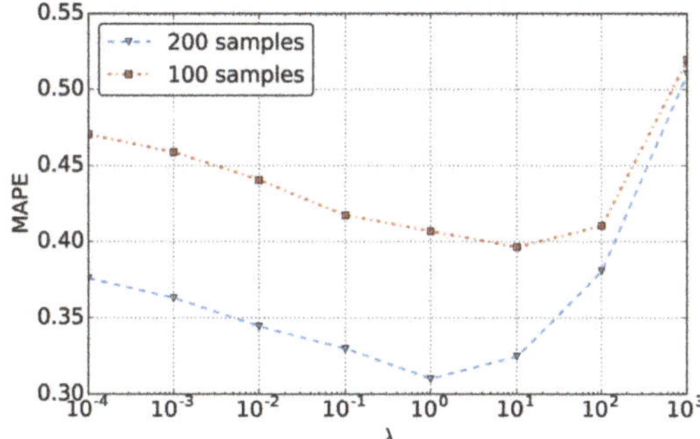

Figure 7. Influence of λ to estimation's Mean Relative Error when $n_{BS} = 200$ and $n_{BS} = 100$ for stress-testing. The figure is based on the November dataset. Results on the December dataset are similar.

6. Related Work

The Telecom Italia Big Data Challenge dataset is widely used to study different problems [2–5,20–24]. However, little research work has been done to estimate the spatial distribution of cellphone activity itself, despite the great value of this problem.

There are various tasks where the key problem is estimating a spatial field over a region based on observations of sampled points, such as house price estimation and population density estimation. Chopra et al. [26] model the underlying surface of land desirability using kernel-based interpolation. However, it is hard to choose the form of kernel functions and tune a large number of hyper-parameters. Spatial Spline Regression technique is applied to the problem of population density estimation in Sangalli et al. [16]. However, in our problem, we only get the accumulated activity density in base stations, rather than real densities in each base station location. In addition, BS locations distribution is highly sparse in our case.

Although a range of kernel-based methods [26,31,32] can be applied to fit a spatial field, the common drawback of these approaches is that, by using uniformly damping weights in distance-based kernels, they tend to link weakly related data points across areas in a non-convex domain. Spatial spline regression [16] on the other hand uses finite-element analysis approach to jointly solve for f and β from the model described by Equation (8) over any irregularly shaped domain Ω.

As it was discussed earlier, the fine-grained data for the distribution of the volume of calls and SMS are not usually available. A common type of data is the data collected by cell phone base stations. Sometimes, cell phone providers interpolate the data collected by the base stations as is discussed in Manfredini et al. [33]. Some researchers interpolate the data to obtain fine grained distributions as in Ratti et al. [29]. However, in Ratti et al. [29], authors do not evaluate the performance of the interpolated distribution. To the best of our knowledge, there is no extensive work done in trying to obtain optimal reconstructions of fine grained cell phone data distribution. We are the first to apply the latest spatial functional analysis techniques to cellphone activity distribution modeling, assuming the activity densities consist of a regression part based on social or demographical statistic features and a spatial field that captures the underlying smoothness property of cellphone activities. In particular, we leverage the idea of spatial spline regression to handle any irregularly shaped geographic regions. We have developed a novel Constrained Spatial Smoothing approach and corresponding training algorithm to recover spatial distribution of cellphone activities from highly sparse observations.

7. Conclusions

In this paper, we study the problem of inferring the fine-grained spatial distribution of certain density data in a region based on the aggregate observations recorded for each of its subregions, which is extremely challenging and seldom visited before, and analyze the challenges of it. We propose the Constrained Spatial Smoothing (CSS) approach that exploits both the intrinsic smooth property of underlying factors and the additional features from external social or domestic statics. We further propose a training algorithm which combines the Spatial Spline Regression (SSR) technique and ADMM technique to learn our model parameters efficiently.

To evaluate our algorithm and compare it with various other approaches, we run extensive evaluations based on the Milan Call Detail Records dataset provided by Telecom Italia Mobile. The simulation results on the dataset show that our algorithm significantly outperforms other baseline approaches by a great percentage. (Note that cross validation and statistical testing are techniques that are usually applied in experiments. However, both techniques require sampling effectively from the sparse spatial data while keeping the intrinsic spatial structure, which is difficult in our problem.) This shows that jointly modeling the underlying spatial continuity and the local features that characterize the heterogeneity of different locations can effectively improve the performance of spatial recovery.

Although we use the data on cell phone activities to illustrate our methodology, our algorithm is not limited to solving the problem of inferring the distribution of cell phone activities, but is also applicable to a variety of problems where estimating an implicit or explicit smooth surface is required, such as inferring the spatial distribution of population densities based on the aggregate population observed at sparsely scattered polling stations, reconstructing a fine-grained geographical distribution of users for an Internet media provider or retailer only from aggregated user counts observed at certain datacenters or points of presence (PoPs), and so on.

Author Contributions: Conceptualization, B.L., B.M., L.K. and D.N.; Data curation, B.L. and B.M.; Formal analysis, B.L., B.M., L.K. and D.N.; Investigation, B.L., B.M., L.K. and D.N.; Methodology, B.L., B.M., L.K. and D.N.; Project administration, B.L., B.M., L.K. and D.N.; Software, B.L. and B.M.; Supervision, B.L., B.M., L.K. and D.N.; Validation, B.L., B.M., L.K. and D.N.; Visualization, B.L., B.M. and L.K.; Writing—original draft, B.L., B.M., L.K. and D.N.; Writing—review and editing, B.L., B.M. and D.N.

Funding: This research received no external funding.

Abbreviations

The following abbreviations are used in this manuscript:

CSS	Constrained Spatial Smoothing
SSR	Spatial Spline Regression
SMS	Short Message Service
ADMM	Alternating Direction Method of Multipliers
QP	Quadratic Programming
CDR	Call Detail Records
PoPs	Presence of Points
CDF	Cumulative Distribution Function

References

1. Barlacchi, G.; De Nadai, M.; Larcher, R.; Casella, A.; Chitic, C.; Torrisi, G.; Antonelli, F.; Vespignani, A.; Pentland, A.; Lepri, B. A multi-source dataset of urban life in the city of Milan and the Province of Trentino. *Sci. Data* **2015**, *2*, 150055. [CrossRef] [PubMed]
2. Cici, B.; Gjoka, M.; Markopoulou, A.; Butts, C.T. On the decomposition of cell phone activity patterns and their connection with urban ecology. In Proceedings of the 16th ACM International Symposium on Mobile Ad Hoc Networking and Computing, ACM MobiHoc'15, Hangzhou, China, 22–25 June 2015.

3. Louail, T.; Lenormand, M.; Ros, O.G.C.; Picornell, M.; Herranz, R.; Frias-Martinez, E.; Ramasco, J.J.; Barthelemy, M. From mobile phone data to the spatial structure of cities. *Sci. Rep.* **2014**, *4*, 5276. [CrossRef]
4. Douglass, R.W.; Meyer, D.A.; Ram, M.; Rideout, D.; Song, D. High resolution population estimates from telecommunications data. *EPJ Data Sci.* **2015**, *4*, 4. [CrossRef]
5. Deville, P.; Linard, C.; Martin, S.; Gilbert, M.; Stevens, F.R.; Gaughan, A.E.; Blondel, V.D.; Tatem, A.J. Dynamic population mapping using mobile phone data. *Proc. Natl. Acad. Sci. USA* **2014**, *111*, 15888–15893. [CrossRef] [PubMed]
6. Liu, X.; Kyriakidis, P.C.; Goodchild, M.F. Population-density estimation using regression and area-to-point residual kriging. *Int. J. Geogr. Inf. Sci.* **2008**, *22*, 431–447. [CrossRef]
7. Cho, E.; Myers, S.A.; Leskovec, J. Friendship and mobility: User movement in location-based social networks. In Proceedings of the 17th ACM SIGKDD International Conference on Knowledge Discovery and Data Mining, San Diego, CA, USA, 21–24 August 2011; ACM: New York, NY, USA, 2011; pp. 1082–1090.
8. Li, X.; Pan, G.; Wu, Z.; Qi, G.; Li, S.; Zhang, D.; Zhang, W.; Wang, Z. Prediction of urban human mobility using large-scale taxi traces and its applications. *Front. Comput. Sci.* **2012**, *6*, 111–121.
9. Lu, X.; Wetter, E.; Bharti, N.; Tatem, A.J.; Bengtsson, L. Approaching the limit of predictability in human mobility. *Sci. Rep.* **2013**, *3*, 2923. [CrossRef] [PubMed]
10. Wang, D.; Pedreschi, D.; Song, C.; Giannotti, F.; Barabasi, A.L. Human mobility, social ties, and link prediction. In Proceedings of the 17th ACM SIGKDD International Conference on Knowledge Discovery and Data Mining, San Diego, CA, USA, 21–24 August 2011; ACM: New York, NY, USA, 2011; pp. 1100–1108.
11. De Domenico, M.; Lima, A.; Musolesi, M. Interdependence and predictability of human mobility and social interactions. *Pervasive Mob. Comput.* **2013**, *9*, 798–807. [CrossRef]
12. Wood, S.N. Thin plate regression splines. *J. R. Stat. Soc. Ser. B Stat. Methodol.* **2003**, *65*, 95–114. [CrossRef]
13. Wood, S.N.; Bravington, M.V.; Hedley, S.L. Soap film smoothing. *J. R. Stat. Soc. Ser. B Pervasiv Methodol.* **2008**, *70*, 931–955. [CrossRef]
14. Ramsay, T. Spline smoothing over difficult regions. *J. R. Stat. Soc. Ser. B Pervasiv Methodol.* **2002**, *64*, 307–319. [CrossRef]
15. Guillas, S.; Lai, M.J. Bivariate splines for spatial functional regression models. *J. Nonparametr. Stat.* **2010**, *22*, 477–497. [CrossRef]
16. Sangalli, L.M.; Ramsay, J.O.; Ramsay, T.O. Spatial spline regression models. *J. R. Stat. Soc. Ser. B Stat. Methodol.* **2013**, *75*, 681–703. [CrossRef]
17. Boyd, S.; Parikh, N.; Chu, E.; Peleato, B.; Eckstein, J. Distributed optimization and statistical learning via the alternating direction method of multipliers. *Found. Trends Mach. Learn.* **2011**, *3*, 1–122. [CrossRef]
18. Parikh, N.; Boyd, S. Proximal algorithms. *Found. Trends Optim.* **2014**, *1*, 127–239. [CrossRef]
19. Bertsekas, D.P. *Nonlinear Programming*; Athena Scientific: Belmont, MA, USA, 1999.
20. Gonzalez, M.C.; Hidalgo, C.A.; Barabasi, A.L. Understanding individual human mobility patterns. *Nature* **2008**, *453*, 779–782. [CrossRef]
21. Csáji, B.C.; Browet, A.; Traag, V.A.; Delvenne, J.C.; Huens, E.; Van Dooren, P.; Smoreda, Z.; Blondel, V.D. Exploring the mobility of mobile phone users. *Phys. A Stat. Mech. Its Appl.* **2013**, *392*, 1459–1473. [CrossRef]
22. Song, C.; Qu, Z.; Blumm, N.; Barabási, A.L. Limits of predictability in human mobility. *Science* **2010**, *327*, 1018–1021. [CrossRef]
23. Blondel, V.D.; Decuyper, A.; Krings, G. A survey of results on mobile phone datasets analysis. *EPJ Data Sci.* **2015**, *4*, 10.
24. Grauwin, S.; Sobolevsky, S.; Moritz, S.; Gódor, I.; Ratti, C. Towards a comparative science of cities: Using mobile traffic records in new york, london, and hong kong. In *Computational Approaches For Urban Environments*; Springer: Berlin/Heidelberg, Germany, 2015; pp. 363–387.
25. Telecom. Telecom Italia Big Data Challenge. 2014. Available online: https://dandelion.eu/datamine/open-big-data/ (accessed on 27 July 2016).
26. Chopra, S.; Thampy, T.; Leahy, J.; Caplin, A.; LeCun, Y. Discovering the hidden structure of house prices with a non-parametric latent manifold model. In Proceedings of the 13th ACM SIGKDD International Conference on Knowledge Discovery and Data Mining, San Jose, CA, USA, 12–15 August 2007; ACM: New York, NY, USA, 2007; pp. 173–182.
27. Hjelle, Ø.; Dæhlen, M. *Triangulations and Applications*; Springer Science & Business Media: Berlin, Germany, 2006.

28. Douglas, J.; Rachford, H.H. On the numerical solution of heat conduction problems in two and three space variables. *Trans. Am. Math. Soc.* **1956**, *82*, 421–439. [CrossRef]

29. Ratti, C.; Frenchman, D.; Pulselli, R.M.; Williams, S. Mobile landscapes: Using location data from cell phones for urban analysis. *Environ. Plan. B Plan. Des.* **2006**, *33*, 727–748. [CrossRef]

30. Lila, E.; Sangalli, L.M.; Ramsay, J.; Formaggia, L. *fdaPDE: Functional Data Analysis and Partial Differential Equations; Statistical Analysis of Functional and Spatial Data, Based on Regression with Partial Differential Regularizations*; R Package Version 0.1-4; The Comprehensive R Archive Network. 2016. Available online: https://cran.r-project.org/web/packages/fdaPDE/index.html (accessed on 26 April 2019).

31. Clapp, J.M. A semiparametric method for estimating local house price indices. *Real Estate Econ.* **2004**, *32*, 127–160. [CrossRef]

32. Caplin, A.; Chopra, S.; Leahy, J.V.; LeCun, Y.; Thampy, T. Machine Learning and the Spatial Structure of House Prices and Housing Returns. 2008. Available online: https://ssrn.com/abstract=1316046 (accessed on 26 April 2019).

33. Manfredini, F.; Pucci, P.; Tagliolato, P. Toward a systemic use of manifold cell phone network data for urban analysis and planning. *J. Urban Technol.* **2014**, *21*, 39–59. [CrossRef]

 *applied sciences*

Article

Landslide Prediction with Model Switching †

Darmawan Utomo, Shi-Feng Chen and Pao-Ann Hsiung *

Computer Science and Information Engineering, National Chung Cheng University, No. 168, Sec. 1, University Rd., Minhsiung, Chiayi 62102, Taiwan; du88@yahoo.com (D.U.); vincet0809@gmail.com (S.-F.C.)

* Correspondence: pahsiung@cs.ccu.edu.tw; Tel.: +886-5272-0411

† This paper is an extended version of paper published in the 2017 IEEE Conference on Dependable and Secure Computing, held in Taipei, Taiwan, 7–10 August 2017.

Received: 30 March 2019; Accepted: 28 April 2019; Published: 4 May 2019

Abstract: Landslides could cause huge damages to properties and severe loss of lives. Landslides can be detected by analyzing the environmental data collected by wireless sensor networks (WSNs). However, environmental data are usually complex and undergo rapid changes. Thus, if landslides can be predicted, people can leave the hazardous areas earlier. A good prediction mechanism is, thus, critical. Currently, a widely-used method is Artificial Neural Networks (ANNs), which give accurate predictions and exhibit high learning ability. Through training, the ANN weight coefficients can be made precise enough such that the network works in analogy to a human brain. However, when there is an imbalanced distribution of data, an ANN will not be able to learn the pattern of the minority class; that is, the class having very few data samples. As a result, the predictions could be inaccurate. To overcome this shortcoming of ANNs, this work proposes a model switching strategy that can choose between different predictors, according to environmental states. In addition, ANN-based error models have also been designed to predict future errors from prediction models and to compensate for these errors in the prediction phase. As a result, our proposed method can improve prediction performance, and the landslide prediction system can give warnings, on average, 44.2 min prior to the occurrence of a landslide.

Keywords: landslide prediction; machine learning; neural networks; model switching

1. Introduction

Landslides are natural disasters which can cause huge damage to properties and severe loss of lives. Many studies have focused on how to detect and monitor landslides. Though landslide detections could be performed in real-time, there might not be enough time to react, so as to save human lives and properties. In order to minimize the losses caused by landslides, an early prediction mechanism, with pre-warnings, is necessary. Once the system can give an alarm in advance, people would have more response time to evacuate before the landslide occurs.

There are several problems in landslide prediction. First, just as in most safety-critical applications, landslide prediction also exhibits the same data imbalance problem, where the class of stable data has much more data than the class of unstable data. Stable data, here, refers to the normal conditions (where there is no landslide), while unstable data represents landslide-related information. Second, a low true-positive rate (TPR) problem is often found in safety-critical applications, because of the interference in learning between two or more classes in the datasets. For example, learning from the normal stable conditions often affects the learning from the unstable (landslide) conditions, thus resulting in a low TPR. Third, predictive applications are often faced with the problem of determining an appropriate prediction horizon; that is, the size of the time window of past history to be used for predictions. Finally, real-time applications face the problem of determining when to re-train the models.

To address the above-mentioned four problems existing in landslide prediction, this work provides a total solution in the form of an early warning system, called the Model Switch-based Landslide Prediction System (MoSLaPS). To address the data imbalance problem, we adapt the popular Adaptive Synthetic Sampling (ADASYN) method [1] to landslide prediction. To address the low TPR problem, we propose a novel event-class model switch predictor design that significantly improves TPR. To address the problem of customizing the prediction horizon, we also propose a novel dynamic tuning method for the prediction horizon, in order to achieve the goal of early warning. To address the problem of determining model re-training time, we propose a novel learning-based re-training method, based on an error model which considers both the long-term and short-term accumulated errors. Errors are also predicted, so re-training can be done earlier in preparation for future data changes; as a result of which, our proposed system can achieve the goal of early warning.

Section 2 introduces some related work. Section 3 presents the proposed model switching method. Section 5 presents and analyzes the experimental results. In Section 6, the conclusions and future work are described.

2. Related Work

Landslide prediction methods can be classified into three types: Image analysis, machine learning, and mathematical evaluation models. Table 1 shows a comparison among these types of methods. First, image analysis uses Geographic Information Systems (GIS), which can collect, store, manage, and analyze geographical data. By analyzing disaster data, such as history of landslides and data on land development for agriculture, the risk of landslides can be predicted. The probability of landslides is variable, as it is based on the number of layers of data used for analysis. Second, machine learning techniques, such as Bayesian networks [2], neural networks [3], or genetic algorithm [4], use computational intelligence to calculate the probability of landslides. These methods incorporate different factors that might cause landslides to evaluate the probability of landslide occurrence. They are not real-time, because they require huge computational times for prediction. Finally, mathematical evaluation models use a single evaluate equation, such as Factor of Safety (FS) [5]. A hazard model is combined with the physical concepts of mechanics and hydrographic data for the stability of slopes. It is easy for simulation and fits a wide range of environments, but it is difficult to obtain the whole hydrographic data as groundwater elevation is difficult to measure.

Table 1. Comparison of Landslide Detection/Prediction Methods.

Types	Methods	Advantages	Accuracies
Image Analysis	Geographical Information System [4]	Suitable for Large Area	Accuracy based on number of layers
Machine Learning	Bayesian Network [2]	Simple Network	75%
	Neural Network [3]	Simple Network	67%
	Genetic Algorithm [4]	Optimal Solution	90%
Mathematical Evaluation	SHALSTAB * [6]	High Accuracy	>90%

* Shallow Landsliding Stability Model.

Landslides occur when the down-slope shear stress is large. As shown in Equation (1) [5], the Factor of Safety (FS) refers to the stability of the soils. It takes physical properties, including rainfall, slope, and soil properties, into consideration. It can easily predict landslides with the trend of each parameter. Therefore, to predict landslides, a FS equation is matched with these attributes. Three regions of the FS value, based on the SHALSTAB model [6], are defined to distinguish between the dangerous levels of a slope, as shown in Table 2. The Stable Region is classified as the stable class and the Marginally Stable and Actively Unstable Regions are classified as the unstable class. Based on the classes, different training samples are used to train multiple neural network predictors.

$$FS = \frac{C + (1 - \frac{R}{T}\frac{\alpha}{sin\theta}\frac{\rho_w}{\rho_s})\rho_s g Z cos^2\theta tan\phi}{\rho_s g Z cos\theta sin\theta},$$

(1)

with

- C: The effective coefficient (kPa);
- R: The rainfall intensity (mm/hr);
- T: The soil transmissivity (mm/hr);
- Z: The soil depth (m);
- ρ_w: The density of water (kg/m^2);
- ρ_s: The density of soil (kg/m^2);
- ϕ: The internal friction angle of the slope material (degree);
- θ: The slope gradient (degree); and
- α: The specific contributing area [5].

Table 2. Classification of Factor of Safety (FS) Levels.

Stable Region	$FS \geq 1.3$
Marginally Stable Region	$1.3 > FS \geq 1$
Actively Unstable Region	$FS < 1$

Of particular mention is the work done by Lian et al. [7,8] on landslide displacement prediction using Prediction Intervals (PIs) and an ANN switched prediction method. The authors employed K-means clustering for dividing the landslide data into two classes; namely, a majority class (stationary points) and a minority class (mutational points). Then, a weighted Extreme Learning Machine (ELM) classifier was used to construct the switch rules. Finally, bootstrap and kernel-based ELMs were applied to construct the PIs. This work was concerned with how the displacements are predicted accurately and early. In contrast, our work is focused on how landslides can be actually predicted accurately and early. Not only is the goal different, the methods or techniques used or proposed are also quite different. We employ a very popular mathematical estimation model for landslide prediction, as defined above (namely, the SHALSTAB model and the factor of safety). We use ANN models for the model switching, as well as for the predictions. We adapt ADASYN for resolving the data imbalance problem. We also propose novel methods for model retraining and prediction horizon tuning. Details are given in the next section.

3. Model Switched Landslide Prediction System

A total solution for landslide prediction with early warnings is proposed in this work. The design of the proposed Model Switched Landslide Prediction System (MoSLaPS) is shown in Figure 1. It consists of two parts; namely, Physical Entities and Computation Elements. In the Physical Entities, environmental data, such as rainfall, soil moisture, and slope, are collected by sensor nodes. Coordinator nodes integrate the sensed data and transmit them to the Computation Elements through Zigbee transmitters. The Computation Elements consists of a SHALSTAB Model, a Switch-based Prediction Model, and an Accurate Early Warning System, as described in the following.

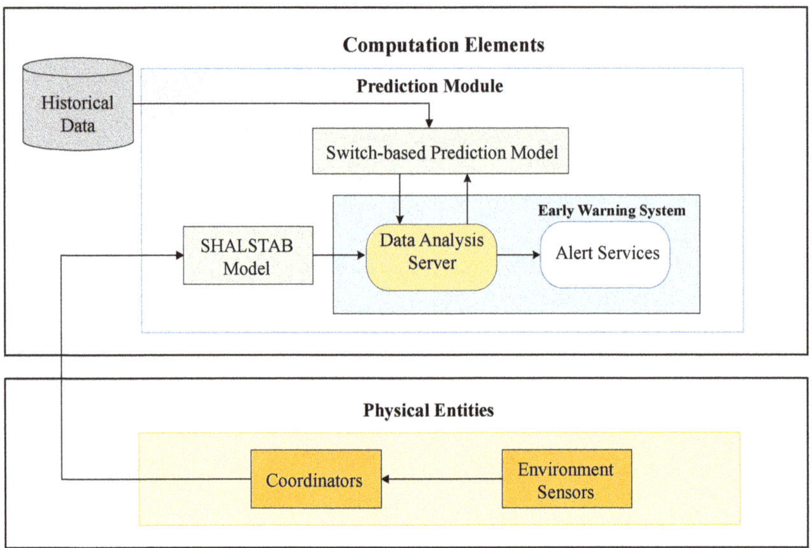

Figure 1. Model switched landslide prediction system architecture.

- SHALSTAB Model
  The SHALSTAB model takes the sensed environmental data, including rainfall, soil moisture, and slope, from the Physical Entities and evaluates the Factor of Safety, using Equation (1). Over time, the FS values are recorded as $FS_{actual} = \{FS_0, FS_1, ..., FS_t\}$, which is the input to the Accurate Early Warning System to predict the occurrence of landslides.

- Switch-based Prediction Model
  From historical environmental data, the proposed system consists of two prediction models to learn two different data patterns; namely, the stable pattern and the unstable pattern. To switch between the different prediction models, a neural network classifier is designed to predict the future class. The Switch-based Prediction Model can improve the prediction accuracy when the neural network classifier switches the prediction models accurately. The detailed technique is described in Section 3.3.

- Accurate Early Warning System
  The accurate early warning system consists of a data analysis server and alert services. The data analysis server uses the above-described switch-based prediction model to predict landslides. The input data, FS_{actual}, is used to predict future FS values, denoted as $FS_{predict} = \{FS_{t+1}, FS_{t+2}, ..., FS_{t+n}\}$. For each predicted FS value, there is a difference between the predicted FS and the actual FS calculated using Equation (1). This difference is called prediction error. The data analysis server will assess the applicability of the prediction model, according to the trend of prediction errors. If the error exceeds a pre-defined threshold, it means the prediction model is not suitable for the environment at that time and the predicted results have large prediction errors. Thus, based on the error measurement and a given error tolerance threshold, the prediction model is re-trained. The entire process will be described in Section 4. If a predicted FS value, FS_{t+k}, is smaller than 1, then it is estimated that a landslide will occur after k time slots [9]. Thus, alert services can send an alert in advance.

The details of the prediction models, model switching, and early warning system will be introduced in the rest of this section.

3.1. Prediction Model Design

To predict the future FS, a feed-forward Back-Propagation Neural Network (BPNN) is employed as the prediction model. A BPNN is a powerful computation system, created by generalizing the Widrow-Hoff learning rules [10] into a multi-layer with a non-linear differential transfer function network. The complex network connections imply that the high learning and reasoning ability of BPNN can be applied to deal with problems with high complexity. Figure 2 shows the framework of a BPNN-based prediction model.

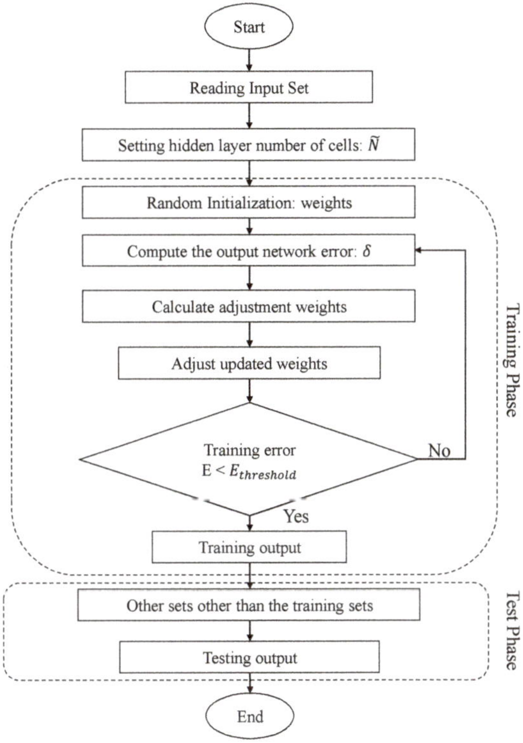

Figure 2. Back-Propagation Neural Network (BPNN) prediction model framework design.

Training Method for Time-Series Based BPNN

The basic computational procedure of a BPNN is explained to provide a basic description of the type of ANN that is implemented here. Figure 3 shows the basic structure of a time-series based BPNN. There are three types of layers: input, hidden, and output layers. In the input layer, time-series input for time slots $t - n$ to t, corresponding to a specific feature, such as factor of safety, rainfall, and soil moisture, is taken as input data. Each pair of nodes in the adjacent layers are linked by a weight. The values in all nodes in a previous layer and the weights are multiplied and accumulated as the input for a node of the next layer. The inputs are, then, given to an activation function to calculate the output value of the node. Repeating the above operations, layer by layer, from input layer to output layer, the final output can be derived.

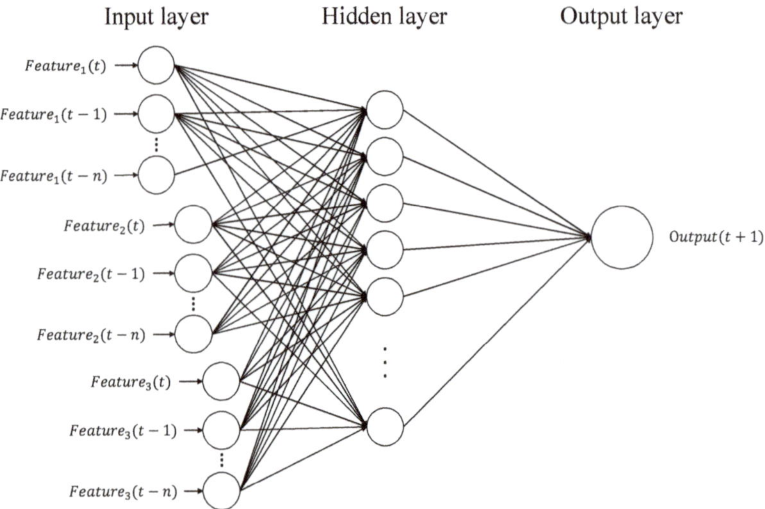

Figure 3. A fully connected feed-forward Back-Propagation Neural Network with time-series.

Algorithm 1 has a complete description as a prediction model; the following equations explain the functions that are used in this algorithm. At first, the previous FS is used to predict the future FS by a BPNN model. To dynamically determine suitable weights for different FS, a back-propagation method is applied to train and update the weights before prediction, in order to completely illustrate the details of back propagation method. Given the training sample, $T_{data} = (x_i, p_i), i = 1, ..., N$, where $x_i = [x_{i1}, x_{i2}, ...x_{in}] \in R^n$ is the impact factor and $p_i = [p_{i1}, p_{i2}, ...p_{ik}] \in R^k$ is the training target, the general mathematical model of a standard single hidden layer feed-forward network with $\tilde{N}$ hidden neurons is shown in Equation (2).

$$o_j = \sum_{i=1}^{\tilde{N}} g(w_i \cdot x_i), \qquad j = 1, ..., N,$$ (2)

where $o_j = [o_{j1}, o_{j2}, ...o_{jk}]$ is the jth output of the BPNN, w_i is the weight of the connection from the input neurons to the ith hidden neuron, and $g(x)$ is an activation function that represents how much adjustment the output should be from the neuron, based on the sum of the input. The activation function used in our BPNN model is depicted in Equation (3); namely, the sigmoid function. The sigmoid function, also called the logistic function, is a commonly-used activation function which has an output range from 0 to 1.

$$g(x) = \frac{1}{1 + e^{-x}}.$$ (3)

The difference between the prediction result and the actual result is called the prediction error. In order to reduce prediction error, the weights need to be updated. The Levenberg-Marquardt (LMA) [11] method has the fastest convergence and the lowest mean square error. Therefore, LMA is selected as a training function to calculate the output network error δ and adjustment weight W. It is depicted in Equation (4):

$$W_{k+1} = W_k - [J^T J + uI]^{-1} J^T \delta,$$ (4)

where W_k represents the weight matrix in the kth iteration, J is the Jacobian matrix [12] that contains the network error for weight and the first-order differential of weight, I is the unit matrix, δ is the network output error, and u is a constant. LMA can dynamically adjust the constant u to reduce the network output error δ.

Algorithm 1: Prediction Model Algorithm.

Input:
T_{data}: Data of FS values used for training BPNN;
P_{input}: Inputs of FS values for prediction;
$E_{threshold}$: Training error threshold;
Output:
P_{output}: Prediction output of FS values;
Variable:
W: Weights of BPNN;
E: Training error between training outputs and target;
δ: Error value for adjusting weights;
c_{sat}: 1: Training cycle is complete, 0: Training cycle is incomplete;

```
1  Set hidden layer number of cells N̂ ; // Equation (6)
2  Set the maximum iteration number for training Epochs;
3  Randomly initialize weight W;
   // Training model
4  i = 0;
5  while (E > E_threshold)&&(i <= Epochs) do
6  |    c_sat = 0;
7  |    while c_sat = 0 do
8  |    |    CalculateBPNN(T_data, W) ; // Equations (2) and (3)
9  |    |    Calculate output network error δ ; // Equation (4)
10 |    |    Calculate adjustment weight W ; // Equation (4)
11 |    |    if all samples are trained then
12 |    |    |    c_sat = 1;
13 |    Calculate training error E ; // Equation (5)
14 |    i + +;
   // Do prediction
15 P_output = CalculateBPNN(P_input, W);
16 return P_output;
```

After the training phase, the training error E is calculated by Equation (5), to determine whether the training step has reached convergence. If it is not less than the training error threshold, $E_{threshold}$, the training phase is restarted.

$$E = \frac{1}{N} \sum_{i=1}^{N} (T_i - O_i)^2, \tag{5}$$

where:

- T_i: Target FS value of training sample i;
- O_i: BPNN output FS value of training sample i; and
- N: Number of training samples.

Considering the training time in our proposed re-training process, the number of outputs, Num_{output}, is set to 1 to avoid a long training time. This means that only one prediction result per iteration will be obtained, by taking the past FS value as input.

The number of neurons in the hidden layer, denoted as Num_{neuron}, is determined based on the following experience rule [13]:

- The number of neurons in the first hidden layer is calculated using (6):

$$Num_{neuron} = \sqrt{Num_{input} * Num_{output}}.$$ (6)

3.2. Imbalanced-Class Prediction Design

Class balance enhancements are needed to handle training samples with an unbalanced class distribution [14]. The Adaptive Synthetic Sampling (ADASYN) method [1] is used here for balancing the imbalanced data (i.e., data is pre-processed using ADASYN), and then the processed dataset are used to train the event-class predictor, which is also a BPNN model. In the following, the ADASYN algorithm is described as follows.

Data Pre-Processing Using ADASYN Algorithm

The ADASYN algorithm can improve the data imbalance problem by synthetically creating new samples from the unstable class by linear interpolation between existing unstable class samples. This approach, by itself, is known as the Synthetic Minority Over-sampling Technique (SMOTE) method [15]. ADASYN is an extension of SMOTE, creating more samples in the vicinity of the boundary between the two classes than in the interior of the unstable class.

To create more synthetic data for the unstable class, FS, rainfall, and soil moisture are used as the training data and the corresponding class label, y_i, is constructed according the classification region. Given the training samples (X_i, y_i), $i = 1, ..., N$, where $X_i = < x_{i,fs}, x_{i,r}, x_{i,sm}, x_{i,slope} > \in R^n$, $x_{i,fs}$ is the FS value, $x_{i,r}$ is the rainfall, $x_{i,sm}$ is the soil moisture, $x_{i,slope}$ is the slope gradient, and y_i is the class label. The training samples, X_i, are classified by Equation (7). For $x_{i,fs} \geq 1.3$, x_i classifies as stable class. On the other hand, if $x_{i,fs} < 1.3$, then x_i classifies as unstable class. After X_i is classified, the class label, y_i, is set by Equation (8). If $X_i \in StableClass$, y_i is set as 0. If $X_i \in UnstableClass$, y_i is set as 1.

$$\begin{cases} \text{if } x_{i,fs} \geq 1.3, X_i \in Stable\ Class \\ \text{if } x_{i,fs} < 1.3, X_i \in Unstable\ Class \end{cases},$$ (7)

$$y_i = \begin{cases} 0 & \text{if } X_i \in Stable\ Class \\ 1 & \text{if } X_i \in Unstable\ Class \end{cases}.$$ (8)

To adjust class balance, the degree of class imbalance is needed to be calculated by Equation (9):

$$d = N_m / N_s,$$ (9)

where

- N_m: The size of unstable class examples; and
- N_s: The size of stable class examples.

Furthermore, the default level, $d_{default}$, which is the threshold for the level of maximum class imbalance tolerated, also needs to be determined beforehand. If the current d is smaller than the threshold degree, $d_{default}$, then the number of synthetic data samples that need to be generated for the unstable class is calculated using Equation (10):

$$R = (N_s - N_m) \times \beta,$$ (10)

where $\beta \in [0, 1]$ is a parameter used to specify the desired balance level after generation of the synthetic data: $\beta = 1$ means a fully balanced data set is created after the generalization process.

For each $X_i \in Unstable\ Class$, K nearest neighbours can be found by using the Euclidean distance. The ratio r_i, defined in Equation (11), which represents the number of stable-classified in the K nearest neighbours, and its normal form, $\hat{r}_i$, defined in Equation (12), are calculated, where $\hat{r}_i$ is called a density distribution of r_i, $(\sum_i \hat{r}_i = 1)$.

$$r_i = h_i / K, \qquad i = 1, \ldots, N_m,$$ (11)

$$\hat{r}_i = r_i / \sum_{i=1}^{N_m} r_i,$$ (12)

where h_i is the number of samples in the K nearest neighbours of x_i that belong to the stable class; therefore, $r_i \in [0, 1]$.

Thus, the number of synthetic data samples that need to be generated for each unstable class sample, $X_{unstable}$, can be calculated by Equation (13).

$$g_i = \hat{r}_i \times R,$$ (13)

where R is the total number of synthetic data samples that need to be generated for the unstable class, as defined in Equation (10).

By random, the program chooses one unstable data sample, X_{zi}, from the K nearest neighbours for $X_{unstable}$ to generate new synthetic samples, sd_{new}, by Equation (14). This procedure is repeated g_i times to produce new synthetic samples.

$$sd_{new} = X_{unstable} + (X_{zi} - X_{unstable}) \times \lambda,$$ (14)

where

- $(X_{zi} - X_{unstable})$: The difference vector; and
- λ: A random number $\lambda \in [0, 1]$.

3.3. Switch Based Prediction Model Design

To address the issue of imbalanced data between the unstable and stable classes, a switch-based neural network prediction algorithm is proposed, as detailed in Algorithm 2.

The environmental factors, including rainfall, soil moisture, and slope gradient, are used to calculate the FS values, $x_{i,fs}$, using the SHALSTAB model. Given the environmental samples $D_{landslide} = \{x_i\}$, where $x_i = <x_{i,r}, x_{i,sm}, x_{i,slope}>$, $x_{i,r}$ is the rainfall, $x_{i,sm}$ is the soil moisture, and $x_{i,slope}$ is the slope gradient, the FS values, $x_{i,fs}$, are calculated given the set of all training samples, T_{sample}, where $T_{sample} = \{X_i | X_i = <x_{i,fs}, x_{i,r}, x_{i,sm}, x_{i,slope}>\}$. Then, the corresponding class labels y_i, $T_{class} = \{X_i, y_i\}$, can be constructed and classified by Equations (7) and (8). To construct the BPNN models for different data patterns, the calculated FS need to be classified in two subsets, as follows:

$$T_{class1} = \{x_{i,fs} | x_{i,fs} \in X_i, x_{i,fs} \in Stable\ Class\},$$ (15)

$$T_{class2} = \{x_{i,fs} | x_{i,fs} \in X_i, x_{i,fs} \in Unstable\ Class\},$$ (16)

where T_{class1} is the set of FS values for $x_{i,fs}$, in $X_i \in StableClass$, and T_{class2} is the set of FS values for $x_{i,fs}$, in $X_i \in UnstableClass$; so that $T_{class1} \cap T_{class2} = \phi$ and $T_{class1} \cup T_{class2} = x_{i,fs}$.

Algorithm 2: Switch-based Neural Networks Prediction Algorithm.

Input:
$D_{landslide}$: $\{x_i | x_i = < x_{i,r}, x_{i,sm}, x_{i,slope} >\}$;
P_{test}: $\{x_t | x_t = < x_{t,fs}, x_{t,r}, x_{t,sm}, x_{t,slope}, y_t >\}$ for prediction;
Output:
P_{output}: $\{FS_{predict}\}$;
Variable:
T_{sample}: $\{X_i | X_i = < x_{i,fs}, x_{i,r}, x_{i,sm}, x_{i,slope} >\}$;
T_{class}: $\{< y_i, x_{i,fs}, x_{i,r}, x_{i,sm}, x_{i,slope} >\}$;
T_{class1}: $\{x_{i,fs} | y_i = 0\}$;
T_{class2}: $\{x_{i,fs} | y_i = 1\}$;
T_{ADASYN}: $\{< x_{new,fs}, x_{new,r}, x_{new,sm}, x_{new,slope}, y_i = 1 >\}$;
P_{class}: $\{y_{predict}\}$;

    // SHALSTAB model
1 $T_{sample} = calculateSHALSTAB(D_{landslide})$; // Equation (1)
    // Classification
2 $[T_{class}, T_{class1}, T_{class2}] = calculateClass(T_{sample})$; // Equations (7) and (8)
    // Data pre-processing
3 $T_{ADASYN} = calculateADASYN(T_{class})$; // Equation (14)
    // Construct each prediction model
4 $Feed - Forward\ BPNN_{Stable} \leftarrow PredictionModel(T_{class1})$;
5 $Feed - Forward\ BPNN_{Unstable} \leftarrow PredictionModel(T_{class2})$;
6 $Event - class\ predictor \leftarrow PredictionModel(T_{class} + T_{ADASYN})$;
    // Model switch
7 $P_{class} = Event - class\ predictor(P_{test})$;
8 **if** $P_{class} == Stable$ **then**
      // Prediction class is stable
9     $P_{output} = Feed - Forward\ BPNN_{Stable}(P_{test})$
10 **else**
      // Prediction class is unstable
11    $P_{output} = Feed - Forward\ BPNN_{Unstable}(P_{test})$
12 **return** P_{output};

Here, the ADASYN algorithm is used to produce new synthetic samples for the unstable class, in order to balance the sizes of the two classes. The processed dataset, T_{ADASYN}, is used to predict the future class using a BPNN model. The event-class predictor can switch between the different models, according to the predicted class. As shown in Figure 4, the steps of the event-class predictor are as follows.

First, T_{class} is selected to construct the synthetic dataset $T_{ADASYN} = \{< sd_{new}, y_{new} >\}$ for balancing class distribution by the ADASYN algorithm, where sd_{new} are the new synthetic samples, $sd_{new} = < x_{new,fs}, x_{new,r}, x_{new,sm}, x_{new,slope} >$, and $y_{new} = 1$ represents that the synthetic class label is unstable class. The synthetic sd_{new} include $x_{new,fs}$, the new synthetic FS value; $x_{new,r}$, the new synthetic rainfall; $x_{new,sm}$, the new synthetic soil moisture; and $x_{new,slope}$, the new synthetic slope gradient. Both of the classes T_{class} and T_{ADASYN} are integrated into the training set of the event-class predictor. Second, the event-class predictor is constructed using the BPNN model. Finally, the event-class predictor is used to predict the future class label, $P_{class} \in \{0, 1\}$, for the testing phase, where $P_{class} = 0$ represents that the prediction class label is stable and $P_{class} = 1$ represents that the prediction class label is unstable.

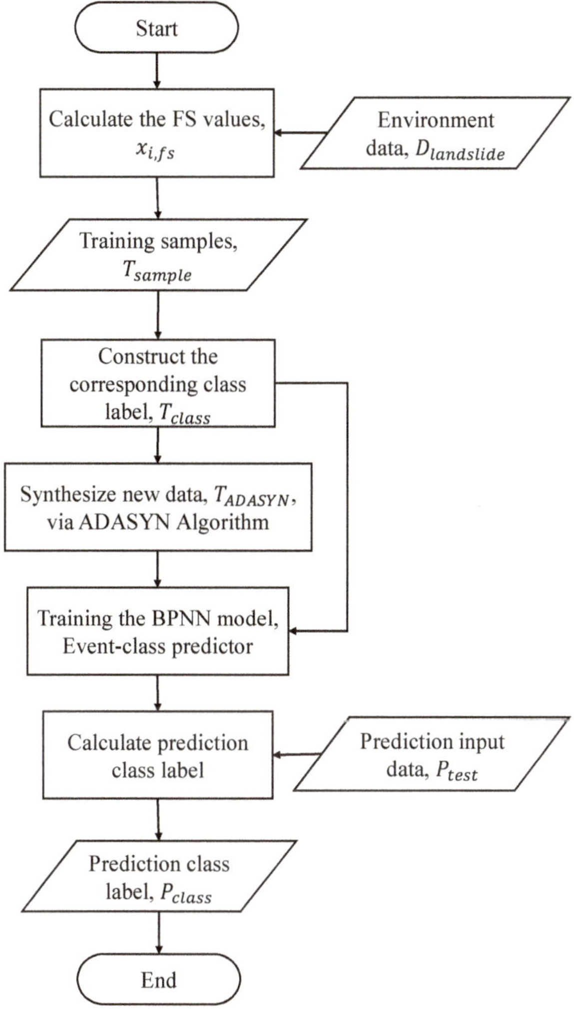

Figure 4. Flow of constructing the event-class predictor.

The aim of our proposed method is to construct different pattern predictors, as shown in Figure 5. The steps of different pattern predictors are as follows. First, T_{class1} is used as training data to train a BPNN model. After training, the BPNN model is the stable pattern of $x_{i,fs}$ (i.e., $Feed - Forward\ BPNN_{Stable}$). On the other side, T_{class2} is applied as training data to train another BPNN model. After training, the BPNN model is the unstable pattern of $x_{i,fs}$ (i.e., $Feed - Forward\ BPNN_{Unstable}$). Thus, two BPNNs are built to deal with different patterns. As shown in Figure 6, this procedure can switch different pattern predictors, according to the predicted class, P_{class}, that is obtained by the event-class predictor. When $P_{class} = 0$, the testing data, $P_{test} = \{x_t\}$, is applied to predict the future FS using $Feed - Forward\ BPNN_{Stable}$, where $x_t = < x_{t,fs}, x_{t,r}, x_{t,sm}, x_{t,slope}, y_t >$, $x_{t,fs}$ is the testing data of the FS value, $x_{t,r}$ is the testing data of the rainfall, $x_{t,sm}$ is the testing data of the soil moisture, $x_{t,slope}$ is the testing data of the slope gradient, and y_t is the testing data of the corresponding class label. On the other side, when $P_{class} = 1$, P_{test} is employed to predict the future FS using $Feed - Forward\ BPNN_{Unstable}$. Finally, the predicted FS, $P_{output} = \{FS_{predict}\}$, can be obtained by using the proposed Switch-based Prediction Model.

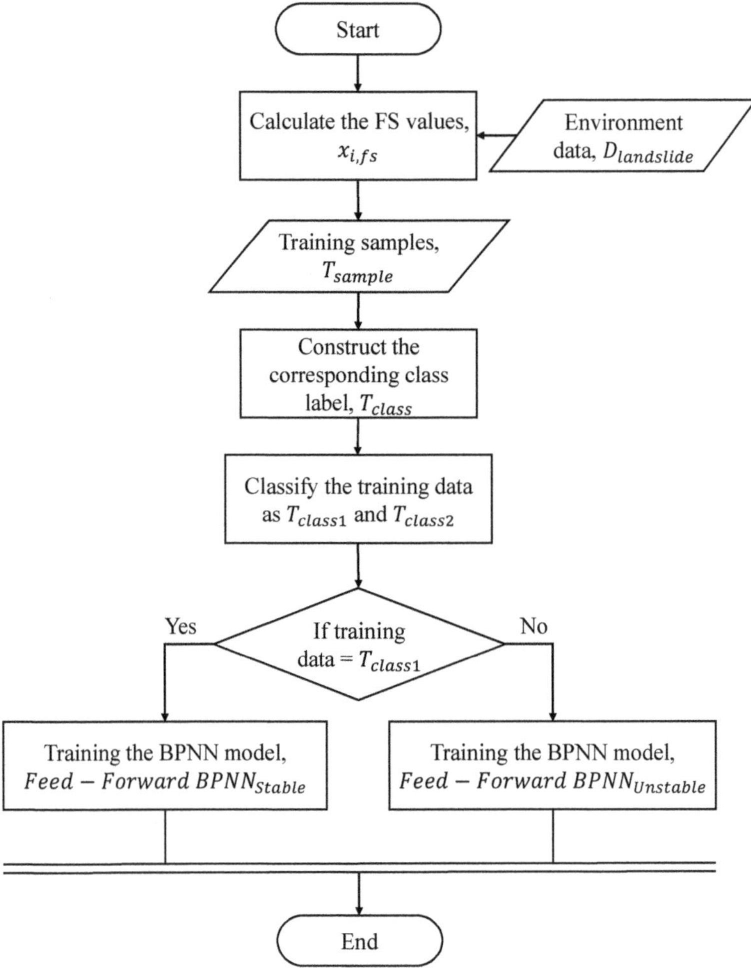

Figure 5. Flow of constructing different pattern predictors.

The main contribution of this work is that the proposed method can make highly accurate predictions, even in the case of highly imbalanced data. Two techniques were employed, ADASYN and an event-class predictor.

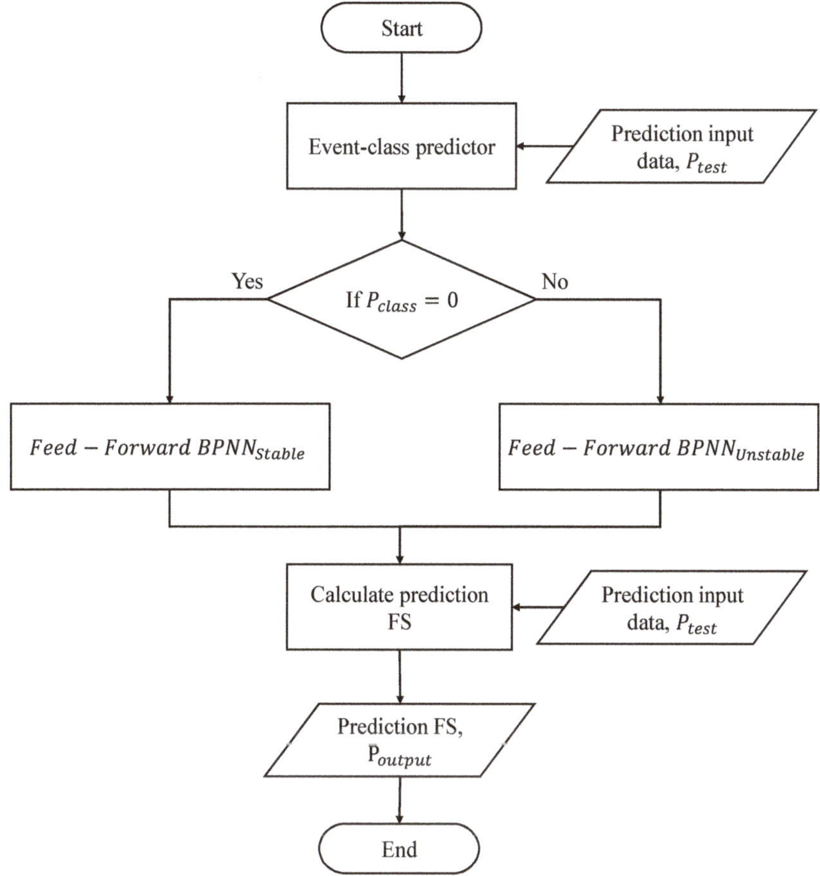

Figure 6. Flow of the switching strategy.

4. Accurate Early Warning System Design

To ensure the switch-based neural networks prediction model can be precise in a changing environment, an accurate early warning system is designed, as shown in Figure 7. It is divided into two parts: A learning-based re-training flow, as described in Section 4.1, and a Prediction Horizon tuning flow, as shown in Section 4.2.

4.1. Learning-Based Re-Training Flow

Figure 8 shows the flow of learning-based re-training. The determination of ret-raining is based on the error estimation. The error estimation process calculates the average error of all (FS_{actual}, $FS_{predict}$) pairs in an error-estimation window (EEW). For each period of the EEW, this procedure compares the average error, $AVGE$, of two error estimation windows, EEW_{now} and EEW_{prev}, and the accumulated error, $ACCE$, to check the two conditions for re-training. Equations (17) and (18) give the two conditions under which the prediction model needs to be re-trained, where C_{IE} is the coefficient of interval error used to specify the short-term tolerable error range, which is equal to the size of the prediction horizon in our work. If the difference of $AVGE$ between two continuous EEW is too large, the prediction model will be re-trained, due to the high variability of the input pattern that the original prediction model could not predict. Here, C_{AE} is the coefficient of accumulated error used to specify the long-term tolerable error range (here, long-term means since epoch). If the *ACCE*, compared to average error of the prediction model, $AVGE_{Model}$, is too large, the prediction model will also be re-trained, as the prediction results are becoming inaccurate. This further implies that the environment is changing with time, and that adaptation is needed. $AVGE_{now}$ and $ACCE_{now}$ are calculated using Equations (19) and (20), respectively.

$$AVGE_{now} > C_{IE} \times AVGE_{prev},\tag{17}$$

$$ACCE_{now} > C_{AE} \times AVGE_{Model},\tag{18}$$

$$AVGE_{now} = \frac{\sum |FS_{actual} - FS_{predict}|}{SIZE_{EEW}},\tag{19}$$

$$\begin{aligned}ACCE_{now} &= ACCE_{prev} + \Delta AVGE\\ &= ACCE_{prev} + \left(AVGE_{now} - AVGE_{prev}\right)\end{aligned}.\tag{20}$$

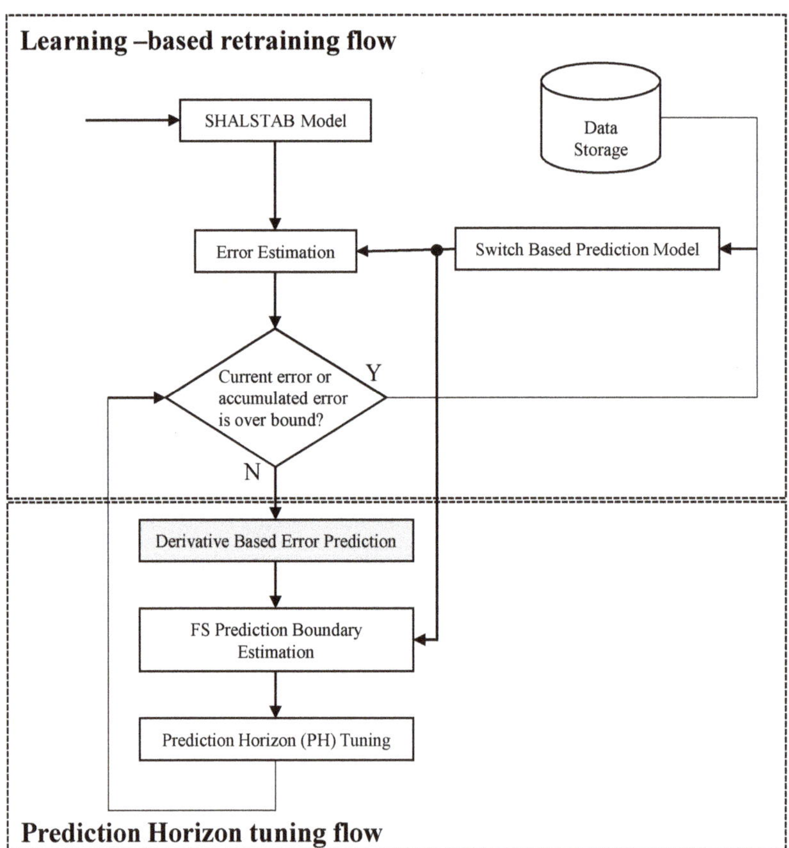

Figure 7. Overall Flowchart of Prediction Model Analysis.

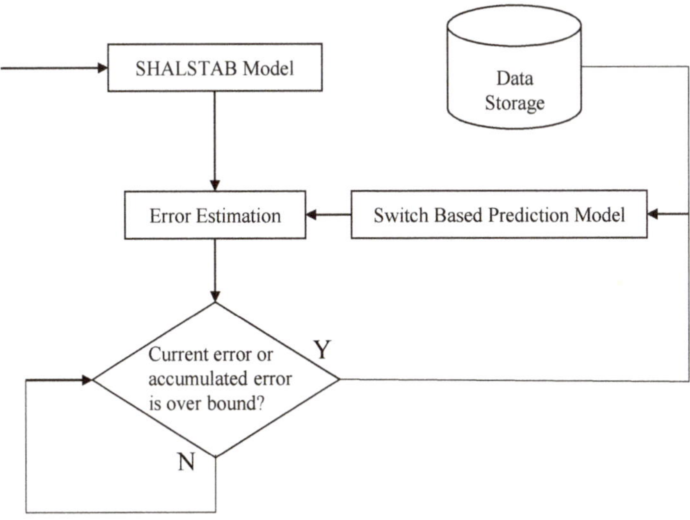

Figure 8. Learning-based re-training flow.

4.2. Prediction Horizon Tuning

In the re-training flow, the error estimation results are utilized further to tune the prediction horizon. The advantages of a variable-length prediction horizon (PH) are as follows:

- The occurrence of landslides can be predicted earlier; and
- The number of false predictions can be reduced.

The prediction horizon tuning flow is shown in Figure 9. Prediction errors are non-linear, and the prediction model is applied (as described in Section 3.1) in order to learn the inherent pattern for predicting future errors, $ERR_{predcit}$. In this way, the size of the PH is set to predict the occurrence of landslides earlier. Then, the future target ranges, P_{future}, can be determined by Equation (21). If the range of $P_{future} < 1$, this system can send alerts in advance.

$$FS_{predict} - |ERR_{predict}| \leq P_{future} \leq FS_{predict} + |ERR_{predict}|. \tag{21}$$

To decide whether the size of the PH is tuned, error boundary estimation is needed. As the program already has the predicted error, $ERR_{predict}$, and the predicted FS, $FS_{predict}$, then the predicted lower bound, $Bound_{low}$, can be estimated by Equation (22). According to the estimated results, the tuning is performed based on the following rules:

- If there is no $Bound_{low}$ lower than the lower bound of Stable Class (i.e., 1.3), for every time point in the prediction horizon, the size of the prediction horizon is increased by 1; and
- If there exists one $Bound_{low}$ lower than the lower bound of Stable Class (i.e., 1.3), for every time point in the prediction horizon, the size of prediction horizon is reset to the default value.

$$Bound_{low} = FS_{predcit} - |ERR_{predict}|. \tag{22}$$

Based on the above rules, we make our method a little more flexible for the landslide prediction scenario with variable-length prediction horizon.

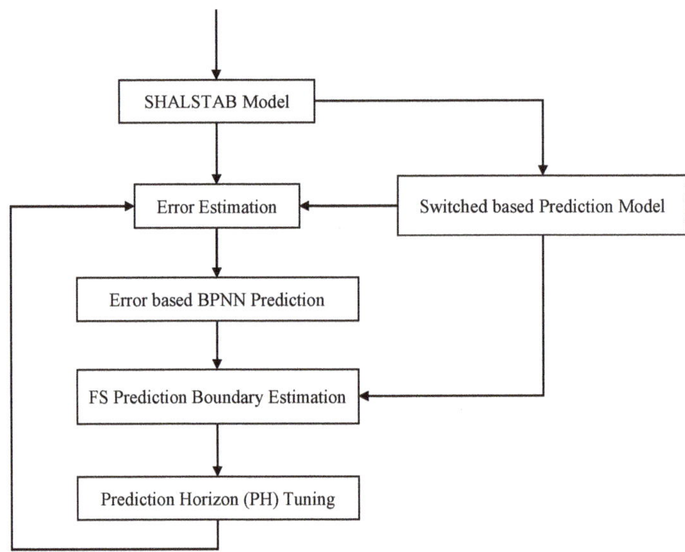

Figure 9. Prediction horizon tuning.

5. Experiments

In this section, evaluations of the proposed method for landslide prediction are presented. First, the experimental datasets used for the experiments are introduced. Then, the experimental results are illustrated. All experiments were carried out using the MATLAB® programming language on a PC with an Intel® Core™ i7-3770 CPU @ 3.40GHz and 16 GB RAM, running the Windows® 10 (64-bit) OS.

5.1. Experimental Datasets

Historical environmental monitoring datasets from the Shen-Mu station [16] were selected as a case study. Landslides are mainly influenced by rainfall and soil moisture. Using the FS, the occurrence of a landslide is estimated. The monitoring dataset of the Shen-Mu station is shown in Figures 10 and 11. Figures 10 and 11 show the relationships between rainfall and FS, and between soil moisture and FS, respectively. When rainfall and/or soil moisture increase, the FS decreases and the slope becomes unstable.

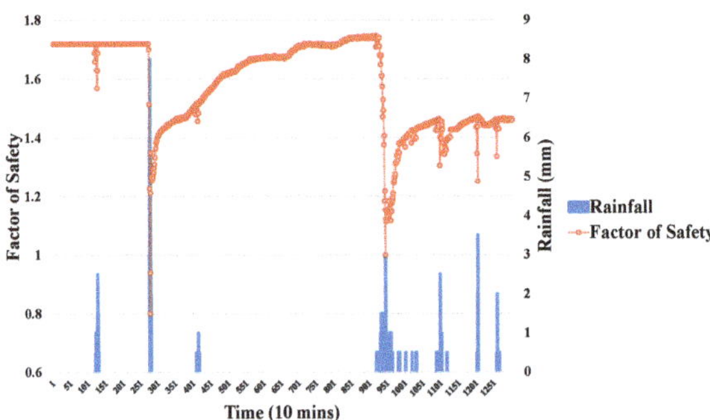

Figure 10. Monitoring curves of Factor of Safety and rainfall at Shen-Mu station.

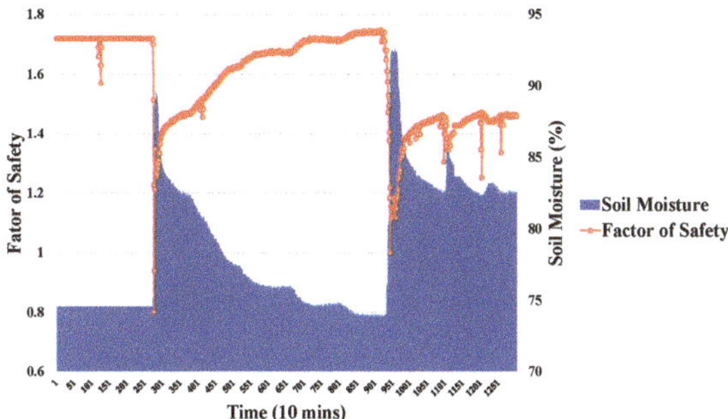

Figure 11. Monitoring curves of Factor of Safety and soil moisture at Shen-Mu station.

In the Shen-Mu datasets [16], data was recorded once per 10 min, and was collected in 2016. The program randomly selected 10 sets of samples, where each set had 1300 samples. Each dataset was further divided into two parts, the training set (75%) and the test set (25%). Note that the original data were used as input; that is, they were not normalized, because we needed to calculate the FS according to the SHALTAB model, as given in Equation (1).

Firstly, an evaluation of imbalanced-class prediction is described. Then, it is shown how the prediction accuracy is increased due to the proposed MoSLaPS.

5.2. Evaluation of Event-Class Prediction

ADASYN [1] was applied for data pre-processing in this program. It synthetically created new samples from the unstable class to balance the distribution of the data, if required. In the ADASYN algorithm, the desired level of balance, β, could be adjusted to control the number of new synthetic samples, which were needed when $0 \leq \beta \leq 1$.

Our event-class predictor used ADASYN [1] for imbalanced data processing. After data pre-processing, the processed dataset was used to train the event-class predictor (i.e., a BPNN model). To evaluate the event-class predictor, several performance indicators were applied and defined, as follows:

- True Positive Rate (TPR) is defined as in Equation (23);
- False Positive Rate (FPR) is defined as in Equation (24); and
- Accuracy (ACC) is defined as in Equation (25).

If the prediction class was unstable and the actual class was also unstable, then the result was said to be a True Positive (TP). If the prediction class was stable and the actual class Was also stable, then the result was said to be a True Negative (TN). If the prediction class was unstable and the actual class was stable, then the result was said to be a False Positive (FP). If the prediction class was stable and the actual class was unstable, then the result was said to be a False Negative (FN). Table 3 shows the classification of the above four different categories.

$$TPR = \frac{TP}{TP + FN} \times 100\%, \tag{23}$$

$$FPR = \frac{FP}{FP + TN} \times 100\%, \tag{24}$$

$$ACC = \frac{TP + TN}{TP + FN + FP + TN} \times 100\%. \tag{25}$$

Table 3. Confusion matrix.

		Actual Class	
		Unstable	Stable
Prediction Class	**Unstable**	True Positive	False Positive
	Stable	False Negative	True Negative

In our experiment, Figure 12 shows the evaluation results of the ADASYN algorithm for different β levels. When β was greater than 0.45, the TPR was more than 0.98. Therefore, the balance coefficient $\beta = 0.45$ was selected. Table 4 shows the number of new synthetic samples. Majority represents the size of the stable class and Minority represents the size of the unstable class.

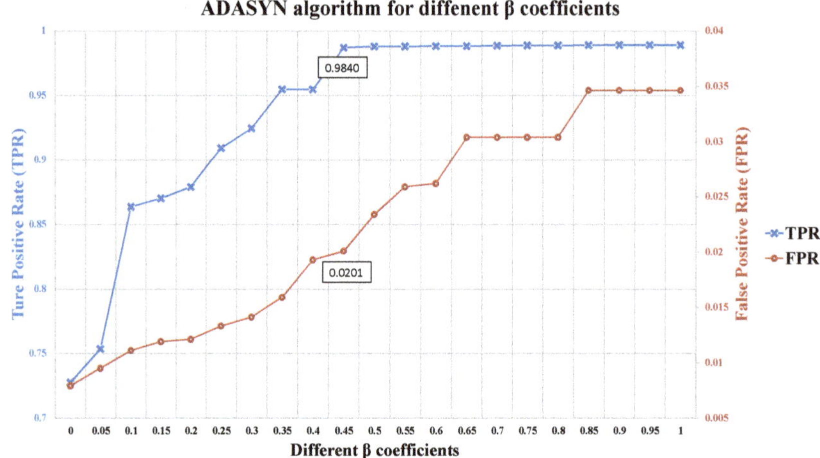

Figure 12. The evaluation result of the ADASYN algorithm for different values of the coefficient β.

Table 4. Number of synthetic new samples generated by the ADASYN algorithm ($\beta = 0.45$).

All Dataset	Majority	Minority	ADASYN
1300	1280	20	449
1300	1199	101	389
1300	1222	78	424
1300	1181	119	374
1300	1199	101	420
1300	1291	9	464
1300	1227	73	414
1300	1224	76	408
1300	1056	244	299
1300	1140	160	341

The evaluation results of the event-class predictor are shown in Table 5. There were 10 sets of testing samples. The average ACC was 97.94%, the average TPR was 98.40%, and the average FPR was 2.01%. Due to the high accuracy of event-class predictor, it can be used to choose a decision to switch between the models of different data patterns.

Table 5. The evaluation results of the event-class predictor.

No.	ACC	TPR	FPR	TP	FP	TN	FN
1	96.43%	100.00%	3.60%	2	9	241	0
2	97.62%	94.44%	2.14%	17	5	229	1
3	98.41%	95.45%	1.30%	21	3	227	1
4	98.41%	100.00%	1.64%	8	4	240	0
5	96.03%	100.00%	3.98%	1	10	241	0
6	100.00%	100.00%	0.00%	16	0	236	0
7	96.03%	100.00%	4.03%	4	10	238	0
8	99.21%	100.00%	0.83%	12	2	238	0
9	98.41%	100.00%	1.68%	14	4	234	0
10	98.81%	94.12%	0.85%	16	2	233	1
Average	97.94%	98.40%	2.01%				

Table 6 shows comparisons of the event-class predictor with other common classifiers, such as BPNN, Support Vector Machine (SVM), and Adaboost. The ACC of all classifiers were greater than 90%. A good classifier needs to have high TPR and low FPR. Although the FPR of our classifier, compared with BPNN and Adaboost, was a little higher, the TPR of our proposed classifier was much higher than that of the others. This means that our classifier exhibited a higher ratio of correct classification.

Table 6. Comparison of event-class predictor with other common classifiers.

Method	ACC	TPR	FPR
Event-Class	97.94%	98.40%	2.01%
BPNN	97.16%	57%	0.42%
SVM	90.42%	78.06%	9.02%
Adaboost	98.27%	75.1%	0.96%

5.3. Evaluation of Model Switched Landslide Prediction System

In the following experiments, the same datasets as used in Section 5.1 were employed to evaluate our landslide prediction model. To evaluate the proposed MoSLaPS model, several performance indicators were used and defined as follows:

- Mean Absolute Percent Error (MAPE) is defined as in Equation (26), where n is the number of predicted data, A_t is the actual value, and P_t is the predicted value.

$$MAPE = \frac{1}{n} \sum_{t=1}^{n} \frac{|A_t - P_t|}{P_t}. \tag{26}$$

- Root Mean Squared Error (RMSE) is defined as in Equation (27), where n is the number of predicted data samples, y_i is the actual FS value, and $\hat{y}_i$ is the predicted FS value.

$$RMSE = \sqrt{\frac{1}{n} \sum_{i=1}^{n} (y_i - \hat{y}_i)^2}. \tag{27}$$

- Normalized Root Mean Squared Error (NRMSE) is defined as in Equation (28), where $\bar{y}$ is the mean of the actual values.

$$NRMSE = \frac{RMSE}{\bar{y}}. \tag{28}$$

Figure 13 shows the actual FS and predicted FS for every 10 min. A BPNN prediction model is not able to learn the pattern of unstable class, as the size of the unstable class is much smaller than that of the stable class. If the original past data were used, the BPNN was not able to predict landslide occurrences. However, after processing the training data using the ADASYN algorithm to re-balance the distribution of classes, a single BPNN prediction model was still not able to learn the pattern of unstable class perfectly. This is because the data of the stable and unstable classes affected each other. As a solution, two BPNN prediction models Were proposed, one to learn the pattern of the stable class and another to learn the pattern of the unstable class. To switch between two BPNN models, an event-class predictor was constructed that can deal with imbalanced data distribution to predict the future class as a decision. Therefore, our proposed MoSLaPS method could learn the patterns of both the stable and unstable classes.

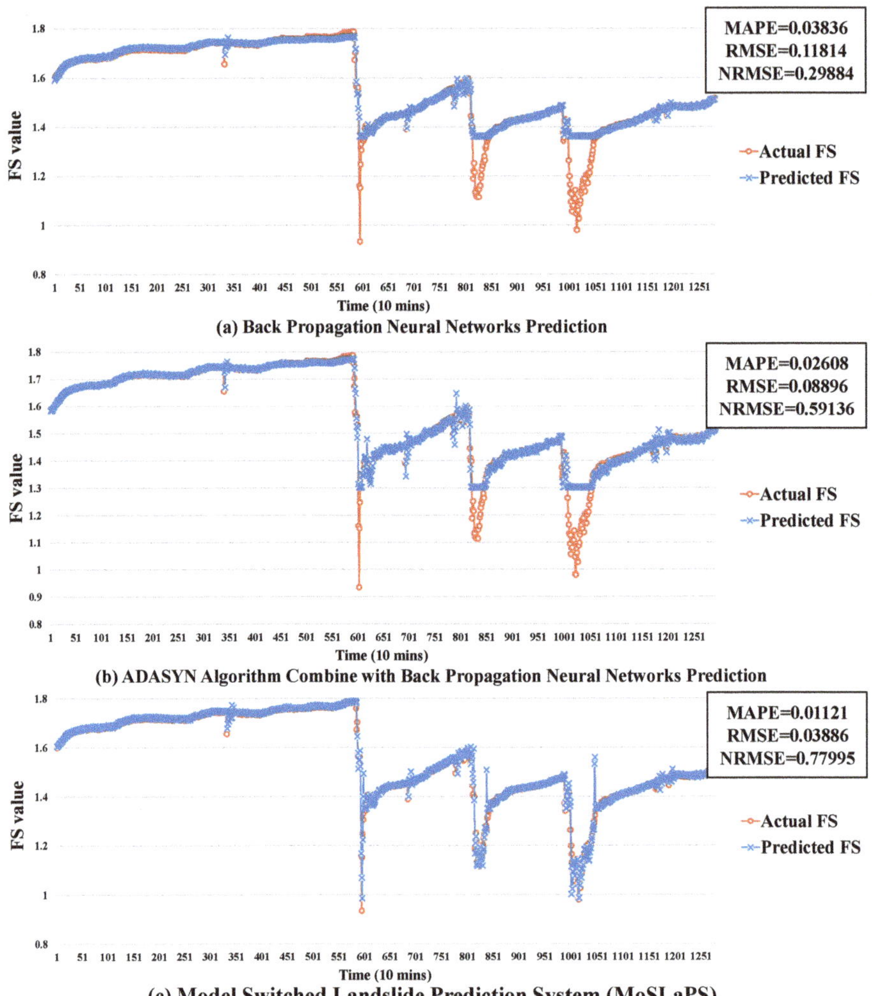

Figure 13. Comparison of the proposed method with other BPNN methods.

We compared our proposed MoSLaPS method with the above-mentioned methods, including a single BPNN and ADASYN+BPNN. A single BPNN was described, in detail, in Section 3.1; and ADASYN+BPNN use the ADASYN algorithm to re-balance the training data. After re-balancing, the processed training data were applied to train the BPNN.

The error metrics were evaluated for MoSLaPS, ADASYN+BPNN, and BPNN. Compared with BPNN and ADASYN + BPNN, our method resulted in much smaller MAPE and RMSE. This means that the prediction accuracy of our method was higher. The NRMSE of our method was also larger than that of the others, which means that our prediction results were closer to the real situation.

Although MoSLaPS was more accurate than the other methods, it requires a little more computation time and resources. As shown in Table 7, the simulation results of the different methods shows that the speed of MoSLaPS was a little slower than that of BPNN and ADASYN+BPNN, and its CPU and memory usages were higher than the others. This is because MoSLaPS took extra computational time and resources to deal with the imbalanced data classification and switching between the different predictors. In our experiment, the time interval was 10 mins; so there was ample time to deal with the process.

Table 7. Comparisons of time consumption and resource usage.

Method	Time (s)	CPU Usage	Memory
MoSLaPS	1.205	23.90%	972 kb
ADASYN+BPNN	0.739	20.30%	192 kb
BPNN	0.639	19.90%	148 kb

5.4. Evaluation of Landslide Pre-Alarm

The same datasets as in Section 5.1 were applied to evaluate our landslide pre-alarm method. Table 8 shows the prediction time (PT) for ten different experiments. For each experiment, the minimum PT, maximum PT, and average PT were recorded. From Table 8, the program was able to observe that the prediction time for dataset ♯5 was the longest (52.2 min); while that for dataset ♯10 was the shortest (38.4 min). As a shorter prediction time indicates that the change of FS is intense and quick, dataset ♯10 represented a higher probability of landslide occurrence. Taking the average of all timings, it can be seen that the proposed MoSLaPS method could warn of the occurrence of a landslide an average of 44.2 min in advance.

Table 8. Prediction time in advance for different datasets.

No.	Min. PT (min)	Max. PT (min)	Avg. PT (min)
1	10	80	40
2	10	70	33.3
3	10	50	35.7
4	20	80	47.6
5	10	80	52.2
6	10	80	51.8
7	10	80	42.6
8	10	70	45.7
9	10	80	43.3
10	10	80	38.4
Avg.	10.7	76.4	44.2

Further, we take a best-case example, to demonstrate how the proposed method can warn of landslide occurrence far in advance. At time point 255, the 8th prediction result is $FS < 1$; that is, a landslide will occur after eight time units, as shown in Figure 14. In our experiments, the interval between two time points is 10 min. Hence, the landslide occurrence could be predicted and warned about 80 min in advance.

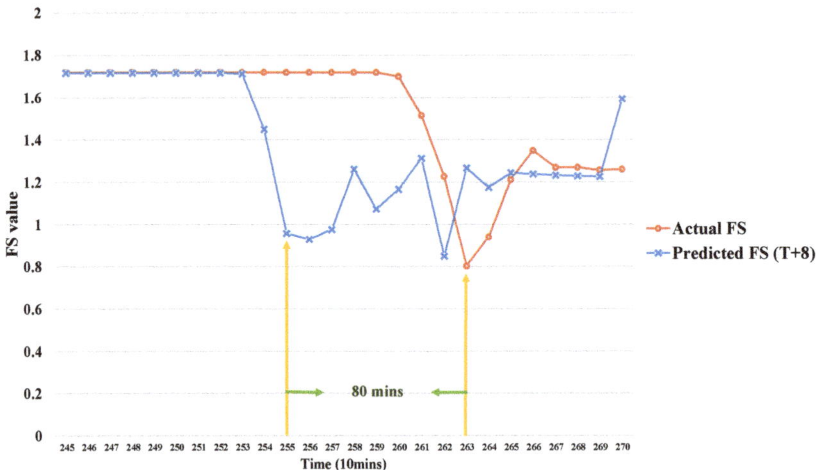

Figure 14. Landslide early warning time point.

6. Conclusions

To address the problems of imbalanced data, low true positive rate for learning, determining the prediction horizon, and the time for model re-training, MoSLaPS has been proposed as a novel method for landslide prediction. MoSLaPS employs the ADASYN method to balance the stable class (no landslide) with the unstable class (landslide), where the classification is based on the factor of safety calculated using the SHALSTAB model. To solve the problem of low true positive rate, a BPNN-based event-class predictor was proposed and two BPNN predictors were designed to learn the stable class pattern and the unstable class pattern. A novel prediction horizon tuning method was proposed, along with a learning-based model re-training technique. All of these optimizations contribute towards the goal of the proposed MoSLaPS; that is, accurate early warning of landslides.

Compared with BPNN and Adaboost, though our event-class predictor has a higher FPR, it also has a much higher TPR of 98.40%. This means our classifier has a higher ratio of correct classification. According to the predicted class, our system can switch between different predictors to adapt to the environmental state. In addition, BPNN is employed to construct the error model to predict the future errors of our proposed MoSLaPS and compensate for these errors in the prediction phase. As a result, MoSLaPS has much lower MAPE and RMSE than BPNN and ADASYN+BPNN, which means that MoSLaPS is more accurate. In addition, the NRMSE of our method is larger than the NRMSE of the other methods, which means that our method is closer to the actual conditions. Statistically, our landslide prediction system could send warnings an average of 44.2 min prior to the actual occurrence of a landslide.

In the future, we will further consider other advanced imbalanced learning algorithms to improve the performance of BPNN-based event-class predictors. For different applications, a larger number of categories (cases) can be considered for model switching. The maximum range of errors allowed in the re-training phase can also be limited, so that prediction models are more stable. As a result, the model switching strategy will be more accurate. Moreover, well-known time-series deep learning technologies, such as Recurrent Neural Networks (RNNs) or Long-Short Term Memory (LSTM) blocks, will be used to combine all three BPNN models into one.

Author Contributions: Conceptualization, P.-A.H.; methodology, D.U.; software, S.-F.C.; validation, S.-F.C. and D.U.; resources, P.-A.H.; data curation, S.-F.C.; writing—original draft preparation, S.-F.C.; writing—review and editing, D.U.; supervision, P.-A.H.; project administration, P.-A.H.; funding acquisition, P.-A.H.

Funding: This research was funded by Ministry of Science and Technology, Taiwan, grant number MOST 140-2221-E-194-064.

Acknowledgments: This research was supported partially by the project grant MOST-107-2221-E-194-001-MY3 from the Ministry of Science and Technology, Taiwan.

Conflicts of Interest: The authors declare no conflict of interest.

Abbreviations

The following abbreviations are used in this manuscript:

ANN	Artificial Neural Network
WSN	Wireless Sensor Network
FS	Factor of Safety
MoSLaPS	Model Switched Landslide Prediction System
LMA	Levenberg–Marquardt algorithm
ADASYN	Adaptive Synthetic Sampling
SMOTE	Synthetic Minority Oversampling Technique
EEW	Error Estimation Window
PH	Prediction Horizon
TPR	True Positive Rate
FPR	False Positive Rate
ACC	Accuracy
TP	True Positive
TN	True Negative
FP	False Positive
FN	False Negative
MAPE	Mean Absolute Percentage Error
RMSE	Root Mean Squared Error
NRMSE	Normalized Root Mean Squared Error
PT	Prediction Time

References

1. He, H.; Bai, Y.; Garcia, E.A.; Li, S. ADASYN: Adaptive synthetic sampling approach for imbalanced learning. In Proceedings of the IEEE International Joint Conference on Neural Networks (IEEE World Congress on Computational Intelligence), Hong Kong, China, 1–8 June 2008; pp. 1322–1328.
2. Jiang, T.; Wand, D. A landslide stability calculation method based on Bayesian network. In Proceedings of the Instrumentation and Measurement, Sensor Network and Automation, Toronto, ON, Canada, 23–24 December 2013; pp. 905–908.
3. Devi, S.R.; Venkatesh, C.; Agarwa, P.; Arulmozhivarman, P. Daily rainfall forecasting using artificial neural networks for early warning of landslides. In Proceedings of the International Conference on Advances in Computing, Communications and Informatics, New Delhi, India, 24–27 September 2014; pp. 2218–2224.
4. Kavzoglu, T.; Sahin, E.K.; Colkesen, I. Selecting optimal conditioning factors in shallow translational landslide susceptibility mapping using genetic algorithm. *Eng. Geol.* **2015**, *192*, 101–112. [CrossRef]
5. Huang, J.C.; Kao, S.J.; Hsu, M.L.; Liu, Y.A. Influence of specific contributing area algorithms on slope failure prediction in landslide modeling. *Nat. Hazards Earth Syst. Sci.* **2007**, *7*, 781–792. [CrossRef]
6. Popescu, M. A suggested method for reporting landslide remedial measures. *Bull. Eng. Geol. Environ.* **2001**, *60*, 69–74. [CrossRef]
7. Lian, C.; Zeng, Z.; Yao, W.; Tang, H. Multiple neural networks switched prediction for landslide displacement. *Eng. Geol.* **2015**, *186*, 91–99. [CrossRef]
8. Lian, C.; Chen, C.L.P.; Zeng, Z.; Yao, W.; Tang, H. Prediction Intervals for Landslide Displacement Based on Switched Neural Networks. *IEEE Trans. Reliab.* **2016**, *65*, 1483–1495. [CrossRef]
9. Huang, J.C.; Kao, S.J. Optimal estimator for assessing landslide model performance. *Hydrol. Earth Syst. Sci.* **2006**, *10*, 957–965. [CrossRef]

10. Hagan, M.T.; Demuth, H.B.; Beale, M.H.; De Jesús, O. *Neural Network Design*; 2014. Available online: http://hagan.okstate.edu/NNDesign.pdf (accessed on 3 May 2019).

11. Levenberg Marquardt Algorithm. 2017. Available online: https://en.wikipedia.org/w/index.php?title=Levenberg%E2%80%93Marquardt_algorithm&oldid=771936514 (accessed on 3 May 2019).

12. Jacobian Matrix and Determinant. 2017. Available online: https://en.wikipedia.org/w/index.php?title=Jacobian_matrix_and_determinant&oldid=776906112 (accessed on 3 May 2019).

13. Palit, A.K.; Popovic, D. *Computational Intelligence in Time Series Forecasting: Theory and Engineering Applications*; Springer Science & Business Media: London, UK, 2006.

14. Ertekin, S.; Huang, J.; Bottou, L.; Giles, L. Learning on the border: active learning in imbalanced data classification. In Proceedings of the 16th ACM Conference on Conference on Information and Knowledge Management, Lisbon, Portugal, 6–10 November 2007; pp. 127–136.

15. Chawla, N.V.; Bowyer, K.W.; Hall, L.O.; Kegelmeyer, W.P. SMOTE: Synthetic minority over-sampling technique. *J. Artif. Intell. Res.* **2002**, *16*, 321–357. [CrossRef]

16. Soil and Water Conservation Bureau Platform for Monitoring Data. 2016. Available online: http://monitor.swcb.gov.tw (accessed on 3 May 2019).

Review

A Critical Review of Spatial Predictive Modeling Process in Environmental Sciences with Reproducible Examples in R

Jin Li

National Earth and Marine Observations Branch, Environmental Geoscience Division, Geoscience Australia, Canberra 2601, Australian Capital Territory, Australia; Jin.Li@ga.gov.au

Received: 22 February 2019; Accepted: 13 May 2019; Published: 17 May 2019

Abstract: Spatial predictive methods are increasingly being used to generate predictions across various disciplines in environmental sciences. Accuracy of the predictions is critical as they form the basis for environmental management and conservation. Therefore, improving the accuracy by selecting an appropriate method and then developing the most accurate predictive model(s) is essential. However, it is challenging to select an appropriate method and find the most accurate predictive model for a given dataset due to many aspects and multiple factors involved in the modeling process. Many previous studies considered only a portion of these aspects and factors, often leading to sub-optimal or even misleading predictive models. This study evaluates a spatial predictive modeling process, and identifies nine major components for spatial predictive modeling. Each of these nine components is then reviewed, and guidelines for selecting and applying relevant components and developing accurate predictive models are provided. Finally, reproducible examples using *spm*, an R package, are provided to demonstrate how to select and develop predictive models using machine learning, geostatistics, and their hybrid methods according to predictive accuracy for spatial predictive modeling; reproducible examples are also provided to generate and visualize spatial predictions in environmental sciences.

Keywords: spatial predictive models; predictive accuracy; model assessment; variable selection; feature selection; model validation; spatial predictions; reproducible research

---

1. Introduction

Spatial predictions of environmental variables are increasingly required in environmental sciences and management. Accurate spatially continuous data are required for environmental modeling, and for evidence-based environmental management and conservation. Such data are, however, usually not readily available and they are difficult and expensive to acquire, especially in areas that are difficult to access (e.g., mountainous or marine regions). In many cases, the spatial data of environmental variables are collected from point locations. Thus, spatial predictive methods are essential for generating spatially continuous predictions of environmental variables from the point samples. Moreover, predictive methods are increasingly being used to generate spatial predictions across various disciplines in environmental sciences [1–6] in parallel to recent advances in (1) computing technology and modeling techniques [7–9], and (2) data acquisition and data processing using remote-sensing techniques and geographic information systems. These advancements resulted in increasingly more environmental variables available for spatial predictive modeling. Consequently, more sophisticated spatial predictive modeling approaches are needed to deal with a large number of predictive variables.

Accuracy of spatial predictive model(s) is critical as it determines the quality of their predictions that form the scientific evidence to inform decision- and policy-making. Therefore, improving the accuracy by choosing an appropriate method and then identifying and developing the most accurate predictive model(s) is essential. It is often difficult to select an appropriate method for any given dataset because spatial predictive methods may be data- or even variable-specific and many factors need to be considered [10,11]. Although the development of hybrid methods of machine learning and geostatistics, and their application considerably improved predictive accuracy, these methods may also be data- or variable-specific [12–14]. For spatial predictive modeling, "no free lunch theorems" [15] are also applicable.

Furthermore, even with the right predictive method, it is a challenging task to identify and develop the most accurate predictive model(s). This is because the spatial predictive modeling process involves many factors or components [10,16,17]. In fact, only a portion of such factors were considered in many previous studies, which often led to sub-optimal or even misleading predictive models [11,18]. This not only presents an opportunity for scientists to develop and improve their predictive models, but also highlights the challenge of selecting relevant predictive variables from a large number of available predictive variables to form the most accurate predictive model. Heavy computations are often involved in identifying and selecting accuracy-improved predictive models for given datasets when the number of predictive variables is large, although high-performance computing facilities may be able to significantly alleviate this challenge.

This study aims to assist spatial modelers and scientists by critically reviewing the spatial predictive modeling process, developing guidelines for selecting the most appropriate spatial predictive methods and identifying and developing the most accurate predictive model to generate spatial predictions. In this study, I focus on spatial predictive models or spatial predictive modeling for generating spatially continuous predictions rather than on other models (e.g., inferential model) as discussed previously [19]. Consequently, the term accuracy in this study refers to the accuracy of predictive model(s) based on validation, and the term uncertainty refers to prediction uncertainty generated by predictive model(s). In this study, the term accuracy is used interchangeably with predictive accuracy. Furthermore, in this study, I mainly focus on the predictive methods for numerical data that are usually encountered in environmental sciences, with a brief discussion on categorical data. In this study, the following nine major components of spatial predictive modeling are identified and reviewed: (1) sampling design, sample quality control, and spatial reference systems; (2) selection of spatial predictive methods; (3) pre-selection of predictive variables; (4) exploratory analysis for variable selection; (5) parameter selection for relevant methods; (6) variable selection; (7) accuracy and error measures for predictive models (numerical vs. categorical); (8) model validation; and (9) spatial predictions, prediction uncertainty, and their visualization. In addition, reproducible examples using *spm* [20], an R package for machine learning, geostatistics, and their hybrid methods, are employed to demonstrate how to select and develop predictive models based on predictive accuracy for spatial predictive modeling; reproducible examples are also provided to generate and visualize spatial predictions in environmental sciences.

2. Sampling Design, Sample Quality Control, and Spatial Reference Systems

2.1. Sampling Design

Although samples are usually collected, stored, and ready to use for spatial predictive modeling, sometimes samples are not available and need to be collected. In the latter situation, a sampling design needs to be produced. In this study, I focus on sampling designs over space. To collect samples from a survey area for a certain survey purpose, a sampling design is an important step and must be created. A good sampling design ensures that data collected from a survey are capable of answering relevant research questions. Better designs, such as spatially stratified sampling designs, will also be as precise and efficient as possible [21,22]. Many methods were developed to generate sampling designs [23–26]. They typically fall into four main categories: (1) non-random sampling design; (2) unstratified random sampling design; (3) stratified random sampling design; and (4) stratified random sampling design with prior information.

The non-random sampling design can be ad hoc sampling based on expert knowledge, purely opportunistic when a certain type of environmental condition becomes available, or systematic sampling. This type of sampling design was applied to many surveys [27–29]. For spatial predictive modeling, this method is not recommended for future studies. However, an interesting comparison of non-random sampling designs was reported for spatial predictive modeling [26]; it may provide some useful clues for sampling designs (e.g., lattice plus close pairs) for spatial predictive modeling.

The unstratified random sampling design is that sampling locations are randomly selected. This can be (1) an unstratified equal probability design, or (2) an unstratified unequal probability design [30,31]. This type of sampling design is not recommended for spatial predictive modeling studies because (1) spatial information is available for sampling design and thus spatially stratified sampling design should be used as discussed below, and (2) it may even be over-performed by the non-random design (i.e., lattice design) [26].

The stratified random sampling design is often used when additional information is available. Such information can be spatial (or location) information, elevation, bathymetry, or geomorphological information. For spatial predictions, such information is important and should be considered when designing a survey for a region. A few recently developed randomized spatial sampling procedures were reviewed and compared using simple random sampling without replacement as a benchmark for comparison [22]. This study provided some empirical evidence for the improvement of sampling efficiency from using these designs and provided some guidance for choosing an appropriate spatial sampling design [22]. Furthermore, some R packages, such as *spsurvey* [31], *GSIF* [32], *spcosa* [33], *clhs* [34], and *BalancedSampling* [35], were developed for this type of sampling design. The stratified random sampling designs with spatial information (i.e., spatially stratified sampling design) are increasingly being used in practice [28,36,37].

The stratified random sampling design with prior information is a new development for sampling over a space. It incorporates the locations of legacy sites into new spatially balanced survey designs to ensure spatial balance among all sample locations [21]. It can be seen as a stratified unequal probability design. An R package, *MBHdesign*, was developed for this method [38].

2.2. Sample Quality Control

Sample quality is vitally important because samples provide the fundamental information for spatial predictive modeling. Many factors may affect sample quality and they are usually dataset-specific [39,40]. Consequently, relevant factors need to be identified for each dataset and then relevant data quality control (QC) criteria need to be developed to QC the dataset [39,40] prior to undertaking spatial predictive modeling. For example, in Geoscience Australia's Marine Samples Database (MARS; http://dbforms.ga.gov.au/pls/www/npm.mars.search), seabed sediment samples were initially quality controlled prior to and after entering the database according to various criteria [12,41]. However, the quality of the samples was still affected by many factors, including data credibility (e.g., non-dredge), data accuracy (e.g., non-positive bathymetry), completeness (e.g., no missing values), etc.; hence, data quality control approaches were developed to QC the samples of seabed mud content and sand content [12,41]. These approaches may provide examples about how to identify relevant factors and develop possible data QC criteria for a given dataset. In some instances, data noise may result from repeated measurements, and certain rules may need to be developed to clean such samples based on professional knowledge [42]. Moreover, exploratory analysis can be used to further detect abnormal samples, as detailed in Section 5.

2.3. Spatial Reference Systems

To generate spatial predictions (i.e., spatially continuous data) for a region using spatial predictive models, two types of georeferenced data are required: (1) point samples of response and predictive variables; and (2) grid data of predictive variables. Such georeferenced data are often stored according to various spatial reference systems [43]. The spatial reference system used to project or store the spatial information is often assumed to have certain effects on the performance of predictive models; thus, in practice, various spatial reference systems were used to minimize such effects [43]. When a study area is small and within one particular UTM zone, spatial data are often projected using either the UTM zone or an appropriate projection system; when the study area is spanning multiple UTM zones, the existing geographic coordinate system (i.e., WGS84) or another appropriate projection system can be used.

Although the spatial reference system by which the spatial information is stored is often considered as a potential source of error for spatial predictive modeling, a series of studies demonstrated that the effects of spatial reference systems on the performance of spatial predictive methods (i.e., inverse distance weighting and ordinary kriging) are negligible for areas at various latitudinal locations (up to 70 dd) and spatial scales (i.e., regional and continental) [43–45]. Therefore, it was recommended that spatial reference system selection and re-projection can be removed for spatial predictive modeling for areas with latitude less than 70 dd, and spatial data can be modeled in WGS84 or the spatial projection system already used for the data. Although new spatial reference systems (e.g., DGGS [46]) may be developed to remedy various limitations of existing ones, the above recommendation may still be applicable as discussed previously on why the effects of spatial reference systems on the predictive accuracy are negligible [43–45].

3. Selection of Spatial Predictive Methods

3.1. Spatial Predictive Methods

For spatial predictive modeling, there are many methods available [3]. Previously, over 20 spatial predictive methods were grouped into (1) non-geostatistical methods (e.g., inverse distance weighting (IDW)), (2) geostatistical methods (e.g., ordinary kriging (OK)), and (3) combined or hybrid methods [10]. Collectively, these methods are largely non-machine learning methods and a small portion of these methods, like regression tree (RT) and linear regression models (LM), use secondary information.

When sufficient secondary information is available, a number of other methods could be used. These methods include traditional statistical methods, machine learning methods, the hybrid methods of traditional statistical methods and geostatistical methods, and the hybrid methods of machine learning and geostatistical methods (Table 1). These methods were applied or compared in various spatial predictive modeling studies [12,41,47–55]. Of these methods, random forest (RF), hybrid method of RF and OK (RFOK), and hybrid method of RF and IDW (RFIDW) were among the most accurate methods in these applications. Generalized boosted regression modeling (GBM), hybrid method of GBM and OK (GBMOK), and hybrid method of GBM and IDW (GBMIDW) showed great potential based on our unpublished study. In the current study, these methods are presented in three main groups: (1) non-machine learning methods; (2) machine learning methods; and (3) the hybrid methods.

Table 1. Spatial predictive methods using predictive variables [6,12,41,48–55].

Non-Machine Learning Method and Hybrid Methods		Machine Learning Method and Hybrid Methods	
Non-machine learning method	Hybrid methods	Machine learning method	Hybrid methods
Generalized additive models		Cubist	Cubist and OK (cubistOK)
Generalized least squares trend estimation (GLS)	GLS and OK	Generalized boosted regression modeling (GBM)	GBM and IDS (GBMIDS)
Generalized linear models (GLM)	GLM and IDW (GLMIDW)		GBM and OK (GBMOK)
	GLM and OK (GLMOK)	General regression neural network (GRNN)	GRNN and IDS (GRNNIDS)
GLM with lasso or elastic net regularization			GRNN and OK (GRNNOK)
	Linear models and OK	Multivariate adaptive regression splines	
	RT and IDS (RTIDS)	Naïve Bayes	
	RT and OK (RTOK)	Random forest (RF)	RF and IDS (RKIDS)
			RF and OK (RKOK)
		Support vector machine (SVM)	SVM and OK (SVMOK)
			SVM and OK (SVMIDS)

3.2. Selecting Spatial Predictive Methods

Selection of appropriate spatial predictive methods for a response variable (or dependent variable, or primary variable in geostatistics) is critical. For data without predictive variables, geostatistical methods are the only methods that can be used. Method selection was discussed and guidelines were developed for using geostatistical methods in various studies [16,56–61]. A decision tree was developed for selecting the most appropriate method from a pool of 25 spatial predictive methods according to the availability and nature of data and the expected predictions, together with the features of each method [10]. However, it was argued that there was no simple answer regarding the choice of appropriate geostatistical methods, because the hallmark of a good geostatistical modeling work is customization of the approach to the dataset at hand [56]. This suggests that "no free lunch theorems" [15] are also applicable for spatial predictive modeling using geostatistical methods. Joint application of two spatial predictive methods might produce additional benefits such as the combined procedures in previous studies [10,62–64].

For data with predictive variables, there are many options available. It is often difficult to select an appropriate method because the performance of spatial predictive methods depends on many factors, including the assumptions and properties of each method, the nature and spatial structure of data for the response variable, sample size and distribution, the availability of predictive variables, availability of software, computational demands, and many other factors [10,11]. All of these factors need to be considered when making an appropriate selection.

Moreover, if more than one method can be applied, model comparison techniques such as cross-validation in combination with the measures of predictive error or accuracy can be used to select a method. This selection technique not only selects the most appropriate method but also the most accurate predictive model that can maximize the predictive accuracy [12,65,66]. This selection technique can be applied to methods irrespective of whether they use predictive variables.

4. Pre-Selection of Predictive Variables

Predictive variables are termed predictor variables, independent variables, predictors, and features. They are also called secondary variables/information in geostatistics. They are essential for spatial predictive methods that use predictive variables.

4.1. Principles for Pre-Selection of Predictive Variables and Limitations

Principles for pre-selecting predictive variables may change with disciplines. For environmental sciences, the main principle is that predictive variables need to be closely related to the variable to be predicted (i.e., the response variable) [67,68]; ideally, they should be causal variables, or variables directly caused by the response variable (e.g., optical reflectance of vegetation types, backscatter of seabed substrates). They are usually identified based on expert or professional knowledge. However, in many cases, it is hard to know what the causal variable(s) is (are) for a response variable. Proxy (or surrogate) variables are often used instead of causal variable for spatial predictive modeling. Again, they are usually identified based on expert or professional knowledge [69]. Certainly, predictive models can use causal variables, proxy variables, or both if causal variables are not all available.

When the accuracy of a resultant predictive model is unexpectedly poor, then we may need to consider that we may have missed some important predictive variables, for which we may have no knowledge or even awareness (e.g., hidden variables [70]). Further actions are required to expand the professional knowledge pool in order to identify such possible predictive variables.

For spatial predictive modeling, the selection of potential predictive variables is even more challenging. This is because the selection could be constrained by certain factors. For example, predictive variables need to be continuously available for a target region. Spatial resolution is also a critical issue as the resolutions of various predictive variables need to meet the desired resolution for the final predictions, although they can be rescaled. Sometimes, even though we know the possible causal predictive variables based on expert or professional knowledge, they may not meet these requirements and cannot be used for spatial predictive modeling. This is particularly true in marine environmental sciences.

4.2. Predictive Variables for Environmental Sciences

For terrestrial environmental modeling, many predictive variables are available. Many previous applications provided examples of variables used for terrestrial environmental modeling [13,42,49–51, 53,67,71,72].

In contrast, the information of predictive variables is often scarce for marine environmental modeling. In many cases, proxy variables are usually used for predictive modeling [69,73]. For example, to predict the spatial distribution of seabed sediments for Australian Exclusive Economic Zone (AEEZ) at a resolution of 0.01 or 0.0025 degrees, only a few predictive variables were available for the whole AEEZ [12,41,74] (Table 2). For spatial predictions over smaller areas, quite often more predictive variables became available at desired resolution such as for seabed sediment [4,75–77], seabed hardness [78–80], and sponge species richness [6,81] (Table 2). Bathymetry and backscatter were also used to predict seabed sediment at local scale [82]. Some derived information may be used as predictive variables. For example, in Table 2, predictive variables 5–13 were derived from bathymetry (bathy), while predictive variables 15–19 were derived from backscatter (bs). Some other variables were used for seabed grain size at small scale [48]. In addition, many variables were reviewed [69,83] and could be used for marine environmental modeling. Fuzzy geomorphic features were also used for spatial predictive modeling at local scales [77,84].

Table 2. A list of predictive variables used for some marine environmental variables.

No	Predictive Variables	Seabed Sediment/Grain Size	Seabed Hardness	Sponge Species Richness	Window/Kernel Size(s)
1	Longitude (long)	yes	yes	yes	
2	Latitude (lat)	yes	yes	yes	
3	Distance to coast (dist)	yes		yes	
4	Bathymetry (bathy)	yes	yes	yes	
5	Local Moran's I from bathymetry	yes	yes	yes	yes
6	Mean curvature	yes			yes
7	Planar curvature	yes	yes	yes	yes
8	Profile curvature	yes	yes	yes	yes
9	Relief	yes	yes	yes	yes
10	Rugosity (or surface, surface complexity)	yes	yes	yes	yes
11	Slope	yes	yes	yes	yes
12	Topographic or bathymetric position index (tpi or bpi)	yes	yes	yes	yes
13	Fuzzy morphometric features	yes			yes
14	Backscatter (bs) 10–36	yes	yes	yes	
15	Entropy from bs			yes	yes
16	Homogeneity from bs		yes	yes	yes
17	Local Moran's I from bs		yes	yes	yes
18	Prock from bs		yes		
19	Variance from bs		yes	yes	yes
20	Suspended particulate matter	yes			
21	Mean tidal current velocity	yes			
22	Peak orbital velocity of waves at seabed	yes			
23	Roughness from bathy *	yes			
24	Roughness from bs *	yes			
25	Sobel filter from bathy #	yes			

* The difference between the minimum and maximum of cell and its eight neighbors [48]. # A directional filter that emphasizes areas of large spatial frequency (edges) running horizontally (X) or vertically (Y) across the image [48].

5. Exploratory Analysis for Variable Pre-Selection

5.1. Non-Machine Learning Methods

Exploratory analysis is often used to detect the relationships between the response variable and predictive variables for non-machine learning methods, such as LM, generalized linear models (GLM), and kriging with an external drift (KED). By applying such analysis, people intend to find data nature and structure [85]. Key issues may include the identification of (1) outliers, (2) homogeneity in variance of response variable, (3) data distribution of response variables including normality, (4) collinearity (i.e., the correlation among predictive variables), (5) relationships or response curves of response variable to predictive variables, (6) how strong the relationships are between response variable and predictive variables, (7) possible interactions among predictive variables, (8) independence of a response variable, whether temporally, spatially, or both, and (9) source of random errors, which may lead to a mixed-effect model or additional predictive variable(s) [71]. On the basis of the above analyses, certain actions can be taken to deal with relevant samples or variables. For instance, some predictive variables can be removed if their correlations with the response variable are low. However, it would be wise to let the variable selection process determine which variables should be removed because some variables may be important predictive variables even with low correlations. Highly correlated predictive variables, usually determined based on correlation coefficient (r) or variance inflation factor (VIF), can also be eliminated to reduce the collinearity, although caution should be taken for this exercise [9,86]. Relevant issues about collinearity were also discussed [87]. Some predictive variables may need to be specified to their second or third orders if a non-linear relationship is detected. Moreover, some interactions may need to be considered and tested.

5.2. Machine Learning Methods and Hybrid Methods

For machine learning methods, exploratory analysis is useful for understanding data and interpreting modeling results [78]. However, some roles of exploratory analysis for non-machine learning methods are no longer needed for machine learning methods. This is because machine learning methods, like RF, are free of assumptions on the data distribution and can handle non-linear relationships and interactive effects [88,89]. They can also handle highly correlated predictive variables [6,47]. Furthermore, the use of highly correlated predictive variables is encouraged for RF because they may be able to make a meaningful contribution to improving predictive accuracy [6].

5.3. Hybrid Methods

For the hybrid methods, exploratory analysis is as useful as for the aforementioned methods. The residuals of a detrending method (e.g., GLM, RF) are assumed to be normal if kriging methods are applied. Thus, the residuals need to be analyzed to check this assumption [12,41].

6. Parameter Selection

6.1. Parameter Selection for Non-Machine Learning Methods

For non-machine learning methods, I mainly focus on two commonly used methods [11], IDW and kriging (e.g., OK). For IDW, it is really dependent on the selection of appropriate values for a power parameter and the number of nearest observations, which can be selected based on their resultant predictive accuracy [41,90]. The smoothness of the estimated surface increases as the value of power parameter increases [91]; however, manipulating the power parameter to smooth the predictions and to produce visually pleasant maps does not warrant the quality of the resultant predictions and is not recommended.

For kriging methods, a number of parameters, including window size, and isotropy and anisotropy of data, need to be considered, as well as the variogram model and its parameters. Data transformation needs to be considered when the data are skewed and anisotropic. Three methods of data transformation (i.e., logarithms, standardized rank order, and normal scores) can be employed to reduce the skewness [74, 92]. Some other methods, such as Box–Cox transformation [82], arcsine [88], square-root transformation [47], and double square root or square root and log [12], can be used to normalize the data. The selection of these transformation and normalization methods is largely data-dependent and careful examination should be taken. The selection can also be determined according to their effects on the predictive accuracy.

In addition, for anisotropy, non-stationary methods like KED should be used in cases with a general anisotropy or trend (i.e., drift) [75]. If different types of non-stationarity exist, application of different spatial predictive methods to each type may improve predictive accuracy [93].

The parameter selection for these methods can be determined according to the predictive accuracy of resultant predictive models. This is demonstrated in Section 11.

6.2. Parameter Selection for Machine Learning Methods

For machine learning methods, RF and GBM are considered. For RF, relevant parameters are mtry, ntree, and so on [94], while, for GBM, these include n.trees, learning.rate, interaction.depth, bag.fraction, and so on [95]. Some commonly used default parameter values can be used as they are quite often optimal [94,96], except the distribution parameter for GBM. The distribution parameter for GBM should be based on data type of the response variable; for example, Poisson should be used for count data. Relevant parameters can also be selected based on cross-validation [41,48,96].

6.3. Parameter Selection for Hybrid Methods

The parameter selections for the above non-machine learning methods and machine learning methods are equally applicable to relevant hybrid methods.

7. Variable Selection

For machine learning methods, variable selection is termed feature selection, while, for non-machine learning methods, it is often called model selection. However, model selection often leads to the most parsimonious fitted model rather than the most accurate predictive model [6]. In this study, we use the term "variable selection" for both non-machine learning and machine learning methods to identify and develop the most accurate spatial predictive model(s).

Variable selection is important for many predictive methods, although it is not required for all methods. For instance, classification and regression trees [97] and LIVES [98] are exempt from variable selection. However, as per all other methods, they assume that the predictive variables used are informative and not misleading because they treat each predictive variable as equally important. Thus, misleading predictive variables may considerably reduce predictive accuracy as discussed previously [98]. The variable selection procedure for machine learning methods and their hybrid methods is fundamentally different from the procedure for non-machine learning methods [47,79,99]. For geostatistical methods like IDW and OK, no variable selection is required, and it is really about the selection of appropriate values for relevant parameters, as discussed in Section 6. In this section, I focus on the following three methods: (1) GLM; (2) RF; and (3) GBM. This is because of their wide applications, robustness, or the recent developments in variable selection techniques for these methods.

For GLM, there are many methods available in R for variable selection [100–103]. These methods may include (1) *stepAIC* or *step*; (2) *dropterm*, *drop1*, or *add1*; (3) *anova*; (4) *regsubsets*; and (5) *bestglm* [100,102,103]. The application of these methods for spatial predictive modeling can be seen in recent studies [6,104]. Variables selected based on these methods may form the most parsimonious model, but the model may have low predictive accuracy or even be misleading [6,104], with the exception that *bestglm* is promising if cross-validation, instead of Akaike's information criterion (AIC) or Bayesian information criterion (BIC), is used for information criteria [104]. Alternatively, the variable selection for GLM can also be based on variables selected by other method such as RF [71]. It was found that traditional variable selection methods are unsuitable for identifying GLM predictive models, and joint application of RF and AIC can select accuracy-improved predictive models [6]. This highlights the importance of differentiating variable selection for predictive modeling [6] from variable selection for hypothesis testing [99] or inferential modeling [19]. The common mistakes associated with incorrectly distinguishing data analytic types were briefly summarized and discussed previously [19].

For RF, variable selection methods may include (1) variable importance (VI) [78], (2) averaged variable importance (AVI) [79], (3) Boruta [105], (4) knowledge informed AVI (KIAVI) [6,79], (5) recursive feature selection (RFE) [106], and (6) variable selection using RF (VSURF) [107]. Of these methods, KIAVI is recommended because it outperforms all other variable selection methods [6,79,104,108].

For GBM, variables can be selected in terms of the relative influence [95,108]. The recursive feature selection [106] can also be used for variable selection.

Two concepts were proposed for variable selection: important and unimportant predictive variables based on the predictive accuracy [6,79]. They were defined as follows:

1. Important variable based on the predictive accuracy (IVPA). This refers to the variable for which exclusion during the variable selection process would reduce the accuracy of a predictive model based on cross-validation. It may be more appropriate to call it predictive accuracy boosting variable (PABV).
2. Unimportant variable based on the predictive accuracy (UVPA). This refers to variables for which exclusion during the variable selection process would increase the accuracy of a predictive model based on cross-validation [6,79]. It may be more precise to call it predictive accuracy reducing variable (PARV).

Application of relevant variable selection methods and concepts can further improve the accuracy of predictive models [6,79]; this is demonstrated in Section 11. Although these concepts were developed based on RF and its hybridization with geostatistical methods, they can be equally applied to any other predictive methods.

8. Accuracy and Error Measures for Predictive Models

8.1. Relationship between Observed, Predicted, and True Values

Predictive accuracy is about the differences between observed and predicted values that are derived based on validation methods [18]. However, it is often questioned what the differences between the predicted values and true values are. Since the true values are mostly unknown, the observed values are used to validate predictive models. For an observed value, it may be again different from its corresponding true value. The difference between the true value and observed value is the error associated with the observed value. Let us refer to this error as an observational error that is the sum of random error associated with observed variable, sampling error, and measuring error. The sampling and measuring errors are the sum of errors resulting from various factors that may affect the accuracy of observation (i.e., measurement) and change with the variable observed. Let us take seabed sediment as an example; the factors may include sampling design, the position accuracy of survey vessel, equipment used for sample collection, field operation, sample storage, sample processing procedure and analysis in laboratory, data entry, etc. However, how much error can be attributed to each of these factors is unknown in most cases. Hence, we have to use observed and predicted values to assess the predictive accuracy in practice.

8.2. Error and Accuracy Measures of Predictive Models

Many error and accuracy measures were developed to assess the accuracy of predictive models for numerical data [65,109,110]. Some of these error and accuracy measures were assessed and their limitations were previously discussed [3,18]. Of these error and accuracy measures, VEcv (i.e., variance explained by a predictive model based on cross-validation) measures how accurate the predictive model is and was proven to be independent of unit, scale, data mean, and variance [18,111], while root mean squared error (RMSE) measures how wrong the predictive model and the resultant predictions can be. Therefore, VEcv and RMSE are recommended for numerical predictions. Legates and McCabe's E1 (E1) was recommended for numerical data as well [111]. The commonly used measure, r or r^2, is not recommended because it is an incorrect measure of predictive accuracy [111].

For categorical data, correct classification rate (CCR), kappa (kappa), sensitivity (sens), specificity (spec), and true skill statistic (TSS) are often recommended [112,113]. RMSE is also used for presence and absence data [114]. One commonly used measure, area under the curve (AUC) (or receiver operating characteristics (ROC)), is not recommended for reasons previously highlighted [112,115].

9. Model Validation

9.1. Model Validation Methods

The accuracy of predictive models is critical as it determines the quality of the resultant predictions. The accuracy is often assessed based on model validation methods that may include the following:

1. Hold-out validation;
2. K-fold cross-validation;
3. Leave-one-out cross-validation;
4. Leave-q-out cross-validation;
5. Bootstrapping cross-validation;
6. Using any new samples that are not used for model training.

In environmental sciences, the most commonly used validation methods are hold-out and leave-one-out [18]. However, of these validation methods, five- or 10-fold cross-validation was recommended [7,116].

9.2. Randomness Associated with Cross-Validation Methods

Although a five- or 10-fold cross-validation was recommended to evaluate the performance of spatial predictive models [116], the datasets are randomly generated for each fold of the cross-validation change when the process is repeated. Thus, the randomness associated with the cross-validation would produce predictive accuracy or error measures that change with each iteration of the cross-validation [47]. To reduce the influence of the randomness on predictive accuracy (i.e., to stabilize the resultant performance measures), the cross-validation needs to be repeated (e.g., 100 times) [47,74,78]. The choice of this iteration number is data-dependent and can be determined based on the method used in previous studies [47,78].

10. Spatial Predictions, Prediction Uncertainty, and Their Visualization

10.1. Spatial Predictions

In spatial predictive modeling, the goal is not only to develop the most accurate predictive model, but more importantly to generate spatial predictions. The spatial predictions are usually produced using the most accurate predictive model developed according to the above procedures. To make predictions, in addition to the spatial predictive model, we need relevant information of each model predictive variable to be available at each grid cell at a desired resolution. When all this information is prepared, the spatial predictions can then be generated. The predictions contain three columns, i.e., longitude, latitude, and predictions. Sometimes, uncertainty of the predictions can be produced.

10.2. Prediction Uncertainty

Prediction uncertainty in environmental modeling may refer to various aspects of the modeling process and is used to encompass many concepts [117–120]. It can result from various sources or factors as previously discussed [17,120,121]. In this study, the uncertainty which is produced by a predictive model is about spatial predictions. Prediction uncertainty is increasingly required for decision-making and many methods are used to produce such uncertainty. In this study, I focus on the uncertainty produced by some commonly used methods: OK, LM, and RF.

For OK, prediction variances can be produced [58]. However, it was shown that the variances produced are independent of the actual predicted values [122]. Thus, the resulting variances should not be used to measure uncertainty, although many studies used them for such purpose. Since they reflect the variations in spatial departures among samples, they can be used as good indicators where samples are sparse and, thus, may provide useful information for selecting future sampling locations.

For LM, prediction uncertainty (i.e., prediction intervals) produced are much wider than confidence intervals of a model fitted [101]. Such uncertainty, however, has little to do with the predictive accuracy of the model. For instance, it was found that the models developed according to goodness of fit could be misleading when they were used as predictive models [6]; hence, its prediction uncertainty could also be misleading, and further studies are recommended.

For RF, many types of uncertainty could be produced, which actually reflect the difference in sampling strategies [123–126]. For example, prediction uncertainty produced for RF in a previous study based on Monte Carlo resampling [127] was, in fact, measuring the variation in predictions among individual trees rather than by RF. A further example for RF is that an ensemble of equally probable realizations was generated and the differences amongst the realizations were used as a measure of uncertainty [128]. This type of uncertainty only measures the differences among the results of various runs of RF, that is, measuring the difference resulted from the randomness associated with each run. Hence, these values do not relate to predictive accuracy and do not measure prediction uncertainty.

In addition, for any of the spatial predictive methods above, predictive errors based on validation can be produced for a predictive model developed. However, this leads to only one error value, and all predictions would have the same uncertainty value if it is used as an uncertainty measurement [129].

It is apparent that the uncertainty values produced above are either not measuring prediction uncertainty, or they depend on various factors as discussed above. This consequently results in the need to question the uncertainty of uncertainty. In short, how to assess prediction uncertainty needs further study. Any uncertainty measures that can incorporate the information of predictive accuracy are worth further investigation and recommended for future studies.

10.3. Visualization

Spatial predictions can be visualized using various tools, most commonly ArcGIS and QGIS. The function, *spplot*, in R is often used to plot the distribution of spatially continuous predictions [59]. The R package, *raster*, can also be used for such purpose [130]. Joint application of R and Google Earth can be used to visualize the predictions. In this study, I demonstrate how to use the latter approach along with *spplot* to visualize the predictions as below.

11. Reproducible Examples for Spatial Predictive Modeling

In this section, reproducible examples using *spm*, an R package, are provided to demonstrate how to select and develop a predictive model according to the guidelines and recommendations provided in the previous sections for spatial predictive modeling in environmental sciences. The predictive model to be used was developed using RFOK [74], where data preparation, including pre-selection of predictive variables, relevant parameter selection, variable selection, and model validation, was detailed. Seabed gravel content samples in the Petrel sub-basin, northern Australia marine margin are used to demonstrate how to select relevant parameters, test the predictive accuracy, and generate and visualize spatial predictions. These examples for RFOK can be easily extended to other predictive methods including IDW, OK, RF, GBM, RFIDW, GBMOK, GBMIDW, RFOKRFIDW, and GBMOKGBMIDW by replacing *rfokcv* and its associated parameters with relevant functions and parameters for these methods in *spm*.

11.1. Accuracy of a Predictive Model for Seabed Gravel Content

In a previous predictive model [74], a spherical variogram model and a searching window size of 12 were used. The accuracy of this predictive model [74] can be shown using the *rfokcv* function in *spm* as shown below. To stabilize the accuracy derived, I repeat the cross-validation 100 times, which can be determined using the methods discussed in Section 9.2.

```
> library(spm)
> data(petrel)
> names(petrel)
[1] "long" "lat" "mud" "sand" "gravel" "bathy" "dist" "relief" "slope"
> set.seed(1234)
> n <- 100
> rfokvecv1 <- NULL
> for (i in 1:n) {
+   rfokcv1 <- rfokcv(petrel[, c(1,2)], petrel[, c(1,2, 6:9)], petrel[, 5], predacc = "VEcv")
+   rfokvecv1 [i] <- rfokcv1
+ }
> mean(rfokvecv1)
[1] 37.44799
```

It suggests that the predictive accuracy is 37.4% in terms of VEcv.

11.2. Parameter Selection

The *rfokcv* function in *spm* is used to demonstrate how to select the best parameters for a predictive model (by using above predictive model as an example), and to check if the parameters used are optimal.

```
> library(spm)
> data(petrel)
> nmax <- c(5:12); vgm.args <- c("Sph", "Mat", "Ste", "Log")
> rfokopt3 <- array(0, dim = c(length(nmax), length(vgm.args)))
> set.seed(1234)
> for (i in 1:length(nmax)) {
+   for (j in 1:length(vgm.args)) {
+   rfokcv1.1 <- NULL
+     for (k in 1:100) {
+     rfokcv1.1[k] <- rfokcv(petrel[, c(1, 2)], petrel[, c(1, 2, 6:9)], petrel[, 5], nmax = nmax[i],
+     vgm.args = vgm.args[j], predacc = "VEcv") }
+     rfokopt3[i, j] <- mean(rfokcv1.1) } }
> which (rfokopt3 == max(rfokopt3, na.rm = T), arr.ind = T)
   [1,]  6 4
> vgm.args[4]; nmax[6]
   [1] "Log"
   [1] 10
```

The results suggest that the model would achieve the best predictive accuracy if "Log" is used for variogram modeling, and the 10 nearest samples are used for nmax. Of course, a different range may be used to choose the best nmax, and other variogram models can also be tested if needed.

We can use *rfokcv* in *spm* to assess the accuracy of RFOK by using the parameters identified above.

```
> library(spm)
> data(petrel)
> set.seed(1234)
> n <- 100
> rfokvecv1 <- NULL
> for (i in 1:n) {
+  rfokcv1 <- rfokcv(petrel[, c(1, 2)], petrel[, c(1, 2, 6:9)], petrel[, 5], vgm.args = "Log",
+  nmax = 10,
+  predacc = "VEcv")
+  rfokvecv1 [i] <- rfokcv1
+ }
> mean(rfokvecv1)
[1] 38.30175
```

This finding suggests that the overall averaged accuracy of the RFOK predictive model for seabed gravel content in terms of VEcv is 38.3%, higher than that of the previous model. This demonstrates that the parameters used previously are not optimal and that parameter selection improved predictive accuracy.

11.3. Predictive Variable Selection

In this study, we use the predictive variables previously identified [74], where the predictive variables were selected based on VI. Since then, more advanced variable selection methods for RF, RFOK, and RFIDW, such as AVI, KIAVI, PABV, and PARV [6,79], were developed. Application of these model selection methods and concepts may further improve the predictive accuracy of the model above. It is apparent that latitude (lat) is a PARV, as shown in the previous study [74]; thus, the removal of lat is expected to improve the predictive accuracy. This can be demonstrated below.

```
> library(spm)
> set.seed(1234)
> rfokvecv1.1 <- NULL
> for (i in 1:n) {
>  rfokcv1 <- rfokcv(petrel[, c(1, 2)], petrel[, c(1, 6:9)], petrel[, 5], vgm.args = "Log",
+  nmax = 10,
+  predacc = "Vecv")
+  rfokvecv1.1 [i] <- rfokcv1
+ }
> mean(rfokvecv1.1, na.rm=T)
[1] 39.00298
```

A further improvement in predictive accuracy is achieved after applying PARV. This further demonstrates the role of variable selection, especially the importance of newly developed variable selection methods.

11.4. Generation of Spatial Predictions

The predictive model developed above can be used to generate spatial predictions. The function *rfokpred* in *spm* is used to produce the predictions.

```
> set.seed(1234)
> library(spm)
> data(petrel); data(petrel.grid)
```

```
> rfokpred1 <- rfokpred(petrel[, c(1, 2)], petrel[, c(1, 6:9)], petrel[, 5], petrel.grid[, c(1,
2)],   +  petrel.grid, ntree = 500, nmax = 10, vgm.args = ("Log"))
> names(rfokpred1)
[1] "LON"  "LAT"  "Predictions"  "Variances"
```

The output dataset has four columns named longitude, latitude, predictions, and variances. Please note that the uncertainty information (i.e., variances) is produced for readers interested; however, be aware of the various limitations as discussed in Section 10 when using such information.

11.5. Visualisation of Spatial Predictions

Joint application of R and Google Earth can be used to visualize the predictions generated above.

```
> library(sp); library(plotKML)
> rfok1 <- rfokpred1
> gridded(rfok1) <- ~ longitude + latitude
> proj4string(rfok1) <- CRS("+proj=longlat +datum=WGS84")
> plotKML(rfok1, colour_scale = SAGA_pal[[1]], grid2poly = TRUE)
```

The resultant map is shown in Figure 1a. One of the advantages of using R and Google Earth is that it can place the prediction map into the context map by Google Earth, which provides additional information to final users. However, the labels of longitude and latitude are hard to place in Figure 1a. If these labels are required, *spplot* can be applied to the above gridded data as shown below (Figure 1b).

```
> par(font.axis=2, font.lab=2)
> spplot(s1, c("Predictions"), key.space=list(x=0.1,y=.95, corner=c(-1.2,2.8)),
+  col.regions = SAGA_pal[[1]], # this requires plotKML
+  scales=list(draw=T), colorkey = list(at = c(seq(0,80,5)), space="right",
+  labels = c("0%"," "," "," ","20%"," "," "," ","40%"," "," "," ","60%"," "," "," ","80%")),
+  at=c(seq(0,80, 5)))
```

With regard to the prediction map, it is obvious that there are artefacts (e.g., sharp vertical changes associated with longitude) in the predictions. These artefacts may disappear or be alleviated if more variables could be used; in other words, different predictive variables should be tested according to the recently development in variable selection [6,79], as discussed in Section 7.

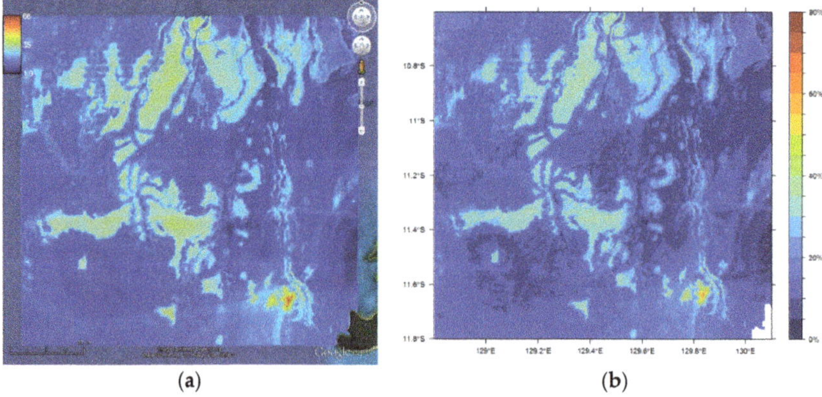

(a)   (b)

Figure 1. Predictions of seabed gravel in the Petrel sub-basin, northern Australian marine margin using a hybrid method of random forest and ordinary kriging (RFOK): (a) *plotKML* (left) and (b) *spplot* (right).

12. Summary

This study reviewed the modeling process for spatial predictive modeling in environmental sciences. The modeling process covers the following nine components:

1. Sampling design and data preparation;
2. Selection of predictive methods;
3. Pre-selection of predictive variables;
4. Exploratory analysis;
5. Parameter selection;
6. Variable selection;
7. Accuracy assessment;
8. Model validation;
9. Spatial predictions, prediction uncertainty, and their visualization.

Each of these components plays a significant role in model development. Incorrect or inappropriate implementation of any components may lead to less accurate or even misleading predictive model(s). To select the most accurate predictive model, all components and relevant requirements and factors for each component need to be considered and carefully implemented by following the guidelines, suggestions, and recommendations provided under relevant components in this study. Reproducible examples were provided to demonstrate how to select and identify the most accurate spatial predictive model using *spm*, and to generate and visualize spatial predictions in environmental sciences. For a predictive model, predictive accuracy is a key criterion for model selection and is critical for subsequent spatial predictions. This modeling process is not only important for spatial predictive modeling, but also provides valuable reference to other predictive modeling fields. Although this study attempts to cover relevant components, which may contribute to the improvement of predictive accuracy, as completely as possible, the spatial predictive modeling field is too broad to allow that to be done comprehensively in this study. This is because different disciplines have their own specific features and requirements. Therefore, further studies are needed to identify factors in relevant components or additional components that can further improve the accuracy of predictive models in various disciplines. This study would be expected to not only boost applications of appropriate spatial predictive modeling processes, but also provide spatial predictive modeling tools for various modeling components to improve the quality of spatial predictions.

Funding: This research received no external funding.

Acknowledgments: I would like to thank Gareth Davies, Peter Tan, Trevor Dhu, Andrew Carroll, and Kim Picard for their valuable comments and suggestions. This study was supported by Geoscience Australia. This paper was published with the permission of the Chief Executive Officer, Geoscience Australia.

Conflicts of Interest: The authors declare no conflicts of interest.

References

1. Marmion, M.; Luoto, M.; Heikkinen, R.K.; Thuiller, W. The performance of state-of-the-art modelling techniques depends on geographical distribution of species. *Ecol. Model.* **2009**, *220*, 3512–3520. [CrossRef]
2. Maier, H.R.; Kapelan, Z.; Kasprzyk, J.; Kollat, J.; Matott, L.S.; Cunha, M.C.; Dandy, G.C.; Gibbs, M.S.; Keedwell, E.; Marchi, A.; et al. Evolutionary algorithms and other metaheuristics in water resources: Current status, research challenges and future directions. *Environ. Model. Softw.* **2014**, *62*, 271–299. [CrossRef]
3. Li, J.; Heap, A. *A Review of Spatial Interpolation Methods for Environmental Scientists*; Record 2008/23; Geoscience Australia: Canberra, Australia, 2008; 137p.
4. Stephens, D.; Diesing, M. Towards quantitative spatial models of seabed sediment composition. *PLoS ONE* **2015**, *10*, e0142502. [CrossRef] [PubMed]

5. Sanabria, L.A.; Cechet, R.P.; Li, J. Mapping of australian fire weather potential: Observational and modelling studies. In Proceedings of the 20th International Congress on Modelling and Simulation (MODSIM2013), Adelaide, Australia, 1–6 December 2013; pp. 242–248.

6. Li, J.; Alvarez, B.; Siwabessy, J.; Tran, M.; Huang, Z.; Przeslawski, R.; Radke, L.; Howard, F.; Nichol, S. Application of random forest, generalised linear model and their hybrid methods with geostatistical techniques to count data: Predicting sponge species richness. *Environ. Model. Softw.* **2017**, *97*, 112–129. [CrossRef]

7. Hastie, T.; Tibshirani, R.; Friedman, J. *The Elements of Statistical Learning: Data Mining, Inference, and Prediction*, 2nd ed.; Springer: New York, NY, USA, 2009; p. 763.

8. Crawley, M.J. *The R Book*; John Wiley & Sons, Ltd.: Chichester, UK, 2007; p. 942.

9. Kuhn, M.; Johnson, K. *Applied Predictive Modeling*; Springer: New York, NY, USA, 2013.

10. Li, J.; Heap, A.D. Spatial interpolation methods applied in the environmental sciences: A review. *Environ. Model. Softw.* **2014**, *53*, 173–189. [CrossRef]

11. Li, J.; Heap, A. A review of comparative studies of spatial interpolation methods in environmental sciences: Performance and impact factors. *Ecol. Inform.* **2011**, *6*, 228–241. [CrossRef]

12. Li, J.; Potter, A.; Huang, Z.; Daniell, J.J.; Heap, A. *Predicting Seabed Mud Content across the Australian Margin: Comparison of Statistical and Mathematical Techniques Using a Simulation Experiment*; Record 2010/11; Geoscience Australia: Canberra, Australia, 2010; 146p.

13. Sanabria, L.A.; Qin, X.; Li, J.; Cechet, R.P.; Lucas, C. Spatial interpolation of mcarthur's forest fire danger index across australia: Observational study. *Environ. Model. Softw.* **2013**, *50*, 37–50. [CrossRef]

14. Tadić, J.M.; Ilić, V.; Biraud, S. Examination of geostatistical and machine-learning techniques as interpolaters in anisotropic atmospheric environments. *Atmos. Environ.* **2015**, *111*, 28–38. [CrossRef]

15. Wolpert, D.; Macready, W. No free lunch theorems for optimization. *IEEE Trans. Evol. Comput.* **1997**, *1*, 67–82. [CrossRef]

16. Burrough, P.A.; McDonnell, R.A. *Principles of Geographical Information Systems*; Oxford University Press: Oxford, UK, 1998; p. 333.

17. Jakeman, A.J.; Letcher, R.A.; Norton, J.P. Ten iterative steps in development and evaluation of environmental models. *Environ. Model. Softw.* **2006**, *21*, 602–614. [CrossRef]

18. Li, J. Assessing spatial predictive models in the environmental sciences: Accuracy measures, data variation and variance explained. *Environ. Model. Softw.* **2016**, *80*, 1–8. [CrossRef]

19. Leek, J.T.; Peng, R.D. What is the question? *Science* **2015**, *347*, 1314–1315. [CrossRef]

20. Li, J. spm: Spatial Predictive Modelling. R Package Version 1.1.0. Available online: https://CRAN.R-project.org/package=spm:2018 (accessed on 17 May 2019).

21. Foster, S.D.; Hosack, G.R.; Lawrence, E.; Przeslawski, R.; Hedge, P.; Caley, M.J.; Barrett, N.S.; Williams, A.; Li, J.; Lynch, T.; et al. Spatially balanced designs that incorporate legacy sites. *Methods Ecol. Evol.* **2017**, *8*, 1433–1442. [CrossRef]

22. Benedetti, R.; Piersimoni, F.; Postigione, P. Spatially balanced sampling: A review and a reappraisal. *Int. Stat. Rev.* **2017**, *85*, 439–454. [CrossRef]

23. Stevens, D.L.; Olsen, A.R. Spatially balanced sampling of natural resources. *J. Am. Stat. Assoc.* **2004**, *99*, 262–278. [CrossRef]

24. Benedetti, R.; Piersimoni, F. A spatially balanced design with probability function proportional to the within sample distance. *Biom. J.* **2017**, *59*, 1067–1084. [CrossRef]

25. Wang, J.-F.; Stein, A.; Gao, B.-B.; Ge, Y. A review of spatial sampling. *Spat. Stat.* **2012**, *2*, 1–14. [CrossRef]

26. Diggle, P.J.; Ribeiro, P.J., Jr. *Model-Based Geostatistics*; Springer: New York, NY, USA, 2010; p. 228.

27. Przeslawski, R.; Daniell, J.; Anderson, T.; Vaughn Barrie, J.; Heap, A.; Hughes, M.; Li, J.; Potter, A.; Radke, L.; Siwabessy, J.; et al. *Seabed Habitats and Hazards of the Joseph Bonaparte Gulf and Timor Sea, Northern Australia*; Record 2008/23; Geoscience Australia: Canberra, Australia, 2011; 69p.

28. Radke, L.C.; Li, J.; Douglas, G.; Przeslawski, R.; Nichol, S.; Siwabessy, J.; Huang, Z.; Trafford, J.; Watson, T.; Whiteway, T. Characterising sediments for a tropical sediment-starved shelf using cluster analysis of physical and geochemical variables. *Environ. Chem.* **2015**, *12*, 204–226. [CrossRef]

29. Radke, L.; Nicholas, T.; Thompson, P.; Li, J.; Raes, E.; Carey, M.; Atkinson, I.; Huang, Z.; Trafford, J.; Nichol, S. Baseline biogeochemical data from australia's continental margin links seabed sediments to water column characteristics. *Mar. Freshw. Res.* **2017**. [CrossRef]

30. Kincaid, T. GRTS Survey Designs for an Area Resource. 2019. Available online: https://cran.r-project.org/web/packages/spsurvey/vignettes/Area_Design.pdf (accessed on 17 May 2019).

31. Kincaid, T.M.; Olsen, A.R. spsurvey: Spatial Survey Design and Analysis. R Package Version 3.3. 2016. Available online: https://cran.r-project.org/web/packages/spsurvey/index.html (accessed on 17 May 2019).

32. Hengl, T. GSIF: Global Soil Information Facilities. R Package Version 0.4-1. 2014. Available online: https://cran.r-project.org/web/packages/GSIF/index.html (accessed on 17 May 2019).

33. Walvoort, D.J.J. Spatial Coverage Sampling and Random Sampling from Compact Geographical Strata. R Package Version 0.3-6. Available online: https://cran.r-project.org/web/packages/spcosa/index.html (accessed on 17 May 2019).

34. Roudier, P. CLHS: A R Package for Conditioned Latin Hypercube Sampling. 2011. Available online: https://cran.r-project.org/web/packages/clhs/index.html (accessed on 17 May 2019).

35. Grafström, A.; Lisic, J. Balancedsampling: Balanced and Saptially Balanced Sampling. R Package Version 1.5.4. 2018. Available online: https://cran.r-project.org/web/packages/BalancedSampling/index.html (accessed on 17 May 2019).

36. Radke, L.; Smit, N.; Li, J.; Nicholas, T.; Picard, K. *Outer Darwin Harbour Shallow Water Sediment Survey 2016: Ga0356—Post-Survey Report*; Record 2017/06; Geoscience Australia: Canberra, Australia, 2017. [CrossRef]

37. Siwabessy, P.J.W.; Smit, N.; Atkinson, I.; Dando, N.; Harries, S.; Howard, F.J.F.; Li, J.; Nicholas, W.A.; Picard, K.; Radke, L.C.; et al. *Bynoe Harbour Marine Survey 2016: Ga4452/sol6432—Post-Survey Report*; Record 2017/04; Geoscience Australia: Canberra, Australia, 2017.

38. Foster, S.D. MBHdesign: Spatial Designs for Ecological and Environmental Surveys. R Package Version 1.0.76. 2017. Available online: https://cran.r-project.org/web/packages/MBHdesign/index.html (accessed on 17 May 2019).

39. Cai, L.; Zhu, Y. The challenges of data quality and data quality assessment in the big data era. *Data Sci. J.* **2015**, *14*, 1–10.

40. Pipino, L.L.; Lee, Y.W.; Wang, R.Y. Data quality assessment. *Commun. ACM* **2002**, *45*, 211–218. [CrossRef]

41. Li, J.; Potter, A.; Huang, Z.; Heap, A. *Predicting Seabed sand Content across the Australian Margin Using Machine Learning and Geostatistical Methods*; Record 2012/48; Geoscience Australia: Canberra, Australia, 2012; 115p.

42. Li, J.; Hilbert, D.W.; Parker, T.; Williams, S. How do species respond to climate change along an elevation gradient? A case study of the grey-headed robin (*Heteromyias albispecularis*). *Glob. Chang. Biol.* **2009**, *15*, 255–267. [CrossRef]

43. Jiang, W.; Li, J. *The Effects of Spatial Reference Systems on the Predictive Accuracy of Spatial Interpolation Methods*; Record 2014/01; Geoscience Australia: Canberra, Australia, 2014; p. 33.

44. Jiang, W.; Li, J. Are Spatial Modelling Methods Sensitive to Spatial Reference Systems for Predicting Marine Environmental Variables. In Proceedings of the 20th International Congress on Modelling and Simulation, Adelaide, Australia, 1–6 December 2013; pp. 387–393.

45. Turner, A.J.; Li, J.; Jiang, W. Effects of spatial reference systems on the accuracy of spatial predictive modelling along a latitudinal gradient. In Proceedings of the 22nd International Congress on Modelling and Simulation, Hobart, Australia, 3–8 December 2017; pp. 106–112.

46. Purss, M. Topic 21: Discrete Global Grid Systems Abstract Specification, Open Geospatial Consortium [OGC 15-104r5]. 2017. Available online: https://www.google.com.au/url?sa=t&rct=j&q=&esrc=s&source=web&cd=4&cad=rja&uact=8&ved=2ahUKEwiHmPmnrqHiAhWFfisKHfTlB18QFjADegQIABAC&url=https%3A%2F%2Fportal.opengeospatial.org%2Ffiles%2F15-104r5&usg=AOvVaw3Ww2TasQntx17y99VlHwig (accessed on 17 May 2019).

47. Li, J. Predictive modelling using random forest and its hybrid methods with geostatistical techniques in marine environmental geosciences. In Proceedings of the Eleventh Australasian Data Mining Conference (AusDM 2013), Canberra, Australia, 13–15 November 2013; Volume 146.

48. Stephens, D.; Diesing, M. A comparison of supervised classification methods for the prediction of substrate type using multibeam acoustic and legacy grain-size data. *PLoS ONE* **2014**, *9*, e93950. [CrossRef] [PubMed]

49. Hengl, T.; Heuvelink, G.B.M.; Kempen, B.; Leenaars, J.G.B.; Walsh, M.G.; Shepherd, K.D.; Sila, A.; MacMillan, R.A.; de Jesus, J.M.; Tamene, L.; et al. Mapping soil properties of africa at 250 m resolution: Random forests significantly improve current predictions. *PLoS ONE* **2015**, *10*, e0125814. [CrossRef] [PubMed]

50. Zhang, X.; Liu, G.; Wang, H.; Li, X. Application of a hybrid interpolation method based on support vector machine in the precipitation spatial interpolation of basins. *Water* **2017**, *9*, 760. [CrossRef]

51. Seo, Y.; Kim, S.; Singh, V.P. Estimating spatial precipitation using regression kriging and artificial neural network residual kriging (rknnrk) hybrid approach. *Water Resour. Manag.* **2015**, *29*, 2189–2204. [CrossRef]

52. Demyanov, V.; Kanevsky, M.; Chernov, S.; Savelieva, E.; Timonin, V. Neural network residual kriging application for climatic data. *J. Geogr. Inf. Decis. Anal.* **1998**, *2*, 215–232.

53. Appelhans, T.; Mwangomo, E.; Hardy, D.R.; Hemp, A.; Nauss, T. Evaluating machine learning approaches for the interpolation of monthly air temperature at mt. Kilimanjaro, tanzania. *Spat. Stat.* **2015**, *14*, 91–113. [CrossRef]

54. Leathwick, J.R.; Elith, J.; Francis, M.P.; Hastie, T.; Taylor, P. Variation in demersal fish species richness in the oceans surrounding new zealand: An analysis using boosted regression trees. *Mar. Ecol. Prog. Ser.* **2006**, *321*, 267–281. [CrossRef]

55. Leathwick, J.R.; Elith, J.; Hastie, T. Comparative performance of generalised additive models and multivariate adaptive regression splines for statistical modelling of species distributions. *Ecol. Model.* **2006**, *199*, 188–196. [CrossRef]

56. Isaaks, E.H.; Srivastava, R.M. *Applied Geostatistics*; Oxford University Press: New York, NY, USA, 1989; p. 561.

57. Hengl, T. *A Practical Guide to Geostatistical Mapping of Environmental Variables*; Office for Official Publication of the European Communities: Luxembourg, 2007; p. 143.

58. Pebesma, E.J. Multivariable geostatistics in s: The gstat package. *Comput. Geosci.* **2004**, *30*, 683–691. [CrossRef]

59. Bivand, R.S.; Pebesma, E.J.; Gómez-Rubio, V. *Applied Spatial Data Analysis with R*; Springer: New York, NY, USA, 2008; p. 374.

60. Lark, R.M.; Ferguson, R.B. Mapping risk of soil nutrient deficiency or excess by disjunctive and indicator kriging. *Geoderma* **2004**, *118*, 39–53. [CrossRef]

61. Huang, H.; Chen, C. Optimal geostatistical model selection. *J. Am. Stat. Assoc.* **2007**, *102*, 1009–1024. [CrossRef]

62. Hernandez-Stefanoni, J.L.; Ponce-Hernandez, R. Mapping the spatial variability of plant diversity in a tropical forest: Comparison of spatial interpolation methods. *Environ. Monit. Assess.* **2006**, *117*, 307–334. [CrossRef] [PubMed]

63. Stein, A.; Hoogerwerf, M.; Bouma, J. Use of soil map delineations to improve (co-)kriging of point data on moisture deficits. *Geoderma* **1988**, *43*, 163–177. [CrossRef]

64. Voltz, M.; Webster, R. A comparison of kriging, cubic splines and classification for predicting soil properties from sample information. *J. Soil Sci.* **1990**, *41*, 473–490. [CrossRef]

65. Bennett, N.D.; Croke, B.F.W.; Guariso, G.; Guillaume, J.H.A.; Hamilton, S.H.; Jakeman, A.J.; Marsili-Libelli, S.; Newham, L.T.H.; Norton, J.P.; Perrin, C.; et al. Characterising performance of environmental models. *Environ. Model. Softw.* **2013**, *40*, 1–20. [CrossRef]

66. Gneiting, T.; Balabdaoui, F.; Raftery, A.E. Probabilistic forecasts, calibration and sharpness. *J. R. Stat. Soc. Ser. B* **2007**, *69*, 243–268. [CrossRef]

67. Austin, M. Species distribution models and ecological theory: A critical assessment and some possible new approaches. *Ecol. Model.* **2007**, *200*, 1–19. [CrossRef]

68. Elith, J.; Leathwick, J. Species distribution models: Ecological explanation and prediction across space and time. *Annu. Rev. Ecol. Evol. Syst.* **2009**, *40*, 677–697. [CrossRef]

69. McArthur, M.A.; Brooke, B.P.; Przeslawski, R.; Ryan, D.A.; Lucieer, V.L.; Nichol, S.; McCallum, A.W.; Mellin, C.; Cresswell, I.D.; Radke, L.C. On the use of abiotic surrogates to describe marine benthic biodiversity. *Estuar. Coast. Shelf Sci.* **2010**, *88*, 21–32. [CrossRef]

70. Huston, M.A. Hidden treatments in ecological experiments: Re-evaluating the ecosystem function of biodiversity. *Oecologia* **1997**, *110*, 449–460. [CrossRef]

71. Arthur, A.D.; Li, J.; Henry, S.; Cunningham, S.A. Influence of woody vegetation on pollinator densities in oilseed *brassica* fields in an australian temperate landscape. *Basic Appl. Ecol.* **2010**, *11*, 406–414. [CrossRef]

72. Elith, J.; Graham, C.H.; Anderson, R.P.; Dulik, M.; Ferrier, S.; Guisan, A.; Hijmans, R.J.; Huettmann, F.; Leathwick, J.R.; Lehmann, A.; et al. Novel methods improve prediction of species' distributions from occurrence data. *Ecography* **2006**, *29*, 129–151. [CrossRef]

73. Miller, K.; Puotinen, M.; Przeslawski, R.; Huang, Z.; Bouchet, P.; Radford, B.; Li, J.; Kool, J.; Picard, K.; Thums, M.; et al. *Ecosystem Understanding to Support Sustainable Use, Management and Monitoring of Marine Assets in the North and North-West Regions: Final Report for NESP d1 2016e*; Report to the National Environmental Science Program, Marine Biodiversity Hub; Australian Institute of Marine Science: Townsville, Australia, 2016; 146p. Available online:

https://www.nespmarine.edu.au/system/files/Miller%20et%20al%20Project%20D1%20Report%20summarising%20outputs%20from%20synthesis%20of%20datasets%20and%20predictive%20models%20for%20N%20and%20NW_Milestone%204_RPv3.pdf (accessed on 17 May 2019).

74. Li, J. Predicting the spatial distribution of seabed gravel content using random forest, spatial interpolation methods and their hybrid methods. In Proceedings of the International Congress on Modelling and Simulation (MODSIM) 2013, Adelaide, Austrialia, 1–6 December 2013; pp. 394–400.

75. Verfaillie, E.; Van Lancker, V.; Van Meirvenne, M. Multivariate geostatistics for the predictive modelling of the surficial sand distribution in shelf seas. *Cont. Shelf Res.* **2006**, *26*, 2454–2468. [CrossRef]

76. Verfaillie, E.; Du Four, I.; Van Meirvenne, M.; Van Lancker, V. Geostatistical modeling of sedimentological parameters using multi-scale terrain variables: Application along the belgian part of the north sea. *Int. J. Geogr. Inf. Sci.* **2008**. [CrossRef]

77. Huang, Z.; Nichol, S.; Siwabessy, P.J.W.; Daniell, J.; Brooke, B.P. Predictive modelling of seabed sediment parameters using multibeam acoustic data: A case study on the carnarvon shelf, western australia. *Int. J. Geogr. Inf. Sci.* **2012**, *26*, 283–307. [CrossRef]

78. Li, J.; Siwabessy, J.; Tran, M.; Huang, Z.; Heap, A. Predicting seabed hardness using random forest in R. In *Data Mining Applications with R*; Zhao, Y., Cen, Y., Eds.; Elsevier: Amsterdam, The Netherlands, 2014; pp. 299–329.

79. Li, J.; Tran, M.; Siwabessy, J. Selecting optimal random forest predictive models: A case study on predicting the spatial distribution of seabed hardness. *PLoS ONE* **2016**, *11*, e0149089. [CrossRef]

80. Siwabessy, P.J.W.; Daniell, J.; Li, J.; Huang, Z.; Heap, A.D.; Nichol, S.; Anderson, T.J.; Tran, M. *Methodologies for Seabed Substrate Characterisation Using Multibeam Bathymetry, Backscatter and Video Data: A Case Study from the Carbonate Banks of the Timor Sea, Northern Australia*; Record 2013/11; Geoscience Australia: Canberra, Australia, 2013; 82p.

81. Huang, Z.; Brooke, B.; Li, J. Performance of predictive models in marine benthic environments based on predictions of sponge distribution on the australian continental shelf. *Ecol. Inform.* **2011**, *6*, 205–216. [CrossRef]

82. Lark, R.M.; Marchant, B.P.; Dove, D.; Green, S.L.; Stewart, H.; Diesing, M. Combining observations with acoustic swath bathymetry and backscatter to map seabed sediment texture classes: The empirical best linear unbiased predi. *Sediment. Geol.* **2015**, *328*, 17–32. [CrossRef]

83. Diesing, M.; Mitchell, P.; Stephens, D. Image-based seabed classification: What can we learn from terrestrial remote sensing? *ICES J. Mar. Sci.* **2016**, fsw 118. [CrossRef]

84. Fisher, P.; Wood, J.; Cheng, T. Where is helvellyn? Fuzziness of multi-scale landscape morphometry. *Trans. Inst. Br. Geogr.* **2004**, *29*, 106–128. [CrossRef]

85. Zuur, A.; Leno, E.N.; Elphick, C.S. A protocol for data exploration to avoid common statistical problems. *Methods Ecol. Evol.* **2010**, *1*, 3–14. [CrossRef]

86. O'Brien, R.M. A caution regarding rules of thumb for variance inflation factors. *Qual. Quant.* **2007**, *41*, 673–690. [CrossRef]

87. Harrell, F.E., Jr. *Regression modelling strategies: with applications to linear models, logistic regression, and survival analysis*; Springer: New York, NY, USA, 1997.

88. Li, J.; Heap, A.D.; Potter, A.; Daniell, J. Application of machine learning methods to spatial interpolation of environmental variables. *Environ. Model. Softw.* **2011**, *26*, 1647–1659. [CrossRef]

89. Cutler, D.R.; Edwards, T.C.J.; Beard, K.H.; Cutler, A.; Hess, K.T.; Gibson, J.; Lawler, J.J. Random forests for classification in ecology. *Ecography* **2007**, *88*, 2783–2792. [CrossRef]

90. Collins, F.C.; Bolstad, P.V. A comparison of spatial interpolation techniques in temperature estimation. In Proceedings of the Third International Conference/Workshop on Integrating GIS and Environmental Modeling, Santa Fe, NM, USA, 21–25 January 1996.

91. Ripley, B.D. *Spatial Statistics*; John Wiley & Sons: New York, NY, USA, 1981; p. 252.

92. Wu, J.; Norvell, W.A.; Welch, R.M. Kriging on highly skewed data for dtpa-extractable soil zn with auxiliary information for ph and organic carbon. *Geoderma* **2006**, *134*, 187–199. [CrossRef]

93. Meul, M.; Van Meirvenne, M. Kriging soil texture under different types of nonstationarity. *Geoderma* **2003**, *112*, 217–233. [CrossRef]

94. Liaw, A.; Wiener, M. Classification and regression by randomforest. *R News* **2002**, *2*, 18–22.

95. Ridgeway, G. gbm: Generalized Boosted Regression Models. R Package Version 2.1.3. 2017. Available online: https://cran.r-project.org/web/packages/gbm/index.html (accessed on 17 May 2019).

96. Elith, J.; Leathwick, J.R.; Hastie, T. A working guide to boosted regression trees. *J. Anim. Ecol.* **2008**, *77*, 802–813. [CrossRef]

97. Breiman, L.; Friedman, J.H.; Olshen, R.A.; Stone, C.J. *Classification and Regression Trees*; Belmont: Wadsworth, OH, USA, 1984.

98. Li, J.; Hilbert, D.W. Lives: A new habitat modelling technique for predicting the distributions of species' occurrence using presence-only data based on limiting factor theory. *Biodivers. Conserv.* **2008**, *17*, 3079–3095. [CrossRef]

99. Johnson, J.B.; Omland, K.S. Model selection in ecology and evolution. *Trends Ecol. Evol.* **2004**, *19*, 101–108. [CrossRef]

100. Venables, W.N.; Ripley, B.D. *Modern Applied Statistics with S-Plus*, 4th ed.; Springer: New York, NY, USA, 2002; p. 495.

101. Chambers, J.M.; Hastie, T.J. *Statistical Models in S*; Wadsworth and Brooks/Cole Advanced Books and Software: Pacific Grove, CA, USA, 1992; p. 608.

102. Lumley, T.; Miller, A. leaps: Regression Subset Selection. R Package Version 3.0. 2009. Available online: https://cran.r-project.org/web/packages/leaps/index.html (accessed on 17 May 2019).

103. McLeod, A.I.; Xu, C. bestglm: Best Subset GLM. R Package Version 0.36. 2017. Available online: https://cran.r-project.org/web/packages/bestglm/index.html (accessed on 17 May 2019).

104. Li, J.; Alvarez, B.; Siwabessy, J.; Tran, M.; Huang, Z.; Przeslawski, R.; Radke, L.; Howard, F.; Nichol, S. Selecting predictors to form the most accurate predictive model for count data. In Proceedings of the International Congress on Modelling and Simulation (MODSIM) 2017, Hobart, Australia, 3–8 December 2017.

105. Kursa, M.B.; Rudnicki, W.R. Feature selection with the boruta package. *J. Stat. Softw.* **2010**, *36*, 1–13. [CrossRef]

106. Kuhn, M. caret: Classification and Regression Training. R Package Version 6.0-81. 2018. Available online: https://cran.r-project.org/web/packages/caret/index.html (accessed on 17 May 2019).

107. Genuer, R.; Poggi, J.M.; Tuleau-Malot, C. VSURF: Variable Selection Using Random Forests. R Package Version 1.0.2. 2015. Available online: https://cran.r-project.org/web/packages/VSURF/index.html (accessed on 17 May 2019).

108. Li, J.; Siwabessy, J.; Huang, Z.; Nichol, S. Developing an optimal spatial predictive model for seabed sand content using machine learning, geostatistics and their hybrid methods. *Geosciences* **2019**, *9*, 180. [CrossRef]

109. Han, J.; Kamber, M. *Data Mining: Concept and Techniques*, 2nd ed.; Elsevier: Amsterdam, The Netherlands, 2006; p. 770.

110. Moriasi, D.N.; Arnold, J.G.; Van Liew, M.W.; Bingner, R.L.; Harmel, R.D.; Veith, T.L. Model evaluation guidelines for systematic quantification of accuracy in watershed simulations. *Am. Soc. Agric. Biol. Eng.* **2007**, *50*, 885–900.

111. Li, J. Assessing the accuracy of predictive models for numerical data: Not r nor r^2, why not? Then what? *PLoS ONE* **2017**, *12*, e0183250. [CrossRef] [PubMed]

112. Allouche, O.; Tsoar, A.; Kadmon, R. Assessing the accuracy of species distribution models: Prevalence, kappa and true skill statistic (tss). *J. Appl. Ecol.* **2006**, *43*, 1223–1232. [CrossRef]

113. Fielding, A.H.; Bell, J.F. A review of methods for the assessment of prediction errors in conservation presence/absence models. *Environ. Conserv.* **1997**, *24*, 38–49. [CrossRef]

114. Thibaud, E.; Petitpierre, B.; Broennimann, O.; Davison, A.C.; Guisan, A. Measuring the relative effect of factors affecting species distribution model predictions. *Methods Ecol. Evol.* **2014**, *5*, 947–955. [CrossRef]

115. Lobo, J.M.; Jiménez-Valverde, A.; Real, R. Auc: A misleading measure of the performance of predictive distribution models. *Glob. Ecol. Biogeogr.* **2008**, *7*, 145–151. [CrossRef]

116. Kohavi, R. A study of cross-validation and bootstrap for accuracy estimation and model selection. In Proceedings of the International Joint Conference on Artificial Intelligence (IJCAI), Montreal, QC, Canada, 20–25 August 1995; pp. 1137–1143.

117. Refsgaard, J.C.; van der Sluijs, J.P.; Højberg, A.L.; Vanrolleghem, P.A. Uncertainty in the environmental modelling process - a framework and guidance. *Environ. Model. Softw.* **2007**, *22*, 1543–1556. [CrossRef]

118. Hayes, K.R. *Uncertainty and Uncertainty Analysis Methods*; CSIRO: Canberra, Australia, 2011; p. 131. Available online: https://publications.csiro.au/rpr/download?pid=csiro:EP102467&dsid=DS3 (accessed on 17 May 2019).

119. Barry, S.; Elith, J. Error and uncertainty in habitat models. *J. Appl. Ecol.* **2006**, *43*, 413–423. [CrossRef]

Appl. Sci. **2019**, *9*, 2048

120. Oxley, T.; ApSimon, H. A conceptual framework for mapping uncertainty in integrated assessment. In Proceedings of the 19th International Congress on Modelling and Simulation, Perth, Australia, 12–16 December 2011.
121. Walker, W.E.; Harremoes, P.; Rotmans, J.; Van der Sluijs, J.P.; van Asselt, M.B.A.; Janssen, P.; Krayer von Krauss, M.P. Defining uncertainty: A conceptual basis for uncertainty management in model-based decision support. *Integr. Assess.* **2003**, *4*, 5–17. [CrossRef]
122. Goovaerts, P. *Geostatistics for Natural Resources Evaluation*; Oxford University Press: New York, NY, USA, 1997; p. 483.
123. Mentch, L.; Hooker, G. Quantifying uncertainty in random forests via confidence intervals and hypothesis tests. *J. Mach. Learn. Res.* **2016**, *17*, 1–41.
124. Slaets, J.I.F.; Piepho, H.-P.; Schmitter, P.; Hilger, T.; Cadisch, G. Quantifying uncertainty on sediment loads using bootstrap confidence intervals. *Hydrol. Earth Syst. Sci.* **2017**, *21*, 571–588. [CrossRef]
125. Wager, S.; Hastie, T.; Efron, B. Confidence intervals for random forests: The jackknife and the infinitesimal jackknife. *J. Mach. Learn. Res.* **2014**, *15*, 1625–1651. [PubMed]
126. Wright, M.N.; Ziegler, A. Ranger: A fast implementation of random forests for high dimensional data in c++ and r. *J. Stat. Softw.* **2017**, *77*, 1–17. [CrossRef]
127. Coulston, J.W.; Blinn, C.E.; Thomas, V.A.; Wynne, R.H. Approximating prediction uncertainty for random forest regression models. *Photogramm. Eng. Remote Sens.* **2016**, *82*, 189–197. [CrossRef]
128. Chen, J.; Li, M.-C.; Wang, W. Statistical uncertainty estimation using random forests and its application to drought forecast. *Math. Probl. Eng.* **2012**, *2012*, 915053. [CrossRef]
129. Bishop, T.F.A.; Minasny, B.; McBratney, A.B. Uncertainty analysis for soil-terrain models. *Int. J. Geogr. Inf. Sci.* **2006**, *20*, 117–134. [CrossRef]
130. Hijmans, R.J. raster: Geographic Data Analysis and Modeling. Available online: http://CRAN.R-project.org/package=raster (accessed on 17 May 2019).

Article

Mapping Areal Precipitation with Fusion Data by ANN Machine Learning in Sparse Gauged Region

Guoyin Xu [1], Zhongjing Wang [1,2,3,]* and Ting Xia [4]

[1] Department of Hydraulic Engineering, Tsinghua University, Beijing 100084, China; xgy13@mails.tsinghua.edu.cn
[2] State Key Laboratory of Hydro-Science and Engineering, Tsinghua University, Beijing 100084, China
[3] State Key Lab of Plateau Ecology and Agriculture, Qinghai University, Xining 810016, China
[4] China Renewable Energy Engineering Institute, Beijing 100120, China; yifuling@126.com
* Correspondence: zj.wang@tsinghua.edu.cn; Tel.: +86-10-6278-2021

Received: 30 March 2019; Accepted: 28 May 2019; Published: 4 June 2019

Abstract: Focusing on water resources assessment in ungauged or sparse gauged areas, a comparative evaluation of areal precipitation was conducted by remote sensing data, limited gauged data, and a fusion of gauged data and remote sensing data based on machine learning. The artificial neural network (ANN) model was used to fuse the remote sensing precipitation and ground gauge precipitation. The correlation coefficient, root mean square deviation, relative deviation and consistency principle were used to evaluate the reliability of the remote sensing precipitation. The case study in the Qaidam Basin, northwest of China, shows that the precision of the original remote sensing precipitation product of Tropical Precipitation Measurement Satellite (TRMM)-3B42RT and TRMM-3B43 was 0.61, 72.25 mm, 36.51%, 27% and 0.70, 64.24 mm, 31.63%, 32%, respectively, comparing with gauged precipitation. The precision of corrected TRMM-3B42RT and TRMM-3B43 improved to 0.89, 37.51 mm, –0.08%, 41% and 0.91, 34.22 mm, 0.11%, 42%, respectively, which indicates that the data mining considering elevation, longitude and latitude as the main influencing factors of precipitation is efficient and effective. The evaluation of areal precipitation in the Qaidam Basin shows that the mean annual precipitation is 104.34 mm, 186.01 mm and 174.76 mm based on the gauge data, corrected TRMM-3B42RT and corrected TRMM-3B43. The results show many differences in the areal precipitation based on sparse gauge precipitation data and fusion remote sensing data.

Keywords: Qaidam Basin; remote sensing; TRMM; artificial neural network

---

1. Introduction

Precipitation is one of the essential links in the water cycle process and varies significantly whether it is spatial or temporal [1,2]. Traditionally, the measurement of precipitation is based on a ground gauge station such as a hydrometric station or meteorological station. The gauge precipitation is identified in terms of both effectiveness and accuracy due to its direct measurement. The spatial distribution of precipitation is mostly interpolated from the gauged data. However, the accuracy of interpolation in the sparse and uneven gauged area is generally not reliable [3]. Therefore, the fusion of remote sensing data and gauged data for evaluation has become a challenging topic [4–8].

There are many high-resolution rainfall products at both the global and regional scales which have been released successively [9,10], such as the Global Precipitation Climate Program (GPCP), Global Satellite Mapping Precipitation Program (GSMaP), Tropical Precipitation Measurement Satellite (TRMM) and Global Precipitation Measurement (GPM) [11–13]. Many remote sensing precipitation products [14–17] are widely used to compensate for the shortage of gauged data areas [18–21].

However, the remote sensing precipitation production is not highly reliable due to its indirect observation which needs adjusting and evaluation [22–25]. There are many achievements published about the evaluation of remote sensing precipitation products. [26–30]. The precision evaluation index of remote sensing precipitation products mainly includes a correlation coefficient, determination coefficient, scatter slope, fuzzy comprehensive score, etc. [31–34]. For those remote sensing precipitation products with low precision, it is necessary to be corrected. The most used method is machine learning, such as the classification and regression tree (CART), random forest (RF), multi-factor data mining set correction, etc. [18,35]. An artificial neural network (ANN) is a powerful machine learning algorithm with a complex network structure formed by the interconnection of a large number of processing units (neurons) [36]. It is an information processing system based on imitating the structure and function of the brain neural network [37]. The theory of an ANN has made significant progress in pattern recognition [38,39], automatic control, signal processing [40], assistant decision-making, artificial intelligence [41], networking and healthcare [42–45]. It has been successfully introduced into the field of hydrology and water resources [46,47].

There are many publications on TRMM remote sensing precipitation products for applicability in specific areas [48–50]. Dominque et al. [51] found that the accuracy of both the total and the monthly precipitation of TRMM in the Amazon basin are high enough. Ji et al. [52] validated the accuracy of TRMM precipitation products and found it has a high accuracy in Sichuan and Chongqing in China on seasonal and monthly scales. Wang et al. [53] analyzed TRMM precipitation products with the observation data of meteorological stations in the Tianshan Mountains and its surrounding areas, and the results showed that the TRMM products had good applicability. A large number of research results showed that the accuracy of TRMM precipitation products was higher on monthly and annual scales [54,55], which could be used for analyzing the dynamic variability of a long-time precipitation sequence [56].

However, Qu et al. [57] evaluated the daily precipitation products of TRMM in the Irrawaddy River basin and found that the remote sensing precipitation and the measured values had a high correlation but a large deviation. Xu et al. [58] evaluated the TRMM precipitation in the southern part of the Qinghai–Tibet Plateau by gauged data from high-density rainfall stations and found that TRMM overestimated the amount of light rains. The altitude, slope, direction, latitude, longitude and other factors impact the accuracy of TRMM precipitation [18]. Therefore, TRMM precipitation products should be corrected before being applied in some areas, especially high mountain areas. Based on the evaluation of TRMM 3B42RT and 3B43 in the Qaidam Basin, northwest of China, this paper fused the gauged data and remote sensing data of precipitation by machine learning and assessed the rainfall resources in the Qaidam Basin. The methodology can be used in other sparse gauged areas.

2. Data and Methods

2.1. Study Area and Data Sources

2.1.1. Study Area

The Qaidam Basin is located in the northeastern edge of the Tibetan Plateau. The geographical coordinates are 34°41′–39°20′ N and 87°48′–99°18′ E, spanning the Gansu Province, Qinghai Province and Xinjiang Uygur Autonomous Region. The vast majority of the Qaidam Basin is in the Qinghai Province with an area of 234.14 thousand km^2. The area in Xinjiang Uygur Autonomous Region is 17.42 thousand km^2 and that in the Gansu Province is 17.89 thousand km^2. The northwest, northeast and south of the Qaidam Basin are surrounded by the Altun Mountains, Qilian Mountains and Kunlun Mountains, respectively, as shown in Figure 1. The Qaidam Basin is the only large plateau inland basin in the world and its elevation ranges from 2653 m to 6748 m. The basin is deep in the mainland and surrounded by mountains. It is hard for the warm and humid airflow from the southwest to reach the basin, forming the typical cold-dry continental climate. Affected by the topography and latitude, the temperature of the basin is high in the central portion, but low all around. The lowest temperature occurs in January with −9.8–−16.1 °C in the basin area and −14.7–−17.2 °C in the mountainous area.

The highest temperature is in July with 13.5–19.2 °C in the basin area and 5.6–10.4 °C in the mountainous area. The annual sunshine duration is generally above 3100 h.

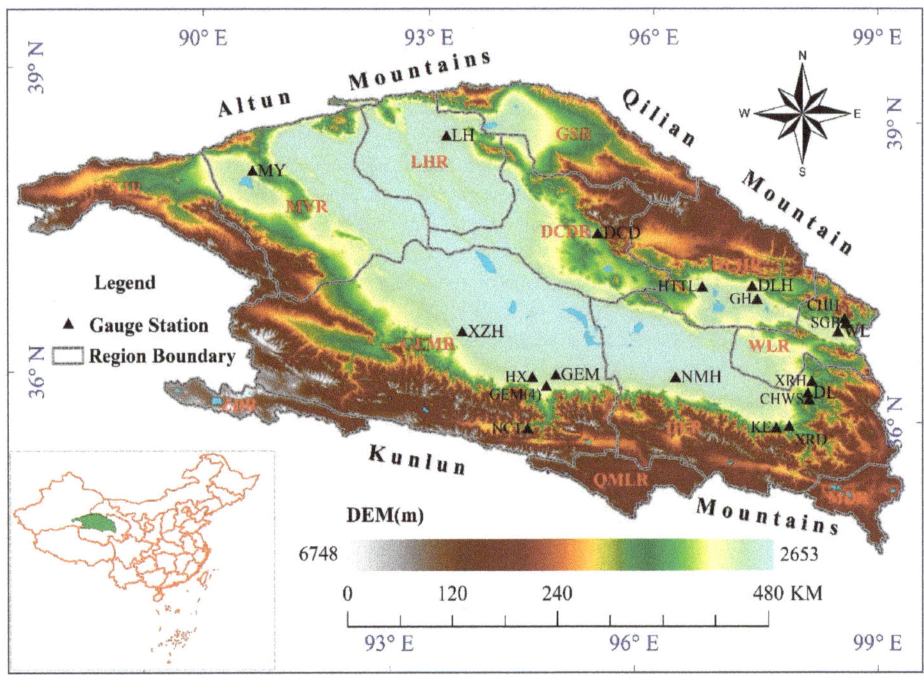

Figure 1. The research area of the Qaidam Basin.

2.1.2. Data Sources

The data used in this study include Digital Elevation Model (DEM), representing ground elevation, TRMM 3B42RT and 3B43, representing remote sensing precipitation, and gauged precipitation data by surface ground meteorological and hydrometric stations. The DEM data is from the Geographical Information Monitoring Cloud Platform (GIM Cloud) [59], and the spatial resolution is 1 km × 1 km. The TRMM 3B42RT and 3B43 are from the NASA website [60] with a spatial resolution of 0.25° × 0.25° and a time resolution of 3 h. The TRMM products were processed by ArcGIS to fit the Qaidam Basin and the data is from 2001 to 2016. In order to consider the effect of elevation, the 1 km × 1 km spatial resolution matched with DEM data was used when resampling the TRMM data. The gauged precipitation data, including 9 meteorological stations and 11 hydrometric stations from 2001 to 2016, was from China Meteorological Science Data Sharing Service Network [61] and Hydrological Red Book of the People's Republic of China [62], respectively.

2.2. Methodology

2.2.1. Evaluation of Remote Sensing Precipitation Precision

(1) The TRMM products' precision in gauged grids:

The correlation coefficient (R^2), relative deviation (Bias) and root mean square deviation (RMSD) were calculated in the grids where a gauge station was located. The data series is from 2001 to 2016 and the temporal resolution is all on an annual scale. The formulas are as follows:

$$R^2 = \frac{\left[\sum\limits_{i=1}^{n}\left(Ps_i - \overline{Ps}\right)\left(Pt_i - \overline{Pt}\right)\right]^2}{\sum\limits_{i=1}^{n}\left(Ps_i - \overline{Ps}\right)^2 \cdot \sum\limits_{i=1}^{n}\left(Pt_i - \overline{Pt}\right)^2}, \tag{1}$$

$$Bias = \frac{\sum\limits_{i=1}^{n}(Pt_i - Ps_i)}{\sum\limits_{i=1}^{n}Ps_i} \times 100\%, \tag{2}$$

$$RMSD = \sqrt{\frac{1}{n}\sum\limits_{i=1}^{n}(Pt_i - Ps_i)^2}, \tag{3}$$

where, Ps is the gauged precipitation by the ground gauge stations (mm); Pt is the remote sensing precipitation retrieved from TRMM products (mm); $\overline{Ps}$ and $\overline{Pt}$ are the average value of Ps and Pt, respectively; n is the number of years (2001–2016). The R reflects the correlation between the gauged precipitation and the TRMM precipitation. The relative deviation (Bias) and root mean square deviation (RMSD) reflect the deviation degree between the gauged precipitation data and the TRMM precipitation.

(2) The TRMM products' precision in ungauged grids:

There are only 20 gauges in the study area which means the gauged grids are equal or less than 20. All the other grids are ungauged. According to Xia's achievements [9], the accuracy of the remote sensing precipitation on the grids without a gauge station can be evaluated by the criteria of the consistency rate (CR). The formulas are as follows:

$$Count_i = \begin{cases} 1, & if\ Pt_i \in D \\ 0, & if\ Pt_i \notin D \end{cases} i = 1, 2, \cdots, N, \tag{4}$$

$$S = \sum\limits_{i=1}^{N} Count_i, \tag{5}$$

$$CR = \frac{S}{N} \times 100\%, \tag{6}$$

where, D is the precipitation-elevation mask (PEM) derived from the relationship of gauged precipitation and the elevation of the gauge stations; N is the total number of grids without gauge stations; S is the total number of grids in which remote sensing precipitation falls into the PEM. If the remote sensing precipitation value falls into the mask, then the remote sensing precipitation is considered consistent with the gauged precipitation at the same elevation region, that is, the remote sensing precipitation is reliable and vice versa.

2.2.2. Correction of Remote Sensing Precipitation by ANN Model

A three-layer ANN model was set up for remote sensing precipitation correction. The model has five input variables (i.e., gauged precipitation (Ps), elevation (DEM), longitude (X) and latitude (Y) of the gauge stations, and TRMM precipitation). Specifically, the Ps is the target value of the model, and the four others are the variables of the model input layer. The output variable is only the corrected precipitation in the model output layer. The hidden layer nodes are set to 20 by the preferred selection. The structure of the ANN model in this paper is shown in Figure 2.

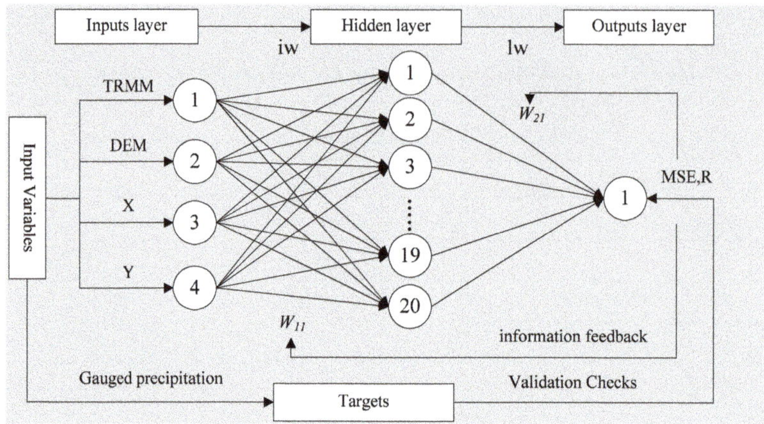

Figure 2. The artificial neural network (ANN) structure diagram.

The model training function is 'trainlm' which updates weight and bias values according to Levenberg–Marquardt optimization. At the same time, we chose the 'tansig' as the model neural transfer function to calculate a layer's output from its net input. The divide function is accessed automatically whenever the network is trained, which is used to divide the data into training, validation and testing subsets. In this study, the net divide function is set to 'dividerand', and the ratios for training, validation and testing are 0.7, 0.0 and 0.3 according to the needs of the training function of 'trainlm'. The prepared sequence data is randomly divided into the training subset (70%) and testing subset (30%) using the division parameters by the divide function. The fusion models for 3B42RT and 3B43 correction completed 103 and 127 times of training, and the convergence error (mean square error) was 0.0036 and 0.0041, respectively.

The process of data fusion mainly includes four steps: (1) the Pt is extracted by combining ArcGIS technology according to the spatial locations of the gauge stations. Moreover, the data of Ps, DEM, X, Y and Pt are stored in a one-to-one correspondence; (2) training and testing the model after setting each function of the artificial neural network. When the parameters such as the R^2 reach certain conditions, the construction of the model is completed; (3) correction remote sensing precipitation based on the fusion model which is established in the previous step; (4) the fused precipitation data are inversed to the research area on a spatial–temporal scale according to the spatial information of the grids.

2.2.3. Assessment of Rainwater Resources

The areal rainwater resources (i.e., precipitation) is evaluated by the mean value. The variation trend of the areal precipitation is predicted by the improved Mann–Kendall method [63–65].

Corresponding to the time series X with n sample sizes, the order column is constructed as follows:

$$S_k = \sum_{i=1}^{k} r_i \quad r_i = \begin{cases} 1 & x_i > x_j \\ 0 & x_i \leq x_j \end{cases} j = 1, 2, \cdots, i. \tag{7}$$

It can be seen that the order sequence S is the cumulative number of values at the ith moment greater than that at the jth moment. Under the assumption that the time series is randomly independent, the statistics are defined as follows:

$$UF_k = \frac{S_k - E(S_k)}{\sqrt{Var(S_k)}} k = 1, 2, \cdots, n, \tag{8}$$

where $UF_k = 0$, $E(S_k)$ and $Var(S_k)$ are the mean and variance of the cumulative number S_k, respectively. When $x_1, x_2, \ldots, x_n$ are independent of each other and have the same continuous distribution, they can be calculated by the following formula:

$$E(S_k) = \frac{n(n+1)}{4} \, Var(S_k) = \frac{n(n-1)(2n+5)}{72}, \tag{9}$$

where UF_i is the standard normal distribution, which is the statistical sequence calculated according to the time series X $(x_1, x_2, \ldots, x_n)$. The significance level α is determined and the normal distribution table is checked. If $|UF_i| > U_\alpha$, it indicates that there is a significant trend change in the sequence.

Repeating the above process according to the time series X in inverse order $x_n, x_{n-1}, \ldots, x_1$, while letting $UB_k = -UF_k$, $k = n, n-1, \ldots, 1$, $UB_1 = 0$. If the calculated value of UB_k or UF_k is greater than 0, it indicates that the sequence has an upward trend; if lower than 0, it indicates a downward trend. When they exceed the critical line, it suggests that the upward or downward trend is significant.

3. Results and Discussion

3.1. Precision of Original TRMM Products

3.1.1. Grids with the Gauge Station

According to the methods above, the precision criteria of original TRMM products on the 20 grids with the gauge station are shown in Table 1. The names of the gauge stations are LengHu (LH), XiaoZaoHuo (XZH), GeErMu (GEM), HeXi (HX), MangYa (MY), NuoMuHong (NMH), GeErMu4 (GEM4), DaChaDan (DCD), HuaiTouTaiLa (HTTL), GaHai (GH), NaChiTai (NCT), WuLan (WL), DeLingHa (DLH), DuLan (DL), ChaHanWuSu (CHWS), XiaRiHa (XRH), XiangRiDe (XRD), KeEr (KE), ShangGaBa (SGB), ChaHanHe (CHH), respectively.

Table 1. The precision of original Tropical Precipitation Measurement Satellite (TRMM) products (2001–2016). LH: LengHu; XZH: XiaoZaoHuo; GEM: GeErMu; HX: HeXi; MY: MangYa; NMH: NuoMuHong; GEM4: GeErMu4; DCD: DaChaDan; HTTL: HuaiTouTaiLa; GH: GaHai; NCT: NaChiTai; WL: WuLan; DLH: DeLingHa; DL: DuLan; CHWS: ChaHanWuSu; XRH: XiaRiHa; XRD: XiangRiDe; KE: KeEr; SGB: ShangGaBa; CHH: ChaHanHe; MAP: Mean Annual Precipitation.

		Gauges				Original 3B42RT			Original 3B43		
No.	Name	X (°)	Y (°)	DEM (m)	MAP (mm)	MAP (mm)	Bias (%)	RMSD (mm)	MAP (mm)	Bias (%)	RMSD (mm)
1	LH	93.33	38.75	2777	19.88	41.10	106.79	23.17	36.19	82.07	18.42
2	XZH	93.68	36.80	2772	30.34	81.60	168.97	57.76	77.49	155.44	52.74
3	GEM	94.90	36.42	2812	47.32	126.01	166.30	81.28	113.31	139.46	68.68
4	HX	94.60	36.38	2822	47.34	121.63	156.91	77.60	111.21	134.89	68.22
5	MY	90.85	38.25	2942	50.16	68.37	34.55	26.85	60.59	19.79	20.07
6	NMH	96.42	36.43	2796	55.54	118.66	113.65	67.30	117.21	111.05	65.48
7	GEM4	94.78	36.30	2957	61.76	126.01	104.05	67.61	113.31	83.48	55.44
8	DCD	95.37	37.85	3190	104.00	122.89	18.17	38.35	119.28	14.69	28.18
9	HTTL	96.73	37.35	2867	107.94	160.30	48.51	65.39	162.39	50.45	63.29
10	GH	97.43	37.23	2877	161.24	157.18	−2.52	32.02	168.12	4.27	26.18
11	NCT	94.57	35.87	3966	182.31	230.40	26.38	70.91	222.21	21.89	65.31
12	WL	98.48	36.92	2959	222.26	157.31	−29.22	83.38	174.37	−21.55	66.26
13	DLH	97.37	37.37	2988	228.68	201.19	−12.02	41.53	203.84	−10.86	35.93
14	DL	98.10	36.30	3190	240.73	183.92	−23.60	73.11	195.62	−18.74	58.43
15	CHWS	98.12	36.23	3273	244.97	211.97	−13.47	54.70	218.89	−10.65	45.53
16	XRH	98.15	36.42	3143	252.91	183.92	−27.28	95.16	195.62	−22.65	81.60
17	XRD	97.87	35.97	3100	285.18	264.48	−7.26	57.26	265.41	−6.93	55.43
18	KE	97.70	35.95	3269	307.43	278.26	−9.49	70.71	277.94	−9.59	70.63
19	SGB	98.58	37.00	3168	332.05	170.04	−48.79	171.22	187.25	−43.61	151.87
20	CHH	98.57	37.05	3351	432.30	257.26	−40.49	189.76	258.21	−40.20	187.08
–	Average	–	–	–	170.72	163.12	36.51	72.25	163.92	31.63	64.24

It is clear in Table 1 that the average mean annual precipitation (MAP) of 3B42RT and 3B43 on the 20 grids are 163.12 mm and 163.92 mm, respectively, which look quite close to that of the gauged precipitation (170.72 mm). However, it can be found that the TRMM precipitation is higher in the low gauged precipitation areas (LGPA), including the stations of LH, XZH, GEM, HX, MY, NMH, GEM4, DCD and HTTL. On the contrary, the TRMM precipitation is lower in the high gauged precipitation area (HGPA), including the stations of WL, DLH, DL, CHWS, XRH, XRD, KE, SGB and CHH. As a result, the average MAP of TRMM is approximately equal to that of the gauged precipitation, but the underestimated HGPA and overestimated LGPA will lead to systematic bias. The systematic error is considered the result of planarization of the original TRMM when calibration was done with too limited gauged data.

The average RMSD of the original 3B42RT and 3B43 in the Qaidam Basin is 72.25 mm and 64.24 mm, respectively. Considering with the bias together, the minimum RMSD (23.17 mm of 3B43RT and 18.42mm of 3B43) relates to the overestimation (bias of 106.79%, 82.07% and value of 21.26 mm, 16.31 mm) of 3B42RT and 3B43 in the LGPA (LH station), while the maximum RMSD (189.76 mm of 3B43RT and 187.08 mm of 3B43) relates to the underestimation (bias of −40.49%, −40.20% and value of −175.04 mm, −174.09 mm) in the HGPA (CHH station). It is obvious that the bias of underestimation is smaller than that of overestimation, but the absolute amount of underestimation is far greater than that of overestimation, as shown in Figure 3. Therfore, it is supposed that the average TRMM precipitation in the Qaidam Basin would be an underestimation overall.

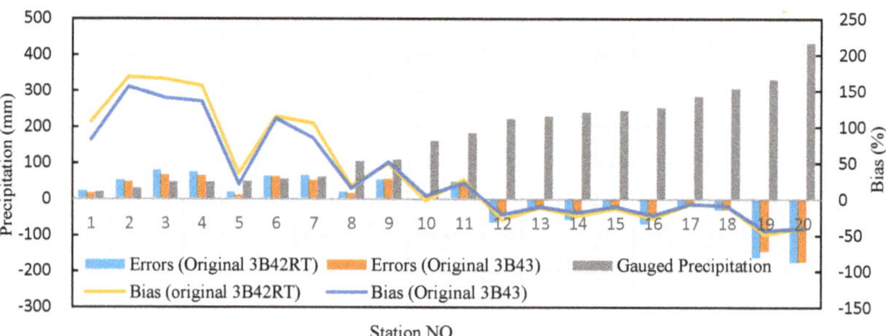

Figure 3. The Original TRMM Precipitation Products Errors.

3.1.2. Grids without a Gauge Station

Most of the grids have no ground gauge so that we could not evaluate the precision by precipitation itself. Some other validation principle is introduced. Here, it is the consistency principle (CR), which is a relationship rule of rainfall and elevation retrieved from the gauged data. The situation of 3B42RT and 3B43 precipitation falling into the PEM is shown in Figure 4.

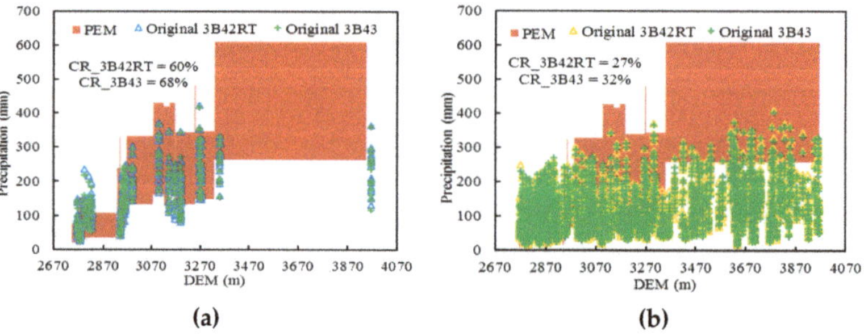

Figure 4. The gauged precipitation – elevation mask (PEM) and original TRMM precipitation filter from 2001 to 2016. (**a**) Grids with gauge station. (**b**) Grids without gauge station.

It can be found that the CR value of original 3B42RT and 3B43 precipitation on the grids with gauge is 60% and 68%, while on the grids without gauge it is only 27% and 32%, respectively. It means the TRMM precipitation on the gauged grids is more reliable than that on the ungauged grids.

3.2. Correction of TRMM Products

3.2.1. Calibration and Validation

The regression machine learning by the ANN model mentioned above was carried out. In the process, we have a total of 20 gauges with 16 years of data sets, of which 224 sets of data are used for the model training, and the remaining 96 sets are used for model testing. The result of a comparison of output and target is shown in Figure 5.

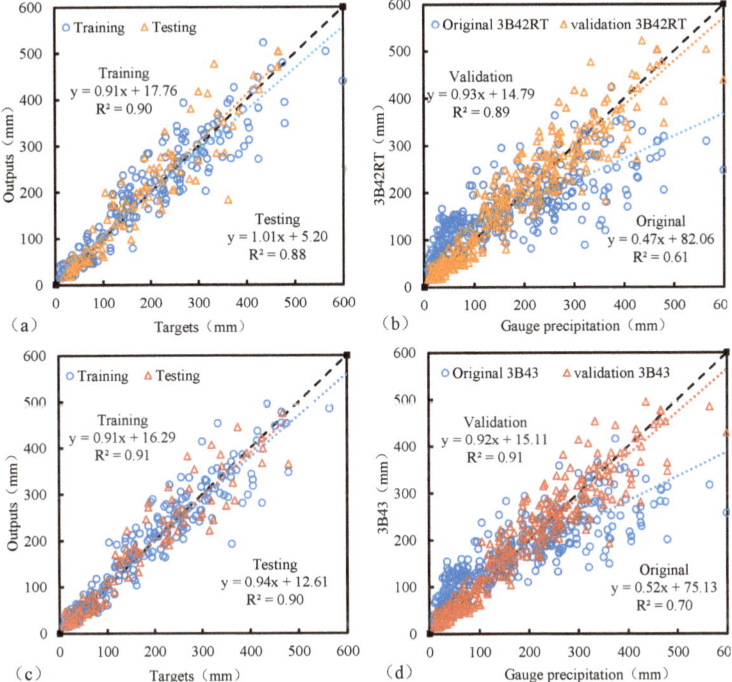

Figure 5. The calibration and validation of TRMM precipitation by artificial neural networks (ANN). (**a**) Training and testing for 3B42RT; (**b**) correction for 3B42RT; (**c**) training and testing for 3B43; (**d**) correction for 3B43.

It can be seen from Figure 5a,c that the training and testing are very good with the determinant coefficients 0.90, 0.91 and 0.88, 0.90, respectively. There was also improvement with the determinant coefficients 0.89, 0.91 of corrected TRMM precipitation compared with the original TRMM determinant coefficients of 0.61, 0.70. It also can be seen that the values of original 3B42RT and 3B43 are all mostly above the 1:1 line when gauged precipitation was less than 100 mm, while a lot of the values of original TRMM precipitation are under the 1:1 line when gauged precipitation was more than 200 mm. This indicates that the original TRMM precipitation in the LGPA was overestimated and the TRMM precipitation in the HGPA was underestimated. On the whole, the dispersion of TRMM precipitation points on both sides of the 1:1 line is large and uneven, the R^2 is only 0.61 and 0.70, respectively. Fortunately, the 3B42RT and 3B43 precipitation improved significantly both in the LGPA and HGPA. They closely dispersed on the both sides of the 1:1 line after the correction by the ANN model. The R^2

increased to 0.89 and 0.91, respectively. It was proven that the fusion model based on an ANN is effective for the correction of TRMM products.

3.2.2. The Precision of Corrected TRMM Products

After the correction, the precision of TRMM precipitation in the Qaidam Basin on the grids with the gauge station is shown in Table 2. It can be found that the average bias is significantly reduced, from 36.51% to –0.08% (3B42RT) and 31.63% to 0.11% (3B43), respectively. It also can be seen from the RMSD that the fusion model has a significant correction effect on TRMM products. The average RMSD value decreased from 72.25 mm and 64.24mm to 37.51 mm and 34.22 mm after correction.

Table 2. The precision evaluation of corrected 3B42RT and 3B43 (2001–2016). - LH: LengHu; XZH: XiaoZaoHuo; GEM: GeErMu; HX: HeXi; MY: MangYa; NMH: NuoMuHong; GEM4: GeErMu4; DCD: DaChaDan; HTTL: HuaiTouTaiLa; GH: GaHai; NCT: NaChiTai; WL: WuLan; DLH: DeLingHa; DL: DuLan; CHWS: ChaHanWuSu; XRH: XiaRiHa; XRD: XiangRiDe; KE: KeEr; SGB: ShangGaBa; CHH: ChaHanHe; MAP: Mean Annual Precipitation.

	Gauges				Corrected 3B42RT			Corrected 3B43		
Name	X (°)	Y (°)	DEM (m)	MAP (mm)	MAP (mm)	Bias (%)	RMSD (mm)	MAP (mm)	Bias (%)	RMSD (mm)
LH	93.33	38.75	2777	19.88	19.03	−12.15	8.21	17.80	−10.44	7.21
XZH	93.68	36.80	2772	30.34	31.60	4.15	11.71	32.25	6.32	12.21
GEM	94.90	36.42	2812	47.32	45.53	−3.78	11.06	44.48	−6.00	10.49
HX	94.60	36.38	2822	47.34	45.53	−3.82	14.92	45.09	−4.77	15.71
MY	90.85	38.25	2942	50.16	48.87	−2.46	22.2	52.54	4.52	22.01
NMH	96.42	36.43	2796	55.54	54.30	−2.23	17.24	53.51	−3.66	17.53
GEM4	94.78	36.30	2957	61.76	55.26	−10.52	13.57	56.97	−7.74	12.57
DCD	95.37	37.85	3190	104.00	98.31	−5.47	29.43	105.38	1.33	23.10
HTTL	96.73	37.35	2867	107.94	123.67	14.57	38.85	123.21	14.15	35.36
GH	97.43	37.23	2877	161.24	160.28	−0.60	31.44	158.83	−1.50	26.12
NCT	94.57	35.87	3966	182.31	182.35	0.02	35.08	185.47	1.74	34.19
WL	98.48	36.92	2959	222.26	243.09	9.38	56.33	231.36	4.10	48.45
DLH	97.37	37.37	2988	228.68	226.20	−1.08	31.62	223.36	−2.32	24.75
DL	98.11	36.30	3190	240.73	246.18	2.26	48.60	247.77	2.93	40.58
CHWS	98.12	36.23	3273	244.97	290.87	18.74	63.93	289.24	18.07	58.15
XRH	98.15	36.42	3143	252.91	247.61	−2.11	66.42	247.41	−2.17	59.82
XRD	97.87	35.97	3100	285.18	271.46	−4.81	54.38	275.56	−3.37	52.80
KE	97.7	35.95	3269	307.43	304.7	−0.89	62.78	308.31	0.29	63.06
SGB	98.58	37	3168	332.05	316.19	−4.78	57.49	305.51	−7.99	51.4
CHH	98.57	37.05	3351	432.3	449.37	3.95	74.96	426.29	−1.27	68.8
Average	–	–	–	170.72	173.02	−0.08	37.51	171.52	0.11	34.22

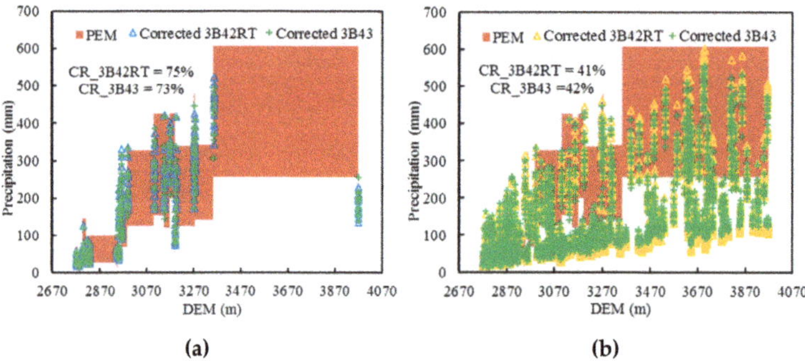

Figure 6. The consistency rate (CR) elevation for corrected TRMM from 2001 to 2016 in the Qaidam Basin. (**a**) Gauged grids. (**b**) Ungauged grids.

For those grids without gauges, the CR values were also evaluated. The CR of 3B42RT and 3B43 on the grids with and without the gauge station improved up to 75%, 73% and 41%, 42% after correction

(Figure 6). This means the precision of TRMM precipitation on grids with and without gauge station are also improved.

3.3. Assessment of Rainwater Resources of the Qaidam Basin

3.3.1. The Average Amount of Precipitation

The spatial distribution of the mean annual precipitation from 2001 to 2016 before and after the correction of 3B42RT and 3B43 in the Qaidam Basin is shown in Figure 7. It can be seen that the precipitation on the eastern, southern and southeastern edges of the Qaidam Basin is high, while in the center and northwest it is small. The original distribution of precipitation in the east and west of the Kunlun Mountains is not even, but the trend of decreasing from east to west is enhanced significantly after the correction. The main reason for this is the increase of precipitation in the eastern Kunlun Mountains after the TRMM products corrected. The precipitation in the Qilian Mountains and the southeastern edges of the Qaidam Basin was also relatively high and increased after correction. At the same time, there was a higher consistency between precipitation and elevation in those regions, that is, precipitation increased with elevation.

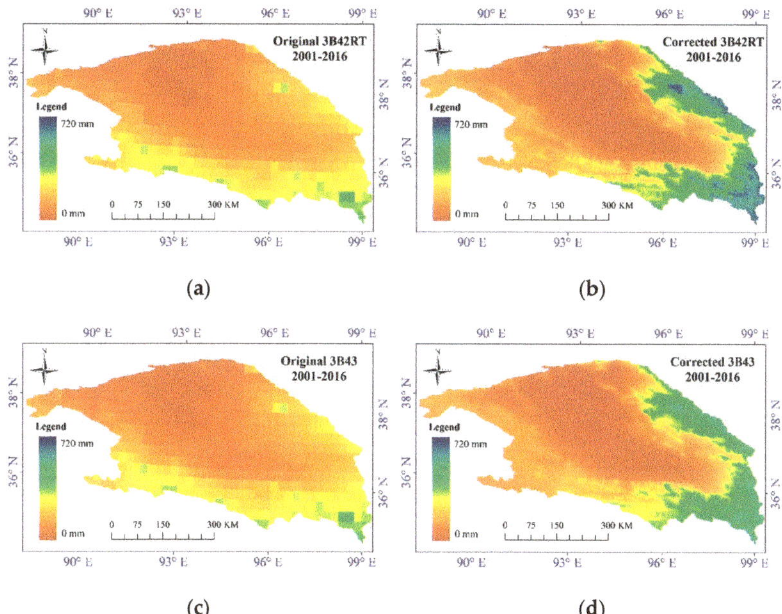

Figure 7. Comparison of spatial distribution of TRMM precipitation before and after correction. (**a**) Original 3B42RT precipitation; (**b**) Corrected 3B42RT precipitation; (**c**) Original 3B43 precipitation; (**d**) Corrected 3B43 precipitation.

For the 12 subregions of the basin, including MA, LE, DCD, DLH, WL, DL, GEM, QML, MD, ZD, XJR and GSR as shown in Figure 1, the areal precipitation in each region is shown in Table 3.

We can see from Table 3 that the average annual precipitation of the Qaidam Basin based on the original 3B42RT and 3B43 are 148.45 mm and 146.26 mm, respectively, while the new assessment value is 186.01 mm and 174.76 mm based on corrected TRMM 3B42 and 3B43. It is obvious that the annual precipitation (104.34 mm) based on the interpolated precipitation by gauges is significantly less than that of original and corrected TRMM products.

The previous studies on precipitation in the Qaidam Basin were mostly based on the gauged precipitation data to get the areal precipitation of the Qaidam Basin according to the traditional interpolation method. The pity is that the gauge distribution is too sparse and uneven, causing an unreliable interpolation result. It is obvious that the precipitation in the area was underestimated in the past. The fusion remote sensing precipitation with local ground information in high mountainous areas is helpful.

Table 3. Assessment of annual precipitation in each region of the Qaidam Basin.

Region Name	Area (10³ km²)	Interpolated Precipitation by Gauges	Original 3B42RT	Corrected 3B42RT	Original 3B43	Corrected 3B43
LE	19.3	29.65	43.79	22.42	39.43	25.3
MY	31.3	42.43	67.72	55.27	62.22	62.48
GEM	69.4	65.64	162.73	118.69	157.5	114.27
DCD	21.3	69.55	100.92	128.46	97.61	126.22
XJ	17.4	74.31	114.1	131.95	108.28	133.53
ZD	4.8	83.98	239.7	171.39	238.79	141.03
GS	17.9	86.57	112.33	221.02	107.09	208.24
QML	6.5	157.57	275.74	363.65	273.22	321.82
DLH	22.7	168.45	190.01	369.43	192.47	343.08
DL	43.8	183.01	203.43	302.37	205.56	273.62
WL	10.4	259.57	171.17	299.41	185.38	288.63
MD	4.7	292.9	354.29	602.71	358.21	523.32
The entire basin	269.4	104.34	148.45	186.01	146.26	174.76

3.3.2. The Precipitation Variation Trend

The annual precipitation time series and its variation trend of the gauged precipitation, original and corrected 3B42RT and 3B43 products at the 20 gauge stations are shown in Figure 8 and Table 4. From Figure 8 we can know that the original 3B42RT and 3B43 precipitation at the GEM, LH, NMH, XZH. HX, NCT and GEM4 stations, where the precipitation is low, is obviously higher than the gauged precipitation. On the contrary, the original 3B42RT and 3B43 precipitation at the CHH and SGB stations, where the precipitation is high, is obviously less than the gauged precipitation. This again indicates that the original 3B42RT and 3B43 are significantly overestimated in the LGPA and underestimated in the HGPA. What is nice is that the 3B42RT and 3B43 decreases or increases significantly and tends to the gauged precipitation values after the fusion model was corrected.

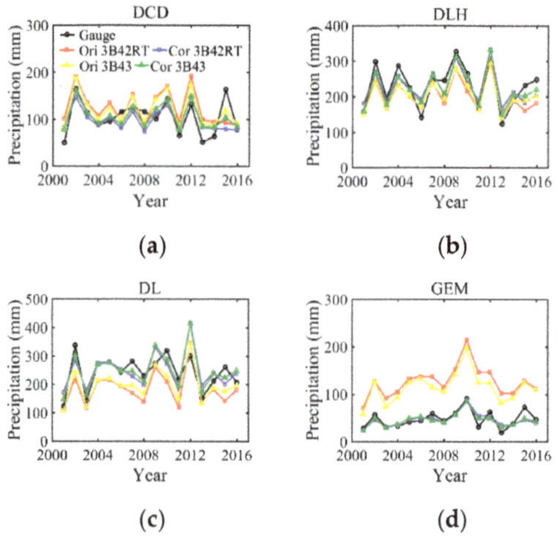

Figure 8. *Cont.*

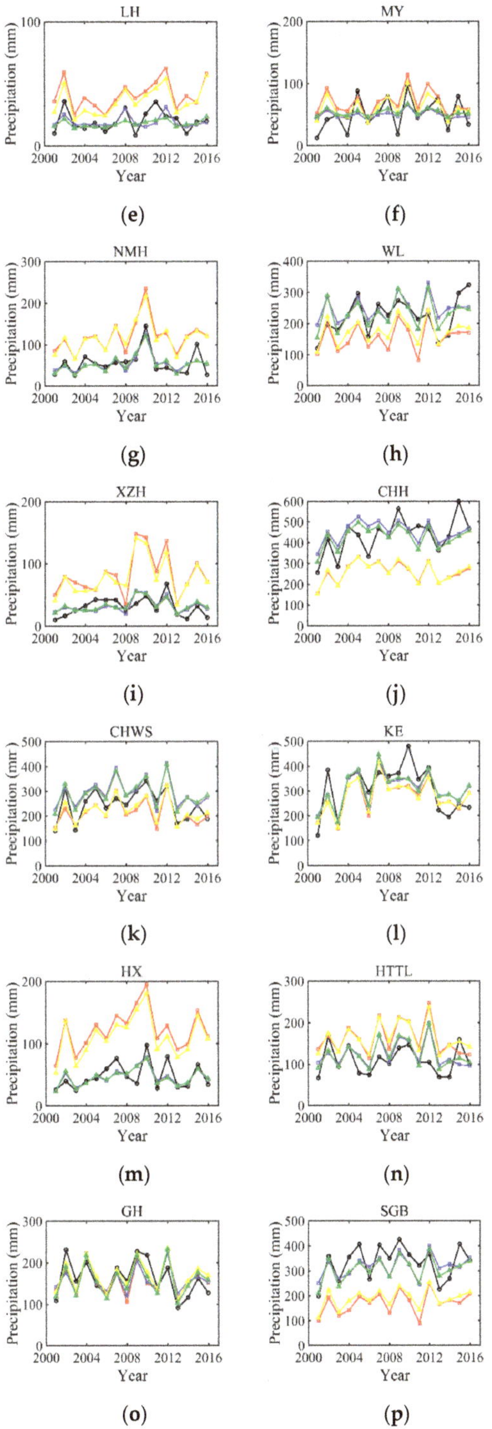

Figure 8. *Cont.*

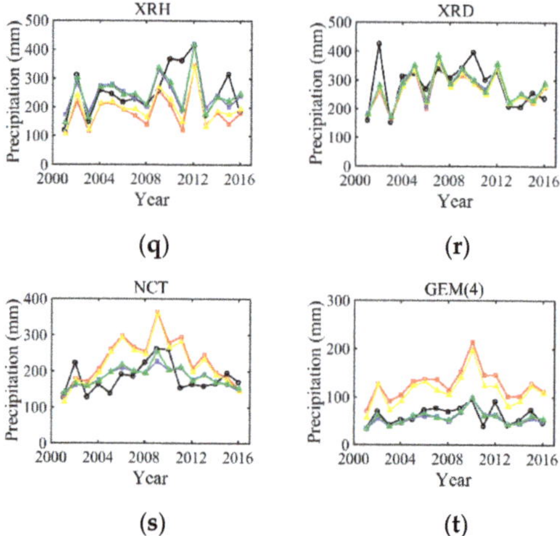

Figure 8. The annual precipitation time series of the gauged precipitation, original and corrected of 3B42RT and 3B43 products at the 20 gauge stations. (**a**) DCD: DaChaDan; (**b**) DLH: DeLingHa; (**c**) DL: DuLan; (**d**) GEM: GeErMu; (**e**) LH: LengHu; (**f**) MY: MangYa; (**g**) NMH: NuoMuHong; (**h**) WL: WuLan; (**i**) XZH: XiaoZaoHuo; (**j**) CHH: ChaHanHe (**k**) CHWS: ChaHanWuSu (**l**) KE: KeEr; (**m**) HX: HeXi; (**n**) HTTL: HuaiTouTaiLa; (**o**) GH: GaHai; (**p**) SGB: ShangGaBa; (**q**) XRH: XiaRiHa; (**r**) XRD: XiangRiDe; (**s**) NCT: NaChiTai; (**t**) GEM(4): GeErMu4.

Table 4. Comparison of the annual precipitation variation trend (2001–2016) between gauged and TRMM products on the grids with the gauge station. LH: LengHu; XZH: XiaoZaoHuo; GEM: GeErMu; HX: HeXi; MY: MangYa; NMH: NuoMuHong; GEM4: GeErMu4; DCD: DaChaDan; HTTL: HuaiTouTaiLa; GH: GaHai; NCT: NaChiTai; WL: WuLan; DLH: DeLingHa; DL: DuLan; CHWS: ChaHanWuSu; XRH: XiaRiHa; XRD: XiangRiDe; KE: KeEr; SGB: ShangGaBa; CHH: ChaHanHe; MAP: Mean Annual Precipitation.

Station Name	Gauged	Original 3B42RT	Corrected 3B42RT	Original 3B43	Corrected 3B43
GH	−2.21	−0.23	−0.18	−0.27	−0.85
XRD	−1.32	0.01	−0.02	−0.06	−0.07
DCD	−0.63	−1.55	−0.75	−1.07	−0.77
KE	−0.62	−0.37	−0.40	−0.47	−0.59
DLH	−0.32	−0.94	−0.91	−0.27	−0.33
HTTL	−0.15	−1.05	−0.63	0.13	−0.10
CHWS	0.09	2.78	2.87	0.58	0.67
LH	0.13	0.91	0.05	1.10	0.26
XZH	0.14	1.44	0.39	1.94	0.57
DL	0.23	0.01	0.02	0.94	1.05
GEM4	0.58	1.87	0.88	1.29	0.58
NMH	0.63	1.91	0.95	2.33	1.29
HX	0.89	2.08	0.69	2.23	0.85
GEM	0.93	1.87	0.84	1.29	0.54
NCT	1.31	0.63	0.32	0.86	0.35
MY	1.45	0.17	0.06	0.41	0.13
SGB	2.43	4.25	3.95	3.22	3.03
WL	4.71	3.14	2.77	2.28	2.69
XRH	3.21	2.21	2.29	2.94	3.06
CHH	10.45	0.19	0.18	1.16	1.78
Average	1.10	0.97	0.67	1.03	0.71

From the average change trends of precipitation at the 20 gauge stations in the three data sources, the annual precipitation of the Qaidam Basin shows an increasing trend. The gauged precipitation at GH, XRD DCD, KE, DLH and HTTL stations showed a decreasing trend, and the others showed an increasing trend in different degrees at the other 14 gauge stations. The change trend direction is basically the same. Therefore, the spatial distribution and temporal variation trend of TRMM products after correction can characterize the spatial–temporal variation characteristics of precipitation in the Qaidam Basin.

According to the corrected 3B42RT and 3B43 precipitation in the Qaidam Basin from 2001 to 2016, the change trend and significance on the grid scale in the study area were calculated, as shown in Figure 9.

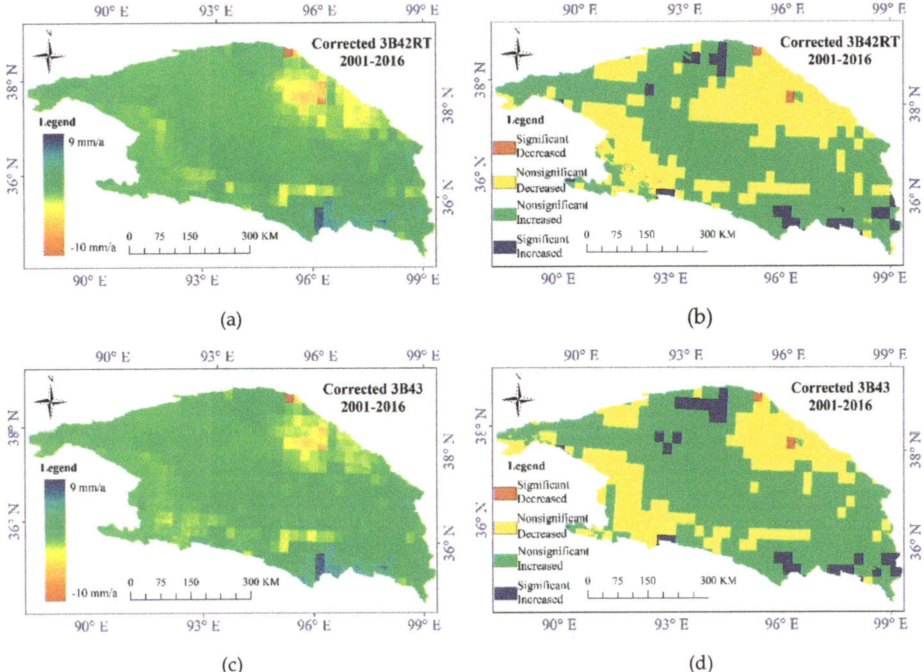

Figure 9. The variation trend and significance of precipitation in the Qaidam Basin. (**a**) Corrected 3B42RT change trend; (**b**) corrected 3B42RT change significance; (**c**) corrected 3B43 change trend; (**d**) corrected 3B43 change significance.

It can be seen from the Figure 9a,b that the precipitation in the northeast of the Qaidam Basin (Qilian Mountain area) showed a decreasing trend, while in the southeastern edge of the basin (east of the Kunlun Mountains) showed an increasing trend based on corrected 3B42RT and 3B43 products. It is obvious the precipitation slightly increased at the center and northwest of the basin. Figure 9b,d showed the significant test results of the precipitation change trend. We know that most areas of the Qaidam Basin were non-significantly increased or decreased, and only a small part of the areas passed the significant change test ($Z \geq 2.58$).

4. Conclusions

(1) The Qaidam Basin is located in an arid and semi-arid region with a dry climate and fragile natural ecological environment. The precipitation process in the Qaidam Basin is significantly different spatially and temporally. Due to the sparse gauge stations and maldistribution, the rainwater resources

are hard to assess. A fusion of remote sensing precipitation of TRMM products and gauged precipitation is helpful.

(2) The average mean annual precipitation is only 104.34 mm by ground gauges interpolation, and 148.45 mm and 146.26 mm by original 3B42RT and 3B43. However, the precision evaluation of TRMM precipitation shows it was overestimated in the LGPA and underestimated in the HGPA. The original TRMM products need a correction.

(3) The correction result shows the average mean annual precipitation is 186.01 mm by 3B42RT and 174.74 mm by 3B43. The average bias of 3B42RT and 3B43 at gauge stations are significantly reduced to -0.08% and 0.11% after being corrected, and the average RMSD is significantly reduced to 37.51 mm and 34.22 mm, respectively. All of those indicated that the precipitation products fusion model based on ANN could effectively work on TRMM products.

(4) The final result shows there are many differences in areal precipitation based on sparse gauge precipitation data, original TRMM data and fusion remote sensing data. The rainwater resources in the study basin have been underestimated in the past year, and both were derived from ground gauge stations and from original TRMM products.

In the future, research needs to test the ANN effectivity on mining the normalized difference vegetation index and rainwater resources in this area.

Author Contributions: Conceptualization, Z.W. and G.X.; methodology, Z.W. and T.X.; formal analysis, G.X.; writing—original draft preparation, G.X.; writing—review and editing, Z.W. All authors read and approved the final manuscript.

Funding: This research was funded by the National Key Research and Development Program (No. 2016YFC0402900), Key R&D and Transformation Projects in Qinghai Province (No. 2017-SF-116), and the National Natural Science Foundation of China (NSFC, Grant No. 41671020).

Acknowledgments: The authors would like to thank all the colleagues who generously provided their dataset. The authors greatly thank the editors and reviewers for providing thorough and constructive comments to improve the manuscript.

References

1. Bohnenstengel, S.I.; Schlünzen, K.H.; Beyrich, F. Representativity of in situ precipitation measurements—A case study for the LITFASS area in North-Eastern Germany. *J. Hydrol.* **2011**, *400*, 387–395. [CrossRef]

2. Marzano, F.S.; Cimini, D.; Montopoli, M. Investigating precipitation microphysics using ground-based microwave remote sensors and disdrometer data. *Atmos. Res.* **2010**, *97*, 583–600. [CrossRef]

3. Zhuoqi, C.; Xiaogu, Z.; Shupeng, Z.; Tao, L. Mapping Daily Precipitation over China Based on TRMM Multisatellite Precipitation Analysis and Gauge Data. In Proceedings of the International Conference on Remote Sensing, Kyoto, Japan, 29 October–1 November 2012.

4. XIE, P.; Arkin, P.A. Analyses of Global Monthly Precipitation Using Gauge Observations, Satellite Estimates, and Numerical Model Predictions. *J. Clim.* **1996**, *9*, 840–858. [CrossRef]

5. Huffman, G.J.; Adler, R.F.; Rudolf, B.; Schneider, U.; Keehn, P.R. Global Precipitation Estimates Based on a Technique for Combining Satellite-Based Estimates, Rain Gauge Analysis, and NWP Model Precipitation Information. *J. Clim.* **1995**, *8*, 1284–1295. [CrossRef]

6. Vila, D.; Goncalves, L.; Toll, D.; Rozante, J.R. Statistical Evaluation of Combined Daily Gauge Observations and Rainfall Satellite Estimates over Continental South America. *J. Hydrometeorol.* **2009**, *10*, 533–543. [CrossRef]

7. Rozante, J.R.; Moreira, D.S.; de Goncalves, L.G.G.; Vila, D.A. Combining TRMM and Surface Observations of Precipitation: Technique and Validation over South America. *Weather* **2010**, *25*, 885–894. [CrossRef]

8. Shen, Y.; Zhao, P.; Pan, Y.; Yu, J. A high spatiotemporal gauge-satellite merged precipitation analysis over China. *J. Geophys. Res. Atmos.* **2014**, *119*, 3063–3075. [CrossRef]

9. Huffman, G.J.; Robert, F.A.; David, T.B.; Nelkin, E.J.; Acheampong, M. *The TRMM Multi-Satellite Precipitation Analysis (TMPA)*; Springer: Berlin, Germany, 2008.

10. Kubota, T.; Ushio, T.; Shige, S.; Kida, S.; Kachi, M.; Okamoto, K. Verification of High-Resolution Satellite-Based Rainfall Estimates around Japan Using a Gauge-Calibrated Ground-Radar Dataset. *J. Meteorol. Soc. Jpn.* **2009**, *87*, 203–222. [CrossRef]

11. Kidd, C.; Huffman, G. Global precipitation measurement. *Meteorol. Appl.* **2011**, *18*, 334–353. [CrossRef]

12. Tapiador, F.J.; Turk, F.J.; Petersen, W.; Hou, A.Y.; García-Ortega, E.; Machado, L.A.T.; Angelis, C.F.; Salio, P.; Kidd, C.; Huffman, G.J.; et al. Global precipitation measurement: Methods, datasets and applications. *Atmos. Res.* **2012**, *104*, 70–97. [CrossRef]

13. Seto, S.; Iguchi, T.; Oki, T. The Basic Performance of a Precipitation Retrieval Algorithm for the Global Precipitation Measurement Mission's Single/Dual-Frequency Radar Measurements. *IEEE Trans. Geosci. Remote* **2013**, *51*, 5239–5251. [CrossRef]

14. Zhu, Z.; Yong, B.; Ke, L.; Wang, G.; Ren, L.; Chen, X. Tracing the Error Sources of Global Satellite Mapping of Precipitation for GPM (GPM-GSMaP) Over the Tibetan Plateau, China. *IEEE J. Stars* **2018**, *11*, 2181–2191. [CrossRef]

15. Guo, H.; Chen, S.; Bao, A.; Behrangi, A.; Hong, Y.; Ndayisaba, F.; Hu, J.; Stepanian, P.M. Early assessment of Integrated Multi-satellite Retrievals for Global Precipitation Measurement over China. *Atmos. Res.* **2016**, *176–177*. [CrossRef]

16. Haile, A.T.; Yan, F.; Habib, E. Accuracy of the CMORPH satellite-rainfall product over Lake Tana Basin in Eastern Africa. *Atmos. Res.* **2015**, *163*, 177–187. [CrossRef]

17. Liu, S.; Yan, D.; Qin, T.; Weng, B.; Li, M. Correction of TRMM 3B42V7 Based on Linear Regression Models over China. *Adv. Meteorol.* **2016**, *2016*, 1–13. [CrossRef]

18. Xia, T.; Wang, Z.; Zheng, H. Topography and Data Mining Based Methods for Improving Satellite Precipitation in Mountainous Areas of China. *Atmosphere* **2015**, *6*, 983–1005. [CrossRef]

19. Zheng, X.; Zhu, J. A methodological approach for spatial downscaling of TRMM precipitation data in North China. *Int. J. Remote Sens.* **2015**, *36*, 144–169. [CrossRef]

20. Tao, Z.; Yuanqing, H.; Jian, M.; Juan, P. Spatial and temporal distribution of precipitation based on corrected TRMM data around the Hexi Corridor, China. *Sci. Cold Arid Reg.* **2014**, *6*, 159–167.

21. Shi, Y.; Song, L.; Xia, Z.; Lin, Y.; Myneni, R.; Choi, S.; Wang, L.; Ni, X.; Lao, C.; Yang, F. Mapping Annual Precipitation across Mainland China in the Period 2001–2010 from TRMM3B43 Product Using Spatial Downscaling Approach. *Remote Sens.* **2015**, *7*, 5849–5878. [CrossRef]

22. Dinku, T.; Chidzambwa, S.; Ceccato, P.; Connor, S.J.; Ropelewski, C.F. Validation of high-resolution satellite rainfall products over complex terrain. *Int. J. Remote Sens.* **2008**, *29*, 4097–4110. [CrossRef]

23. Bitew, M.M.; Gebremichael, M. Assessment of satellite rainfall products for streamflow simulation in medium watersheds of the Ethiopian highlands. *Hydrol. Earth Syst. Sci.* **2011**, *15*, 1147–1155. [CrossRef]

24. Condom, T.; Rau, P.; Espinoza, J.C. Correction of TRMM 3B43 monthly precipitation data over the mountainous areas of Peru during the period 1998–2007. *Hydrol. Process.* **2011**, *25*, 1924–1933. [CrossRef]

25. Darand, M.; Amanollahi, J.; Zandkarimi, S. Evaluation of the performance of TRMM Multi-satellite Precipitation Analysis (TMPA) estimation over Iran. *Atmos. Res.* **2017**, *190*, 121–127. [CrossRef]

26. Romilly, T.G.; Gebremichael, M. Evaluation of satellite rainfall estimates over Ethiopian river basins. *Hydrol. Earth Syst. Sci.* **2011**, *15*, 1505–1514. [CrossRef]

27. Moazami, S.; Golian, S.; Kavianpour, M.R.; Hong, Y. Uncertainty analysis of bias from satellite rainfall estimates using copula method. *Atmos. Res.* **2014**, *137*, 145–166. [CrossRef]

28. Wang, Z.; Zhong, R.; Lai, C.; Chen, J. Evaluation of the GPM IMERG satellite-based precipitation products and the hydrological utility. *Atmos. Res.* **2017**, *196*, 151–163. [CrossRef]

29. Lu, X.; Wei, M.; Tang, G.; Zhang, Y. Evaluation and correction of the TRMM 3B43V7 and GPM 3IMERGM satellite precipitation products by use of ground-based data over Xinjiang, China. *Env. Earth Sci.* **2018**, *77*, 209. [CrossRef]

30. Lekula, M.; Lubczynski, M.W.; Shemang, E.M.; Verhoef, W. Validation of satellite-based rainfall in Kalahari. *Phys. Chem. Earth Parts A/B/C* **2018**, *105*, 84–97. [CrossRef]

31. Duan, Z.; Bastiaanssen, W.G.M. First results from Version 7 TRMM 3B43 precipitation product in combination with a new downscaling–calibration procedure. *Remote Sens. Env.* **2013**, *131*, 1–13. [CrossRef]

32. Guofeng, Z.; Dahe, Q.; Yuanfeng, L.; Fenli, C.; Pengfei, H.; Dongdong, C.; Kai, W. Accuracy of TRMM precipitation data in the southwest monsoon region of China. *Appl. Clim.* **2017**, *129*, 353–362. [CrossRef]

33. Jia, S.; Zhu, W.; Lü, A.; Yan, T. A statistical spatial downscaling algorithm of TRMM precipitation based on NDVI and DEM in the Qaidam Basin of China. *Remote Sens. Env.* **2011**, *115*, 3069–3079. [CrossRef]

34. Liu, S.; Yan, D.; Wang, H.; Li, C.; Qin, T.; Weng, B.; Xing, Z. Evaluation of TRMM 3B42V7 at the basin scale over mainland China. *Adv. Water Sci.* **2016**, *27*, 639–651.

35. Seyyedi, H.; Anagnostou, E.N.; Beighley, E.; McCollum, J. Satellite-driven downscaling of global reanalysis precipitation products for hydrological applications. *Hydrol. Earth Syst. Sci.* **2014**, *18*, 5077–5091. [CrossRef]

36. Agatonovic-Kustrin, S.; Beresford, R. Basic Concepts of Artificial Neural Network (ANN) Modeling and its Application in Pharmaceutical Research. *J. Pharm. Biomed. Anal.* **2000**, *22*, 717–727. [CrossRef]

37. Partridge, D.; Rae, S.; Wang, W.J. Artificial neural networks. *J. Roy. Soc. Med.* **1999**, *92*, 385. [CrossRef]

38. Lyons, W.B.; Flanagan, C.; Lewis, E.; Ewald, H.; Lochmann, S. Interrogation of multipoint optical fibre sensor signals based on artificial neural network pattern recognition techniques. *Sens. Actuators A Phys.* **2004**, *114*, 7–12. [CrossRef]

39. Lyons, W.; Fitzpatrick, C.; Flanagan, C.; Lewis, E. A novel multipoint luminescent coated ultra violet fibre sensor utilising artificial neural network pattern recognition techniques. *Sens. Actuators A Phys.* **2004**, *115*, 267–272. [CrossRef]

40. Nissar, A.I.; Upadhyaya, S.J. Fault Diagnosis of Mixed Signal VLSI Systems Using Artificial Neural Networks. In Proceedings of the Southwest Symposium on Mixed-Signal Design (SSMSD 99), Tucson, AZ, USA, 11–13 April 1999; IEEE: Piscataway, NY, USA, 1999; pp. 93–98.

41. Huesken, D.; Lange, J.; Mickanin, C.; Weiler, J.; Asselbergs, F.; Warner, J.; Meloon, B.; Engel, S.; Rosenberg, A.; Cohen, D.; et al. Design of a genome-wide siRNA library using an artificial neural network. *Nat. Biotechnol.* **2005**, *23*, 995–1001. [CrossRef]

42. Aceto, G.; Ciuonzo, D.; Montieri, A.; Pescapé, A. Mobile Encrypted Traffic Classification Using Deep Learning. In Proceedings of the 2018 Network Traffic Measurement and Analysis Conference (TMA), Vienna, Austria, 26–29 June 2018; IEEE: Piscataway, NJ, USA, 2018; pp. 1–8.

43. Aceto, G.; Ciuonzo, D.; Montieri, A.; Pescape, A. Mobile Encrypted Traffic Classification Using Deep Learning: Experimental Evaluation, Lessons Learned, and Challenges. *IEEE Trans. Netw. Serv. Manag.* **2019**. [CrossRef]

44. Rucco, R.; Sorriso, A.; Liparoti, M.; Ferraioli, G.; Sorrentino, P.; Ambrosanio, M.; Baselice, F. Type and Location of Wearable Sensors for Monitoring Falls during Static and Dynamic Tasks in Healthy Elderly: A Review. *Sensors* **2018**, *18*, 1613. [CrossRef]

45. Gardner, G.G.; Keating, D.; Williamson, T.H.; Elliott, A.T. Automatic detection of diabetic retinopathy using an artificial neural network: A screening tool. *Brit. J. Ophthalmol.* **1996**, *80*, 940–944. [CrossRef]

46. Hsu, K.; Gupta, H.V.; Sorooshian, S. Artificial Neural Network Modeling of the Rainfall-Runoff Process. *Water Resour. Res.* **1995**, *31*, 2517–2530. [CrossRef]

47. Hopfield, J.J. Neural networks and physical systems with emergent collective computational abilities. *Proc. Natl. Acad. Sci. USA* **1982**, *79*, 2554–2558. [CrossRef]

48. AghaKouchak, A.; Nasrollahi, N.; Habib, E. Accounting for Uncertainties of the TRMM Satellite Estimates. *Remote Sens.* **2009**, *1*, 606–619. [CrossRef]

49. Heidinger, H.; Yarlequé, C.; Posadas, A.; Quiroz, R. TRMM rainfall correction over the Andean Plateau using wavelet multi-resolution analysis. *Int. J. Remote Sens.* **2012**, *33*, 4583–4602. [CrossRef]

50. Liu, Z.; Ostrenga, D.; Teng, W.; Kempler, S. Tropical Rainfall Measuring Mission (TRMM) Precipitation Data and Services for Research and Applications. *B Am. Meteorol. Soc.* **2012**, *93*, 1317–1325. [CrossRef]

51. Courault, D.; Seguin, B.; Olioso, A. Review on estimation of evapotranspiration from remote sensing data: From empirical to numerical modeling approaches. *Irrig. Drain. Syst.* **2005**, *19*, 223–249. [CrossRef]

52. Ji, T.; Yang, H.; Liu, R.; He, T.; Wu, J. Applicability analysis of the TRMM precipitation data in the Sichuan-Chongqing region. *Prog. Geogr.* **2014**, 1375–1386.

53. Wang, X.; Liu, H.; Bao, A. Applicability Research on TRMM Precipitation Data in Tianshan Mountains. *J. China Hydrol.* **2014**, 58–64.

54. Yongqing, B.; Juanle, W.; Yujie, W.; Xuehua, H.; Tsydypov, B.Z.; Ochir, A.; Davaasuren, D. Spatio-Temporal Distribution of Drought in the Belt and Road Area During 1998–2015 Based on TRMM Precipitation Data. *J. Resour. Ecol.* **2017**, *8*, 559–570. [CrossRef]

55. Tian, Y.; Peters-Lidard, C.D.; Choudhury, B.J.; Garcia, M. Multitemporal Analysis of TRMM-Based Satellite Precipitation Products for Land Data Assimilation Applications. *J. Hydrometeorol.* **2007**, *8*, 1165–1183. [CrossRef]

56. Bookhagen, B.; Burbank, D.W. Topography, relief, and TRMM-derived rainfall variations along the Himalaya. *Geophys. Res. Lett.* **2006**, *33*.
57. Qu, W.; Lu, J.; Song, W.; Zhang, T.; Tan, Y.; Huang, P. Research on Accuracy Validation and Calibration Methods of TRMM Remote Sensing Precipitation Data in Irrawaddy Basin. *Adv. Earth Sci.* **2014**, 1262–1270.
58. Xu, R.; Tian, F.; Yang, L.; Hu, H.; Lu, H.; Hou, A. Ground validation of GPM IMERG and TRMM 3B42V7 rainfall products over southern Tibetan Plateau based on a high-density rain gauge network. *J. Geophys. Res. Atmos.* **2017**, *122*, 910–924. [CrossRef]
59. Geographical Information Monitoring Cloud Platform. Available online: http://www.dsac.cn/ (accessed on 25 May 2018).
60. National Aeronautics and Space Administration. Available online: https://mirador.gsfc.nasa.gov/ (accessed on 17 March 2018).
61. National Meteorological Information Center. Available online: http://data.cma.cn/ (accessed on 9 May 2018).
62. Qinghai Water Conservancy Bureau. *Annual Hydrological Report P. R. China: Hydrological Data of Inland Rivers and Lakes*, 1st ed.; Journal of Qinghai Water Conservancy Bureau: Xining, China, 2016; pp. 64–73.
63. Marden, J.; Kendall, M.; Gibbons, J. Rank Correlation Methods (5th ed.). *J. Am. Stat. Assoc.* **1992**, *87*, 249. [CrossRef]
64. Mann, H.B. Nonparametric Tests Against Trend. *Econometrica* **1945**, *13*, 245. [CrossRef]
65. Burn, D.H.; Hag Elnur, M.A. Detection of hydrologic trends and variability. *J. Hydrol.* **2002**, *255*, 107–122. [CrossRef]

MDPI
St. Alban-Anlage 66
4052 Basel
Switzerland
Tel. +41 61 683 77 34
Fax +41 61 302 89 18
www.mdpi.com

Applied Sciences Editorial Office
E-mail: applsci@mdpi.com
www.mdpi.com/journal/applsci